NOBEL PRIZES
and
Notable Discoveries

NOBEL PRIZES
and
Notable Discoveries

Erling Norrby

The Royal Swedish Academy of Sciences, Sweden

We World Scientific

NEW JERSEY · LONDON · SINGAPORE · BEIJING · SHANGHAI · HONG KONG · TAIPEI · CHENNAI · TOKYO

Published by

World Scientific Publishing Co. Pte. Ltd.
5 Toh Tuck Link, Singapore 596224
USA office: 27 Warren Street, Suite 401-402, Hackensack, NJ 07601
UK office: 57 Shelton Street, Covent Garden, London WC2H 9HE

Library of Congress Cataloging-in-Publication Data
Names: Norrby, Erling, author.
Title: Nobel prizes and notable discoveries / by Erling Norrby.
Description: New Jersey : World Scientific, [2016] | Includes bibliographical
    references and index.
Identifiers: LCCN 2016027969| ISBN 9789813144637 (hardcover : alk. paper) |
    ISBN 9789813144644 (pbk. : alk. paper)
Subjects: | MESH: Nobel Prize | Neurophysiology--history |
    Molecular Biology--history | History, 20th Century
Classification: LCC QP360 | NLM WL 11.1 | DDC 612.8--dc23
LC record available at https://lccn.loc.gov/2016027969

British Library Cataloguing-in-Publication Data
A catalogue record for this book is available from the British Library.

Preface

In my writing about Nobel Prizes I profit from access to the unique Nobel Archives as they become available for scholarly analysis after the 50 years' secrecy period. They represent a world unique real time analysis of the advance of natural sciences as recognized by the prizes. The nominations vary in length, but not infrequently they are quite informative. However, the most valuable part of the archives is the thorough reviews, at the time made almost exclusively by members of the committees. Thus these analyses have been made by the best scientists in the discipline in Sweden and get their strength from the evaluators' accumulated insights into their science furthered by their participation in the work of the Nobel committees and their capacity for objective evaluations. Potentially they may also have weaknesses reflecting subjectivity of analysis and also difficulties in detaching oneself from prevailing opinions and appreciating the paradigmatic quality of new knowledge. The emergence of dramatically new insights often offers surprises and requires a major reassessment of accepted facts. The advance of knowledge represents one of the most, if not *the* most important quality in the progress of human cultures and civilizations. Presenting the story of the advance of science has many challenges. To what degree is it possible to popularize the complex discoveries made for a broader audience? Like any writer I always have an invisible reader or listener when the text is being formulated.

In the second book of Cicero's *De Oratore* there is a formulation that appeals to me. It reads in Latin *Neque ab indoctissimis neque a doctissimis legi vellem*, which freely translated and turned into a positive formulation might read "I should like to be read, not by the ignorant, nor yet by the very learned," which may be interpreted to mean the educated public. This obviously is not a well defined readership, but it is my hope that the joy of advancing knowledge might be contagious — possibly I am tainted by my background as a virologist in using this term. Reading the books of life, unraveling the remarkable inventiveness of evolution, is a uniquely enriching experience. It represents a never-ending journey that in an exceptional way is illustrated by the discoveries recognized by the Nobel Prizes in natural sciences. These prizes can be used as a guide to a journey through the continuously advancing field of medical sciences and of life sciences in general. Learning about the history of the human species and about our relations to the rest of nature is absolutely critical for developing health care that gives the best possible quality of life for each individual. It also provides guidance as to how we can best carry out our responsibility as stewards of Earth.

A book of the present kind obviously contains a lot of facts and hence cannot be read like a well-crafted novel. Certain sections like the recording of the way different Nobel committees have handled individual nominees by necessity include names of many nominators and of reviewers. The author's ambition is to extract from these rich materials the essentials in recognition and acceptance of the new discoveries. In addition it is important to note to what extent priorities in the advance of a field have been allocated appropriately. It is not possible to write in a neutral and detached way about new discoveries, since they are made by people. Hence a good part of the stories presented concerns the individuals, the committed scientists, who are instrumental in the progression of the scientific enterprise. But it is also about the environment in which they develop their successful science. Thus the major emphasis in these books is on individuals and institutions, with hindsight that we may learn something about the nature of the remarkable individual, and to a certain extent collective creativity that is the source of the relentless progression of knowledge. Having said this it is important to emphasize that science is not a predictable venture. This is of course the reason why it is so exciting to follow, but at the same time politicians who provide the major part of the resources that allow scientists to conduct their free science might wish that the results deriving from the efforts could to a larger extent be foreseen.

In this, my third book on Nobel Prizes, I have written about fields which are more distant from those with which I am most familiar, like infectious diseases, immunology and molecular biology. Thus the first three chapters deal with the exciting advances in the field that came to be called neurophysiology or neurobiology. The focal point is the prize in 1963 to John Eccles, Alan Hodgkin and Andrew Huxley, who discovered the ionic mechanisms involved in excitation and inhibition of the nerve cell membrane. I have tried to put this prize into a context by also presenting the previous development of the field as illustrated by earlier Nobel Prizes and to a certain extent also later developments as highlighted by ensuing prizes.

The following chapters 4 to 6 use the 1964 Nobel Prize in physiology or medicine to Konrad Bloch and Feodor Lynen as a second focal point. They described the complex synthesis of the biologically very important molecule cholesterol, a process involving more than 30 steps. Again prior knowledge in the field is illustrated with reference to earlier prizes in both chemistry and in physiology or medicine. The later developments leading to the introductions of so-called "statins" for prevention of atherosclerosis is emphasized. The last three chapters of the book take us back to an analysis of the development of the field that came to be called molecular biology. This has been discussed at length in my two previous books, *Nobel Prizes and Life Sciences* (2010) and *Nobel Prizes and Nature's Surprises* (2013). In this book one chapter each is devoted to André Lwoff, Jacques Monod and François Jacob, the colorful French recipients of the 1965 prize in physiology or medicine.

The writing of this book was started in early 2014 when I had the privilege of spending seven weeks at the Stellenbosch Institute for Advanced Studies (STIAS) in South Africa. Its director Hendrik Geyer and his senior administrative officer Maria Mouton provided the best of conditions for the work. The stay allowed very stimulating contacts with other visiting scientists, representing many different disciplines. I also had a generous offer to return to STIAS in early 2016, but regrettably I could not accept this for family reasons. However, what made the 2014 stay unique was that three other scientists, friends of mine for more than 40 years in the field of virology, also stayed at STIAS during the same time. A remarkable friendship was nursed by the contacts at work and at play. The three friends Marian Horzinek from the Netherlands, Frederick Murphy from the U.S. and Marc van Regenmortel from France and South Africa have given valuable advice at any time I have been writing about viruses, which, when relevant, I of course have a tendency to do. In particular they contributed to Chapter 7 which ended with a virocentric view of biology, a theme that is likely to re-emerge in possible forthcoming books.

As mentioned I needed a lot of advice when writing about the field of neurophysiology. Sten Grillner, an authority in the field and a good friend since my later time at the Nobel committee at the Karolinska Institute made a very thorough review of the three first chapters in their close to final form. I am very grateful for this analysis and for his recommendations for modifications.

Others who inspired me in the writing were Derek Denton, Torsten Wiesel and Peter Århem. They all gave valuable suggestions. Erik Kandel provided me with very useful reading material. Thomas Starzl provided important information on his early interaction with Horace Magoun, a near prize recipient presented in Chapter 2. Bodvar Vandvik gave valuable information on Fridtjof Nansen's contributions as a neuroscientist including a reference (in Norwegian). Nils Uddenberg's recent books on the history of medicine (in Swedish) gave some additional inspiration to the text and also provided suggestions for some pictures.

Chapter 4 starts with a description of the development of chemistry, later biochemistry, at the Karolinska Institute. Jan Trofast, a specialist on Berzelius, gave valuable comments on this remarkable pioneer in the field of chemistry. Thomas Tydén and Ami Ekman allowed me to learn more about the fascinating Hammarsten family and in particular the chemist Einar Hammarsten, who played a central role in the impressive growth of the discipline of chemistry at the Institute during the first half of the previous century. Thomas Lindahl, Hammarsten's last student kindly provided personal comments on their interaction, cited in the book. In the presentation of Hugo Theorell, the first Nobel Prize recipient at the Karolinska Institute (see front cover), I received valuable advice from his sons, in particular Henning Theorell, who himself has written about his father. My cousin Lars-Johan Norrby, an emeritus physical chemist at Stockholm University, made valuable remarks on the general text of the chapter and in particular on his father Johannes. Ragnar Björk, who himself has written in Swedish about Theorell, also gave valuable advice on this chapter.

In the autumn of 2014 Theodore Friedman arranged a conference in La Jolla to commemorate Frederick Sanger. Many of Sanger's students and collaborators participated in this meeting. Because Sanger, a unique two-time Nobel prize recipient in chemistry, had died in 2013 the archives relating to his first Nobel Prize had become available for reviewing in 2014. I presented a summary of their content at the enriching meeting and this gave an incentive to write Chapter 5. This chapter also discusses Sanger's second Nobel Prize which recognized his development of a technique for sequencing of DNA, which has completely revolutionized the field of molecular biology. I received valuable comments on the impressive developments of this technique from Craig Venter and Mark Adams at the J. Craig Venter Institute in La Jolla. I have a close connection to this Institute as the Vice-Chairman of its Board of Trustees. It allows me to get access to books of value for my writing about Nobel Prizes and this is arranged by contacts with Julie Adelson, the legal advisor to the Institute. Chapter 5 also presents Arne Tiselius, a Swedish Nobel laureate (see cover picture) and for decades a very central figure in the Nobel Prize work. Tiselius made a major impression on me already during the writing of my two previous books on Nobel Prizes. Contacts with his son Per Tiselius, a physician in the charming city of Sigtuna, provided precious personal information. His private clinic in this city is in part a museum commemorating his father.

Chapter 6 highlights the 1964 prize to Bloch and Lynen and to write this I needed to learn more about sterols, a large family of biomolecules of wide-ranging importance in biology and medicine. K. C. Nicolaides made recommendations of general textbooks on the subject and later read some of my texts and proposed corrections. Invaluable contacts were established with Joe Goldstein, who together with Michael Brown received the 1985 Nobel Prize in physiology or medicine for their elegant and medically very important studies of cholesterol metabolism. I received many valuable review articles and also learnt a lot about the field from the email contacts. In 2036 or later the Nobel archives will be available for historians of science who want to write the full story of research on cholesterol that led to the introduction of critical modes of preventive treatments markedly reducing the impact of cardiovascular diseases.

Chapters 7 to 9 brought me back to more familiar grounds of research, but still I relied on many valuable contacts for advice. First and foremost I would like to mention Georg Klein. He represents a unique source of information, because of his exceptional memory and also the simple fact that he has been involved in the Nobel Prize work at the Karolinska Institute since 1958 and for many subsequent years. Another important source of information from the same time is Peter Reichard. I also received valuable help and good reading material from Simon Wain-Hobson and in particular Agnes Ullmann at the Pasteur Institute. It was a joy, but also a daunting challenge, to write about the three French intellectual giants Lwoff, Monod and Jacob.

I am grateful for the full support given to me by World Scientific Publishing in the development of this, the third of my books on Nobel Prizes. In particular I would like to thank its Chairman Professor Kok Khoo Phua and the highly competent editor Kim Tan who with great patience has given support and advice in the processing of the large material for the book. Pleasant working conditions have been provided for me at the office in Singapore and in this city I have also received support from various colleagues, in particular a long-time friend, the President of Nanyang Technical University, Professor

Bertil Andersson. We were simultaneously members of the Board of the Nobel Foundation for a number of years.

Not having English as my native language I need assistance to ensure that the English used is idiomatic. As in the case of my previous books Harry D. Watson has provided expedient and careful reviewing of the texts. Sven and Dagmar Salén Foundation generously covered the costs for this work.

I have received regular support from many different parts of the Nobel system. The Director of the Nobel Museum Olof (Olle) Amelin has shared his interest in my work. Nobel media gave access to the 2000 interview of Andrew Huxley by Joanna Rose. At the Karolinska Institute the secretary of the Nobel committee for physiology or medicine, Göran Hansson, and later his successor Urban Lendahl, have given me permission to examine the archives that progressively became available. During my January visits and also on other occasions Ann-Mari Dumanski and Tatiana Goriatcheva have looked after me well.

For many years I have had my office at the Center for the History of Science at the Royal Swedish Academy of Sciences. This location is ideal because the Nobel archives for physics and chemistry are only 30 seconds away and in particular because of the very attractive working environment created by all the colleagues sharing this milieu. The Director of the Center Karl (Kalle) Grandin has generously made the resources of the institution available to me and in addition he has personally spent many hours with me working on the picture material for the book. Furthermore when I have become stuck at my computer he has swiftly helped me out. My other colleagues Maria Asp, Anne Miche de Malleray, Jonas Häggblom and Åse Frid have shared their enthusiasm and knowledge and innumerable topics of comprehensive or smaller dimensions have been aired at the breaks from my work at the desk. Very special thanks go to my next door neighbor at the Center, Bengt Jangfeldt. He is a highly qualified Slavist and humanist who has published a number of high quality books. Hence we can share the joy and challenge of writing, but also discuss essentially any kind of intellectual problem, in particular if it transcends the border between the humanities and natural sciences.

At the end of this Preface I would like to thank my family. My wife of 57 years, Margareta, is a remarkable life partner, who by now knows all my strengths and weaknesses. She intuits what kind of encouragement or restrain I need. Our three children, to whom I dedicated my previous book, provide continuous support and encouragement of my endeavors. Time passes on and our five grandchildren, at the time of publication of this book, have reached the age of 15–23 years. I would like to dedicate this book to them all — Heimir, Smári (Samuel), Sindri, Henrik and Lisa. They represent the future and will have opportunities to follow the advance of science and share in the application of new unexpected knowledge in the shaping of an ever better world.

Contents

Chapter 8
A Scientist of Many Talents

Chapter 9
From Heroic War Efforts to Intellectual Battles

Chapter 1

Neurophysiology, a Discipline Developing Consciousness

BILLIONS OF NERVE CELLS
UNCOUNTABLE NETWORK CONTACTS
OUR BRAIN, A WONDER

Once in the dawn of modern big time there was a human primate ancestor living in the heart of Africa who looked at his face in a pond. He reflected on the fact that what he saw must be his own image. He further pondered on this particular moment of his existence. His thinking led him to deduce that this existence extended from the time when he was born to the present fleeting moment. It also extended into a future time of unpredictable length. As his thoughts meandered further they came to focus on the — irrelevant? — question, if his existence as an individual could possibly have a meaning. Thus was born the human primate consciousness about consciousness.

Of course in reality, emergence of consciousness was an incremental process. Throughout eons of evolution there has been a progress from perceived awareness of ever increasing complexity towards the advanced form of (self-)consciousness enriching the life of us humans. The uniqueness of human consciousness is the emphasis on the subject, the capacity to reflect, to plan and to act. Not surprisingly attempts to understand consciousness has challenged the most brilliant inquisitive minds of humans. Naturally it is tantalizing to think about and examine the organ with which we think — to be aware of the existence of possibilities for self-reflection. There is, however, often a pronounced difference between how natural scientists and philosophers/humanists conceptualize this state and which qualities of mind they emphasize.

Before approaching the epitome of integration of nerve functions, consciousness, it was imperative to understand the basic functions of the nervous system. This included interpreting its anatomy. The function of individual specialized cellular components needed to be elucidated. The critical cell, the *neuron*, with its particular extensions, was identified. These extensions were found to be of variable length and to carry different functions. The shorter extensions, the *dendrites*, generally, communicate electrical signals from the periphery to the cell body. These signals are then processed and transmitted in an integrated form from this body to the periphery by *axons*. The axons form bundles which are referred to as *nerves*. The nerves may stretch over long distances, in the human body maximally from the spinal cord to our big toe.

The nervous system has two major parts, the central and the peripheral structures. The central part includes the brain and the spinal cord. The brain can be subdivided into the large brain, *cerebrum*, and the small brain, *cerebellum*, which are connected via their underlying structures, the *brain stem*, and the *spinal cord*. The stem includes several distinguishable parts. These will be presented in more detail below. Throughout evolution it is the central nervous system that has shown the most dramatic development in multicellular organisms. In general there is proportionality between body size and brain size, but in primates the brain is relatively larger compared to other vertebrates (elephants and whales have the heaviest brains of any animal). In primates it is in particular the convoluted cortex, looking like a cauliflower which added to the weight. Over about three million years the hominin brain has increased in volume some three times to its present size in modern humans. It might be noted that Neanderthals, who became extinct some 30,000 years ago and with whom we had joint offspring, had somewhat larger brains than the modern *Homo sapiens*.

The right half of the primate large brain is generally larger than the left. Furthermore individual parts of the brain can increase in size after prolonged and intense training. The brain size of men on average is about ten percent larger than the brains of women. However, it should be emphasized that it is not the size of the brain that decides its fully developed capacity to integrate information. The critical factor in the brain is the propensity for development of synaptic contacts and these in turn are consequential to exposure to appropriate environmental stimuli. There are many stages in the development of the embryonic brain to that of the full grown individual. New networks of connections are continually established and trimmed. Cells are retained or removed by programmed cell death, during some periods of the young life at an impressive speed. The flexibility in functions of the brain is remarkable. It is per weight the most energy-consuming organ in the body. That is the reason for the fact that about 20% of all circulation of blood passes through the brain, which weighs only 1.3 to 1.4 kilos in adults. Since it needs to be quick in action it gains calories from using the rapidly metabolized carbohydrates, mostly glucose converted into lactate, in the blood. The mobilization of regionalized brain activity can be readily detected by measuring the local blood flow using magnetic resonance imaging (MRI) to be discussed further in Chapter 3.

The peripheral part of the nervous system is represented by pair-wise nerves. Twelve cranial nerves communicate with the brain structures directly or via their relay stations, except the first two pairs, including the olfactory nerves, which end up directly within the brain stem. In all vertebrates there are *efferent* nerve fibers which transport electrical signals from a central organ like the spinal cord to for example a muscle allowing the initiation of a controlled contraction of this structure. However, nerves generally represent a two-way form of communication allowing sense organs in muscles to inform central structures about the conditions of tension of the tissues. The latter kind of nerve fibers is called *afferent*. There are many kinds of afferent fibers, which transport signals from the different sense organs — registering pain, touch, taste, light, sound, etc. — to the spinal cord and possibly further to higher centers in the brain or directly to the latter organ. Whereas efferent nerves act directly on their target cells, like muscle cells, the afferent sensory nerves pass via a collection of cell bodies outside the scull — a structure referred to as *ganglion* — which serves as a relay station. The central nervous organs are highly complex integrating centers which orchestrate input and output signals. The intricate nature of this system increases with the life form and in vertebrates the highest complexity of the brain is found in humans. Thus a roundworm, *Caenorhabditis elegans*, a popular experimental animal introduced by Sydney Brenner (recipient of the shared 2002 Nobel Prize in physiology or medicine[1]) has 302 nerve cells which form about 7,500 contacts, whereas humans have about 85 billion nerve cells which establish of the order of 10^{14} to 10^{15} contacts. Incoming signals may elicit a wide range of reactions, like activation of different categories of muscles in a flight reaction. The prime functions of centers at different levels with the most advanced integrating functions located to the brain cortex are to allow the animal to manage its environment, to survive. To secure the capacity to reproduce is equally important.

Attempts to decipher the function(s) of the brain, from its simplest to its most advanced forms have been the focus of many studies of animals and man. Neurobiological research has provided deep insights into the operations of the nervous system, not least due to the development of remarkably refined techniques for measurements of the electrical nerve activities. This development has been dependent on the introduction of new physical methods for examination of weak currents, and later for examination of the genetics and chemical dynamics of the functions of the brain by use of molecular techniques. However, studies of the brain started way before the present time. In fact it does not suffice to go back to the ancient Greeks.

The first mention of the brain is in a text using hieroglyphs. It was an Egyptian battle surgeon who in the 17th century BCE noticed that damage to one side of the brain led to symptoms also from one side, a phenomenon referred to as *lateralization*. Lack of speech after damage and seizures were also mentioned in this remarkable text. More than a thousand years later the Greeks dominated speculations about the function of the brain. However, because they considered the human body sacred they generally did not perform dissections. Although some thought that the brain was a center for registration of sensations and for intelligence, Aristotle's school favored another view. It located intelligence in the heart and saw the brain as a cooling device. And of course we still learn things by heart today. The situation changed in the first half of the third century BCE. Herophilus and Erasistratus in Alexandria made dissections of the dead and even vivisections on prisoners. The anatomy of the brain began to be revealed. A distinction was made between the large brain and the underlying small brain. Furthermore the presence of liquid-filled spaces, the *ventricles*, was also recorded. Nerves were shown to be distinct from ligaments and tendons and functionally they were separated into motor and sensory nerves. Nerves of the former kind transmitted signals to the muscles a concept consolidated some five hundred years later by Galen's vivisection experiments on animals.

Progressively the anatomy of the brain was revealed and in the 16th century it became known in its main structures by the remarkable atlases developed by Andreas Vesalius. Two hundred years later the physicist, natural scientist and physician Luigi Galvani stumbled on the discovery that electricity played a central role in the contraction of the leg of a dead frog. His serendipitous finding was made by use of the following experimental setup. A dissected frog was suspended in a copper wire attached to an iron tripod. As the muscles of the frog relaxed the leg it came to touch the iron foothold of the tripod. The muscle then contracted and became disconnected from the foothold. Repeated contractions could be registered by this experimental arrangement and we can now interpret this to be due to the fact that a current could flow when the foot contacted the tripod. Galvani interpreted the observed phenomenon differently. He believed that electricity was generated in the body of the animal. Many scientists became interested in the effect of electricity on the body and speculated on the possibility that it could be used

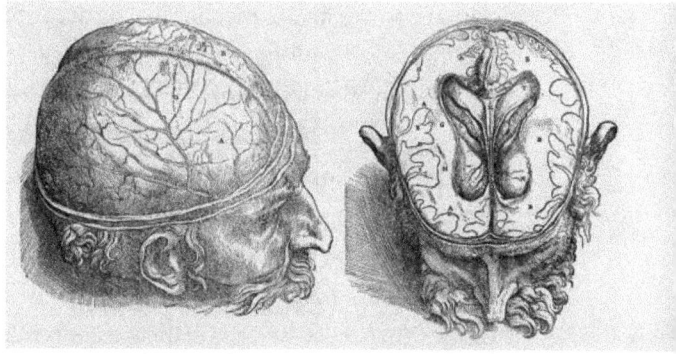

The human brain as illustrated by A. Vesalius. [Courtesy of the Hagströmer Library, Karolinska Institute.]

to heal certain disease conditions. Among those can be mentioned the natural scientist and statesman Benjamin Franklin and the co discoverer of oxygen Joseph Priestley. Carl Linnaeus also became fascinated by what he called *spiritus animalis* and another famous Swede, who we will meet in Chapter 4, Jacob Berzelius, wrote his thesis reflecting on the therapeutic use of electricity in the treatment of different diseases.

Thus electricity could play some kind of role in biological systems, including in the brain. Other important speculations on the latter organ were made by Descartes in the same century. He proposed the theory of dualism possibly to conform to the theocratic conditions of his time. Dualism implied that functions of the body and the mind were organically separated in the brain. He even suggested that a possible site for the mind was the pineal gland, a small cone-like structure at the bottom of this organ. Only humans had a soul and animals were automata, controlled by reflexes only. The Cartesian dilemma has haunted scientists way into the mid-twentieth century and we will return to this in the next chapter.

Phrenology, a term derived from Greek words meaning knowledge of the mind, was a pseudoscience at the turn of the 19th century and it was highly popular for some 50 years. The idea was that character, thoughts and emotions could be mapped to different distinct regions of the brain. During the early part of the twentieth century the Italian physician Cesare Lombroso reintroduced the concept for use in investigations of evolution, anthropology and criminality. Personal traits were considered to be reflected in the detailed structure of the skull. In the spirit of the time it was argued that Europeans were superior to other lesser races. The use of slavery was encouraged. The movement slowly died out when more serious studies were made of the effects of localized damage to the brain, mechanical or because of a bleeding or of blood clotting, thrombosis. In pioneering studies the French physician and anatomist Paul Broca identified a location in the frontal lobe, that, when damaged, led to the loss of capacity to form articulated language, expressive *aphasia*, from Greek for speechlessness. The site is now referred to as Broca's area. Similarly, the German physician and neuroanatomist Karl Wernicke wanted to examine the effect of certain brain lesions on speech and language. He observed that not all deficits of language could be explained by damage to Broca's area. There were other situations when the critical region was located further back in the brain, in a structure referred to as left posterior, superior, temporal gyrus. Damage to this region, Wernicke's area, led to receptive aphasia. In extended studies there has been an expanded mapping of the allocations of many different functions to the different regions of the brain. In right-handed people Broca's and Wernicke's areas are found almost exclusively on the left hemisphere, exemplifying the phenomenon lateralization introduced above.

There were a number of important steps in the growing insight into the networks formed by nerves in the brain. One of the pioneers in this growing field was the Czech anatomist Jan E. Purkyně. Originally most of his work concerned our senses in particular vision but he was also interested in our sense of balance. His studies on vision focused on how we interpret colors and in this context he befriended Goethe, the father of his own *farbenlehre*. Beginning in the early 1830s Purkyně started to use an achromatic microscope to examine the brain. Large cellular structures could be identified, and the exceptionally large cells in the cerebellum are still today called the Purkinje cells. Progressively histological techniques were improved. The texture of the soft tissues could be fixed, thin sections could be prepared and most importantly effective stains for different purposes were developed. The detailed anatomy of nervous tissue could be revealed by use of a staining technique employing silver chromate introduced by the Italian scientist Camillo Golgi. His staining technique was developed further and systematically employed by the Spanish scientist Santiago Ramon y Cajal. These two scientists shared the 1906 Nobel Prize in physiology and medicine. We will soon meet them again.

The discovery that a unique kind of cell in the brain, the neuron, was the central actor in the transmission of signals, for example between the brain and muscles, revolutionized the field. The term neuron for the nerve cell was introduced by the German late nineteenth century anatomist Heinrich W. G. von Waldeyer-Hartz in an excellent synthesis of the pioneering discoveries by Golgi, Cajal and others. He postulated that neurons had a central role in the function of nerve tissues and that they had the particular quality of being electrically excitable. His proposal was referred to as the *neuron doctrine*. The word neuron has a Greek origin and was used to depict a string. It was used for the first time in Homer's *Iliad*. On the side it can be mentioned that Waldeyer-Hartz also coined the word *chromosome*, for the structures carrying the genetic material. The network of fibers connecting cells in the brain fascinated scientists from the first moment of its discovery and into the present time. An example taken from the olfactory epithelium of a dog illustrates the intricate networks of nerve cells and fibers, as a reminder seen only in two dimensions.

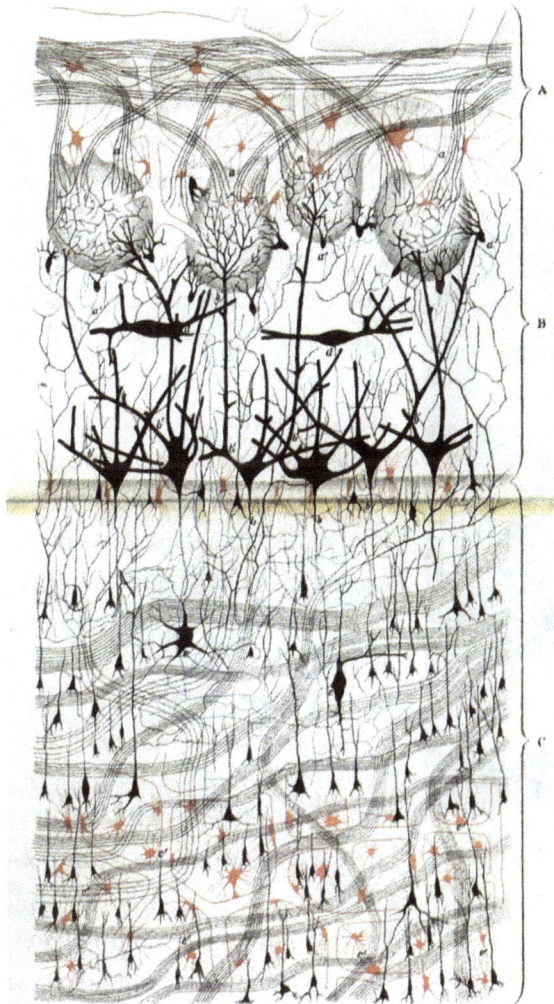

The networks of nerve cells and fibers in the olfactory epithelium of a dog. [Courtesy of the Hagströmer Library, Karolinska Institute.]

After the cellular anatomy of the brain had been clarified the field of neurophysiology has taken many quantum leaps highlighted by Nobel prizes in physiology or medicine. Prior to the prize which is the main focus of the two following chapters, the 1963 Nobel Prize in physiology or medicine to John C. Eccles, Alan L. Hodgkin and Andrew Huxley, there were seven prizes in this field (Table 1.1, p. 14). The first prize given was not concerned directly with discoveries concerning the structure of the nervous tissue, but instead its functional role in the process of digestion.

A Scientist Who Loved Dogs

In 1904 a Nobel Prize in physiology or medicine recognized the work by Ivan P. Pavlov from Russia. This event was commemorated 100 years later when a statue of him was erected. The prize motivation was "in recognition of his work on the physiology of digestion through which knowledge on vital aspects of the subject has been transformed and enlarged." Selection of food and its ingestion provides a good illustration of a remarkable coordination of sensory signals and activation of different functions in the alimentary tract. One critical sense is olfaction and we sort out edible and non-edible potential food materials by our sense of smell. But we also register many other qualities such as the color and general appearance, the texture, the temperature and not least the taste. The brain integrates all these impressions to decide whether we should eat the food or not. Once we decide to do this a large number of signals from the brain activate different parts of the alimentary canal — the oral cavity, the pharynx, the esophagus, the stomach, and the small and large intestines. Pavlov was interested in understanding how these different parts were activated and prepared to process the food. In particular he was interested in the role of nerve regulations.

Ivan Pavlov, statue erected in 2004. [Photo from A. D. Nozdrachov.]

Pavlov was born in 1849 in the city of Ryazan in the Russian Empire as the eldest of eleven children. His father was a parish priest and Pavlov studied at the local church school and later even entered the theological seminary. However, he did not finish his studies at the seminary but instead took up studies in the natural sciences at the University of St. Petersburg. With time he became increasingly involved in studies of physiology and decided to move to the Academy of Medical Surgery. Pavlov studied the circulatory system for his medical dissertation and in 1879 he received his M.D. from the Military Medical Academy with a gold medal award for scientific excellence. Continued research led to a Ph.D. dissertation on "The Centrifugal Nerves of the Heart" in 1883. Thereafter he spent two years abroad in well-established German laboratories. In 1895 he became professor and chairman in physiology at the Military Academy. At that institution he built a very strong and internationally respected group for research in many aspects of physiology.

Pavlov's Nobel Prize in physiology or medicine was the first one in which the discovery did not have an immediate clinical application. Most prizes of this kind during the first ten years were awarded in the fields of microbiology or immunology. They concerned discoveries of obvious practical importance. Pavlov's prize by contrast was given to an experimentalist and theorist, a physiologist. This special character of Pavlov's prize was emphasized by the professor of chemistry and pharmacy, Count Karl A. H. Mörner (p. 193), vice chancellor of the Karolinska Institute, in his introductory speech at the prize ceremony[2]. He said:

> The aim of science is the acquisition of knowledge, the value of which should not be measured by the ease with which it can be brought immediately into practical usefulness. Examples of this can be seen in various accounts of scientific developments which have given their originators a prominent place in the history of Medicine. One may point to Vesalius and Harvey. When Vesalius, in spite of the personal risks to which he exposed himself, through his masterly research opened the way to the study of human anatomy, he was impelled by his desire to carry the torch of science through the covering veil of prejudice and authoritarian belief. When Harvey through long years of investigations and deep study was able to prove the circulation of the blood, it was his thirst for truth which spurred him on in his work; to satisfy it was his reward.

Pavlov was a very impressive physiologist and his research introduced major new concepts in the field of digestion physiology. This became clear from his Nobel lecture[3]. He demonstrated by elegant experiments in dogs that the quality of the food had an influence on the quantity and quality of the digestive gastric juices produced. However, a production of juices could be induced even in the absence of contact with food by repeated exposure to the sight of food. This discovery of the existence of so called *conditioned reflexes* for the first time allowed objective studies of psychic activities. It was demonstrated that the expression of reflexes of importance for digestion, be they direct or conditioned, demanded intact nerve connections between the site of elicitation of the reflex, via the brain and the target organ. Thus was proven the critical role of nerve transmissions in this particular realm of physiology. However, even the sun has its spots. In 1889 Pavlov had made experiments with dogs conducted so that the food could not reach their stomach. Since the gastric juice was still stimulated to run Pavlov concluded that this was due exclusively to nerve signaling via the brain. This conclusion later on turned out not to be the full explanation. As discussed previously[1] the discovery of William M. Bayliss and Ernest H. Starling demonstrated that besides electrical signaling by nerves there were also important chemical messengers. They identified a substance, named *secretin*, the first hormone discovered, which carried signals from the intestine to the pancreatic gland releasing products of importance for digestion. The signaling in this case was not nervous but instead mediated by a chemical substance. It has been recorded[4,5] that Pavlov had one of his assistants repeat the Bayliss and Starling experiments, confirming that they were correct. Regrettably this confirmatory finding was not referred to in his Nobel lecture.

Pavlov was disappointed by his own mistake and this had consequences for his future research. It became oriented towards less groundbreaking endeavors. His close Finnish friend Robert Tigerstedt, a world-renowned sensory physiologist tried in vain to encourage Pavlov to resume some of his original studies (see Ref. 5). Thus he came to have a somewhat reduced influence on the field of physiology, but perhaps more in formulation of psychological concepts. Tigerstedt had a close relationship with Pavlov and the biography of Pavlov in *Les Prix Nobel* is an abbreviated version of a text written by him for a Festschrift in 1904. Disregarding the reduced momentum in Pavlov's work due to the fact that he overlooked the existence of humoral factors it must be concluded that Pavlov by his early work was the founder of a school of research. His discoveries of some basic laws governing the function of the cortex of the hemispheres attracted many physiologists to the field. He was even nominated for a second Nobel award in 1925 and later. However, since no additional important discovery could be identified, no supplementary special investigation was made.

It is of interest that there was such a high regard for Pavlov's and his colleagues' work that their laboratory activities remained untouched by the October Revolution in 1917. Lenin signed a special decree in January 1921 to secure the continuation of Pavlov's pioneering work. This is all the more surprising since Pavlov was very outspoken in his criticism of the young and militant communist regime, not least when it came to the persecution of his fellow

Ivan Pavlov. [Painting by Mikhail Nesterov.]

scientists. In fact in 1927 he even wrote to Stalin strongly protesting about the treatment of Russian intellectuals and stated that he was ashamed of having a Russian nationality. He also wrote several letters to Molotov to save persons he knew from the mass persecutions that followed the murder of Sergei Kirov. Pavlov had a long life and finally died of pneumonia at the age of 86. His study and laboratory are preserved as museums.

Pavlov's Nobel lecture was a miracle of clarity and simplicity[3]. He succinctly outlined his identification of the role of different stimuli and the conditions for development of reflexes by the complex nervous apparatus governing the production of gastric juices. Pavlov had particular consideration for his experimental animals. Thus in one of the paragraphs of his lecture he said:

Now I shall allow myself to leave my main theme for a moment. Cutting of the vagus nerve (this nerve serves the vascular /respiratory and the gastric systems and is one of the cranial nerves connecting organs directly with the brain, my remark) in animals has been practised already for a long time and constituted an absolutely fatal operation. In the course of the 19th century physiologists learned about numerous influences exerted by the vagus nerves on the different organs and their respective investigations revealed at least four disorders in the organism occurring after severing these nerves, each of which was lethal by itself. In our dogs we took appropriate measures against each of these disorders, one of which concerns the digestive system, and due to this procedure the animals whose vagus nerves were cut enjoyed a healthy and happy life. Thus four simultaneously acting lethal causes were deliberately eliminated. A striking proof of the power of science that regards the organism as a machine!

In the next paragraph of his lecture he said:

Some ten years ago the great man to whom the annual science festival in Stockholm owes its existence honored me and my friend the late Professor Nencki with a letter enclosing a considerable donation for the benefit of the laboratories under our direction. In that letter Alfred Nobel expressed his keen interest in the physiological experiments and proposed that we should try several highly instructive projects concerning the supreme tasks of physiology, the problem of the organisms ageing and dying off.

It was in 1893 that Nobel had given a grant of 10,000 rubles, a sizable sum at the time. It allowed Pavlov to expand his laboratory work, but it is unlikely that the proposals in Nobel's late life letter had any influence on the chosen directions of Pavlov's research. The aging Nobel presumably did not receive any advice on how he might prolong his own life.

The last paragraph of Pavlov's lecture summarized his scientific credo in a charming way. It read:

Essentially only one thing in life interests us: our psychic constitution, the mechanism of which was and is wrapped in darkness. All human resources, art, religion, literature, philosophy, and historical sciences, all of them join in bringing light in this darkness. But man has yet another powerful resource: natural science with its strictly objective methods. This science, as we all know, is making huge progress every day. The facts and considerations which I have placed before you at the end of my lecture are the result of numerous attempts to employ a *consistent*, purely scientific method of thinking in the study of the mechanism of the highest manifestations of life in the dog, the representative of the animal kingdom that is man's best friend.

Pavlov was held in deep admiration by many of his colleagues. One of his admirers was Ragnar Granit, the neurophysiologist and Nobel Prize recipient to be, who we will meet repeatedly in this and the coming two chapters. In his 1941 partly autobiographical book in Swedish "Ung mans väg till Minerva (A young man's way to Minerva)" he referred to Pavlov's advice to budding young scientists, a credo Granit had framed and hung on the wall in the laboratory where he was working. This credo summarized the advice to young people contemplating entering science in three points. The first was to *develop slowly*, to give each new awareness and knowledge the time needed. Pavlov emphasized that there is no shortcut, only hard work will do. Using a metaphor he noted that the wing of the bird cannot raise the animal without the support of air. Existing facts are the air of the scientist. However, the scientist should be more than a gatherer of facts, an archivist. He needed to have an open mind and remember with humility and unpretentiousness the fact that there are limitations to available knowledge, that it has large gaps. Thus the second point is the *curiosity that drives the unbiased seeking of new knowledge*. The final and third point emphasized the need for a *total commitment*, the obsessive unrestricted dedication and devotion to the problem selected for examination.

Aggressive Defense of a Doctrine

Two years after Pavlov's 1904 Nobel Prize the founding father of the neuron doctrine, Cajal, was recognized, together with the discoverer of the critical silver staining technique for the elucidation of the structure of nervous tissues, Golgi (Table 1.1, next page). The prize to Golgi and Cajal was awarded "in recognition of their work on the structure of the nervous system." Again it was Mörner who gave the award ceremony speech[6]. In fact he chaired the Nobel Committee for physiology or medicine until his death in 1917 (14 prizes were awarded during this time) and he gave this speech in each of the first eleven years, except in 1907. The reason that another professor of the Karolinska Institute substituted for Mörner in 1907 was that the latter at that time was also the President of the Royal Swedish Academy of Sciences and, being a chemist, as a consequence had been selected to give the introduction to the Nobel Prize in chemistry awarded to Eduard Buchner, the discoverer of cell-free fermentation[1].

Table 1.1. Nobel Prizes in physiology or medicine (1901–1962) recognizing advances in neurophysiology.

Year	Awardee(s)	Motivation
1904	Ivan P. Pavlov	in recognition of his work on the physiology of digestion, through which knowledge on vital aspects of the subject has been transformed and enlarged
1906	Camillo Golgi Santiago Ramón y Cajal	in recognition of their work on the structure of the nervous system
1932	Charles S. Sherrington Edgar D. Adrian	for their discoveries regarding the function of neurons
1936	Henry H. Dale Otto Loewi	for their discoveries relating to chemical transmission of nerve impulses
1944	Joseph Erlanger Herbert S. Gasser	for their discoveries relating to the highly differentiated functions of single nerve fibers
1949	Walter R. Hess	for his discovery of the functional organization of the interbrain as a coordinator of the activities of the internal organs
	Antonio C. A. Moniz	for his discovery of the therapeutic value of leucotomy in certain psychoses
1957	Daniel Bovet	for his discoveries relating to synthetic compounds that inhibit the action of certain body substances, and especially their action on the vascular system and the skeletal muscles

In his introductory speech to the award of the 1906 Nobel Prize in physiology or medicine, Mörner started by recognizing the unique features of the anatomy of the nervous system. He said in the beginning of his presentation:

> The importance of the field they have undertaken to explore is obvious since it concerns the nervous system, an organic structure of such paramount importance to the most delicately organized of all living creatures. It is this system which brings us into relation with the outside world, be it that we receive impressions from it which act on our sensory organs and from there transmit themselves to the nervous centers, or be it that by movements or other forms of activity we intervene in the environmental phenomena. This same organic structure provides the basis and instrument for the highest form of activity of all, intellectual work.

He then discussed the complexity of the system and the difficulties of studying it. In a particular paragraph he referred to consciousness and said "But a far

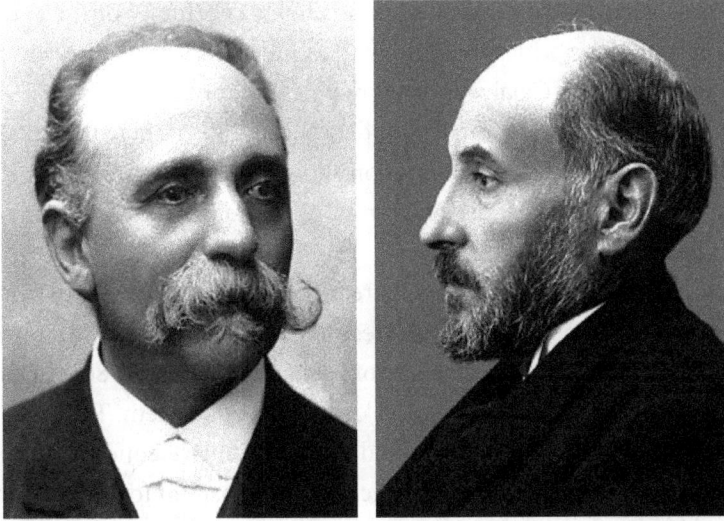

Camillo Golgi (1843–1926) and Santiago Ramón y Cajal (1852–1934), recipients of the 1906 Nobel Prize in physiology or medicine. [From *Les Prix Nobel en 1906*.]

greater complexity appears if the impulse continues to be transmitted and reaches the centers of consciousness. The impulse progresses along nerve tracts which follow complex pathways until it reaches the surface of the brain, i.e. the cerebral cortex. For consciousness — in man at least — is exclusively located in this area." As we shall see interpretation of consciousness over time has come to include many hypotheses. The highlight of Mörner's speech was of course the description of the silver impregnation technique introduced by Golgi and improved by Cajal, which "must be considered as a fundamental discovery in the field of nerve anatomy." The technique allowed for the first time a visualization in more detail of nerve cells and their extensions. These applications to brain studies were developed in depth by Cajal and the two of them were referred to as the "standard bearers of the modern science of neurology." Mörner did not in his speech touch upon the sensitive issues concerning the relative role of the nerve cells and of their extensions in the communications, but the laureates did in fact do this!

The two recipients of awards had quite divergent opinions on the impact of their contributions and polemics were introduced into their Nobel lectures. The conflict concerned fundamental aspects of how the neurons operated and related to each other. Cajal emphasized in his later writings that nerve elements possess reciprocal relationships *in contiguity* but not *in continuity*. This is the critical distinction that separates the school of *neuronism* — the contiguity of neurons as independent cellular units — from that of *reticularism* — a postulated intercellular continuity of the cytoplasm among neurons in a widespread network. Golgi was a spokesperson for the latter theory whereas Cajal was a staunch defender of neuronism. The sharp difference in the interpretation of their data became apparent in their respective Nobel lectures. The intense polemics on show maintained the courtesy forms used for example in the House of Commons in the British Parliament — The Right Honorable, etc.

The backgrounds of the two laureates showed certain similarities. The Italian Golgi, born in Corteno near Brescia in 1843 was the son of a physician. He studied medicine at the University of Pavia, graduating in 1865. Inspired by one of his teachers he initiated studies of the nervous system. The histological studies of this system were conducted under primitive conditions during a time when he was Chief Medical Officer at the Hospital for the Chronically Sick at Abbiategrasso. Golgi then returned to the University of Pavia where he succeeded his mentor as Chair of General Pathology in 1881. Besides his pioneering work on the cellular structures of the nervous system he also made seminal discoveries regarding different forms of malaria parasites and various types of fever. However, his main contribution was the "black reaction" he developed to stain individual nerve and cell structures. The weak solution of silver nitrate employed was found to be particularly valuable in tracing fine processes and ramification to and from cells. Throughout his career he continued to develop and apply this technique. As a person Golgi was modest and reticent. He was highly appreciated by the scientific community and became a member of the Royal Swedish Academy of Sciences in 1910.

Golgi was nominated in a handwritten letter, referring to his "revolutionary" contribution to the understanding of nerve cell anatomy, for this membership by the famous Swedish neurologist Salomon E. Henschen. He was a clinical professor in medicine at the Karolinska Institute from 1900, and had a particular interest in neurological diseases. He made important contributions to studies of aphasia and also brain damage which impairs mathematical abilities, *dyscalculia*. To digress it can be mentioned that Salomon had a son, Folke, who followed in his father's footsteps and became a famous pathologist at the Karolinska Institute. In 1924 the two of them collaborated on the autopsy of Lenin's brain. This happened after the Soviet leader, who was only 54, had had his third and fatal stroke. A sample of the brain was shipped to a famous neurologist at the Kaiser Wilhelm Institute in Berlin, Oskar Vogt[7]. Vogt had a particular interest in localizing the origins of traits indicating "genius" in the brain. He thought he could identify certain unique large pyramidal cells in this organ as a marker of advanced intelligence. Towards the end of the Second World War Russian troops managed to retrieve Lenin's brain from the Kaiser Wilhelm Institute, where it still remained, before the Americans could grab it. For some time it was on display in Lenin's mausoleum at the Red Square of Moscow, but presently it can be viewed at the Moscow Brain Institute.

The family name Vogt has a particular resonance for virologists. Oscar Vogt and his equally well-known neuroscientist French wife had two daughters, both of whom became famous scientists. One of them, Marguerite published important work on the poliovirus together with Renato Dulbecco, who in 1975 received the Nobel Prize in physiology or medicine for his studies of tumor viruses[1]. Like Dulbecco, she became a faculty member of the Salk Institute for Biological Studies in La Jolla, CA, and made important investigations of tumor viruses. The other daughter had an equally successful career as a pharmacologist in Great Britain and was elected a Fellow of the Royal Society. Although not a Jew she left Germany as a protest against Hitler. After her arrival in London she first worked for some time with Henry Dale, who we soon will meet again. But it is now time to return to Henschen's nomination of Golgi for the membership in the Royal Swedish Academy of Sciences. This recognition was only one of many that Golgi received. At the Historical Museum of the University of Pavia a full hall is dedicated to him. In this room can be seen his more than 80 official recognitions in the form of honorary degrees, diplomas and awards. Golgi ended his long life in Pavia in 1926 at the age of 83 years.

Cajal had another Latin country of origin, Spain. He was born in 1852 at Petilla de Aragón. During his younger years he served apprenticeships both as a barber and a cobbler. However, his father, who was a Professor of Applied Anatomy at the University of Zaragoza, convinced him that he should study medicine. He finished his medical studies in 1873, after which he participated in an expedition to Cuba and contracted tuberculosis and malaria. Back in Spain he received his degree as doctor of medicine in Madrid and was appointed Professor of Descriptive and General Anatomy at Valencia in 1883. Together with a colleague Eduardo Garcia Sola, a professor of histopathology and microbiology, he published in 1885 some important work on cholera, examining the pathogenic bacillus and attempting to develop a vaccine. The provincial government wanted to express its gratitude for these important findings and presented Cajal with a ZEISS microscope. With the help of this new instrument it was possible for him to initiate groundbreaking studies of cells, as expressed by his own words "to attack the delicate problems of the structure of the cells without misgivings and with the requisite efficiency." Later Cajal was appointed to positions in General Anatomy at the universities at Barcelona and Madrid, and finally he became the Director of the National Institute of Hygiene in the latter city. He published a number of high-quality scientific articles and books in Spanish and French and became a highly respected member of the international scientific community. He was a foreign member of several international societies, not, however, including the Royal Swedish Academy of Sciences. Among his books can be mentioned *Textbook on the Nervous System of Man and Vertebrates* (in Spanish) published in 1897–99. Cajal used Golgi's staining technique in his work, but he did improve it by a modification referred to as "double impregnation". In this form it is still in use today (see figure, p. 7). Like Golgi he had a long life and died in 1934, 82 years old in Madrid.

Paradoxically Golgi started his Nobel lecture[8], entitled (in English) "The Neuron Doctrine — Theory and Facts" by acknowledging that he had always been opposed to the neuron theory, which, he appreciated, had its origin in work using the staining technique he had developed. He even indicated that he felt that this theory seemed to be going out of favor at the time. The Nobel lecture in many ways was an attack on Cajal's supportive interpretations of the neuron doctrine. Golgi first summarized the doctrine in the following way:

> The fundamental points of the doctrine may be summed up like this: the transmission of nerve impulses is conducted from the protoplasmic extensions and the cell body towards the nerve extension; consequently, each nerve cell possesses a receiving apparatus constituted by the body and the protoplasmic processes, a conducting apparatus — the nerve process — and a transmitting or discharging organ. The protoplasmic processes should, therefore, act as conductors towards the cell body, the nerve process should act as a conductor away from it.

Thus like Cajal, Golgi accepted the fundamental concept formulated by the late nineteenth century German anatomist Waldeyer that the cell he named a *neuron* was central to the functions of the nervous system. But they diverged on their interpretation of how the impulses were communicated.

After his introduction Golgi goes on to criticize Cajal. Sentences in the written text read "I shall therefore confine myself to saying that, while I admire the brilliance of the doctrine which is a worthy product of the high intellect of my illustrious Spanish colleague, I cannot agree with him on some points of an anatomical nature which are, for the theory, of fundamental importance, for example that the peripheral branch of spinal ganglion cells must be identified with a protoplasmic process, since one must consider the myelin sheath as an absolutely secondary event, for it is only necessitated by the length of the process. Similarly, I cannot accept as a good argument in support of the theory the statement which, however, is its starting point, that says ....." The diatribes continued and Golgi followed up his critique over 28 pages with 19 illustrations. In the middle of his lecture he made the following reference:

> In this connection I recall that (Fridtjof)Nansen (this is the thesis presented at Bergen University in Norway[9] by the forthcoming world-renowned polar explorer and 1922 Nobel Peace Prize recipient, my remark), for example, referring to his work on lower animals, had already thought that the true organ of specific nervous activity was the fibrillar network rather than the ganglion cells. This idea, as is well-known, has recently been confirmed by Bethe who has no hesitation in writing "the doctrine attributing to cells the role of being the centre of specific nervous activity is only a morphological speculation which is supported by no convincing proof, while there are several facts which are decidedly against it."

This formulation indicated a skewed interpretation of Nansen's data. Later judges have concluded that he emphasized the role of the nerve cell and did not believe that there were direct connections between the nerve extensions[10]. As a conclusion to his highly polarized and emotional lecture Golgi used the following citation by Nobel:

> Each new discovery leaves in the brains of men seeds which make it possible for an ever-increasing number of minds of new generations to embrace even greater scientific concepts. My wish is that these new anatomical studies, to which this Institute, at such a high intellectual level, has wished to draw attentions of the world, may represent a new element of progress for humanity.

Cajal mobilized a massive defense of the neuron doctrine over 32 pages and with 23 figures in his Nobel lecture which is entitled (in English) "The Structure and Connexions of Neurons[11]." It may appear a bit less polemic than Golgi's presentation, but it is sharp in its argumentation. Cajal in an early paragraph reemphasized that "… nerve elements possess reciprocal relationships in contiguity but not in continuity." As already emphasized this is the critical distinction that separates neuronism — the contiguity of neurons as independent cellular units, also supported for example by Nansen's findings — from reticularism — the intracellular continuity of the cytoplasm among neurons in a widespread reticulum. Cajal mobilized many arguments supporting his neuronism theory, but only two will be mentioned here. One was his studies of the development of the immature brain of the embryo and the other the regeneration of nerve functions by outgrowth of new axons after a trauma. Observations in both these situations emphasized the independent central role of neurons and he summarized this in his lecture in the following way:

> The results obtained demonstrate almost beyond doubt that at no moment of evolution can the axons be taken as cellular chains, or as discontinued axon cylinders as is supposed by the anti-neuronists: on the contrary, and agreeing with the doctrine of His and Kölliker, the new fibres are produced following the budding of the axons, and are in perfect continuation with the motor or sensitive neurons, in the embryo as well as in the regenerating nerves.

Although less confrontational than Golgi, Cajal also spoke his mind, for example in the following statements:

> Besides, we believe that we have no reason for skepticism. While awaiting the work of the future, let us be calm and confident in the development of our forthcoming work. Let us recall that these terminal dispositions, which modern neurology has discovered in the axons, have been established by the concordant revelations of several methods. If future science reserves big surprises and wonderful conquests for us, it must be supposed that she will complete and develop our knowledge indefinitely, while still starting from the present facts.
> Like many scientific errors professed in good faith by distinguished scientists the link theory (reticularism, my remark) is the result of two conditions: one subjective, and the other objective. The first is regrettable but inevitable with certain impatient minds, to reject the use of elective methods, such as those of Golgi and of Ehrlich which do not lend themselves easily to improvisation; the second is the exclusive

application of processes simple and convenient, but without a specific action on axons, and as a consequence incapable of presenting clearly the neuronal expansions and their peripheral ramifications.

And in the penultimate paragraph:

To sum up: from the entirety of the observations which we have just shown, and from many others about which we have not the time to talk, the doctrine of neurogenesis of His is clearly revealed as an inevitable postulate. We mourn this scientist who, in the last years of his life so well filled, suffered the injustice of seeing a phalanx of young experimenters treat his most elegant and original discoveries as errors.

Seven months after his Nobel lecture Cajal again had to defend his neuron theory. This time he mobilized all his talents for vindictive rhetoric. His rebuttal was published in four different journals, two Spanish and one Argentinian (English translation in Ref. 12). He had been attacked by the Rector (Vice Chancellor) of Granada, his friend Eduardo Garcia Sola, who had given a lecture on "The Decadence of the Neuron." It was together with Sola that Cajal had earlier made the important studies of cholera mentioned above. Sola had been inspired to attack the neuron theory based on some recent interpretations of nerve generation experiments by a German psychologist Albrecht von Bethe from Strassburg (presently Strasbourg, France) and the Hungarian histologist Stephan von Apáthy in Kolozsvár (today Cluj-Napoca, Romania). Cajal attacked their results with full emotional force and ripped apart their structurally based objections interpreted to indicate the existence of intercellular anastomoses. To amplify his argumentation he used abusive formulations like "... they often fall for the unhealthy temptation of doing negative work, discrediting doctrines and tarnishing reputations ...", "... just mention one revealing fact of the arrogant egotism and anarchistic rebellious-ness concealed in the depths of reticularism", "the precarious, difficult and inconsistent methods used ...", "... answer those who attack with the visor down and hidden in the shadows," and so on. The final somewhat facetious paragraph read.

Returning to the issue of neuronism, I am afraid that the neuron will be around for a while, and in my opinion the meritorious colleagues I alluded to should calm their nerves. Yes dear colleagues: *la neurona* or *el neurona* (Cajal preferred the feminine form) will outlast us, and in its march toward the future the neuron will see new sunrises and sunsets.

And so it was.

It would indeed have been very interesting to be personally present during the two consecutive lectures given by the laureates prior to the Nobel Prize ceremony. The continued historical developments of course verified that Cajal was right[13]. The neuron is the central cell in nerve tissues. It is the particular cell that can be electrically excited and transmit information by electrical and chemical signals. Neurons are the core components in the central nervous system and also in the ganglia of the peripheral nervous system, but this insight was only the very beginning of the fascinating journey unraveling their multiple functions and activities to be discussed in the following. This journey will include a discussion of the means of transport of electrical signals at the synapses. As we shall see they are in general chemical, but in certain cases the electrical signal is transmitted directly between cells. In these later cases one can functionally interpret the cells to be a part of a network, as proposed by the reticularists. Maybe nothing in biology is truly black and white. Electrical synapses will be further discussed in the next chapter.

Reflecting on the Reflex and Synthesizing the Synapse

It would take 26 years until another Nobel Prize would be given for studies of the brain and the nervous system. In 1932 a prize in physiology or medicine was given to the two British scientists Charles S. Sherrington and Edgar D. Adrian "for their discoveries regarding the functions of neurons" (see table, p. 14). There is a reason for the fact that the laureates were not placed in alphabetic order. Sherrington was one generation ahead of Adrian. He was born in 1857 and thus had to wait until he was 75 years old before he received his Nobel Prize in physiology or medicine. As a consequence he came to be one of the oldest Nobel laureates of his time as discussed in one of my previous books[1]. One wonders why he had to wait for such a long time. He was the most dominant scientist among the neuroscientists of his generation and he had made several important discoveries. The particular situation regarding Sherrington was discussed by Göran Liljestrand in his 1972 review of Nobel Prizes in physiology or medicine[14]. He started by noting that Sherrington had already made important contributions towards the end of the nineteenth century and then continuously over some four decades. He summarized the contributions in the following way "Sherrington has contributed more than any one else to our knowledge about the integrative functions of the nervous system. The exceptional importance of his contribution was early recognized. In many of the nominations, submitted altogether by 134 persons representing thirteen countries, this was stressed with an emphasis unusual even in such cases." Thus the matter was not a lack of nominations. It was a matter of interpreting the term "discovery", specified by Nobel in his will to be the single criterion for a prize in physiology or medicine. Even now the Nobel committee at the Karolinska Institute needs to return to the consideration of this matter.

Charles S. Sherrington (1857–1952), co-recipient of the 1932 Nobel Prize in physiology or medicine. [From Ref. 5.]

It was Johan Erik (Jöns) Johansson who originally reviewed Sherrington's work. Johansson was professor of physiology at the Institute between 1901 and 1927 and he was referred to repeatedly in my first book on Nobel Prizes[1]. The reason was that he as a young scientist worked with Nobel for five months starting in October 1890. He was an important person in furthering contacts with the Karolinska Institute and hereby presumably influencing the formulations in the final will. In 1910 he, for the first time, became a member of the Nobel committee at the Institute. This was also the year when he made his first review of Sherrington. Subsequently Johansson was frequently a member of the committee and between 1918 and 1927 he was its chairman. In later reviews in 1912 and 1915 Johansson came to the conclusion that Sherrington deserved a prize for his discovery of the reciprocal innervations of the antagonistic muscles. However, this did not carry in the committee, which pointed out that partly similar data had been presented much earlier by other researchers.

As chairman Johansson sharpened his interpretation of the will and accentuated an emphasis on the word *discovery*. He wanted to make sure that the text of the will was taken literally. In 1918 the word was used for the first time in a prize motivation. Hereafter all motivations of prizes awarded

under his aegis as chairman included the word discovery in its presentation of the prize. Johansson's attitude most likely was of importance for the absence of any prize to Sherrington. When Johansson retired in 1927 the attitude of the committee might have changed progressively so that Sherrington, pulled along by Adrian, could finally receive his prize. It would have been interesting to have achieved some insight into Liljestrand's importance for the final developments. As we shall see neurophysiological research did not start to develop at the Karolinska Institute until the beginning of the 1930s. The fact that it was the secretary of the committee, the pharmacologist Liljestrand, who gave the award ceremony speech in 1932 may indicate that he could have been influential in swinging the attitude of the committee in favor of Sherrington. It was, by the way, Sherrington who was one of the most influential nominators of Adrian.

Sherrington was one of the outstanding, if not the most outstanding figures in neurosciences of his generation. His family origin apparently is not properly given in his official biography. Considering that motherhood is a matter of fact and fatherhood a matter of opinion, it can be noted that his father was not James Norton Sherrington, from whom his family name was derived. Charles was born 9 years after the death of his presumed father. Instead Charles and his two brothers were the illegitimate sons of Caleb Rose, a highly regarded Ipswich surgeon. Rose was noteworthy not only as a physician but also as a classical scholar and archeologist. His home was much frequented by scholars and artists. In this environment Charles was stimulated to wide ranging involvements in academic pursuits, including also the arts. Late in life Sherrington, inspired by Goethe's writings, published *The Assaying of Brabantius and Other Verse*, a collection of wartime poems. Rose encouraged Sherrington to study medicine.

He started his studies at St. Thomas's hospital. After three years he went on to Cambridge studying physiology under Michael Foster, the "father of British physiology." At a medical congress in London Foster discussed some recent studies of the consequences of local cerebral damage and the development of specific symptoms. Sherrington became absorbed by neurological work which came to dominate his professional scientific career. In 1884–85 he performed experiments on dogs together with Friedrich Goltz in Strasbourg. This experience influenced his attitude to science. Goltz's credo was "among good things only the best is good enough." Sherrington was also exposed to high class microbiological research, in particular concerning cholera by visits to

Rudolf Virchow's and Robert Koch's laboratories. In 1895 Sherrington obtained his first post as a full professor. He was appointed to the Holt Professorship of physiology at Liverpool. Here he started his groundbreaking work on reflexes. By about 1900 he had developed the major tenets for which he later received his Nobel Prize. Already in 1906 he summarized his synthetic perspective on the function of synapses in *The Integrative Action of the Nervous System*[15]. He accepted the Waynflete Chair of Physiology and moved to Oxford. His laboratory became a Mecca for budding neurophysiologists. Sherrington became famous for his teaching at Oxford. A textbook *Mammalian Physiology: A Course of Practical Exercises* was published immediately after the First World War. During the war Sherrington had started to fight for the right of women to enter medical schools. In the early 1920s he was the President of the Royal Society.

Three future Nobel Prize recipients — John Eccles, Ragnar Granit and Howard Florey — were trained by Sherrington. We will meet Eccles at length in the next chapter, as a strong candidate for a Nobel Prize, and Granit will be mentioned frequently at a later stage in the present chapter and the following two chapters. Granit became a central figure in the development of neurophysiology at the Karolinska Institute. He was a frequent reviewer of candidates in the field for two decades starting after the Second World War. Finally he himself became a strong candidate for the prize which he eventually received in 1967. Florey who received his Nobel Prize for the chemical synthesis of penicillin, was discussed previously[1]. Granit wrote his autobiography in later life[5] and in this book he frequently returned to Sherrington as an unforgettable mentor and friend. In an attempt to describe Sherrington's personality he cited from an obituary written by the renowned neurologist Derek Denny-Brown. The text read:

> He was short in stature, about 5 feet 6 inches, very precise and neat in his movements, and he tended to peer through his rimless spectacles though not severely shortsighted. He had lively, humorous grey eyes and a light, easy, friendly manner. He was one of the mildest men I have ever known, rarely vexed and at most saying, "Dear me" or "That is most annoying."

Granit on an earlier occasion wrote a book about this titan of neurophysiology[16]. In a presentation of Sherrington's masterly management of the experimental laboratory environment he cited Wilder G. Penfield who had studied with

Sherrington in Oxford and he later became a world-famous brain surgeon. This is how he recorded his impressions:

> No one who had the privilege of seeing Sherrington in the laboratory could fail to be impressed by the amount of detail he noticed in the movement of an animal, in some piece of tissue under the microscope or in the general run of an experiment. He had a capacity for translating observation into problem, a capacity he retained to the end of his life as an experimenter. It is better called by its right name, a sense of wonder, a precious gift that distinguishes the truly creative from the merely talented.

Like Golgi, Sherrington also became a foreign member of the Royal Swedish Academy of Sciences. This happened in 1926. He was introduced in a hand-written letter of seven pages by a professor of pharmacology at the Karolinska Institute, Carl Gustaf Santesson. The letter which was supported by F. Henschen and others, gave a good summary of Sherrington's seminal contributions. The second paragraph presented a summary preamble "S:s exceedingly important publications deal with the physiology of the nervous system in its broader context and should be regarded as principally fundamental for our understanding of this field." Among Sherrington's many epoch-making findings Santesson emphasized characterization of reflex phenomena, including the mechanism behind the classical patellar reflex, but also other reflexes. The signaling via the spinal cord and the role of sensory and motor nerves in counteracting (antagonistic) muscles were elucidated for the first time. Similar studies were also performed in monkeys of the control of the orientation of the pair-wise orientation of the eyes. In further experiments animals that had had their brain stem cut were examined. Eliminating the control of the centers in the forebrain leads to decerebrate rigidity, an uncontrollable change of posture including a rigid extension of the limbs — *opistothonus*. This phenomenon can be seen in patients with severe brain damage. Sherrington could study the relative role of different antagonistic muscles in this condition and thereby obtained a deeper insight into reflex phenomena. The studies also allowed a distinction between signals coming from the environment, such as light, sound, touch, etc. Such signals triggered *exteroceptive* reflexes. By way of contrast signals emanating from the organ themselves, via proprioceptive receptors caused *interoceptive* reflexes. Finally Sherrington also studied more complex integrated mechanisms of standing and walking and also complicated reflexive defense phenomena when animals showed either an aggressive attitude or prepared to depart — fight or flight. Maintaining and changing posture of course is much more than keeping balance and moving by walking or running. The posture provides a lot of information on the nonverbal communications and emotional cues of individuals, be they animals or humans. When Sherrington died in 1952 he was succeeded as a foreign member of the Royal Swedish Academy of Sciences, on a proposal by Granit, by his one generation younger co-laureate Adrian.

There must have been something special about Sherrington's capacity to integrate knowledge and for example animate and make understandable the function of synapses. Granit has reflected on this kind of contribution in the final chapter of his own biography[5]. He argued that the concept discovery might be liberally extended to include contributions made by the rare breed of scientists who can integrate observed phenomena and turn them into a meaningful whole. He drew a parallel with Darwin's elucidation of the role of evolution in the establishment of a new species. In the case of Sherrington, Granit proposed that the world of integrative concepts developed by him allowed an improved definition of the individual components in the intricate nervous system. Such contributions are irreplaceable in the progress of science and therefore in the argumentation by Granit, might be stretched to represent a form of discovery. The discussion of the meaning of the term discovery will — and should — continue.

One day in October in 1932 Granit and Eccles who both were working with Sherrington at the time, learned from the BBC news that the Nobel Prize in physiology or medicine had been awarded to Sherrington and Adrian. They called Lady Sherrington and were informed that Sir Charles had quietly left for a dinner at Merton College, sparing the other guests the news. Granit and Eccles gathered all the friends from the laboratory that were available and when Sherrington came home to 9 Chadlington Road there was an improvised party. He was often referred to as "Old Sherry" and this alcoholic nectar was therefore used for the many toasts. Everybody was in high spirits and Sherrington muttered discretely "These Swedes — splendid fellows." He thoroughly enjoyed his visit to Stockholm, receiving his Nobel Prize from the hands of His Majesty the King.

Sherrington summarized his life achievements in the field of neurophysiology in his Nobel lecture[17]. He took the neuron concept of Cajal and Golgi to a higher level. The question was how the body integrates signals in the spinal cord. An incoming signal from the periphery is converted into a signal that directs a muscle to contract strongly, weakly or not at all. The signals in the incoming — afferent — nerves provided a reflection guiding the activity of the muscle as determined by the outgoing — efferent — nerves. Hence it was referred to as a *reflex*. The spinal reflex first studied by Sherrington represented

Sherrington receives his Nobel Prize. [From Ref. 35.]

a very simplified situation. As his studies advanced it came to be understood that a reflex is a very complex phenomenon. Although there can be a direct link between the afferent and efferent neuron, there can also be intermediate neurons and since the response of e.g. contraction can be the result of the signaling from many senses, an integration of signals in the spinal cord and brain may be required. One can just reflect for a moment on the mass of data that need to become integrated in a situation when the former Swedish tennis star Björn Borg, at the height of his career was running at maximum speed, managing to reach the ball and return it by a top spin drive delivering it at a few centimeters within the critical line on his opponent's side. What Sherrington found was that the tonus of a muscle is continuously monitored — there are sense organs in the muscle — and kept at a certain level by both activating and inhibiting nerves. Furthermore a single muscle, or rather its individual bundles of muscle fibers, do not act in an independent fashion. Often there are one or more antagonistic muscles and their activity needs to be controlled in parallel. Sherrington's contributions to the development of the field of neuroscience cannot be emphasized enough.

Sherrington finally retired from Oxford in 1936 but lived for 16 more years active in publishing books of lasting influence. He was remembered as the possessor of an impressive memory and his personality was colored by humility

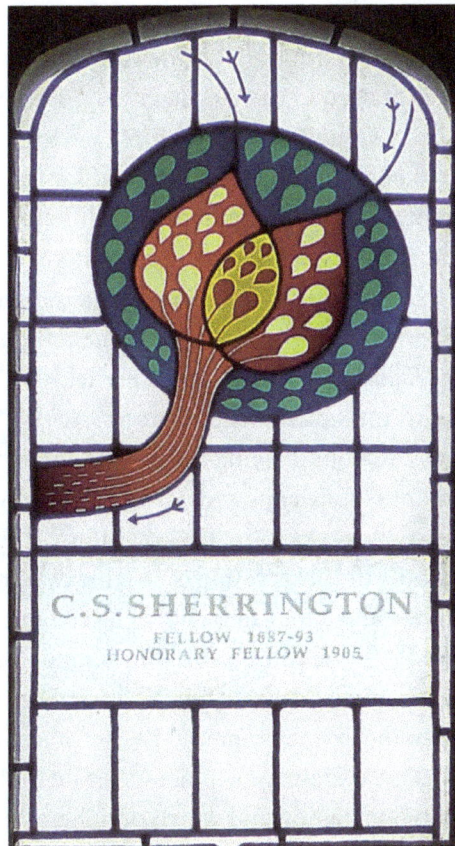

Stained glass window commemorating Sherrington in the dining hall of Gonville and Caius College, Cambridge, UK. [Image by User Schutz. CC BY–SA 2.5]

and friendliness. He was generous in sharing his knowledge. Sherrington's contributions to the development of the field of neurobiology deserve to be repeatedly highlighted. He gave a name to some specific reflexes and formulated two laws relating to the concept of a dermatome — each nerve root innervates a certain area of the body (the virus disease shingles generally extends over a single dermatome) — and to the reciprocal innervations of muscles. Unsurprisingly it was he who came to develop the use of the term *synapse* for the contacts between extensions of a nerve cell and other cells. The term itself had already been introduced in 1897 by the already mentioned English physiologist Foster using a word combined from the Greek *syn* — together, and *haptein* — to fasten, on a suggestion of Arthur W. Verrall, an English classical scholar. Sherrington is commemorated by a stained glass window in the dining hall of Gonville and Caius College, in Cambridge.

Towards the end of his long life he summarized his wise perspectives in two books. One was called *Man on His Nature* in which he reflected on the culture of science. It is praised in the biography by Granit which has already been mentioned[16]. In this book Sherrington is said to have remarked that besides the capacity to observe there is also a need for the scientist to evolve a sense of the quality of wonders, like the curiosity inherently associated with the state of childhood. This engrained feeling can potentially become satisfied in the Eureka moments of major discoveries. At that moment the discoverer is the first and only person in the world to have the unique insight. Another book was entitled *The Endeavour of Jean Fernel*. Fernel was a sixteenth-century French physician who introduced the term *physiology* to the study and description of normal functions of living things. The antonym of physiology is *pathology*, the insight into knowledge about facts pertaining to the origin and development of disease. It serves to remind that normal functions are often identified in their parts by the observation of specific dysfunctions. Pathology is a mirror of physiology and a lot can be learnt also about normal functions at the bedside of patients. Local tumors, bleedings/emboli and loss of myelin can demonstrate the location of specific functions. Some viruses can travel in the axons of neurons both from the center to the periphery and the other way around. Thus the location of damage can demonstrate both where the nerve departs from the brain or ganglion and which peripheral region it serves. We will return to this theme.

Inside Information on Nerve Electricity

Adrian was one generation younger than Sherrington. He was born in 1889. Cambridge University, where he studied physiology, was his alma mater. In 1913 he was elected a Fellow of Trinity College because of his meritorious discovery of the "all or none" principle of nerve electrical discharges. Ten years later he was elected a fellow of the Royal Society and after a further two years he initiated his pioneering studies of sense organs by use of newly developed physiological methods. In 1937 he became the Professor of Physiology at Cambridge, a position he held until his retirement in 1951. There were many older and still very active eminent scientists in the department during his time. Adrian was a very industrious person and demonstrated considerable energy throughout his life. He had more than one string to his bow. In his day he was a daring cyclist, a skilled fencer and a good mountaineer. He also liked sailing. Like Sherrington he contributed several seminal books on the developing field on neurophysiology like *Mechanisms of Nerve*[18] and *The Physical Basis of Perception*. He received many other honors and had a long life, dying in 1977.

The two 1932 prize recipients were introduced at the ceremony by Liljestrand[19]. His opening sentence was "Within the domain of physiology and medicine probably few spheres will be calculated to attract to themselves attention to the same extent as the nervous system, that distributor of rapid messages between the various parts of the body and, beyond that, the material foundation of mental life." He then introduced the neuron concept mentioning Golgi and

Edgar D. Adrian (1889–1977), co-recipient of the 1932 Nobel Prize in physiology or medicine. [From *Les Prix Nobel en 1906*.]

Cajal. After presenting Sherrington's scientific achievements he summarized his immense contributions in the following way:

> I must content myself with this short indication of Sherrington's considerable contributions. His discoveries have ushered in a new epoch in the physiology of the nervous system. On the firm foundation he has laid, many have already built further — among them should be mentioned particularly Magnus's and Kleyn's brilliant work on the posture of the body, how it is assumed and maintained. But Sherrington's work has already partly passed through the ordeal of fire which lies in its application to pathological conditions; it has shown itself to be of great importance for the understanding of certain disturbances within the nervous system, and, certainly, matters here are still in their infancy.

In the presentation of Adrian's discoveries Liljestrand mentioned some developments in the 19th century and referred to "our fellow-countryman" Frithiof Holmgren. He had pioneered measurement of electrical responses in the retina of the eye, leading to the technique of registering the electroretinogram (ERG). However, others had made even more important fundamental discoveries. Among these were the demonstration of the *action potential* by the German physiologist Emil du Bois-Reymond and the first measurements of the speed

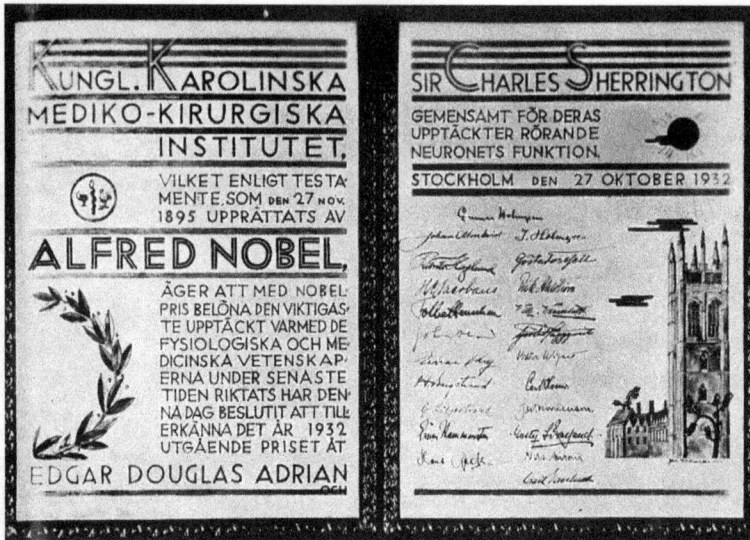

The shared Sherrington-Adrian Nobel Prize diploma. [From Ref. 35.]

of transmission of this potential by his colleague Hermann von Helmholtz, another giant in the study of human physiology.

In contrast to Sherrington Adrian was only 43 years old when he joined him in receiving the Nobel Prize. By a particular arrangement he shared his Nobel diploma with Sherrington. Adrian's groundbreaking discovery concerned the nature of the electrical activity in neurons. It had been known since studies during the nineteenth century that activities elicited in an organ like a muscle were accompanied by electrical changes. With time the techniques to measure weak currents were markedly improved and Adrian was the pioneer of this field as evidenced by his Nobel lecture[20]. He described the importance of his mentor and collaborator Keith Lucas, who died 1916 in mid-career in a wartime airplane crash, only 37 years old. In studies of nerves causing contraction of muscle fibers Lucas had demonstrated that with an increasing stimulus the contraction increased in sudden steps. The number of steps was never greater than the number of fibers in the sample. Hence it could be deduced that muscle fibers and their associated nerves follow an all-or-nothing rule. Exploiting the potential of the cathode ray tube, the capillary electrometer and by the use of so-called thermionic valves to amplify electrical impulses Adrian was able to increase the sensitivity of the recordings 5,000 times. The dramatic improvement in the possibility of studying the function of neurons has been compared to the consequences of the introduction of the light microscope on morphological studies. The highly sensitive techniques developed, with contributions by many scientists, including Herbert S. Gasser at St. Louis, who we will soon meet, could detect as small a change as one or two microvolts. Adrian made the critical finding that the firing of a single neuron, as already suggested by Lucas, was an all-or-nothing phenomenon. Adrian said "In my own work I have tried to follow the lines which Keith Lucas would have developed if he had lived, and I am happy to think that in honoring me with the Nobel Prize you have honored the master as well as the pupil." Somewhat later Adrian in his lecture presented a crucial step forward in the application of the technique. He said:

> The problem was then to limit the activity to only one or two nerve fibres. In this I was happy to have the cooperation of Dr. Zotterman of the Caroline Institute. We found that the sterno-cutaneous muscle of the frog could be divided progressively until it contained only one sense organ; this could be stimulated by stretching the muscle, and we could record the succession of impulses which passed up the single sensory nerve fibre.

Yngve Zotterman was introduced in one of my previous books[21], since he had a major role in the Nobel Prize in physiology or medicine in 1961 to Georg von Békésy. He was also important in the early phase of establishing the field of neurophysiology at the Karolinska Institute, as we shall see.

In his continued work Adrian found that once a cell had fired there was a refractory period during which it could not fire again. So by which mechanism could a signal be graded? One way was of course to vary the frequency of firing, with consideration taken to the existence of the phenomenon of refractoriness. Since a nerve contains many thousands of neuron extensions, the axons, another way of grading a signal was to mobilize different fractions of the available activating neurons or for that matter the already mentioned inhibitory neurons. It was an accidental discovery in 1928 that led Adrian to understand the presence of electricity within nerve cells. In his own words:

> I had arranged electrodes on the optic nerve of a toad in connection with some experiments on the retina. The room was nearly dark and I was puzzled to hear repeated noises in the loudspeaker attached to the amplifier, noises indicating that a great deal of impulse activity was going

on. It was not until I compared the noises with my own movements around the room that I realized I was in the field of vision of the toad's eye and that it was signaling what I was doing.

Adrian studied these fundamental phenomena in pain reflexes and also in the olfactory system. In the practical situation recording of nerve impulses was expressed both visually by the oscilloscope and also audibly over a loudspeaker. Adrian demonstrated this during his lecture accompanying the performance with the following comments "Since rapid potential changes can be made audible as sound waves, a gramophone record will illustrate this. You will be able to hear the two kinds of gradation, the change in frequency in each unit and in the number of units in action." Adrian mentioned in his lecture that besides the large fibers supplying muscle spindles, studies of other sensory nerve fibers which are much smaller had also been possible. He referred to recent work by Joseph Erlanger and Gasser, both of whom we will soon meet. He also discussed the difficulties that he and his collaborators had had in registering meaningful signals from the retina in the eye. The retina seemed to have an overly complex structure. However, the problems were later to be overcome by Haldan K. Hartline, who was to join Granit and George Wald as recipients of the 1967 Nobel Prize in physiology or medicine. Adrian's final paragraph in his lecture foretold new knowledge to come. It read:

Analogies of this kind suggest that sense organs and nerve cells send out impulses because some part of their surface has become depolarized. There are certain difficulties to be faced before this can be treated as more than a crude working hypothesis, but it is one which has important consequences. If the regions from which the discharge originates remain partly or wholly depolarised as long as they are excited, it should be possible to detect potential changes of relatively long duration in sense organs and in the motor nerve centres. Such changes are well known to occur in the eye, and they have been found in the vertebrate brain stem and in the nerve ganglia of insects. Unfortunately the structures in which they occur are so complex that it is difficult to be sure of their interpretation, but at least they suggest the possibility of obtaining direct records of the activities of the grey matter. To extract much information from such records is likely to be a far harder task than it has been in the case of peripheral nerve. In the latter our chief concern is to find out what is happening in the units and this turns out to be a fairly simple series of events. Within the central nervous system the events in each unit are not so important. We are more concerned with the inter-actions of large numbers, and our problem is to find the way in which such interactions can take place.

Hans Berger, Discoverer of the Electroencephalogram

Besides his impressive work described above, in 1934 Adrian together with Brian H. C. Matthews made an important confirmatory identification of spontaneous electrical activity of the brain. This led to the electroencephalography (EEG), first identified by Hans Berger in 1924 eventually being accepted as a valid observation by the scientific community and coming into practical clinical use. The story of Berger and EEG has many dimensions. He was trained as a psychologist and served in this function during the First World War. On one occasion when Berger was close to being run over by a horse-drawn cannon his sister sensed that he was in danger and encouraged their father to send a telegram. Berger, as a trained psychologist thought that this spontaneous telepathy was an example of "psychic energy." This led him to search for a possible spontaneous electrical activity in the brain. In connection with a neurosurgical operation on a 17-year-old boy assisted by his technician and by his wife Baroness Ursula von Bülow he recorded the first EEG in 1924. He was very uncertain about his findings and in fact had very poor insights into mechanics and electricity. It therefore took him five years to publish his findings. He could distinguish alpha and beta waves. Some additional different kinds of waves, considered by Berger possibly not to be true phenomena, later on were found to also be of clinical relevance. Berger's results were originally met with ridicule and skepticism. He is said to have had a low standing among his professional colleagues. The question is whether this generalizing is correct. He received three strong nominations for a Nobel Prize in

Hans Berger (1873–1941).

physiology or medicine in 1940, two years after he had been forced to retire. One was from Adrian. He wrote:

> In 1929 Prof. Berger discovered the human electroencephalogram, the characteristic series of potential oscillations produced by the cerebral hemispheres and appearing in records from the surface of the head. He made a full analysis of the conditions under which the waves appeared, and proved their cerebral origin. For some years his work passed unnoticed or was thought to be erroneous, but it is now fully accepted and his discovery has led to a great deal of valuable knowledge about the activity of the brain. In particular it has led directly to a clearer understanding of major and minor epilepsy (reference), to the changes which occur during sleep (reference) and during attention. I would have nominated Prof. Berger before, but did not do so as I thought he might have difficulty in accepting — the award to Prof. Domagk encourages me to make the present nomination.

The reference to the 1939 prize in physiology or medicine is interesting. When Adrian wrote his letter on November 24 he was probably not aware of the fact that Domagk, after first having accepted the prize was eventually forced by the German Government to decline it. As described previously[1] Hitler became so angry when Carl von Ossietzky received the Nobel Peace Prize in 1936 that he forbade German citizens to accept Nobel Prizes.

The other two nominations for a prize to Berger came from the famous Harvard physiologist Walter B. Cannon, and Tracy J. Putnam, both at Cambridge, Mass. The committee took a serious view on the nominations and asked von Euler to make a full investigation. He made a very thorough analysis over 12 pages discussing the different studies made by Berger of the brain waves under various physiological (wakefulness, sleep, etc.) and pathological conditions. The observations of age-dependent variations of the wave pattern and the influence of different kinds of drugs, including those used for narcosis, were discussed. Von Euler's conclusion read:

> This (the observations just mentioned) and the development of a technique to register the electroencephalogram directly from the skin of the head and the critical analyses of the many different sources of errors, which may affect the results, secure a very high scientific ranking of Berger's observations. Berger's priority regarding (the

discovery of) the encephalogram in humans is uncontestable and is expressed by the fact that Adrian and other prominent researchers use the expression "Berger rhythm" for the slower waves, the so-called alpha-waves. Although the mechanism for the emergence of the wave potentials still remains to be explained, Berger's observations on human E.E.G have already increased our knowledge about the activity of the cortex and seem to be acquiring an increasing importance for clinical diagnosis of different pathological conditions concerning the cortex. In view of these facts it can be considered that the results obtained by Berger are of the kinds that deserve to be recognized by a Nobel Prize in physiology or medicine.

It is not difficult to agree with von Euler that the discovery of the electrical activities of the brain measured by the non-invasive method EEG could have been recognized by a Nobel Prize. EEG has survived into modern times in spite of the fact that other very important methods of non-invasive investigation of the physiology and pathology of the brain have been introduced. However, in 1940 the committee, in spite of von Euler's complimentary conclusions, decided that Berger's work was not worthy of a prize. In addition, because of the war conditions, it was agreed that no prize at all should be awarded in 1940. One wonders if Berger ever found out that he had been nominated for a Nobel Prize. He was forced to discontinue his work on the EEG phenomena after his retirement. In 1941 a very depressed Berger committed suicide by hanging himself in the hospital environment where he had worked.

The Role of Chemical Messengers at Synapses

Only four years after 1932, the field of neurobiology was recognized again by a Nobel Prize in physiology or medicine (Table 1.1, p. 14). It was awarded to Henry H. Dale and Otto Loewi "for their discoveries relating to chemical transmission of nerve impulses." The recommendation of the committee to award these two scientists the prize was not unanimous as we shall see in Chapter 4. Dale and Loewi had focused on another critical step in the functions of nerves. What happens when the electrical signal transported in a nerve reaches the effector organ such as a muscle? Does the electrical signal directly cause a contraction of the muscle fiber or could there be a chemical intermediary? The new finding was that there were chemical intermediaries.

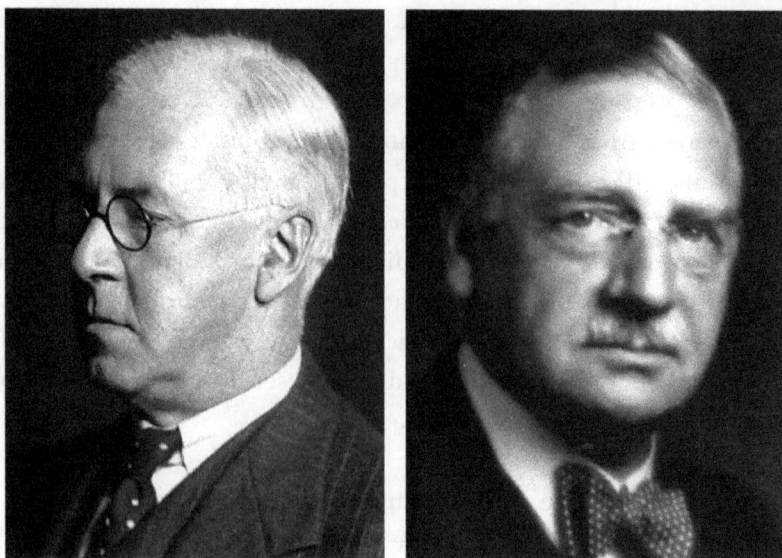

Sir Henry H. Dale (1875–1968) and Otto Loewi (1873–1961), recipients of the 1936 Nobel Prize in physiology or medicine. [From *Les Prix Nobel en 1936*.]

Just as in 1932 these prize recipients were honored by an award ceremony speech given by Liljestrand. As I have already mentioned repeatedly in my two previous books on Nobel Prizes[1, 21] he had a major influence on the Nobel prize work at the Karolinska Institute, being the secretary of the Nobel Committee for more than forty years, 1918–1960. Liljestrand was not a neurobiologist, but a pharmacologist. Still his deep insight into biomedical research allowed him to speak on essentially any subject within the field of biomedicine. He was so familiarized with the advances in the different fields of physiology or medicine that it is said that he wrote the protocols of the meetings with the Nobel Committee before the actual meeting, but that is probably only a tall story. Anyhow, in the absence of a specialist on neurophysiology/biology in the College of Teachers, he introduced for a second time Nobel Prize recipients in this field[22]. His presentation was also very informative to the lay public.

Liljestrand's point of departure was a fable in Livy's (Titus Livius) work on the history of Rome, which stressed the importance of the need for cooperation between all parts in the interest of the whole — "consensus partium." The body fluids, in particular blood, are critical in the distribution of materials and also in the removal of waste products. A particular form of signaling is by use of specific chemical compounds, hormones, produced in a certain organ and active over distance. In both these cases the effect is relatively slow and extends over time. There is also a general system which

Hilding Bergstrand (1886–1967), the vice-chancellor of the Karolinska Institute, member of the Board of the Nobel Foundation, to the left, and Göran Liljestrand (1886–1968), long-time secretary of the Nobel committee at the Karolinska Institute. [From Ref. 35.]

exchanges rapidly transported messages, the nervous system. As we have seen a sensory signal may result in a rapid contraction of a muscle. This is effected by the *motor nervous system*. And there is yet another component of the peripheral nervous system, the so-called *autonomic nervous system*. As its name implies it is not under the control of our will. It is an autonomous system controlling the actions of our many internal organs. There are two components of this system as can be illustrated by its action on the heart. The *sympathetic* system accelerates the heart beats and the *parasympathetic* system reduces its rate of contraction. The effect of, for example, strenuous physical work varies; the cardiovascular system is activated whereas the activities of other organs may be suppressed.

Originally it was thought that electrical impulses traveling in the nerves acted directly on the muscle or, for example, the gland they were in contact with. However, already in the early twentieth century evidence was presented that chemical intermediary messengers could also be involved. The substance *adrenalin* was discovered in extracts of the adrenal glands — riding on top of the kidneys —, hence their name, and it was found to bring about the same effect as activation of the sympathetic nervous system. The release of adrenalin at sympathetic nerve terminals was suggested to lead to contraction of smooth muscle cells. As we shall see this interpretation was corrected in 1946 by Ulf von Euler, who demonstrated that the active substance was noradrenaline and not adrenaline. The activation of adrenaline in situations of fight or flight as the signaling substance of stress was introduced by the abovementioned Cannon. He had a very broad influence on his discipline and was the scientist who popularized the term serendipity in medical science[1]. He is also known outside science by the fact that one of the major mountain peaks in Glacier National Park in the U.S. has been named after him. Ten years after the original discovery of adrenaline it was found by Dale in 1915 that acetylcholine could mimic the effect of the parasympathetic system. It took some time to demonstrate the role of these substances in the intact animal. This was done by Loewi in elegant experiments. They have been described already in one of my previous books about Nobel Prizes[21] and used as an example of a truly paradigmatic discovery. In essence Loewi used a simple experimental approach. He stimulated the heart in one intact experimental animal via the sympathetic nerves and hereafter he collected the fluid surrounding the heart. This fluid was transferred to the medium surrounding a heart from which the nerve connections had been cut off. The fluid accelerated the rate of heart beats! When fluid connected from the first heart after stimulation of the parasympathetic (vagus) nerve was transferred, it instead decelerated the rate of the heartbeat. This is the kind of experiment that attracts the attention of Nobel Prize committees. Loewi was also eventually able to document that it was acetylcholine that was one of the critical chemical messengers and adrenalin the other. Interestingly acetylcholine which has an inhibitory effect on the heart muscle has an excitatory effect at the neuromuscular junctions in skeletal muscle.

It now remained to investigate whether *all* the effects elicited by nerve signals were dependent on the release of a chemical messenger or if there were cases when an effect was elicited directly by the current traveling in the nerve. The "war" between those who believed that a chemical intermediary was essential — the *humoralists (also soupers)* — and those who believed that the electrical current sufficed by itself to elicit an effect — the *ionists (also sparkers)* — raged for many years. Eccles as we shall see in the following chapter was one of the last to accept that in most cases the release of chemical transmitters at the nerve endings was important.

Dale focused on the extension of knowledge about chemical transmission as a critical step in transmission of nerve impulses. He described how it applied to the parasympathetic system but also to the activation of muscle fibers by motor neurons. It was noted that at the time of his Nobel lecture[23] data had been published indicating that the mobilization of sodium ions was of importance in transmission of the nerve impulse, a harbinger of new insights to be gained. Dale was one of the outstanding figures in the field of neurobiology and in pharmacology in general. He was an appraised academic leader and he has been cited on several occasions in my previous books on Nobel Prizes. The rights and responsibilities of a Nobel Prize recipient to nominate future candidates to the prize have always been emphasized in various contexts by Nobel committees. Dale took this task very seriously, as already mentioned[21].

Dale was born in 1875. After basic studies of physiology under John N. Langley — the father of the drug receptor theory — at Trinity College in Cambridge, he received his medical training at St. Bartholomew's Hospital in London, receiving his M.D in 1909. He then joined the eminent laboratory of physiology at University College headed by the abovementioned Starling. In that laboratory he met Loewi, who was to become his life-long friend and co-recipient of the Nobel Prize. Dale also had experience from working abroad. He spent four months with the famous laureate Paul Ehrlich in Frankfurt. After this, in 1914, he joined the Department of Biochemistry and Pharmacology at the National Institute for Medical Research in London, as departmental director. Fourteen years later he became director of the whole Institute, in which post he remained until his retirement in 1942. Following this he became heavily involved in the influential Wellcome Trust. He was chairman of its board from 1938 to 1960.

In 1914 he was elected a Fellow of the Royal Society, and served as its secretary from 1925 to 1935 and its president from 1940 to 1945. He received many honors and was a member of many national academies, including the Royal Swedish Academy of Sciences. He was elected a foreign member of this academy in 1932 based on a proposal which emphasized his discovery of the dual effects of adrenaline, contractive in the small blood vessels causing an increased blood pressure, but also relaxing on smooth muscle cells. The nominator also highlighted Dale's studies of the substance histamine, which plays a special role in anaphylactic shock and in general in allergic reactions. In 1913 Charles Richet had received a Nobel Prize in physiology or medicine for the discovery of the latter phenomenon[21]. Dale had a long life, remained a member of the Academy for 36 years and died in 1968 at the age of 93 years. Fifteen years before his death he published two influential books; *Adventures in Physiology* and *An Autumn Gleaning*.

Loewi was born two years earlier than Dale and he became almost as old, dying at the age of 88 years in 1961. His life took many turns. He was born in 1873 in Frankfurt-am-Main. After high-school studies he registered as a medical student at the Universities of Munich and Strassburg (at the time a part of Germany), but he spent most of his time at lectures in the philosophical faculty. He then became interested in research and completed his doctoral thesis working under Oswald Schmiedeberg, often referred to as the "Father of Pharmacology." After this he received a training in inorganic chemistry and finally became assistant to Hans H. Meyer, a renowned pharmacologist. In 1909 he received his own professorship, the chair of pharmacology in Graz and from there on he developed his academic career in Austria. He was an admired lecturer and physiologists from all over the world came to listen to his lectures. In 1902 he spent some time in Starling's laboratory[1]. In this laboratory the first experiments suggesting the role of chemical messengers were made, as has already been referred to. A substance called secretin was identified. Loewi got the inspiration for his groundbreaking experiments during this visit and he also met Dale, as mentioned. It was in 1921 that Loewi discovered the chemical transmission of nerve impulses. The concept that there might exist chemical messengers had been brewing in his mind for 17 years. The story goes[24] that he woke up one night during the week before Easter with the idea of the existence of chemical messengers. He wrote himself a note, which he could not decipher when he woke up. The following night the idea came back to him and then he went straight to the laboratory and made the experiment. This work culminated in his finding that the parasympathetic transmitter substance was acetylcholine and that a substance closely related to adrenaline played the corresponding role at the sympathetic nerve endings.

Loewi was the one to give his Nobel lecture[25] first in 1936. It presented the whole history of progressive insights into the role of different humoral compounds at the nerve endings. He concluded:

> In conclusion a word or two on the question of how the neurochemical mechanism fits into the connecting pattern of cells. With the discovery that its influence comes about through substances which are released by the nervous system itself, we have the first proof that the nervous system is not only one effector organ for chemical influences from the outside, and not only a participant in general metabolism, but that it has itself a specific chemical influence upon happenings in the organism. On a closer examination this is not surprising.

In nerve-free multicellular organisms, the relationships of the cells to each other can only be of a chemical nature. In multicellular organisms with nerve systems, the nerve cells only represent cells like any others, but they have extensions suited to the purpose which they serve, namely the nerves. Accordingly it is perhaps only natural that the relationships between the nervous system and other organs should be qualitatively of the same kind as that between the non-nervous organs among themselves, that is to say, of a chemical nature.

It is of course always tempting to simplify and generalize. In later studies it has been found that there are in fact certain synapses at which the electrical charge directly releases the effect, as we shall see later, and in the non-multicellular world, bacteria in some cases can communicate directly by wires transmitting electrical signals.

Dale in his Nobel lecture[23] presented data on ongoing experiments and extended the picture given by Loewi. In his final summary Dale stated:

You will see that we are thus led to the conclusion that nearly all the efferent neurons of the whole peripheral system are *cholinergic* (use acetyl choline, my remark); only the postganglionic fibres of the true sympathetic system are *adrenergic* (use adrenaline, my remark), and not even all of these.

He finally commented that when it comes to the chemical transmission in the central nervous system essentially nothing was known at the time. It remained to find techniques to define the range of different transmitter substances used and topographically define the presence of each one of them in different parts of the brain.

In 1938 the Germans invaded Austria. Because of his Jewish ethnic background Loewi was at threat, although he tried to suppress this fact since he was deep into some important experiments. The day of the Nazi invasion, March 12, storm troopers broke into his home and arrested him at gunpoint. His wife and two sons were put under house arrest. Besides his concerns for his family he felt an urge to submit some recent results before he was to be murdered, as he anticipated. A prison guard was persuaded to take a postcard on which he had summarized his seminal findings and mail it to the journal Die Naturwissenschaft! The international community of physiologists reacted strongly to Loewi's imprisonment. Leading American and British delegates, including Cannon and Dale made it clear to the organizers of a physiology congress at the time in Switzerland that if Loewi was not released there would not be any American and British participation in the congress. Two weeks later Loewi was released and was able to leave Austria. However, this was possible only after he had instructed a Swedish Bank, in which he had deposited his Nobel Prize money, to transfer this material component of the reward to a bank controlled by the Nazis. Loewi first went to England where his friend Dale cared for him, as he did for a number of other refugees. Loewi's goal was to continue to the United States, where he wanted to take up a position he had been offered at New York University. In order to travel he needed an American visa. This turned out to be somewhat complicated since the U.S. consul in London did not accept Loewi's letter of dismissal from Graz University as a verification of his teaching capacity. Loewi then suggested that the consul contacted Dale. In order to testify to the consul that Dale was a famous person a copy of *Who's Who* was produced. This satisfied the consul and a visa was granted. When gratefully leaving the consul Loewi's farewell remark was "By the way, do you know who wrote this paragraph in *Who's Who*. I did. Goodbye Mr. Consul." And off Loewi went to a future life in the U.S.

He enjoyed life in his new home country very much. He could develop his science and also his keen interests in the humanities to the full. Like a number of his colleagues he loved to spend summers at the marine biology station at Woods Hole, enjoying its enriching intellectual atmosphere. Besides the Nobel Prize, Loewi received many other prizes and academic honors and he became a Fellow of the Royal Society in London. Throughout his life he retained a deep interest in the humanities. This gentle and modest scientist was very thankful in the end for his change of homeland. He said "I am happy and deeply grateful to the fate that transported me to this country — where I continue to enjoy the stimulating, almost rejuvenating effect of new friendships and the wealth of new impressions and experiences[24]."

The Role of Critical Collaborators

Over the years Dale had many collaborators in his growing research enterprise. There were often many coauthors on his publications and on some important papers he even left his own name out. Deciding on priorities is one of the critical responsibilities of Nobel committees and the task is often challenging. There may also have been a shift in time on the relative emphasis of the role of a leader of a group of scientists and his collaborators. In the first part of the twentieth century a leader of the group may have had a relatively more dominant and authoritarian position than in the period after the Second World War. It was after this that a democratization of research environments developed, starting in the U.S. A particular case to be mentioned is Wilhelm Feldberg[24].

Feldberg was born into a well established bourgeois Jewish family in Germany. He trained in Berlin and when he had obtained his medical degrees he turned to neurophysiology. He received some experience from working abroad with the professor of neurophysiology in Cambridge J. N. Langley. When Langley died he continued his work with Dale. He then returned to the Institute of Physiology in Berlin. Shortly after Hitler had come to power in 1933 the director of the institute called on Feldberg and told him that he had been dismissed and had to leave the same day. After some time Feldberg discovered that there was a representative of the Rockefeller Foundation in Berlin and he managed to contact him. He was very sympathetic but explained by way of caution that he had so many requests and that therefore the chances that he would be able to help a young person like Feldberg were very small. He then asked again what Feldberg's name was and when he had it spelled out he said "I must have heard about you. Let me see." He then looked for a page in his diary and when he found it he said "Here it is. I have a message for you from Sir Henry Dale whom I met in London about a fortnight ago. Sir Henry told me, if by chance I should meet Feldberg in Berlin, and if he has been dismissed, tell him I want him to come to London to work with me. So you are all right," he said affectionately. "There is at least one person I need not worry about any more." Feldberg promptly left for London and was soon joined by his family. He became an important participant in Dale's research group and stayed for two years before he left for Australia to work with a cancer researcher C. H. Kellaway in Melbourne. Due to shortage of funds he returned to England more specifically to a readership in Adrian's department in Cambridge. He was a very much appreciated teacher, but finally he returned to London to head a division at the National Institute. He became a Fellow of the Royal Society. During his two years with Dale important findings were made and Feldberg's name appears on many publications from the group. He himself wrote in a historical review of the time published in 1976[26]:

Between 1933 and 1939, the index volume of Journal of Physiology for volumes 61 to 100 lists, from Dale's laboratory, fourteen publications, printed communications of full papers, on acetylcholine in ganglionic and neuromuscular transmission, and another ten publications of transmission by acetylcholine in other tissues: in the suprarenal medulla, stomach wall, sweat glands, salivary glands and central nervous system. I cannot help it, but my name appears a co-author on all twenty-four of them. What is the explanation? And what was my contribution which resulted in this embarrassing fact, that I was an eye-witness in all of them?

He then goes on to argue that a particular method of using eserine as a blocker of the attachment of acetylcholine to tissues, allowing its identification, was critical for the discoveries made at the time. Thus although Feldberg's life has been summarized to be highly productive and positive due to his courage and optimism[24] there was as so often in human lives a flip side of remaining regrets.

Neither Feldberg nor any of the other collaborators in Dale's group had been nominated for a prize at the time. In his report of August 1936 Liljestrand wrote:

As far as Dale is concerned, I brought forward the previous year that his achievements in the present field partly include the fundamental and meticulous analysis of the action of acetylcholine, which has allowed a closer characterization and expansion of Loewi's discovery (especially Dale's discovery of the conditions for the preganglionic fibers), partly by the demonstration of the biological occurrence of acetylcholine. During the year that has passed Dale in one particular context enlarged our knowledge further, and that concerns the question about the role of acetylcholine in motor nerves. Hereby it seems to me that his part in the fulfillment of the discovery has increased markedly. It should perhaps be mentioned that Dale personally has not always appeared, among the numerous collaborators from his laboratory (especially Gaddum, Euler, Feldberg, Guamarais, Vogt, Brown, Vartiainen) who have often published papers about problems relating to the field, without Dale's name being present among the authors. It is however without doubt that he has had a considerable responsibility for these publications too.

After this he concluded that Dale and Loewi should receive the 1936 prize. Since it was Dale's emphatic considerations which had saved Feldberg in the critical situation in 1933, there was of course a special resonance in their professional relations. Feldberg was in fact finally nominated for a Nobel Prize in physiology or medicine in 1954. The nomination came from B. Mendel in Amsterdam. He wrote:

> Following his important work with Dale on the physiological role of acetylcholine as the transmitter of impulses at the synapses of autonomic ganglia and at neuro-muscular junctions, Feldberg showed that acetylcholine is a synaptic transmitter also in the central nervous system where cholinergic and noncholinergic neurons alternate.
>
> His most recent work is concerned with the introduction of drugs through a permanently implanted cannula into the lateral ventricle of conscious animals. By administering drugs this way, Feldberg was able to elicit an epileptic-like syndrome with tubocurarine, anesthesia with adrenaline and noradrenalin and catatonic stupor with large doses of acetylcholine or with an anti-cholinesterase. These experiments provide a basis for a new approach to the understanding of mental diseases.

Von Euler reviewed the new data and concluded that it was too early to discuss the potential prizeworthiness of the cited publications. This was also the conclusion of the committee. There were no more nominations for Feldberg.

Nerve Signals Travel at Different Velocities

In his Nobel lecture Sherrington referred to some recent work by Joseph Erlanger and Herbert S. Gasser. Their work led to one more Nobel Prize in physiology or medicine in the field of neurobiology. It was awarded during the cataclysmic later part of the Second World War in 1944. This year was the first one in which the Nobel awarding institutions resumed their recognitions of prize recipients. The work by the committees had continued during the war and potential prize recipients had been identified, but not awarded due to the abyss of the world. For natural reasons there was no possibility in 1944 for the prize recipients to come to Stockholm and no Prize ceremony was arranged. However, there was one exception and that concerned the recipient of the prize in chemistry for 1943 (awarded in 1944). It recognized the work by George de Hevesy. His prize has been described in depth in one of my preceding books on Nobel Prizes[21]. At the time Hevesy, who had fled from Denmark and moved his research activities from the Bohr Institute in Copenhagen to Stockholm University, was recognized by an improvised and unique Nobel Prize ceremony at the Royal Swedish Academy of Sciences. This was arranged in connection with a regular meeting of its members.

The other prize recipients in the natural sciences at the time had their domicile in the United States. A reception was arranged by the Swedish Ambassador in New York and speeches and greetings, also from the Swedish Crown Prince, were exchanged over the radio. The Americans Erlanger and Gasser received their prize "for their discoveries relating to the highly differentiated functions of nerve fibers"(see table, p. 14). The award ceremony presentations this year were also made as broadcast speeches. The prize in physiology or medicine on this occasion was not presented by Liljestrand as a "stand-in" neurobiologist, but by a person who was developing a powerful school for the neurosciences at the Karolinska Institute. It was Granit, of Finnish origin, already mentioned above in the presentation of Sherrington. He was also introduced briefly in my preceding Nobel Prize book[21], since he was a critical reviewer of Georg von Békésy, the hearing physiologist, who received the 1961 Nobel Prize in physiology or medicine. However, since Granit played such a central role in the evaluations of candidates in neurobiology during the 1950s and the early 1960s it is due time to recapitulate his background.

Joseph Erlanger (1874–1965) and Herbert S. Gasser (1888–1963), recipients of the Nobel Prize in physiology or medicine in 1944. [From *Les Prix Nobel en 1940–1944.*]

Granit was born in 1900. He took an active part in the Finnish war of liberation in 1918. Thereafter he pursued his university studies at the Turku (Åbo) Academy in Finland. He wanted to study experimental psychology but was persuaded by his uncle to also acquire a medical degree. Soon he decided to focus on physiology, in particular the mechanisms of vision. He got his Ph.D. in 1927 and spent half a year as a post-doc in Sherrington's laboratory. After some two years in the United States as a fellow of the University of Pennsylvania he returned to Sherrington's laboratory where he spent the years 1932–33 on a Rockefeller Foundation scholarship. Back in Helsinki

Ragnar Granit (1900–1991), recipient of the 1967 Nobel Prize in physiology or medicine. [From Ref. 5.]

he was appointed full professor in 1937. He became involved in the Winter War between Finland and Russia, but in 1940 he received an offer to move to Harvard University and a separate invitation to come to the Karolinska Institute. In spite of the former flattering offer he decided to choose Stockholm. Upon his arrival a new era in neurophysiology was initiated at the Institute which will be further discussed at the end of this chapter. Major support was received from the Knut and Alice Wallenberg Foundation. Granit became professor of neurophysiology at the Institute in 1947. His first provisional laboratory in 1945 was upgraded to a section of the Medical Nobel Institute, and very soon a new building was erected for him and his group. In 1947 they moved into a separate Nobel Institute for neurophysiology. Granit became heavily involved in the work with the Nobel Prizes and made several reviews for the committee. Surprisingly he never became full member of the committee, but he was an adjunct member in 1952, 1953, 1955 and in 1961–63. From 1964 and onwards he distanced himself from the committee, probably because the strength of his own candidature increased. Granit shared the 1967 Nobel Prize in Physiology or medicine with Haldan K. Hartline and George Wald. There will be reasons to return to this prize when the archive materials are released for historical studies in 2018.

Granit's award ceremony speech at the 1944 prize events was given as a broadcast lecture on December 10 to reach the prize recipients present at the

New York ceremony[27]. It was short. He described three milestone advances in the field of neurophysiology. The first was the discovery in the mid-nineteenth century that the nerve impulse was an electrical wave (Du Bois-Reymond, Helmholtz). The second according to his judgment was Adrian's demonstration that the impulse was all-or-nothing and that varying the frequency of firing was a way of quantifying the strength of the signal. The discoveries of Erlanger and Gasser were presented as the third step forward. Nerve fibers were found to have different cross sections, ranging from 0.001 to 0.020 mm. Development of new methods allowed a measurement of the speed with which the electrical signal travels in a nerve fiber. By use of an elegantly developed technique — an adaptation of the cathode-ray oscillograph — Erlanger and Gasser could show that there are three major groups of nerve fibers. They were referred to as A, B and C. The thickest fibers (A) could conduct impulses with a speed of 5 to 100 meters per second. B fibers showed an intermediate impulse velocity, 3 to 14 meters per second and the C group finally only allowed a transmission speed of below 2 meters per second. Studies involving the in- and outgoing fibers of the spinal cord showed that for example pain (sensory neurons) was transmitted by the slowly conducting fibers whereas fibers responsible for signaling muscle activities and touch were rapid transmitters. Granit finished his brief presentation in the following way:

> When today Erlanger and Gasser receive the 1944 Nobel Prize for their discoveries concerning the highly differentiated properties of single nerve fibres, it might be pointed out that their achievement was not born, fixed and armored, in the manner of the birth of Pallas Athena. But no sooner had their first results given them the key word than discovery followed hard upon discovery until their colleagues everywhere in the world came to realize that a great synthesis had been born to nerve physiology. This synthesis is based on new facts, well-hardened by a masterly technique cementing them into a groundwork on which will be erected whatever structure that future has in store for the physiology of the central and peripheral nervous system.

Erlanger was the senior scientist in the team. He was born in 1874, whereas his pupil Gasser was born in 1888. They joined forces when Erlanger was a professor at the University of Wisconsin and also later when he moved to Washington University in St. Louis. Erlanger had received training as chemist at the University of California. He then went to Johns Hopkins Hospital and continued from there to the University of Wisconsin, where the joint work with Gasser was performed. The development of the use of the cathode-ray oscilloscope was critical in their joint study. Erlanger finally settled at the Washington University in St. Louis. Gasser studied physiology at the University of Wisconsin, where he also obtained his M.D. in 1915. He then followed Erlanger to St. Louis. After having worked for some time with Dale in London, Gasser became professor at Cornell University in New York and finally he was selected to become a very successful leader of the impressive Rockefeller Institute from 1935 to 1953. Gasser was the coauthor of a book *Electrical Signs of Nervous Activity* published in 1937. He received many academic honors including a foreign membership in the Royal Swedish Academy of Sciences and finished his productive life in 1963, 75 years old. Erlanger in fact survived him by two years, dying at the age of 91.

Gasser came to Stockholm after the war had ended to give his Nobel lecture on December 12, 1945. The title was "Mammalian Nerve Fibers[28]." He provided a background to the extensive work that led to the identification of A, B and C fibers. It is of interest that in three places he refers to work by Zotterman. Thus the reference is not only to Zotterman's collaboration with Adrian, but also to the fact that Zotterman preceded Gasser in studies of signal velocities in mammalian nerve fibers. He was using frog nerve fibers. Gasser's lecture ended:

> The more one sees of the exquisite precision with which events take place in the central nervous system the more one is impressed by it. The more the idea of timing grows in meaning content the more it becomes a directive for future exploration. Differential axonal velocities must play their part in the mechanism. Be this their only contribution to integration, it is still a large one.

Erlanger's lecture was presented two years later, also on December 12. Its title was "Some Observations on the Responses of Single Nerve Fibers[29]." It focused on the studies of the particular electrical signal that travels along the axon of nerve cells and it emphasized the importance of technological developments allowing the measurement of the electrical currents in nerves. A critical change occurred in 1932 when the amplification could be increased from 100,000 times to 2,000,000 times. It was first then that it became possible to measure the signal in individual nerve fibers. The nature of the so-called *action potential* was discussed and towards the end of the lecture the early pioneering work

Erlanger in Stockholm in 1947 to give his Nobel lecture. On the left is Bergstrand and on the right Bernardo A. Houssay (1887–1971), the recipient of half a Nobel Prize in physiology or medicine in 1947. [From Ref. 35.]

by Alan Hodgkin was referred to. Both Erlanger and Hodgkin were interested in understanding the nature of the electrical current in nerves. It would take a few years before Hodgkin could describe the nature of the transmission of current along the axon, to be discussed in Chapter 3.

There are various different reasons for the different rates of conduction of nerve impulses. The most important factor in facilitating a rapid and efficient transmission of them is the existence of an insulating layer around the axon. This layer is formed by a composite material dominated by lipids. Not all axons are wrapped in a lipid-containing sheath. An example is neurons representing the sensory nervous system. The insulating material is called *myelin* and this fatty material is produced by different kinds of cells in the central and in the peripheral nervous system. Its main constituent is cholesterol (see Chapter 6). In the brain the peripheral parts (the cortex) are grayish — the grey matter, whereas the central parts are white — the white matter — due to the predominance of myelin. The presence of myelin isolation can increase the speed of transmission of nerve impulses tenfold. The myelin sheath is not continuous. It is divided into sections separated by the so-called Nodes of Ranvier. Louis-Antoine Ranvier was a renowned French histologist of his time.

The myelin sheath was discovered by the famous German anatomist Rudolf Virchow in 1854, but it was Ranvier, working in the laboratory of the most famous physiologist of his time Claude Bernhard, who a few decades later discovered the nodes that were later given his name. Ion channels can only become activated at the Nodes of Ranvier. The current generated then extends over the neighboring node and jumps to the next node. This is referred to as the *saltatory* conduction. Thus the action potential, to be discussed further below, moves in discrete jumps along a myelinated axon. The moving signal is boosted at each consecutive node. In essence the presence of myelin has two important consequences. It increases the rate of conduction of the signal and it saves energy.

Although it was proposed that the transmission of the nerve impulse was saltatory already in 1925 it was not until 1939 that it was experimentally proven by Ichiji Tasaki, who was mentioned towards the end of Erlanger's Nobel lecture. Ten years later A. Huxley experimentally consolidated this concept. We will meet him later in Chapter 3 and also learn that nerve myelination, which essentially only occurs in vertebrates, is not the only way to increase the speed of the nerve impulse in the axon. Size also matters. After the Second World War Tasaki moved to England and Switzerland and continued to make high quality studies of nerve fibers. Eventually he settled in the U.S. where he made important contributions to the field of audiology. Together with his colleagues he revealed how mechanical vibrations in the cochlea of the ear could be converted into electrical signals for the brain to interpret. These were essential follow-up studies in the field of hearing physiology established by Georg von Békésy, who received a Nobel Prize in physiology or medicine in 1961[21]. Tasaki continued his studies together with his wife throughout his long life. When he died in 2009 he was almost a hundred years old.

A Controversial Nobel Prize

Before we come to the Nobel Prize in physiology or medicine in 1963, there are two more prizes in the field of neurophysiology and in neurosurgery to be discussed (see table, p. 14). The first one is the prize in 1949 to the Swiss scientist Walter R. Hess and the Portuguese neurologist António Egas Moniz. The motivation for Hess's prize was "for his discovery of the functional organization of the interbrain as a coordinator of the activities of the internal organs." This prize will be described in the next chapter together with a discussion of Horace

W. Magoun, who narrowly missed out on a Nobel Prize. Both these scientists investigated integrated systemic functions located in the basal older parts of the brain. Moniz received his prize "for his discovery of the therapeutic value of leucotomy in certain psychoses." The part of the 1949 prize recognizing Moniz has been extensively debated. Nobel archival studies have been carried out and in the light of later historical developments it has been debated whether or not he was worthy of a Nobel Prize[30].

The prize recipients were presented at the award ceremony by the professor of neurosurgery at the Karolinska Institute, Herbert Olivecrona[31]. His contact with neuroanatomy derived from his experiences of consequences of removing tumor tissues from different parts of the brain. Olivecrona gave the following introduction to Moniz's contribution:

> The lines of thought along which Antonio Egas Moniz has advanced to the discovery of the prefrontal leucotomy refer primarily to the localization of certain psychic functions in the brain. It has long been known that the frontal lobes are of great importance for higher cerebral activity, especially in regard to the emotions, and that the destruction of the frontal lobes, by bullet wounds or brain tumours, leads to certain typical changes of the personality, primarily on the affective plane, but sometimes also affecting the intellect, especially highly integrated intellectual functions such as power of judgment, social adaptability, and the like. The American physiologist, Fulton, and his collaborators have proved by experiments on anthropoid apes that neuroses caused experimentally disappeared if the frontal lobes were removed and that it was impossible to induce experimental neuroses in animals deprived of their frontal lobes.

The first and most famous documentation of the role of the prefrontal cortex in coordinating behavior and emotion is the case of Phineas Gage. The story has been told many times. When he was a railroad construction foreman in 1848 he accidentally released an explosion when clearing a roadbed for a new railway line. The explosion drove a thirteen pound iron rod through the front of his skull. Remarkably he survived, but the destruction of his left prefrontal cortex led to major personality changes. Originally he had been a person with excellent knowledge and judgment, but after the accident he became prone to erratic and irresponsible behavior. He had lost the balance of his mind.

Olivecrona continued his laudation of Moniz's work in the following way:

When it is remembered that other methods of treatment have failed or have been followed by recurrence of the disease, it is easy to understand the immense importance of Moniz's discovery for the problems of psychiatric treatment. As was expected, the results are best for the non-schizophrenic groups, that is to say, among those suffering from depression, obsessive neurosis, and the like, where the great majority of patients operated upon have recovered and become capable of working. Within the schizophrenic group, where the disintegration of the personality has often advanced very far, the prospects are less favorable, but even in this group quite a few cases can be released from the mental hospitals, some of them after having fully regained the capacity for work. In other less favorable cases, the nursing problem will be much simplified by the fact that the patient, after the operation, can be kept in a 'quiet' ward.

Moniz, born in 1874, had an aristocratic background. He was trained as a physician and became a specialist in neurology. In 1911 he moved from a professorship at Coimbra to a new chair in neurology at Lisbon. His main contributions in science concerned cerebral angiography, visualization of the main blood vessels in the brain by X-ray after injection of a solution providing contrast, and lobotomy. His first nomination for a prize was in 1928 and the proposal cited his development of the angiographic technique. The level of this discovery was not considered sufficiently high to be recognized by a prize. Instead it was the controversial

António Egas Moniz (1874–1955), recipient of half a Nobel Prize in physiology or medicine in 1949, to the right. [From Ref. 35.]

technique of lobotomy that later on caught the interest of the committee(s). This procedure has been severely criticized and it has been extensively discussed: firstly, whether Moniz really had priority in the discovery and secondly, whether the often dramatic changes of personality including many unwanted side-effects could justify the procedure.

It was after his participation in a Congress of Neurology in 1935 that Moniz initiated a systematic study of lobotomy. There was at the time accumulated evidence that an operation on chimpanzees and also in some cases on humans causing a disconnection of the frontal lobes made the subject docile. Moniz was not a neurosurgeon and in addition he had severely crippled hands due to gout that prevented his active participation in surgery. Therefore the first twenty operations were performed by his colleague Dr Almeida Lima. Together they published their findings that the outcome was positive in 16 of these patients. By this they meant that the patients became calmer and more peaceful. They could be cared for more readily. Moniz then coined the term *psychosurgery* and used this as a title for a monograph in 1942.

With time the procedure became accepted for use not only in Portugal but also in many different countries. For example by 1951 20,000 lobotomies had been performed in the U.S. One patient was President Kennedy's sister Rosemary who underwent lobotomy in 1941 at the age of 23. She remained permanently incapacitated after the operation. During the 1950s various antipsychotic medications were progressively introduced and fortunately the surgical procedure was rapidly phased out. The introduction of different neuropharmacologically active substances, to be further discussed below, completely changed the methods of caring for psychotic patients. Eventually the big enclosed mental hospitals could be closed and the care of patients mostly be managed on an outpatient basis.

The era of lobotomy has been the subject of repeated discussions. It needs to be judged with considerations taken to the prevailing value systems of its time. In Sweden the discussions continued into the 1990s and comparisons were made with forced sterilization, which was not outlawed in Sweden until 1976. Lobotomy has been referred to as medical barbarism disregarding patients' rights but in the early phase it also had its defenders who stressed the functional pacification achieved by the intervention. But did the zombie state often induced represent a life of dignity? How far have we advanced from the paternalistic and sometimes even eugenic attitudes of the middle of the previous century? In the 1962 novel *One Flew Over the Cuckoo's Nest* by Ken Kesey, lobotomy was discussed as a punitive treatment. The message of the book was widely disseminated by Milos Forman's movie starring Jack Nicholson.

In his introductory speech in 1949 Olivecrona did not hesitate to state that frontal leucotomy as he called it "must be considered one of the most important discoveries ever made in psychiatric therapy …." He did take notice though that it seemed not to work as well in schizophrenic patients. The frontal lobe operation was associated with certain risks and in some cases the patient died on the operating table. This led one Swedish psychiatrist to make the following comment in 1997[30] "When the trumpets sounded in the concert hall in 1949 for the medical prize for lobotomy, Sigrid Hjertén had been dead for a year. She died on the operation table in 1948, while undergoing lobotomy." Hjertén was a very well-known and radical Swedish artist who was associated in a complex marriage with another famous Swedish artist Isaac Grünewald. Her personal art showed an influence in the use of contrasting colors from Henri Matisse, with whom she studied and from Paul Cézanne. In the simplified outlining of her motifs, she anticipated German Expressionists. She is represented at the Modern Museum in Stockholm. In spite of this critique it should be added that each historical era deserves to be judged on the basis of prevailing cultural conditions at the time. This obviously also applies to lobotomy.

Moniz could not be personally present at the prize ceremony and his insignia were received by the chargé d'affaires of the Legion of Portugal. Moniz had been an invalid since the beginning of the Second World War. A paranoid patient — not lobotomized — had shot him with eight bullets hitting his hand, chest and back. He survived the attack but became crippled afterwards. Furthermore the formally-required Nobel lecture was never given by Moniz, probably for health reasons, and apparently no text was ever submitted for *Les Prix Nobel*. He died in 1955. Moniz' list of references is somewhat unusual. He was a true renaissance man, who among other things liked fashion and designed his wife's wardrobe. He wrote about many subjects; the physiological and pathological aspects of sex life, the neurology in war, the history of hypnotism, the history of playing cards and his own years of politics. Moniz was appointed Minister for Foreign Affairs in 1917 and he headed the Portuguese Delegation at the Versailles Peace Conference the following year.

The Emerging Field of Neuropharmacology

In his comprehensive 1972 review of the prizes in physiology or medicine Liljestrand[14] also discussed the emerging research area of neuropharmacology.

Daniel Bovet (1907–1992), recipient of the 1957 Nobel Prize in physiology or medicine. [From Ref. 35.]

This field was highlighted by the 1957 Nobel Prize in physiology or medicine to Daniel Bovet from Rome (see table, p. 14). He was recognized "for his discoveries relating to synthetic substances, and especially their action on the vascular system and the skeletal muscles", a complicated prize motivation. Bovet was born in 1907 in Fleurier, Switzerland. He studied at the University of Geneva, where he presented his thesis on zoology and comparative anatomy in 1929. After this he moved to the Pasteur Institute (see Chapter 7) in Paris to work in the field of therapeutic chemistry under the famous scientist Emile Roux. His main mentor however was Ernest Fourneau. Bovet stayed at the institute for almost 20 years, after which he moved to Istituto Superiore di Sanitá in Rome to head the laboratory for therapeutic chemistry. Thus most of the work for which he received the Nobel Prize was done during his stay at the Pasteur Institute. The new principle he and his collaborators introduced was dependent on the discovery of different kinds of chemical transmitters. Once a new kind of transmitter had been identified and chemically characterized, attempts were made to produce compounds with a related molecular structure. The potential capacity for these different compounds to mimic or interfere with the activity elicited by the original transmitter was then evaluated. He pioneered a number of important discoveries. Some of them concerned inhibitors of the transmitter substances identified by Loewi, Dale and von Euler, that we learned about above. Substances that could dampen the effects of the different amines, histamine — a substance released in allergic reactions — acetylcholine, adrenaline and noradrenaline were identified. Bovet together with his wife summarized the advances made in a book in French with the English title *The Chemical Structure and Pharmacodynamic Activity of Drugs on the Vegetative Nervous System*.

In his work he could take advantage of empirical knowledge accumulated way back in human history. A central part of our historical relationship with Nature has been to test the usefulness to us humans of products that are provided by plants we have found in our environment. When Linnaeus was sent by the Swedish Parliament to make a complete inventory of all the plants on the island of Öland off the south-east of Sweden in 1741 he was to separate all plants into one of three categories — edible, useful for medical purposes and useless! No doubt an anthropocentric perspective which has its limitations. What about beauty? It has been empirically known for a long time that plant extracts can have a wide range of effects on the different functions of our body. As exemplified by Börje Uvnäs, professor of pharmacology at the Institute who gave the introduction to Bovet at the 1957 prize ceremony[32] Italian ladies already in the 16th century were increasing their attraction to the opposite sex by dilating their pupils. For this purpose they used extracts from the plant *Atropa belladonna*, which has received its name from the Italian word for beautiful lady. The alkaloid atropine in the plant inhibits the effects of acetylcholine which explains the dilution of the pupils. Another inhibitor of this transmitter substance has been known since the days of hunter-gatherer societies. This is the substance curare discovered by, for example, South American Indian populations. It was used for hunting. When an animal was hit by a poisoned arrow muscular functions were immediately interrupted. It has later been shown that acetylcholine can use two different receptors; muscarinic receptors in the case of atropine and nicotinic receptors in the case of curare. Bovet examined many chemical variants of curare and was able to develop compounds that were useful in connection with operations to achieve complete muscular relaxation. He also made many other contributions, and not only to the field of neuropharmacology as summarized in his extensive Nobel lecture[33]. Already in 1937 he produced the first blockers of the activity of histamine a critical transmitter of allergic reactions. These compounds were found to be of considerable use in such conditions.

Uvnäs finished his presentation by indicating the potential for the future development of drugs that can modify the symptoms of psychotic diseases. His prediction has indeed come true. There has been considerable progress in the field of psychopharmacology. Today there is a rich arsenal of drugs that can be used to manage mood disorders. As Uvnäs expressed it "to influence human mental life." However, this may serve to remind us that there are no shortcuts to a happy life. The increased use at the present time of drugs to influence mood disorders may be deceptive. Drugs may certainly be useful in managing situations of crises but in the end it is a task for each one of us to search for harmony in our daily round. We are social animals and the better we learn to manage our interactions with other people and the more we learn to know ourselves by use of our reflections in other people and enjoying the richness of good literature — Nobel understood this — the better our chance of leading a happy life.

Establishment of Neurophysiological Research at the Karolinska Institute

Neurophysiological research was introduced at a relatively late stage at the
Karolinska Institute but once it got started
it evolved to become one of the strengths
of the Institute. It was at the beginning of
the 1920s that changes started to occur
in the Department of Physiology. Fresh
young scientists introduced novel fields
of research. One of them was Zotterman,
who was briefly introduced in my previous
book[21] because he played a major role in
the 1961 prize in physiology or medicine
to von Békésy as already referred to. A
new branch of physiological science —
neurophysiology — was introduced at

Yngve Zotterman (1898–1982). [Courtesy
of the Karolinska Institute.]

the Institute in the 1930s. Zotterman received his M.D. in 1925 and after that
he was trained to become a neuroscientist in Adrian's laboratory in Great
Britain. He published an important paper together with the latter scientist
entitled "Impulses from a single sensory end organ," which Adrian referred
to in his Nobel lecture[20]. After returning to the Institute Zotterman presented
his thesis "Studies in the peripheral impulses from a single sensory end organ"
and received his Ph.D. in 1933. He demonstrated that pain inflicted by tissue
damage, so-called nociceptive stimulus, was transmitted by nerve fibers
with a relatively slower rate of transmission, later named C fibers by Gasser.
Zotterman continued to work in the expanding department of physiology
and became associate professor in this discipline in 1940. In 1945 he moved
to the School of Veterinary Sciences and became full professor in physiology
and pharmacology at that institution.

In 1933 Ulf von Euler, mentioned briefly above, had his readership (in
Swedish docent position) transferred from the discipline of pharmacology
to physiology and five years later he became chairman of the department of
physiology at the Institute. Von Euler was born to become a scientist. His father
was Hans von Euler-Chelpin (see picture on front cover), who received the Nobel
Prize in Chemistry in 1929, and his mother Astrid Cleve was also a successful
scientist in the field of botany and geology. During his post-doctoral years Ulf

von Euler worked with four scientists who had received or were to receive Nobel Prizes; Dale already introduced above; Corneille J. F. Heymans, physiology or medicine 1938; Archibald V. Hill, physiology or medicine 1922 (awarded in 1923); Bernardo A. Houssay, physiology or medicine 1947 (see picture, p. 52).

Ulf von Euler came to play a major role in the Nobel work during the 1940s through the early 1970s, surprisingly less often as a reviewer, but in many other functions. He was an adjunct member of the committee for physiology or medicine from 1948–51 and a full member between 1953 and 1960, the last year as its chairman. He then became the secretary of the committee, a task he carried out for five years, but in 1963 and 1964 he is also listed as an adjunct member. Thereafter in 1965 he became the Chairman of the Board of the Nobel Foundation and while serving in this function for 11 years he received the Nobel Prize in physiology or medicine in 1970. Even during the years of his chairmanship of the board of the Foundation he participated in the Nobel work at the institute as adjunct member of the committee in 1967, 1968, 1970 — the year of his prize! — and in 1971. He shared his 1970 prize with Bernard Katz and Julius Axelrod and the motivation was "for their discoveries concerning the humoral transmittors in the nerve terminals and the mechanism for their storage, release and inactivation." It was more than two decades earlier that von Euler had made his major discoveries. We will meet him again in Chapter 6 and there will be reasons to return to Ulf von Euler's own Nobel Prize when the relevant archives are fully released in 2021.

The question if a Swedish scientist developing to become a strong candidate for the prize becomes challengeable and hence disqualified to participate in the work has been handled differently by different candidates throughout time. As mentioned above Granit generously withdrew from the work three years prior to the year when he received his prize. Ulf von Euler behaved somewhat differently. Not only was he extensively involved in the Nobel committee work for many years prior to his prize, but he also remained an adjunct member even in the year of his

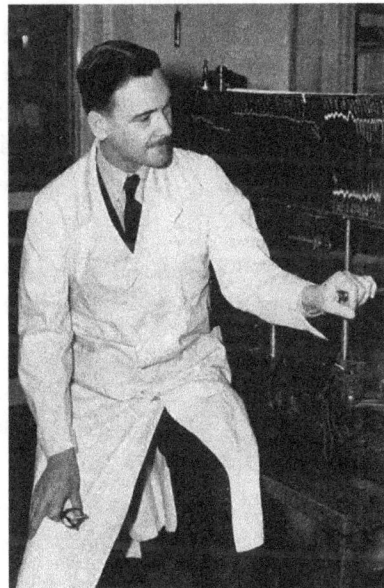

Ulf von Euler (1905–1983), recipient of a shared Nobel Prize in physiology or medicine in 1970. [Courtesy of the Karolinska Institute.]

prize. This, however, is not unprecedented. Both The Svedberg and Arne Tiselius were members of the committee in the years at the time when they received their 1926 and 1948 prizes, but Tiselius withdrew from its work in the second year of his nomination, as discussed in detail in Chapter 5. In the first year of his nomination, in 1947, he declined to stand as a candidate. This was also the position of Svedberg in 1926 and he served as the prime reviewer of Richard A. Zsigmondy, who received the prize for 1925 awarded in that year. The fact that Svedberg received the 1926 prize was due to a coup by the members of the Academy against the committee at the meeting at which the decisions were taken, as previously described[21]. The way Hugo Theorell, the first Nobel Prize recipient at the Karolinska Institute and also Sune Bergström, the joint 1982 prize recipient handled their relations to the Nobel committee will be discussed in Chapters 4 and 6 respectively. Bengt Samuelsson, who shared the prize with Sune Bergström and John Vane "for their discoveries concerning prostaglandins and related biologically active substances," remained unconnected with the prize work until after the award. After that I had the pleasure to collaborate with him in this work for a decade. When reflecting on the ability for impartibility and challenge ability of members of a prize committee it should be remembered that the standards of conduct required have become more rigorous over time. In an earlier phase honorable and fair conduct even in matters involving oneself might have been a more natural feature of the prevailing culture. Today, reviewers have to specify their professional and personal relationships to a candidate whom they have been selected to evaluate.

Returning to Ulf von Euler it should be noted that it was while he was working in Dale's laboratory in the early 1930s he became the co-discoverer of substance P — a peptide later shown to be composed of 11 amino acids serving as a neurotransmitter. After returning home he discovered prostaglandin, to be further mentioned in Chapter 6, and vesiglandin (1935), piperidine (1942) and noradrenaline (1946). Throughout these years the general theme of his science was to search for normal physiological compounds of hormonal nature or with other signaling functions (autopharmaca). All of these findings revolutionized pharmacology and not least neurophysiology. Ulf von Euler became a true school-builder at the Institute. He and his collaborators pioneered studies of different catechol amines, a family of compounds derived from the amino acid tyrosine by addition of different side groups. Important compounds of this kind are adrenaline, noradrenaline and dopamine. Thus for example the mechanisms of functions of noradrenaline in physiological and pathological situations were extensively researched. In collaboration with a colleague at

Nils-Åke Hillarp (1916–1965) *(right)* together with his student Bengt Falck. [Image generously provided by the South Swedish Society for Medical History.]

Lund University, Nils-Åke Hillarp it was found that the transmitter substances at nerve endings were stored in subcellular particles. Much later Hillarp came to be another very critical catalyst for the development of neurobiological research at the Institute. Together with his collaborator in Lund, Bengt Falck, he had developed a very useful fluorescence staining technique, which made it possible to separately label different transmitter substances like dopamine, noradrenaline and adrenaline and to demonstrate their presence in nerve tissues. Hillarp became professor in histology at the Karolinska Institute in 1962. During a few years he was to train and stimulate the enthusiasm of a whole group of promising young neuroscientists, which put their imprint on developments in the field of neurobiology at the Institute for decades to come. This field became one of the Institute's major strongholds for the future. Hillarp died in 1965 of an aggressive malignant melanoma while only 49 years old. *Ars longa, vita brevis.*

In 1947 a second chair in physiology and a separate section of the department of physiology was established at the Institute. Professor Carl-Gustaf Bernhard was appointed to this position. Like Zotterman, we have met him in one of my previous books[21]. In the newly established section II of the Department of Physiology various branches of neurophysiological research were also developed. This new institution was, like the one headed by von

Carl-Gustaf Bernhard (1910–2001). [Courtesy of the Karolinska Institute.]

Euler, developed by use of external grants, not least from the Rockefeller Foundation. The majority of the projects pursued at this new section of the physiology department were devoted to neurophysiology. The projects focused on a range of different problems and many different scientists were involved in the research. The structure and function of the spinal cord, the brain stem and centers higher up, like the pyramidal system and the cortex of the brain were examined. Other projects concerned peripheral nerves and the fundamental processes of synaptic transmissions of nerve signals. The functional development of the brain system during the embryological differentiation was also studied. In spite of all these dynamic projects in Bernhard's section of the physiology department there was another initiative that played an even larger role for the development of the field of neurophysiology at the Institute. This was the recruitment of Granit to the Karolinska Institute in the early part of the Second World War, as previously mentioned. He was the person who, besides von Euler, had the major catalytic effect on the developments in neurobiological sciences at the Institute. Like Zotterman and Bernhard he was briefly introduced in my previous book[21], but he was also presented at length earlier in this chapter. A decisive factor in tempting him to accept the offer from the Karolinska Institute was the possibility of establishing a separate Nobel institute.

The Nobel Foundation has very exact rules concerning activities it might want to initiate, as discussed earlier[1]. All of them should strictly relate only to the management of the different Nobel Prizes. Besides securing money for the prizes the Foundation should support all the work connected to the process of selecting new recipients at the prize-awarding institutions, the committee work and the associated administration at the four Prize-awarding institutions. The support for the committee work also includes special expenses incurred when their members travel to develop the width and depth of their knowledge. There is also a certain amount of support to conferences and symposia arranged under the auspices of members of the committees, Nobel Symposia and Conferences.

All these different compensations for activities initiated by the committees or their members require to be used with careful judgment so as not to reveal the secrecy regarding ongoing discussions about prize candidates or particularly promising fields of science for recognition by a prize.

The Foundation also carries the responsibility for various legal aspects of the Nobel Prize work. It also of course has to manage its own administration, cover costs for hosting the Nobel Prize recipients and also for the illustrious banquet at the City Hall of Stockholm and the corresponding costs for the Nobel Peace Prize recipient(s) in Oslo. However, the Foundation does not have to cover any expenses for the Prize Ceremony in the Concert Hall in Stockholm, since this is generously made available free of charge.

Over time the Nobel Prizes have developed to represent a uniquely strong trademark, in this context a rather trivial modern term. Thus it would be tempting for the Nobel Foundation, and of course also to the City of Stockholm to initiate activities directed for example towards the public and the business sector, but the Foundation is not formally allowed to do this. Thus activities like the Nobel Museum in Stockholm and the media activities of the Foundation are legally and financially managed separately from its formal responsibilities for the prizes. Nobel Media AB is responsible for spreading "… knowledge and inspiration about achievements recognized by the award of a Nobel Prize by producing and developing high quality productions within broadcast and digital media, events and publishing." A new initiative since 2012 has been to arrange a Nobel Prize Dialogue with participation of previous prize recipients during the Nobel week. The theme in 2014 was aging.

Nobel institutes were originally planned to be used for controlling the validity of proposed discoveries. They never came to serve this function, possibly with a few exceptions as we shall see. It was tempting for the prize-awarding institutions to aim at establishing such institutes, since they could further the advance of Swedish science. This might always be excused since only if the quality of members of prize committees were of high international standard would they be expected to manage their demanding and delicate work effectively. For a number of years no prizes were awarded, not only during the time of the First and Second World Wars, but also at a few other earlier times. The money saved could be invested in Nobel institutes. However, even with this extra support the amount of potential resources for research institutions by the Foundation was very limited. In a historical perspective it can be noted that the few institutes that were established, with time came to essentially be taken over and financed by the State. The first initiative to establish a Nobel

The building for the Division of Neurophysiology of the Noble Medical Institute. [Courtesy of the Karolinska Institute.]

Institute at the Karolinska Institute was taken in 1918. Fortunately the proposed Institute for Racial Hygiene and Genetics was defeated by a narrow majority of the College of Professors. In 1935 a new initiative was taken. Plans were made for the establishment of a Nobel Institute to be headed by the promising biochemist Hugo Theorell and in the same building a cell biology division to be led by another promising scientist, Torbjörn Caspersson. This institute will be further discussed in Chapter 4. It was decided to attach one more building to the newly erected Nobel Institute buildings. This third section of the institute focused on neurophysiology and was to accommodate the activities of Granit and collaborators. These new buildings were inaugurated in 1948.

The preamble to Granit's move to the Karolinska Institute was an initiative taken in 1940. Prior to this time Bernhard had worked in Granit's laboratory in Helsinki. When he returned to Stockholm he contacted some influential professors at the Institute. Eventually the Vice-Chancellor of the Institute Gunnar Holmgren became involved. He mobilized some influential politicians and also activated the Knut and Alice Wallenberg Foundation. A package deal was designed and this made Granit choose the Karolinska Institute instead of Harvard University. The first profiled institution for neurophysiology at the Institute became established, originally in provisional localities in the center of Stockholm. Bernhard and Carl Rudolf Skoglund became Granit's assistants. Eventually the activities were established as the Division of Neurophysiology of the Nobel Institute of the Karolinska Institute. Over the years Granit trained a number of qualified scientists and hosted many visiting scientists. One important collaborator was Bernhard Frankenhaeuser, also of Finnish origin. He learnt the technique of voltage clamp as applied to nerves from squids of the genus

Loligo in Cambridge. We will meet him again in the context of the discussion of Alan Hodgkin in Chapter 3. Granit himself mostly developed his scientific involvement in studies of the neurophysiology of the retina. Originally the studies concerned the source of the electrical impulses demonstrated by electroretinogram measurements — a technique for studies of the electrical activity of the retina already developed by the Swede Holmgren as early as in 1865, as mentioned above — but later the main focus was on the mechanisms by which the eye can interpret colors and convey this information to the brain. The advances in this research field have been summarized in consecutive reviews[34]. In parallel with Haldan K. Hartline at the Rockefeller Institute many fundamental discoveries were made. A number of other sensory organs were also investigated with the aim of understanding their neurophysiological background and particulars of their innervation. Another field of particular interest concerned the activation of muscles and the control of their tension. There are sensory organs in muscles controlling their tonus, as discovered by Sherrington. Within the belly of muscles there are muscle spindles also called proprioceptors, from Latin *proprios*, one's own, sensors which register the degree of stretching of the muscle. Many Ph.D. theses at the Institute focused on this field of research at the time.

Another profile of the work at Granit's institution concerned central nervous mechanisms and homeostasis, the control of balanced conditions in the body. Examples were identification of centers in the brain measuring the osmolarity — the salt levels — in the blood, or the thermostat functions controlling the level of temperature in the body. These studies formed a bridge to the general network systems in certain parts of the basal brain described by Magoun and collaborators. This will be the starting theme of the next chapter.

Chapter 2

Testing of Nobel Prize Patience and a Near Prize Recipient

"SOUPERS" OR "SPARKERS"
INTENSIVELY ENGAGED SCIENTISTS
NO ABSOLUTE TRUTH

Due to special circumstances, one of the recipients of the 1963 Nobel Prize in physiology or medicine, John C. Eccles, had to wait a number of years for his award. Some critical interfering events have been described previously[1]. In 1960 the 12 members of the Nobel Committee could not agree on a single proposal for the prize recipient(s). Seven members of the committee headed by the reviewers of F. Macfarlane Burnet and Peter B. Medawar, Sven Gard and Berndt Malmgren, recommended that the prize should be given to these two candidates, but five of the members led by Ulf von Euler wanted to give the prize to Eccles and Horace W. Magoun. The College of Teachers agreed with the thin majority of the committee. In 1961 the committee was again split. There was a majority of seven members who recommended a prize in the field of neurophysiology, but now the proposed candidates had changed. Magoun had been left out and instead Eccles was proposed for a prize together with Alan L. Hodgkin and Andrew F. Huxley. The candidate recommended by the minority was the hearing physiologist Georg von Békésy. Interestingly the College of Teachers this time supported the minority five committee members and decided that von Békésy should receive the prize.

In 1962 there were particular developments which overshadowed any considerations of a prize in neurophysiology. The Swedish chemist and Nobel laureate Arne Tiselius set the agenda[1]. He was a member of the Nobel committee for chemistry and its proposal for a Nobel Prize in this discipline to the protein crystallographers Max F. Perutz and John C. Kendrew was accepted by the Royal Swedish Academy of Sciences. Tiselius had arranged that once this was decided the chemistry committee would share its documentation about its second strongest candidates with the Nobel committee for physiology or medicine. The chemistry committee had in fact shared available information on their deliberations in this matter much earlier. The runners-up were Francis H. C. Crick, James D. Watson and Maurice H. F. Wilkins, the discoverers of the structure of DNA. This led to the professors of the Karolinska Institute, in a delayed decision, voting to recognize these three scientists. So it was not until 1963 that the field of neurophysiology could at last be recognized. Since Magoun was so close to receiving half a Nobel Prize in 1960 his candidature will be reviewed separately in the following. However, Hess who received half a prize in 1949, will be presented first since his investigations preceded some of Magoun's studies in focusing on the integrative functions of the basal parts of the brain, the brainstem gateway.

Studies of the Interbrain and the Midbrain

Walter R. Hess (1881–1973), recipient of half a Nobel Prize in physiology or medicine in 1949. [From Ref. 28.]

Hess's field of interest as cited by Olivecrona in his presentation speech[2] was the so-called mid brain, the mesencephalon. This is not a fully correct description since Hess in his studies focused on more frontal parts of the basal structures of the brain. In fact the motivation for his prize was "for his discovery of the functional organization of the diencephalon as a coordinator of the activities of the internal organs." As can be seen in the drawing the spinal cord continues into the hindbrain, including the bridge — *pons* — and the small brain — *cerebellum* — followed by the midbrain, the *mesencephalon*, and finally in front of this the parts of the forebrain, the *diencephalon*, connecting to the increasingly more complex structures higher up, crowned by the cortex

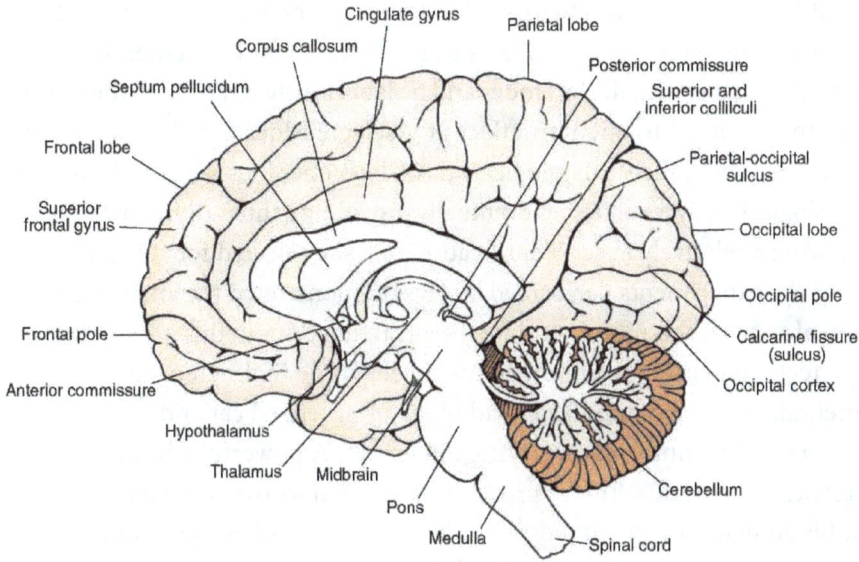

The basal parts of the brain.

of the human brain. The mesencephalon and diencephalon, which represent the older parts of the brain, control different central functions. The structures of the latter part of the brain stem are central contributors to keeping the coordinated harmony of the body — the homeostasis — by monitoring the basic needs of hunger, thirst and the daily rhythm. The central structure in the diencephalon is hypothalamus, and its surrounding structures like a part of the pituitary gland and the pineal gland, the organ where Descartes believed that the soul resided. His dualistic thinking influenced one of the 1963 laureates, Eccles, as we shall see.

Hess was born in 1881 in Frauenfeld, Switzerland. His father, a physics teacher encouraged him to go into science and instructed him in conducting experiments in a physics laboratory. He started to study medicine in Lausanne and continued his studies at other universities, finally receiving his medical degree at the University of Zürich. After training as a surgeon he became an ophthalmologist, but left a successful practice in this specialty at the age of 31 for a new career in science. He developed his approach to research under the physiologist Justus Gaule and presented his Habilitation Thesis in 1913. His studies were mostly concerned with blood flow and respiration. In 1917 Hess succeeded Gaule as Professor and Director of the Physiological Institute at the University of Zurich. In the 1930s he initiated his neurophysiologic studies concerning the function(s) of the diencephalon. At heart he was an old-fashioned generalist in physiology, but he was quite innovative in approaching the complex problems he had selected for his studies. He learned to work with very small electrodes and he laboriously established the minimal current required to elicit an effect at a selected location. Furthermore he developed very precise histological atlases by meticulous serial sectioning and histological analyses of tissues eventually providing a three-dimensional picture allowing a relatively exact identification of a site selected for stimulation.

The experiments performed by Hess demonstrated the anatomical location of some of the abovementioned central functions in this part of the brain. The technique used was to introduce an electrode in different sections of the forebrain in the immobilized head of an anesthetized cat and note the effect of electrical stimulation of varying intensities. A powerful stimulation led to destruction of tissue. This approach was referred to as a "stereotaxic scalpel."[3] In his somewhat convoluted Nobel lecture he described how stimulation of various parts of the diencephalon led to release of visceral activities, those controlled by the sympathetic and parasympathetic systems, but also of particular patterns of somatomotor activities. Thus complex patterns of behavior such as stress and defense reactions or those connected with defecation or micturition could be released. In summary he identified the presence within the diencephalon of stereotyped modules — motor programs — that combine instinct-driven behaviors, appropriate autonomic patterns to accompany them and the relevant emotion. Although not frequently cited — writing in German during the mid-1930s prewar time dominated by Nazi Germany did not favor the circulation of his findings — Hess work has influenced experimental studies of modular emotional/autonomic/behavioral action patterns into modern time. However, it is not easy to identify a single discovery for which he was recognized by his Nobel Prize.

Like Hess, Magoun was involved in studies of integrative centers in the brain, located in the grey substances in its bottom parts. However, his interest primarily concerned structures in the mesencephalon. Structures in this region have been found to be involved in directing very central functions like vision, hearing, motor control, sleeping/waking, alertness (arousal) and temperature regulation. In addition it is responsible for maintaining consciousness and interpreting pain signals. Many different signals are integrated in this part of the brain, which is very old in evolutionary terms. Magoun discovered certain structures that were of particular importance in tuning and modulating the mass of signals passing from the periphery via the spinal cord and the cranial nerves to the higher brain centers. Certain previously unrecognized structures in the midbrain were identified. They were referred to as the *reticular formation*. The background to the development of this concept illustrates the moving frontiers of neurophysiological science at the time.

Magoun grew up in New England and received a M.S. degree in Zoology at Syracuse University. He then became involved in neuroanatomical work at the Northwestern University Medical School in Chicago under Stephen W. Ranson. Ranson pursued studies of problems very similar to those examined by Hess. Already from the beginning the work focused on different integrating func-

Horace W. Magoun (1907–1991).

tions of the brain stem, such as postural tonus, various visceral functions and emotions. In most of these experiments cats were use as experimental animals. When Ranson died in 1942, Magoun suddenly had to create his own scientific environment. He received a grant from the National Foundation for Infantile Paralysis. The purpose was to study a very serious condition called bulbar poliomyelitis. This is a special form of the disease, which in most cases can be explained as a result of tonsillectomy in an individual who coincidentally has an ongoing unidentified symptomless infection with poliovirus. The operation laid bare nerve endings which allowed the virus to enter severed axons and become transported in the nerve back to the brain. This backwards — retrograde — neurogenic spread resulted in a local replication of the virus in very critical centers in the brain destroying nerve cells, some of which were in control of respiration. Because tonsillectomy was a very common procedure at this time many such cases were seen, not least in connection with the large epidemics of polio in the U.S. and other industrialized countries during the 1940s and early 1950s.

The spread of viruses in nerves represents a very important phase in the dissemination of the infectious agent in certain contagious diseases. The classical case is rabies. A dog infected with rabies virus has a modified behavior because of the infection in its brain. This leads to aggressiveness and an urge to bite. Since the virus also simultaneously infects the salivary glands the bite allows an efficient spread of virus to the bitten animal or person, enhancing the possibility for survival of the virus in nature. A rabid dog who bites a human being deposits the virus locally. The virus enters severed nerve endings and is transported inside nerves to the brain where further replication generally leads to death. Another example of a neurogenic spread of viruses is shingles (zoster) as discussed in my previous book[1]. In connection with the disease varicella, the virus from local lesions on the skin has been transported backwards in nerves and settled in a dormant form in some of the nerve ganglia in the body. Many decades later this quiescent virus can become activated and again spread in the nerves, but in this case in the other direction, from the central ganglion structure to the periphery. The spread might occur in one of the two nerves in a pair among the 31 spinal or certain cranial nerves, each one of which conveys the motor, sensory and autonomic signals from the different regions of the body. Each nerve serves only one section of the skin, a dermatome, and this is why the spread of the virus leads to a band-like representation of skin lesions. These skin lesions are a potential source of infectious varicella virus. Contacts between an elderly person and his or her grandchild may lead to the spread of the virus and development of a full-blown varicella in the child. This is a clever way for the virus to secure its survival in nature. It should be noted that this spread in nerves and local replication as in shingles may occur even in the presence of a partial immunity caused by a previous exposure to the virus. A virus that resides in the nerves cannot be attacked by the immune system. Viruses thus use many tricks to spread in the body and to facilitate their survival and one of them is to use mechanisms for transport of matter in nerves both towards and away from the central nervous structures.

A characterization of the phenomenon of bulbar polio fitted well into Magoun's and his collaborators particular focus on the brain stem. However, his most seminal contributions were made in 1949 and they concerned the conditions for sleep and wakefulness. The technique used was to place fine electrodes in predetermined parts of the brain of experimental animals, usually cats, by use of an external frame. Hess had used a similar approach as mentioned above. The frame was attached to fixed points of the head of the anesthetized animal. By use of the electrodes it was possible to destroy specific sites in the brain stem — the stereotactic surgery already referred to — and record the consequences. In collaboration with a visiting Rockefeller fellow, the scientist Giuseppe Moruzzi from Pisa, Italy, the studies were enlarged to also include a recording of the electrical activity of the selected region under different conditions. At the time it was believed that most activities of the central nervous system were initiated in the surface layer of the brain, the cortex, and controlled from there, for example in the case of peripheral

Magoun (middle) and Giuseppe Moruzzi (1910–1986) on his right.

muscular activity. However, it was observed that local lesions to various regions of the folded grey matter representing the outermost layer of the brain — the cerebral cortex — did not directly influence its central control of complexity epitomized by the state of consciousness. The critical region for maintenance of wakefulness was in the middle parts of the brain stem, the mesencephalon and this was clarified by the important studies by Magoun and Moruzzi.

Moruzzi contributed neurophysiological knowledge to the anatomical insight previously provided by Magoun. Fundamental new insights into states of sleep and wakefulness were developed. Magoun and Moruzzi showed that besides signals going directly to the brain surface there was a second parallel system which used synaptic relay stations in the brain stem. It was demonstrated that this second system had an overriding control function. It served as a unifier or organizer. And it censored the signaling between functions of the small brain and the motor cortex of the brain surface in their contacts with the many different peripheral organs of the body. Incoming signals, for example for pain and touch, and outgoing signals, for activation of relevant muscles, were controlled in a coordinated fashion in this part of the brain. Stimulating a selected area in the brain stem might cause an animal to become awake whereas incapacitating the functions of the same area would induce

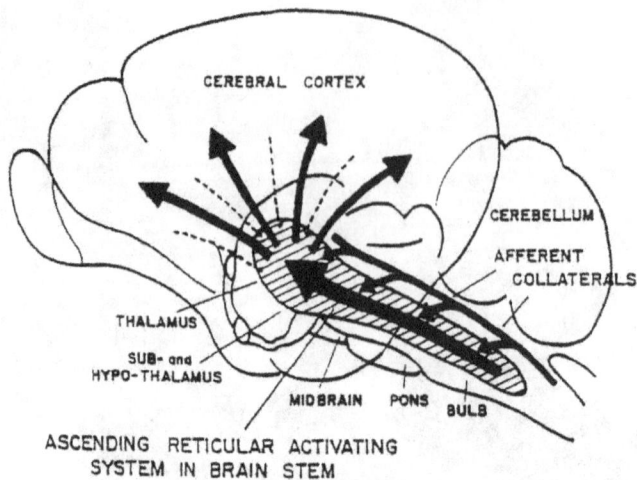

CEREBRAL CORTEX

CEREBELLUM

AFFERENT
COLLATERALS

THALAMUS

SUB- and
HYPO-THALAMUS

MIDBRAIN   PONS   BULB

ASCENDING RETICULAR ACTIVATING
SYSTEM IN BRAIN STEM

Fig. 8. Outline of brain of cat, showing distribution of afferent collaterals to
ascending reticular activating system in brain stem.

Reticular activating system (RAS). [From Ref. 4.]

sleep. Extensive damage might lead to an irrevocable coma. Based on these and additional findings the term (ascending) *reticular activating system* (RAS) was introduced. The newly identified structures were also referred to as the extralemniscal sensory system from a Greek word *lemniscus* for ribbon or band. In summary the system was concluded to be associated "with alert wakefulness as a background for sensory perception, higher intellectual activity, for voluntary movements and behaviors, and to provide insights about brain and mind."[4]

Genetically controlled urges like thirst, hunger, need for air, pain, etc. can be referred to collectively as primordial emotions. All of them are controlled in an integrated way in the basal part of the brain. Processing these emotions represent an intergrated part of the stream of consciousness. The need for controlled, goal directed behavior date back to the emergence of the earliest mammals. During the evolution of ever more complex animals additional, secondary and tertiary layers have been added to the full consciousness of mammals. We keep on learning more and more about the complex networks that integrate information in conscious and subconscious centers to the full and elusive self-consciousness in humans. We will return to this issue.

Later studies have extended the appreciation of the brain stem as a gateway for signaling from the periphery to higher centers and also emphasized the two-way signaling that constantly occurs. The multilane highway is nowadays generally described as including the following prominent brain structures: periphery — spinal level — brainstem gateway — thalamus — prefrontal cortex — orbitofrontal cortex — limbic system and then back again to the brainstem gateway. The role of the frontal cortex in deciding our personality traits and modes and integrating our hedonistic experiences was emphasized in the discussion of lobotomy in the previous chapter.

Sleep and the Active Brain

The role of the brain stem in sleep control was originally identified by early experiments in which the effect of cuts across different parts of this structure was determined. A cut across the top end led to the persistent drowsiness of the animal. The state of the electrical processing in the brains in animals with and without such a cut was studied by the new technique of EEG, introduced by Berger, as presented in the previous chapter. The intricate electrical wave pattern of the brain during wakefulness and sleep were markedly different. Originally it was believed that sleep was a state that evolved in the brain in the absence of stimuli and that it allowed the organ to rest. Studies by Berger and many later studies by others have demonstrated that sleep is a condition of special activities of the brain and that these serve many, only partly known, important functions. Thus a well balanced sleep improved learning and memorizing. The wave pattern of the EEG, not surprisingly is different between wakefulness and sleep, but there are many different distinct states of sleep. Without going into any details of these different conditions of sleep, a particular emphasis needs to be given to the 15–25% of the time we sleep, which is devoted to dreams. Because of the unique influence our dreaming has on the movement of our eyes, the term used is "rapid eye movement" (REM) sleep. The REM sleep apparently has a particular importance since this active processing of impressions (sorting, classification, reinforcement) — conscious and unconscious — stored in the brain occurs in humans from birth, in other vertebrates and in certain forms also in birds and possibly some forms of insects.

It is not clear what the function(s) of sleep may be, although there have been much speculation. It has been documented that a synthesis of proteins is important for the development of our long-term memory, but not for our short-term memory. Such a synthesis also occurs during our dream sleep. It is obvious that our sleep is influenced by many factors in our daily life like stress, excitement and the use of alcohol or drugs. Waking up in the middle of the night, sometimes referred to as the "hour of the wolf" is a common problem. The difficulties posed by events that have occurred and worries about future engagements may become exaggerated. However, sometimes the wake-up phases may also be useful. It happens to me that thoughts about what to include in a chapter of this book may come to me in the middle of the night. It is then important to take a note of the idea that has emerged and hope that the scribbling is decipherable the following morning. The person who has to make sacrifices for this uncontrollable behavior is my wife — "You should sleep and not write in the middle of the night!" Still, also in general, sleep can be a problem for people. We have all had our nightmares and experienced dreams that recur and haunt us. Our dreams involve many sensory experiences using all the different senses, except perhaps smell. This might be due to the olfactory nerves connecting to centers in the brain that are located higher up than the RAS. Although smell may not play a role in our dreams it plays a very important role in our world of memories. One need only remember the effect of the smell of the Madeleine cookies in Marcel Proust's *À la recherche du temps perdu*. Fortunately our dreams as a part of our sleep are associated with a general loss of muscle tone. If that were not the case our nights might be characterized by uncontrollable and unpredictable physical activities. Recognizing the importance of a balanced sleep is nothing new. Shakespeare lets Caesar say to Antony "Let me have men about me that are fat, sleek-headed men and such as sleep o'nights."

Sleep deprivation can be very harmful and has also been exploited in brainwashing procedures. Already Pavlov had tested the effects of alternative positive and negative signals for his dogs. It was possible to induce a state of restlessness and nervousness. These kinds of conditions can become accentuated in the absence of sleep and this is discussed for various conditions in an interesting book by William Sargent[5]. He compared the loss of control of mental conditions in for example the situations of religious conversions, combat exhaustion and various techniques of brain-washing. He also discussed the use of psychoanalysis, shock treatments and leucotomy, the latter of which was already described in the previous chapter. It can be noted as a truism that there are many ways in which our complicated brain functions can be influenced. In fact each day of our life uniquely offers both the encouragements and the challenges which are the hallmark of living.

It has not been clarified how our dreams interact with and possibly retrieve material from the subconscious activities of our brain. There are certain reaction patterns of more general or individual occurrence that definitely have a historical evolutionary background. Examples are our reaction to fire — a friend and a foe — spontaneous feelings of dizziness at a precipice — the risk of our monkey ancestors falling out of trees — our preference for the open landscape — the opportunities and dangers of developing a ground-associated existence instead of living in trees. There are also certain individuals who have a strong emotional reaction towards spiders and snakes, again an ancient appreciation of major threats in nature. However, our subconscious mind is not only a mirror of our ancestors' experiences way back in time, it must also reflect experiences we ourselves have gained. This treasure of the subconscious helps us in managing our daily lives, often in ways that are not apparent to us. Studies of brain functions reveal that choices can be made in its unconscious centers a distinctly measurable time before the decision arrives in its conscious parts. Of course the mechanisms of unconscious processing of the brain, like an exposure to a picture for a too brief time to allow it to be consciously memorized, can also be exploited by skilful advertisers, causing us to develop an irrational spontaneous purchasing behavior.

The qualitative state of sleep can be analyzed not only by EEG measurements, but also by simpler techniques of measuring the degree of oxidation of blood and the pulse in recordings from the finger tips. This is due to the fact that REM sleep is associated with an activation of the autonomous nervous system with consequences for the pulse and for the efficacy of oxidation. One common problem in elderly people is snoring and associated intervals of stopping breathing, so-called "apneas." Temporary lack of breathing immediately leads to a reduction in the oxidation of blood. This has critical effects on the brain cells, which have the most pronounced dependence of any part of our body on oxygenation and active metabolism. One way of effectively managing the emergence of apneas during sleep is to use a mask covering nose and mouth to ensure a continuous positive airway pressure, abbreviated CPAP. The use of this device ensures an even and efficient oxygenation of the blood during the night. This is associated with a normalization of the periods of REM sleep, leading to an improved feeling of well-being during the state of wakefulness.

Specific attacks by viral agents sometimes can provide an insight into the physiological role of certain structures, also in the brain, as mentioned above. Between 1915 and 1926 there was an epidemic caused by an unidentified infectious agent which led to severe damage to the midbrain structures controlling wakefulness. The disease was referred to as *encephalitis lethargica* — a brain infection causing a morbid and prolonged sleep (stupor). It is also referred to as von Economo disease from its discoverer, the Austrian neurologist of Greek origin, Constantin von Economo. The infection left the patients in a speechless and motionless state, "frozen" to a statue-like condition. This condition might last for decades. In his book *Awakenings*[6] Oliver Sacks has described how treatment with the drug L-DOPA in some cases has led to an awakening from the deep sleep and a return to a society which because of the time that had elapsed has undergone dramatic changes. It must be a very puzzling experience to wake up after ten or more years of sleep. Von Economo speculated that there might be more than one center in the basal part of the brain regulating, respectively, sleep and wakefulness. Later studies have identified the existence of different centers in the hypothalamus for these functions.

The brain is not only protected against its surroundings by the solid bone case that encloses it. In addition all the brain structures are enclosed in soft membranes which control the possibility for substances in the circulating blood to reach the brain tissues. This so-called *blood-brain barrier* normally allows the passage only of low molecular weight substances. Hence for example antibodies, representing large protein structures, cannot penetrate into the brain. This limits the important immune defense in attempts to control infections established inside the brain. It is therefore referred to as an immunologically "privileged" site. Of course the barrier also represents a hindrance for infectious agents, be they parasites, bacteria or viruses, to infect the brain. And still in particular cases infectious agents may spread to the brain and cause encephalitis.

The existence of the blood-brain barrier was identified about 100 years ago by Edwin E. Goldmann, born in South Africa but professionally active in Germany[7]. He used a stain, trypan blue, which originally was developed in 1904 by Paul Ehrlich, the recipient of the 1908 Nobel Prize in physiology or medicine. This stain belongs to a group of "vital stains" which means that, because of its chemical structure, it cannot penetrate into a living cell, but can stain a dead cell. Goldmann found that when the stain was administered into the blood stream some cells in different tissues of the body were stained with the exception of the brain tissue — "white as snow." However, when the stain was injected directly into selected localities connecting to the brain, also some cells in this organ were stained, demonstrating the existence of a barrier. Ehrlich had developed this stain to protect against one class of infectious diseases collectively referred to as trypanosomiasis. This kind of diseases is caused by a related group of protozoans, which in a certain form (*Trypanosoma brucei*) have developed a capacity to penetrate into the brain in the later stages of an infection. This may cause "sleeping" sickness. There were severe epidemics of this disease in sub-Saharan Africa at the beginning of the previous century. The protozoan was given its name because of the characteristic screw-like movements seen through the microscope — Greek *trypano*, borer, and *soma*, body. The name of the parasite in its turn gave name to the stain developed by Ehrlich. Regrettably Ehrlich's "magic bullet" did not cure the infection, but it allowed Goldmann to discover the blood-brain barrier. Trypanosomiases still occurs in different forms in sub-Saharan Africa and in Latin American and certain drugs interfering with these infections have been developed.

It has been speculated that the cause of encephalitis lethargica might have been the influenza virus strain H1N1 involved in the devastating Spanish flu pandemic in 1917–18 and that the mechanism of damage possibly was an autoimmune reaction. It is known from later observations that infections and also immunization with influenza virus antigens may elicit autoimmune reactions. In 1976 it was deemed that there was an increased risk for a recurrence of disseminating infections with a swine-flu-related virus, an H1N1 virus strain. President Ford gathered the foremost experts in the field in the White House and based on a consensus of those present, a vaccination program was initiated in the U.S. It was soon discovered that a small fraction of those immunized a certain time after vaccination developed an illness called *Guillain-Barré disease*. This rare disorder has an autoimmune background. The patient's immune system attacks their own nerve cells, which may lead to muscle weakness and even paralysis. Fortunately these symptoms generally fade with time. When this complication became known the immunization program was rapidly suspended.

Recently a similar rare complication was seen after large-scale immunization with a particular kind of influenza vaccine in Sweden and Finland in 2009. The vaccine recipients showed an increased frequency of *narcolepsy*. This is a neurological disorder that affects our control of sleep and wakefulness. It is our circadian, from Latin *circa* meaning round and *diem* meaning day, rhythm which regulates the day and night pattern in us and also in many other forms of life including even certain bacteria. The molecular clock that emerged at the dawn of evolution has been characterized in some detail[8]. In mammals the brain center controlling the circadian "clock" is located in hypothalamus in nerve cells populating the suprachiasmatic nucleus in this part of the brain in front of the brain stem region harboring RAS. It has been proposed that it is pathological autoimmune reaction in patients with narcolepsy that disturb their regular sleep pattern. However, it remains an open question if alternatively narcolepsy may result from a direct attack by an infectious agent, possibly influenza virus. Vaccine recipients developing the complications may have had a symptomless infection prior to immunization. Patients with narcolepsy have uncontrollable episodes of falling asleep during the daytime. Their nightly sleep also shows abnormalities. Normally we develop REM sleep after a period of deeper sleep, but patients with narcolepsy initiate this form of sleep promptly after falling asleep. Patients developing this kind of vaccine-associated complication seem to have a certain make-up of antigens on white blood cells, the so-called HLA antigens. The increased rate of developing the two autoimmune diseases discussed is one additional case of Guillain-Barré disease per 100,000 and 3.6 additional cases of narcolepsy per 100,000 vaccinated individuals above background levels.

No Nobel Prize has been given for discoveries concerning our dreams, but clearly many scientists have had dreams about receiving a Nobel Prize. *Why did Freud never receive the Nobel Prize?* has been used as a title of one essay[9].

Instruction or Selection

As mentioned the brain is an immunologically privileged site. The blood-brain barrier restricts the local mobilization of the full immune armamentaria of the body such as movement of cells participating in inflammatory reactions into this organ. This may change under conditions of a viral, bacterial or parasitic infection in the brain when there is a leakage of the barrier. In addition there are disorders of autoimmune origin that may cause severe dysfunctions of the brain. One of these diseases is *multiple sclerosis*. This is the most common autoimmune disorder of the central nervous system. The inflammatory process of this disease leads to a local destruction of the isolating layer, the abovementioned myelin, causing an array of different symptoms. The mechanism of the disease process remains unknown although it has been documented that both genetic and environmental factors play a role. Over the years a number of drugs that reduce the symptoms and their frequency of occurrence have been introduced, but a specific treatment has yet to be found. Inside the brain there is a production of certain immunoglobulins, which appear as bands on electrophoresis. The significance of such an oligoclonal IgG was discussed in my previous Nobel book[1]. In connection with acute or persistent virus infections in the brain a specific oligoclonal antibody response directed against various parts of the invading infectious agent can be demonstrated. During the 1970s we tried, in collaboration with many other researchers, in particular Bodvar Vandvik, to determine the specificity of the oligoclonal populations of antibodies produced in the brain of patients with multiple sclerosis, but in vain. Some of them had specificity for the antigens of certain viruses, like measles virus, but that seemed to represent a so-called by-stander activation of antibody-producing cells. The majority of the antibodies were directed against some unknown antigen(s).

The nervous system and the immune system share characteristics of being the two major adaptive overriding physiological systems in the body. Their respective functions are influenced by our exposure to the environment. As discussed at length earlier[1] it was long debated how it would be possible for the immune system to recognize in principle any foreign substance, referred to as "antigen" in this context. We can mobilize an immune attack against invading viruses, bacteria and parasites that we have never encountered before. The first attempt to explain this was provided by Paul Ehrlich, who received a Nobel Prize in physiology or medicine in 1908 together with Ilya I. Mechnikov "in recognition of their work on immunity." Ehrlich used the analogy of a lock and key. His idea was that in some way the antigen could induce immune cells to synthesize a matching structure, a lock, an idea he also applied to his studies of drugs. However, it was very difficult to understand how information could be transferred for such a phenomenon to occur, how an instruction allowing a key to cause a synthesis of its matching lock might work.

The great virologist and biologist Macfarlane Burnet thought differently. He adopted an evolutionary perspective and postulated that by some mechanism the body had developed a capacity to recognize in principle any kind of foreign structure. When a new antigen was introduced certain clones of cells capable of recognizing this antigen became activated. This was presented as *the clonal selection theory*. He was of course pleased when he was recognized for another of his theories in immunology — acquired immunological tolerance — together with Peter Medawar in 1960, but he remarked that this was only his second best theoretical contribution to immunology. When Niels Jerne was awarded a prize 24 years later in physiology or medicine in recognition of his contribution to the clonal selection theory, Burnet sent him a telegram commenting that they might have shared the prize. But what did it matter, they both had become Nobel laureates!

Their theory of *selection* resolved the big problem of trying to conceptualize mechanisms of an *instruction* of cells by influence from the environment. It can be noted that from an economy of resources point of view the prize paid is an enormous waste in the immune system. Only a very small minority of clones of immune cells will ever encounter an antigen that will activate them. Waste, however, is not an uncommon hallmark of biological phenomena. One need only think of the spawning of salmon or the fact that a single human egg cell is fertilized by one out of more than 20 million spermatozoa in a single ejaculation. Regarding clonal selection it should be added that the first encounter between a dormant clone and a new foreign antigen leads to a rather weak binding — the clones are somewhat imprecise in their capacity to interact with a specific antigen and hence to some degree promiscuous. Once the first binding has taken place the clone of triggered cells evolves through a maturation process including several successive cell divisions to provide a progressively improved binding. This is evolution in action on a small scale.

The understanding of the selective mechanisms of action of the immune system immediately leads to the question of the conditions for environmental interaction of the nervous system. By exposure to the environment, using all the different kinds of senses, the brain continuously accumulates different kinds of memory functions. This memory can be short-lived or more long-lasting. Since it is not possible for us to synthesize by some advanced combinatorial process all the potential conscious and subconscious memories accumulating from the incipient early awareness of life forms to the end of time for complex life on earth, to sample from, there must be some form of *instruction* process! Furthermore, in contrast to the immune cells there is to my knowledge no proof of restructuring of DNA in nerve cells, although it has been looked for. The potential role of epigenetic phenomena in controlling and storing information needs to be further examined. It should be recalled that the incoming information is transmitted by electrical impulses carried by nerves and subsequently by a transmitter substance. As has been briefly mentioned protein synthesis is required for the establishment of long-term memory. Information flow between proteins is the focus of a rapidly developing field of research. There are many kinds of protein aggregation diseases, not least in the brain[10, 11]. They have their origin in the development of an aberrant conformation of a special kind of host-encoded proteins. These proteins are referred to as "prion-like." They can occur in two different stable forms, one physiological and one pathological, and hence potentially might serve as a protein switch. This kind of phenomenon, a bi-stable protein molecular switch potentially could be used to encode and store information. However, there remains the problem of the degradation of proteins in the process of their continuous turn-over. Most of them have a very short half-life.

Thus biochemical processes may be of importance in memory functions, although their nature remains to be determined, but electrochemical processes also need to be taken into consideration. During later decades the developments of electronic media has been very impressive. After the switch from analogous to digital processing there has been a dramatic increase in the speed of processing of data and in the volume of information that can be retained. Today there is almost no limit to the amount of data that can be stored and retrieved. This has had dramatic consequences for developments in modern science and it has allowed the establishment of the field of artificial intelligence. The question is to what extent new insights in this field can help us to comprehend the instructive memory functions of our brain. In this context it might be appropriate to cite Gerald Edelman who emphasized that "the brain is not a computer but a jungle."

The unique adaptability of brain structures to environmental exposure should be emphasized. Newborn kittens which are blindfolded for some time after birth cannot later in life develop normal sight as demonstrated by David Hubel and Torsten Wiesel, who we will meet in the next chapter. The development of sense organs is dependent on opportunities to use them. Taxi drivers in London have a certain enlargement of parts of the brain apparently involved in memorizing addresses in the complex street layout of the city. There remains a lot to be learnt about the underlying mechanisms of processing and maturation of all the pieces of memory that we store. It would take too long to discuss in this context the different theories of memory but we will return briefly to this issue later in this chapter. However, it could be added that there are close and only partly-elucidated interactions between the immune system and the nervous system. It has been discovered that there are several ways in which these two systems can communicate and influence each other in managing the extremely complex homeostasis of our whole body. There is a growing involvement in scientific studies in the fields of neuroimmunology and even psychoneuroimmunology. The brain and the peripheral nervous system in consort have many different ways of "talking to each other." There are several examples on how various transmitter substances can act on selected immune cells and reciprocally how various cytokines can influence nerve functions.

Before leaving the remarkable capacity of our brain to accumulate, process and store memories a few comments on the emergence of information from chaos might be appropriate. Since the end of the 19th century it has been a challenge to scientists and philosophers to explain how information may accumulate in the perspective of the second law of thermodynamics. This law specifies that in any thermodynamic system there is a development towards maximal spatial homogeneity, expressed in the term *entropy*. The opposite of entropy is information (see also Chapter 8). It is obvious that under the restrictive conditions of the law it is difficult to understand how life ever could have arisen on our planet. The way out of this dilemma was highlighted by observations recognized by a Nobel Prize in chemistry in 1977. This prize was given to Ilya Prigogine "for his contributions to non-equilibrium thermodynamics, in particular the theory of dissipative structures." In essence this theory means that within a large system in which entropy continues to increase there can be "pockets" within which there is an energy-driven increase of information. We should be grateful for these exceptions because without them life could never have evolved on Earth. Such anti-entropic phenomena are particularly well illustrated by the functions of our brains. In an anti-entropic fashion we use our whole life span to accumulate and integrate information. It is just too bad that it all disappears at the time of our death. Since we do not know the mechanisms of memory storage it is difficult to conceptualize how we might collect the life-time memory of an individual before his or her demise. Still, as discussed in the section *Seeds and Deeds* in my previous book[1] we can let some of our particular life experiences survive by the legacy we may possibly leave behind.

It is now high time to get back to Magoun.

A Student Who Did Not Become a Neurophysiologist

The important and dynamic phase of research leading to the breakthrough by Magoun and Moruzzi has been described by a close witness Thomas E. Starzl. Later in life he developed into one of the leading transplant surgeons in the world. In his autobiography[4] Starzl gives the following description of his early career. His studies at the medical school of Northwestern University were extended from four to five years because of his involvement in research. During his sophomore year he attended to lectures by the professor of neuroanatomy, Magoun. They were well-prepared extemporaneous performances providing both substance and style. Magoun offered Starzl a student scholarship which he decided to accept instead of looking for another temporary summer job. When the next semester was about to start, this involvement in exciting science made Starzl cancel his registration for the Fall medical school. He learned

Thomas E. Startzl (in the middle). On his left is Joseph E. Murray (1919–2012), the recipient of a shared Nobel Prize in physiology or medicine in 1990 for organ (kidney) transplantation and on his right Norman Shumway, the father of heart transplantation, who did not receive any Nobel Prize. [From Ref. 4.]

the techniques used in Magoun's laboratory, but added some improvements of his own. The electrodes used for inducing focal stimulus in the basic parts of the brain were now also used in a reverse fashion to record electrical responses. New information was retrieved. In 1950 Magoun left Northwestern to occupy the first chair of anatomy at the University of California at Los Angeles (UCLA). A large part of his time became devoted to administration, but he also developed a laboratory. Starzl followed him to UCLA, but then needed to return to Northwestern to finish his studies. Magoun had arranged a scholarship for him to join Granit in Stockholm, but Starzl's urge to become a clinically active physician was too strong and they went their separate ways. Magoun's belief in Starzl's future potential was graciously expressed in his final letter of recommendation addressed to Alfred Blalock, the heart surgeon at Johns Hopkins University — another near Nobel Prize candidate. Starzl's development into a transplantation surgeon is another story. Briefly he is best known for pioneering liver transplantation, but he has made an impressive number of seminal contributions to the whole field of transplantations in humans. He is one of the most prolific writers of scientific articles in the world, co-authoring more than two thousand articles. In a recent review of the history of transplantation[12] he is praised for his early improvement of the immunosuppressant regime and also for the insights he has provided into the potential establishment of leukocyte dimerism — a takeover of the immune characteristics of an organ recipient from the donor. Starzl's name must have been discussed in the preparation for the 1990 Nobel Prize in physiology or medicine to Joseph E. Murray and E. Donnall Thomas, which has been described earlier (see Ref. 1, Chapter 4).

At UCLA Magoun developed his observations further and wrote a very influential monograph, *The Waking Brain*[13]. Magoun has been referred to as a "warm, modest, honest, generous and rather shy person with a monumental intellect." His nickname was Tid. To further cite Starzl it was by "swimming under his giant (shark) fins that everyone flourished." Further on in his career he turned to university administration and proudly created the Brain Research Institute (BRI) at the UCLA campus. However, this brainchild of Magoun was left to others to manage. Two people with entrepreneurial abilities, a handsome and charming neurosurgeon Jack French and a psychologist Donald Lindsley carried these responsibilities. French had a major social talent, at one time courting the famous actress Ava Gardner and later marrying an opera star, Dorothy Kirsten. Through the networks he established he was successful in raising large sums of money from philanthropists. Sadly in the end French developed Alzheimer's disease and after his death his widow established the French Foundation to support the institute. This influential institute originally used Magoun's schematic picture of the RAS as its symbol (p. 76). It remains highly active into the present time. Magoun retired from UCLA in 1972, after which he spent two years as Director of the National Research Council's Fellowship in Washington and then returned to UCLA as emeritus professor. One of his last acts was to create the UCLA Neuroscience History Archives, located at the BRI. He died 84 years old in 1991. It was a long and rich life[14].

Magoun was nominated for a Nobel Prize in physiology or medicine many times and he was reviewed by the committee on several occasions.

The Nobel Committee Reviews Magoun's Work

The first nomination of Magoun was by the famous American neurophysiologist John F. Fulton in 1954. By way of digression I can mention that his name has a certain significance in my life as a medical student. Fulton was the editor of a well-known teaching material, *A Textbook of Physiology*, which had been published in its 17th edition in 1955. This was the recommended textbook for the teaching of physiology in the spring semester of 1958 at the Karolinska Institute which I participated in. The book was very comprehensive, not least in the field of neurophysiology. About one-third of the content excelled in action potentials, presented in the next chapter. However, we did not have to absorb all this information. When I started to study medicine I took it very seriously. My attitude was that it was a privilege to participate in the academic study of medicine and I needed to prove to myself that I could manage it and hence deserved the opportunity given to me. Therefore I phased out various activities that previously had occupied most of my time, like playing Dixieland music and being a member of the, at the time, elite Stockholm team AIK of the popular Swedish sport handball. The final proof that I could not only manage my studies, but also excel in them came in the final test of the course in physiology. I had produced some voluntary additional material writing a minor essay about the neurons in the lobster *Homerus americanus*. As a result of this I was the only student in the course who was given a mark with a star. This did wonders for my self-confidence, but it did not motivate me to become a neurophysiologist. Instead, already during the subsequent semester I had become attracted to research in virology. The course of my professional life was now assured.

After this digression let us return to Fulton. It can be seen from his letter of nomination that in a preceding letter he had nominated Granit and also mentioned Magoun. This nomination was rejected by the committee, since Granit was a member of the committee. The problem concerning whether a Swedish candidate, as his qualifications improved should dissociate himself from the committee was discussed in the previous chapter (p. 61). In the case of Granit he decided to withdraw from the committee work three years before he received his own prize, but in 1963 he was to have the most important influence on developments. Fulton's nomination contained an enclosure outlining Magoun's most important contributions. In addition to this nomination there were two nominations from Japanese colleagues and one from an Italian colleague. The committee asked Granit to do a review. He praised the work, but also noted the importance of contributions by others. In addition he referred to certain technical deficiencies in the experiments performed. It can be noted that Hess in his Nobel lecture stated "… because in these particulars American investigators, whose merit is beyond doubt (especially Ranson and Magoun) have followed another path, beset with various avoidable experimental errors." In spite of these comments the review expresses a distinct respect for Magoun's work in particular the collaboration with Moruzzi. In the conclusion of the 16-page long report it was stated that Magoun without doubt was deserving a prize. The committee, however, came to the conclusion that Magoun's contributions at the time should not be considered for a prize.

In 1956 there was another nomination of Magoun, this time by J. Auer, an anatomist from Leiden. Another evaluation, also including a second candidate Frederic Bremer, was made by Bernhard. In the final paragraph he, like Granit, concluded that Magoun without doubt was worthy of the prize. At this time the committee agreed and declared that his contributions were worthy of a prize. The nominations continued. In 1957 there were 6 proposals, most of them from Southern Europe and often including Moruzzi, and in 1958 one more. In the latter year the professor of neurology of the Institute Eric Kugelberg made a third evaluation. Like the previous two reviewers he came to the conclusion that Magoun was without doubt worthy of a prize. This was confirmed by the committee. Six more nominations were given for Magoun in 1959, half of them together with Moruzzi. In 1960, besides one nomination from Japan, also including H. Jasper, there was a campaign from Turin in Italy with seven nominations for Magoun and Moruzzi together. It was in this year that Magoun and Eccles together were supported by the strong minority of five members of the committee to get the prize. Since the College of Teachers supported the majority of the committee members and gave the prize to Burnet and Medawar, this was the closest that Magoun ever came to receiving the prize. Thereafter his place was taken by Hodgkin and Huxley and the considerable number of nominations of Magoun in the ensuing years did not influence reviewers and the committee.

In fact in 1961 there were eight nominations of Magoun, one in combination with H. Jasper and the rest in combination with Moruzzi. Again the majority of nominations came from the University of Turin. Granit in his supplementary review in this year continued to support Magoun's candidature, but he now favored a combination of Eccles with Hodgkin and Huxley, as we shall see. Magoun remained worthy of a prize and Moruzzi was declared for the first time to also be this. In 1962 there were as many as 17 nominations of Magoun, but this time only two of them also mentioned Moruzzi. It is worth mentioning that in this year there were no nominations for Hodgkin and Huxley, a remarkable fact we will return to. The nominations also continued in 1963 of either Magoun alone or in combination with Moruzzi. Paradoxically the committee now considered Magoun not worthy of a prize and Moruzzi not at the time worthy of a prize. There was one more nomination in 1964 of the two of them together with F. Bremer. Again Moruzzi was considered not worthy of a prize at the time and for some reason Magoun is not even mentioned in the protocol of the committee. It would seem that once he was out of the picture in 1961, the committee more or less completely lost interest in his — and consequently in Moruzzi's — candidatures. There was a free fall. The discoveries made by the other candidates apparently were interpreted to be so much more impressive and important. This is not an uncommon development with a candidature. In general a candidate only remains close to the group of top nominees for a few years, although there are exceptions, like Sherrington, described in the previous chapter and Peyton Rous, the recipient of the shared Nobel Prize in physiology or medicine in 1966, briefly mentioned in the following chapter.

Enter Eccles

In the early 1950s there were major changes in the methodologies used for neurobiological research. The wave of electricity traveling along axons, the action potential already introduced, came into the focus as a topic of interest. All membranes of cells maintain a voltage difference between their outside and inside. It had become apparent that this voltage difference was maintained by channels through the membranes and molecular pumps that lead to a difference in representation of different ions on the inside and outside of the cell. This is of particular importance in the nervous system because a traveling action potential, which is an all-or-nothing phenomenon, upsets the ion balance. Sodium rushes into the interior of the cell and thereafter potassium leaks out to the surrounding milieu. There are also other modes, as for example in heart muscle cells in which the primary leakage concerns calcium in addition to sodium. These advances in the beginning of the 1950s were dependent on developments of new experimental tools. The goal was to dissect the details of the action potential and in order to do that it was necessary to find a way to register events in single neurons. This was made feasible by improving the speed and sensitivity of the electronics used and by making it possible to produce ever finer glass electrodes, which could be placed *inside* a cell body or an axon without upsetting the balance between their outside and inside milieus. It was these advances that the three prize recipients in physiology or medicine in 1963 pioneered by a series of discoveries that they made. The first one to enter the stage is Eccles.

John C. Eccles (1914–1997). [From Ref. 28.]

Eccles (in the center) together with Stephen W. Kuffler (1913–1980) to his right and Bernhard Katz (1911–2003) to his left in Australia during the 1940s. The latter received a shared Nobel Prize in 1967.

Eccles was one of the leading figures in neurobiological research[15]. He was a man of boundless energy and inspired many in his research group by his example. Born in 1903 in Victoria, Australia he was educated in Warrnambool and Melbourne High Schools. After finishing his medical studies at the University of Melbourne he went to Oxford as a Rhodes Scholar. Since he had already become intrigued by the brain-mind problem he chose the renowned neurophysiologist Sherrington as his mentor. He received his Ph.D. and participated in the publication of a major monograph, *Reflex Activity of the Spinal Cord* in 1932, which, as previously mentioned, had Sherrington as its last author (at this time authors were most often listed alphabethically). Eccles also became involved in studies of synaptic transmissions in autonomic ganglia and the heart. In the mid 1930s there was intense discussion about whether the transmission was due purely to electrical phenomena or if its nature was chemical. In the terminology of the time, Dale, who we met in the previous chapter was the leader of the "soupers" whereas Eccles, a stubborn proponent of electrical transmission, headed the "sparkers." Eccles' preference was based on the rapidity of the process. It took a long time for him to acknowledge that he was wrong. By reason of his authority he may even have held back the field. The central part of Eccles' forthcoming work concerned excitatory and inhibitory synaptic transmission in the central nervous system. In 1935 he returned to Australia and carried on his laboratory activities in the Kanematsu Memorial Institute of Pathology at Sydney Hospital. Together with Bernhard Katz and Stephen W. Kuffler, who had fled wartime Europe, and who we will meet again later, he examined the transmission of signals at the neuro-muscular junction. The latter two scientists accumulated evidence of the importance of acetylcholine even in muscle synapses. As Kandel says[16] "they became newly minted 'soupers'." Eventually Eccles progressively had to give up his position as he accumulated more experimental data.

In 1943 Eccles accepted an appointment to a Chair of Physiology at the Medical School in Dunedin, New Zealand. In this new environment it did not take long until he had again built a strong group of researchers. In Dunedin he got to know the philosopher Karl Popper, at the time a member of the faculty of Canterbury University College in Christchurch. They had long discussions about the mechanism of signal transmission and also about the relationship between body and soul. Eccles was a Catholic Christian and a dualist. Like Descartes he believed that the soul was a separate entity. We will return to his philosophical thinking later. Popper is famous for emphasizing that a scientific theory cannot be experimentally proven to be correct, only to be incorrect. Possibly this helped Eccles later to reconcile himself with the fact that his belief that neural transmission categorically was

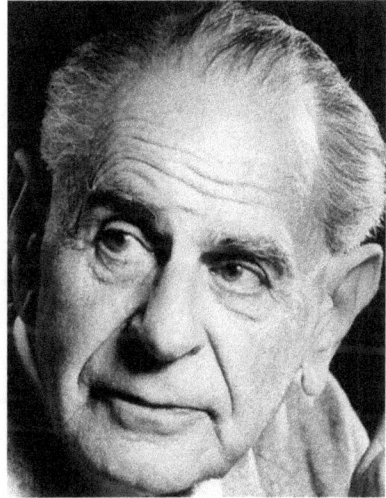

Karl R. Popper (1902–1994).

an electrical phenomenon progressively became more difficult to defend as new knowledge accumulated during the 1940s. Still there are more general problems associated with discarding one's favorite hypothesis. It was said already in the later part of the 19th century by Thomas H. Huxley, who we will meet briefly in the next chapter, that "The great tragedy of science — the slaying of a beautiful hypothesis by an ugly fact." One might also quote the philosopher Alfred N. Whitehead who said "It is more important that an idea is fruitful than that it is correct." It was in 1951 that Eccles eventually had to change his opinion on the nature of nerve transmission. He had then moved back to Australia accepting the Chair of Physiology at the John Curtin School of Medical research in Canberra. It was experiments with electrolyte-filled fine glass electrodes placed inside cells that eventually made it clear to him that the transmission must be chemical in nature.

The newly-introduced techniques in the early 1950 allowed important advances. It was understood for the first time that it was increases in conductivity that was the cause of excitation and inhibition. The group in Canberra continued to make excellent contributions during this and the coming decade elucidating mechanisms for control of signal integration in the dorsal column of the spinal cord[17], the thalamus, the hippocampus and the cerebellum. In October 1955 Eccles was invited to give the Herter lectures at the John Hopkins School of Medicine in Baltimore. He used the four lectures he gave to write a monograph entitled *The Physiology of the Nerve Cell.* It became one of the most influential books in neurobiology at the time. When he was approaching retirement in the late 1960s he became concerned about possibilities to further exploit his boundless energy in science and decided to leave Australia. This cannot have been an easy decision. His wife preferred to stay behind with their large family and the marriage was dissolved in 1968. In that same year he married a neurophysiologist collaborator, Helena Táboríková. He had four sons and five daughters in the marriage with his first wife, and one of his daughters developed to be one of his close collaborators. Together with her he published a number of important papers starting in 1956. She did not join him in his move to the U.S.

In 1966 Eccles moved his operations to the U.S., first of all to Chicago, to a newly-established Institute for Biomedical Research, but since he was not satisfied with the conditions provided he moved on to the State University of New York at Buffalo. Together with a smaller-size group of associated scientists he examined the role of the cerebellum in the control of movements. In 1975 he finally left active science and moved to Switzerland, but he continued to be very active in writing essays and books and to lecturing on a wide range of topics. His view of brain-mind relations to which he frequently returned was idiosyncratic and generated considerable controversy, a theme we will return to.

The Australian Academy of Sciences was founded in 1954 and Eccles was one of the scientists responsible for this event and thus became one of its first fellows. He had already been a Fellow of the Royal Society of London since

1941. In the Australian academy he served on the Academy council and he was also the Academy's second President between 1957 and 1961. Eccles received numerous academic honors and was knighted in 1958, elected Australian of the Year in 1963 and finally appointed a Companion in the Order of Australia in 1990. He had a very long life, dying at the age of 94 in Switzerland.

A Frequently Nominated and Reviewed Candidate

Eccles was nominated for a Nobel Prize in physiology or medicine for the first time in 1953. After that he received one or more nominations every year until he received his prize in 1963. The nominator in 1953 was Paul Hoffman, a professor of physiology at Freiburg. He emphasized the importance of findings made by Eccles in the early 1950s using Hodgkin's technique to perform intracellar measurements of action potentials. Already in this first year of nomination it was considered important to carry out a review. It was made by Granit. As early as the end of the first paragraph of the 12-page review it was stated that the new findings to be discussed were of the highest class. It is mentioned that Eccles was very slow to accept the theory of chemical transmission, but this is interpreted to mean that he preferred to form independent opinions. The conclusion of the review was that Eccles should be declared worthy of a prize. It was also suggested that he should be combined with Hodgkin, since they had used the same technique in their separate studies of each end of the neuron. Hodgkin was examined by Granit in a separate review. As we shall see he praised Hodgkin's work, but raised the issue of whether or not A. Huxley should be included and therefore the committee delayed taking a decision on whether the two nominated candidates were worthy of a prize. In 1954 Eccles was again proposed, this time together with Hoffman, who had been a nominator for him the previous year. No further analysis was made.

In a nomination for 1955 Eccles was combined with Ulf von Euler, who was a member of the committee. Von Euler declared that he did not want to be a candidate and stayed on the committee, whereas Eccles was reviewed by Bernhard. He was somewhat more hesitant than Granit in his evaluation and referred to the fact that critical information on the central regulation was still missing. Thus he proposed that Eccles should not be declared worthy of a prize for the time being. The committee agreed with Bernhard, but at the same time it declared Hodgkin and Huxley worthy of prizes based on a review by Granit as we shall see. The nominations of Eccles continued in 1956–58 but it was first in the latter year that one more review was made. It can be noted that as mentioned above Granit stayed out of the committee between 1956 and 1960, but he remained active in his writing of reviews. The committee asked Granit to update the evaluation he had made five years earlier. During this time Eccles active group had added a number of important data. Granit's 1958 review of Eccles mentioned his important monograph *The Physiology of Nerve Cells* and it also referred to the new data that Eccles had generated in collaboration with his daughter Rosamund and the Swedish scientist Anders Lundberg. In this part of the review Granit referred to work developed in the Nobel Institute headed by himself. It was argued that the results obtained at the Institute had been essential for the interpretations of recent results obtained by Eccles' group. The conclusion of the review was that Eccles should be combined with Hodgkin and Huxley, who had been previously reviewed by Granit and also reviewed by Frankenhaeuser, another member of Granit's team at the Institute, for a prize in physiology or medicine the same year. The committee agreed with the reviewers and the three candidates were declared worthy of a prize

The following year, 1959, Bernhard carried out a second review. He emphasized the importance of the earlier technical developments by Hodgkin and Huxley. He cited nominations specifying "The study by Eccles of the cellular mechanisms taking place during synaptic transmissions in term of permeability changes and ion(m)ic fluxes would not have been possible without the pioneer work of the Cambridge school under the leadership of A. L. Hodgkin" and with another formulation "Indeed, it (Hodgkin's contribution) has provided the stimulus and the framework for the interpretation of Eccles' studies on the motoneurone." Bernhard summarized his review as follows:

> The strength of Eccles contributions in particular seems to me to be that he has successfully managed to apply the micro technology to the central nervous system, and that he has shown by presentation of an overwhelming amount of scientific data that the neurons of the central nervous system are amenable to analysis, which with regard to exactness comes close to that of studies of the peripheral nerves and muscles.

Thereafter he declared Eccles worthy of a prize with reference to his comprehensive contributions to our knowledge about the processes of the synapses. The committee agreed and it also included Hodgkin, Huxley and Katz among the deserving candidates. The prize in physiology or medicine in 1959 went to the pioneers in studies of nucleic acid replication, Severo Ochoa and Arthur Kornberg[9].

In 1960 there were four comprehensive nominations of Eccles, one from Hess in Zurich and three from colleagues in Freiburg. In none of these was Eccles combined with another candidate. No new review was done this year, but as in the previous year Eccles as well as Hodgkin and Huxley and also Katz were declared to deserve a prize. The committee was split as already referred to above. Seven members voted for Burnet and Medawar, but five including Bernhard and von Euler, the chairman, voted for Eccles and Magoun. The College of Teachers agreed with the majority. This could have been the first year that rumors might have reached Eccles that he was a strong candidate for a prize, but since the prize went to Burnet and Medawar he had to uncool the champagne.

The following year again there were two strong nominations for Eccles, one from a fellow Australian, N.N. Robinson, and another from Hartline in New York, who later shared the 1967 prize with Granit. The position of Eccles in the latter nomination was presented in a way that may be representative of his standing in the field. The last summarizing paragraph read:

"Eccles' experimental analysis has provided an explicit basis for the understanding of neural integration at the cellular level. His concepts are the outgrowth of earlier, more general ideas; he has utilized the results of others as well as his own; he has changed his views on details as experimental evidence required. Many details of his theories will undoubtedly be altered as new understanding develops; new principles are certain to be added to those he has established. Nevertheless, essential principles he has developed by his original experimental work are firmly established. The new facts he has found and the synthesis of knowledge he has made constitute a true discovery of lasting value in neurophysiology."

The committee decided to ask Granit to carry out one more review. His 1961 analysis of the four candidates Eccles, Hodgkin, Huxley and Katz was very comprehensive and covered 30 pages. The common core of the work by these four scientists was the physiology of the nerve membrane in different contexts; Hodgkin and Huxley have defined fundamental properties of membranes of separated axons; Eccles has focused on membranes of the cells in the anterior part of the spinal cord and Katz on the membranes in the motor end plate where the nerve meets the muscle fibers. In the following we will focus on Granit's comments on Eccles and in the next chapter return to his discussion of the other three scientists. However, it is appropriate to note that Eccles' major contributions in the early 1950s were dependent on the pioneering work by Hodgkin and Huxley, which as a consequence were discussed first in Granit's review. On the other hand it is equally true that Eccles and colleagues already in the laboratory in Dunedin in 1951/52, had been the first scientists to apply the electrodes to cells in the spinal cord[17].

In his analysis of Eccles, Granit referred to his two earlier reviews done in 1953 and 1958. The major discoveries from Eccles' group dated from the early 1950s and were summarized in the already mentioned book *The Neurophysiological Basis of Mind*, a title Granit justifiably called inadequate. It is not the mind that is studied, it is the transfer of signals in nerves which have their cell body in the anterior part of the spinal cord, and whose axons end at the motor endplates where the electrical signal causes contraction of muscles fibers. Granit discussed the critical question of the transfer of signals at these endplates. Although there was evidence that direct electrical transfer of signals occurs only in certain particular contexts, as will be discussed later, the transfer of signals at the endplate in general turned out to be dependent on the release of chemical messenger. As demonstrated later these are packaged into small membrane-bound vesicles. There are two kinds of chemical amplifiers. One of these acts directly on the muscle fiber, and the excitation signal leads to muscle contraction. The other instead has an inhibiting effect, but this exerts its effect on the motor neuron within the spinal cord. Granit emphasized that the concept of inhibiting nervous signals had already been introduced by Sherrington, and that it was Gasser and Granit himself who had shown that this was dependent on transmission of electrical signals in nerves. However, Eccles was given the credit for having substantiated this concept by his use of intracellular electrodes in the spinal cord. This contribution was described as a discovery.

It was also discussed how Eccles with double intracellular electrodes had presented the role of different ions in the process of their transport, in particular sodium and potassium through the membrane due to a change of permeability. Granit referred to Eccles own summary of his findings in the book, *The Physiology of Nerve Cells*. He noted that Eccles had mastered a "heroic experimental fight" and commented that the cell body of the axon is a more challenging object to study than the axon. One hotly debated subject was whether there could be simultaneous activating and inhibiting influences on a single neuron. Granit's judgment was that Eccles had the last word on this issue at the time. Inhibitory mechanisms are expressed via an intermediate neuron. In his summary Granit discussed the relative weight of the contributions of the four candidates and we will return to this later. The first conclusion was that Eccles could not be given an award before the pioneering contributions by Hodgkin and Huxley had been recognized by a prize. Since the practice that had developed at this time was that at the most three candidates could be recognized simultaneously, the question was whether Eccles or Katz should be included. Both of them were considered worthy of a prize. However, Granit mentioned that Katz was younger and only had been proposed for two years. He therefore argued that Katz could wait.

It should be noted that Granit had worked side by side with Eccles in Sherrington's laboratory in the early 1930s as already described. Eccles, as repeatedly noted, had been nominated several times since the early 1950s whereas Katz was a newcomer as a candidate. Still it was Katz and Kuffler working with Eccles during the Second World War who managed progressively to demonstrate that the "soupers" were right and the "sparkers," for a long time spearheaded by Eccles, were wrong. Thus it could be argued that already in 1963 Katz should have been considered as a stronger candidate than Eccles, but because of the way the evaluations of the candidates in the field had evolved, the committee was not receptive to a discussion of this matter. Katz was to receive his prize seven years later together with Ulf von Euler and Julius Axelrod "for their discoveries concerning the humoral transmittors in the nerve terminals and the mechanisms for their storage, release and inactivation."

In its final judgment the committee had six neurobiologists worthy of a prize to choose between. Following Granit's recommendation (he was a member of the committee this year) it selected Hodgkin, Huxley and Eccles, leaving out Katz as well as Magoun and Moruzzi. As already mentioned the latter two scientists were never to be recognized by a prize. The proposal of Hodgkin, Huxley and Eccles had a majority in the committee (7 votes), but there were five members who preferred Georg von Békésy. As described in my previous book on Nobel Prizes[1], the College of Teachers followed the minority. Thus Eccles again had to uncool the champagne bottles waiting for the anticipated event.

In 1962 there were six nominations for Eccles, including one from Magoun, which also included many other leading neurobiologists. In fact the latter nomination was completely useless since as many as 32 scientists in the field were proposed. It is a puzzling fact that in 1962 there were no nominations for Hodgkin and Huxley and hence the trio of neurophysiologists agreed on after the review by Granit could not be considered this year, as will be further discussed in the next chapter. No additional review was carried out and following the recommendations by the chemist and Nobel laureate Tiselius at the Royal Swedish Academy of Sciences, Crick, Watson and Wilkins, the discoverer of the structure of DNA, swept the floor and were unanimously awarded the prize in physiology or medicine. The neurophysiologists had to wait for one more year.

In 1963 there were again repeat nominations of all three candidates. It can be added at this point that throughout the years Eccles had occasionally been proposed to be combined with other candidates. However, there had never been a proposal for the combination of the three neurophysiologists eventually chosen by the committee. There had been one proposal submitted in 1957 and repeated in 1958 to combine Eccles with Hodgkin. In 1963 there were two nominations for Eccles. One thorough nomination for him was submitted by Dale, who also had provided a nomination the previous year, as mentioned. It was very extensive and covered seven pages. Dale exemplifies a Nobel laureate who during his post-prize time was one of the most active nominators, as was already mentioned above and in my previous books[1, 10]. Dale who identified the chemical transmission at the motor endplate of course was particularly well qualified to nominate Eccles. The indefatigable Granit, who was serving the last of his six years on the committee, made one more final extensive review. It covered 46 pages, but out of these, eleven were the review proper and the rest was a chapter out of a forthcoming book for which he had just submitted the manuscript. As mentioned in the previous chapter he had been invited to write a book in English in a series *British Men of Science* for a general audience. The main character of this book was Sherrington and chapter VI which was enclosed had the title "'Inside' information on synaptic action." In the middle part of his review, Granit noted that "At the present time there is not a single biological process, within the whole field of physiology, which has been analyzed with the same degree of precision and still I have in the text above only presented the main features of these masterly investigations and discoveries, which presently represent the foundation of the physiology of activating signals."

The general review started by stating that at the time there was no need for one more review of the three candidates Eccles, Hodgkin and Huxley. Granit repeated in his presentation the peaks in Eccles' publications, with particular emphasis on the use of the technique with two microelectrodes. His conclusions were that a prize for Hodgkin and Huxley needed to have priority, but that a joint prize with Eccles was a very attractive solution since, as previously mentioned, two different functions of the neuron would be recognized, one that concerned its axon and one its soma (body).

In spite of Granit's powerful review, it was not possible to form a unanimous committee in 1963. In that year there were nine members on the committee and out of these six supported a prize in the field of neurophysiology, whereas three members instead preferred the chemists Konrad Bloch and Fedor Lynen as the prize recipients. The College of Teachers supported the majority of the committee and Eccles was finally able to uncork his champagne bottles. The citation given for the prize to Eccles, Hodgkin and Huxley was "for their discoveries concerning the ionic mechanisms involved in excitation and inhibition in the peripheral and central portions of the nerve cell membrane." Bloch and Lynen had to wait one more year (Chapter 6).

Humoral or Electrical Transfer of Signals between Neurons

As we have learnt there were originally very intense debates between the reticularists arguing that neurons formed a continuous network, and the neuronists who held the view that nerve cells acted as independent units. With time it became apparent that the neuron is the autonomous central module of the nervous system. The next, equally intense debate concerned the modes of transmission of signals between two neurons. Eccles was a "sparker" and remained one of the scientists who held out the longest in his belief that electrical signals were transferred directly between nerve cells at the synapse. His argument was that only an electrical signal could achieve the speed of transmission observed. Eventually he had to sacrifice his belief and accept that the predominant form of transmission was humoral. And yet, very often in Nature two equivalent explanations to a particular phenomenon can be found. There are in fact both humoral and electrical synapses, although the former dominate. The electrical synapse was first detected in crayfish in the late 1950s, but then discovered to exist throughout the animal kingdom. They occur in all different parts of the human brain.

The electrical synapse has a number of unique characteristics. The two nerve cells forming this special form of synapse are in close contact and can exchange cytoplasmic material even including middle-sized molecules like the second messengers to be introduced in the following chapter. The means of communication is thus not by ion channels in the cytoplasmic membrane, but by other much larger structures called gap junctions. The advantage of the electrical synapse is the high speed of transfer of the signal as argued by Eccles. Thus it is used by primitive animals like the sea hare *Aplysia*, which we will also meet again in the next chapter, to scare off enemies. The prompt release of large quantities of ink blurs the vision of nearby predators. The function in the human brain of the electrical synapses is only partly known. The speed of transfer of the impulse can be used to synchronize the action of a group of connected cells. The hippocampus of us mammals contains cells which secrete hormones. A group of such cells can be induced to release their hormone simultaneously. In this particular situation the nerve cells truly form a continuum, as was argued by the reticularists. Nature apparently is never dogmatic and exploits different alternative ways of solving a problem depending on the demands of the situation.

Eccles and a Scientific Approach to Dualism

Discussions of consciousness under the rubric the philosophy of mind has led to a wide array of different interpretations. Descartes put the theory of dualism on a firm footing although the incentive to do this most likely was to accommodate to the prevailing theocratic situation. Mind and body were two separate entities and there was speculated that the pineal gland was the place where the mind resided. This small organ at the bottom of the brain is shaped like a small pine cone — hence its name. Today it is known to have endocrine functions. It produces melatonin, a derivative of serotonin which influences our sleep pattern, the circadian rhythm. Thus it can be stated that this gland influences our awareness, since it monitors whether we are asleep or awake, but it is not the place for the soul. It is not "the Cartesian theatre." That has to be looked for somewhere else, but a search for it is more a matter for philosophical/theological speculation and belief, and not for examination by hypothetical-deductive science. However, Eccles, during the later part of his life argued that special anatomical and physical properties of the mammalian cerebral cortex provided the origin for consciousness.

In 1977 Eccles and Popper summarized some of their shared philosophical thoughts in a book entitled *The Self and Its Brain. An Argument for Interactionism*[18]. In addition to this jointly-authored book Eccles further elaborated his thinking on this subject in a later book of his own. In 1990 he published another book on this subject, *The Evolution of the Brain's Creation of the Self* and he also presented an article published by the Royal Society entitled

"A unitary hypothesis of mind-brain interaction in the cerebral cortex." How may his characteristic mode of thinking have influenced the way he conducted his science? His belief in dualism and his tinkering at the controversial interface between phenomena that can be experimentally analyzed and those which cannot, provided a potential ground for dissent with many of his colleagues.

In their joint book Popper provided the more philosophical dimensions. In a charming book[19] presenting a selection of Max Perutz's letters some interesting comments were also given on his contacts with Popper. In 1986 Perutz listened to a presentation by the 83-year-old Popper at the Royal Society on *A New Interpretation of Darwinism*. To his surprise Perutz heard Popper stating that biochemistry could not be reduced to physics and chemistry. Popper's answer to Perutz's discussion question, "If you think about it for an evening you will understand the reason" was not very helpful. Perutz wrote a critique of Popper's argumentation in *New Scientist*, which led to the latter inviting him for a discussion. The point Popper wanted to make was that biological reactions *in vivo* take on *a purpose*, which they don't have in the test tube in the laboratory. By the same reasoning combustion of gas would be different from the same reaction in the laboratory, because again it has a purpose. At this point Perutz concluded that the point made was philosophical rather than scientific.

Eccles made serious attempts to manage the concept of mind and he resorted to applications of theories formulated by physicists to interpret the phenomena of quantum physics, to understand the sometimes contradictory behavior of the elementary particles, the building stones of atoms. However, there is a difference in the range of discernible irregularities and irrationalities depending on the perspective taken. The seemingly lack of logic in the behavior of elementary particles — mass or wave-like character; differences in activity when observed and when not observed; presence or absence in a given space — is not apparent in the world of atoms. The kind of instability observed at this level limits itself to movements of electrons between the different shells leading to a change from one isotope to another. Eventually a stable state of a particular kind of atom is reached. Generally it is the more stable forms of atoms that are used in building the molecules of life. The importance of radioactive isotopes and their practical use was discussed in the chapter on George de Hevesy in my previous book[1].

The personal form of dualist theory that Eccles presented in his writings at the end of the 1980s is not very easily accessible even for experienced experimental scientists. He introduced a new concept, the *dendron*. This term is meant to represent a conjectural bundle of dendrites or processes of some 200 neurons at the lower layers of the cortex. Borrowing insights from physics it was proposed that the microsites of dendrons should operate in analogy with the probability fields of quantum physics, having neither mass nor energy, following Heisenberg's uncertainty principle. In some mysterious way the mind, in a non-energy consuming way, would initiate the release of transmittor substances thereby initiating new events. The theory of dendrons was criticized because it did not separate chance and choice. The attempt to provide a holistic feat of reconstructing the picture was not accepted by the scientific community even though the theory emanated from a respected neurophysiologist. Günter Stent summarized in his 2002 book *Paradoxes of Free Will*[20] his view on Eccles' and Popper's book in the following way: "Insofar as *The Self and Its Brain* is still known at all, it is remembered as a philosophical *folie-à-deux* of a pair of éminences *grises*." This may be too harsh a comment. Popper's arguments still stimulated discussion and served as a contrast to Ludwig Wittgenstein's mathematically framed argumentation in his *Tractatus Logico-philosophicus* that it all boils down to linguistic misunderstandings.

Throughout the second half of the twentieth century there have been major advances in possibilities for studying the brain and very important new techniques have been introduced that allow us to examine many real-time events by non-invasive methodology. Many new discoveries are being made by the use of these and still there are many aspects of brain functions that we cannot study today because of the immense complexity of the central nervous system and the lack of techniques allowing a sufficiently high resolution. However, the absence of techniques does not permit us as scientists to combine speculations with available facts. Every scientist needs to keep a watertight distinction between what they believe and what they can study by controllable techniques. In true science there is no room for metaphysics or teleology. The fact that Eccles was a dualist might have influenced his performance as a scientist. However, it seems that he managed to keep apart his strict experimental contributions, for which he deservedly received his Nobel Prize, and his philosophical speculations. Each individual of course has to take his own personal approach to resolving the challenging existential problems. Not surprisingly such reflections always seek their nourishment in the concept of consciousness.

Few themes have attracted so many general biological and philosophical thinkers as the mind-body problem. In pre-Enlightenment times it was natural to see the two as separate functions and this was formulated most succinctly by

Descartes in the 17th century. We had a soul but animals were automata. The more science has advanced the more it has become clear that it is the remarkable similarities rather than particular differences that characterize our relation to animals. Gradually it came to be understood that living entities are composed of a number of different chemical components, but these components could be present both in organic and in inorganic materials. The question then arose if the chemistry of living material was of a particular nature. With time this *vitalism* concept has been refuted by science. The first blow to the concept was given by Friedrich Wöhler when he unintentionally demonstrated that the organic compound urea could be synthesized from inorganic components. The full conceptual impact of this information remained hidden not only to him but also to generations of scientists following him. So the concept has survived but in various new disguises.

In the early 1930s the giant of physics Niels Bohr gave a lecture on *Light and life*, which was also published in the journal *Nature*. His idea was that like in physics when light could be interpreted to appear either as particles or as a wave-like phenomenon there might also be some unique difference in the quality of relationships between atoms when they are represented in inorganic and in organic material. With time it has become clear that the complexity of all the chemical reactions that maintain the balance in the living organisms, homeostasis, depends on interactions between different elements represented in the periodic table. There are no unique qualities in the way the atoms interact in living material. The last vestige of the mindset that claims there is some unique force apart from those defined by the available insights into the natural sciences concerns consciousness. Many people find it difficult to accept that our moods and fluctuating feelings, like when we fall in love, can be explained by chemical reactions. Our intuitive feelings, executed by chemical reactions, suggest to us that our dreams and subconscious conceptions must have some source of unknown origin. And still we notice that the function of our brain can readily be influenced by alcohol and drugs. Another important aspect of this discussion is if the categorical dependence on chemical reactions in the function of our brain limits our possibilities to make choices, to have a free will. However, in the real situation we have the opportunity to make choices between different alternatives. In order to discuss these problems in a more penetrating way we need to attempt define the concept of consciousness.

Before that, however, a few comments on the possibility of spontaneous emergence of information. Alan Turing, the first person to conceptualize machines allowing an advance from analogous to digital mathematics and also a critical code breaker during the Second World War[21] reflected on the early phases of embryogenesis. As the fertilized egg divides, the cells of the early divisions remain identical and each one of them can give rise to a complete embryo. The cell ball, the *morula*, then becomes hollow, and at this stage there is an orientation of the structure meaning that all the cells are no longer identical to one other. There is a formation of an opening at this *blastula* stage which later on either becomes the mouth or the anus depending upon which group of animal is involved. It seemed to Turing that in the formation of an opening in the ball-like structure there was an emergence of new information. The molecular mechanism that decides that this irrevocable step is taken remains open to speculation. Could it be oscillation in the concentration of a certain chemical, known to occur from other systems, potentially amplifying or suppressing connected molecular events? Be that as it may, it remains difficult to conceptualize emergence of information from non-information. And still it can be noted that random molecules in a solution may form crystals, which have an orientation in space. There is room for the vague speculation that in our brains too there are seeds of new thoughts, with emerging information possibly generated randomly under the influence of molecular or electrical oscillations. Could this be the substrate for a free will?

Consciousness Revisited

The first paragraph of this book introduced the concept of consciousness. Someone has said that in a cosmological perspective there are three outstanding miracles, the origin of the universe, the origin of life and the origin of consciousness. Thus there are reasons to return repeatedly to this concept. In simple terms one may state that anything we are aware of and experience at a given moment represents a part of our consciousness, which makes this experience at the same time both the most familiar and most mysterious. The modern concept of consciousness was introduced by the philosopher John Locke in 1690. The concept is derived from Latin, *con* "together" and *scio* "I know". Over time the meaning of this concept has been discussed by philosophers as well as behavioral and natural scientists. Not surprisingly the interpretations vary widely. The concepts of wakefulness and sleep were introduced above. However, consciousness, both quantitatively and qualitatively, is so much more than just a state of wakefulness. It is the integrated processing, the summarized perceptions, of external and internal signaling to the brain, which endows us with a sense of selfhood. Descartes emphasized the importance of our thinking for the definition of our individual existence — *cogito ergo sum*. There is a first-person point of focus. Consciousness means the development of a unique identification of us as a subject with the possibility to discuss our own person in a multi-person scenario. The philosopher Isaiah Berlin in his writings has emphasized the paradox that the mind becomes the subject and the object of itself. As we think we modify our thought processes and the use of different brain functions. We are no automata responding in a predictable way.

This feeling of self can be discussed in many contexts; do animals have consciousness, when does the newborn child develop full consciousness; what happens in cases of general anesthesia when a patient during his arousal passes through a continuum of different levels of wakefulness towards full consciousness, can a robot (an object equipped with means for development of artificial intelligence) develop consciousness etc. In the case of animals accumulated evidence shows that there are several cases of display of behaviors that are related to objects that the animal cannot perceive directly. Two different kinds of test have been used. One approach to studying the self-awareness phenomenon is by applying a spot of stain to the skin or the fur and then analyzing if the painted individual, recognizing the spot, attempts to remove it. This test has been documented to be passed by humans (older than 18 months) and other great apes, bottlenose dolphins, pigeons and elephants. The other kind of test is to use a mirror. A subject who can correctly interpret that what he sees in a mirror is an image of himself has some degree of consciousness as introduced in the first paragraph of this book. This capacity develops in children when they reach about 20 months of age. Interestingly there are only certain kinds of primates, the great apes, chimpanzees, gorillas and orangutans, which show a capacity to develop this kind of consciousness. Recently it has also been shown to apply to dolphins. The latter animals have a relative brain size that is second only to man. They have a superb memory and use a great sonar system for communications by signals that humans have not managed to decipher.

Karl Erik Fichtelius was an idiosyncratic Swedish professor of histology at Uppsala University. He made important contributions to the field of immunology but he was also interested in the intelligence of other mammals with big brains. Once he made an application for a grant to build a typewriter at the delphinarium at Kolmården Wildlife Park outside Norrköping. He did not receive the money requested, so we still do not know if it would be possible to communicate with dolphins by this approach. Many dog owners would of course object to the exclusion of canines from animals with an advanced awareness and argue that at least their pet has a certain consciousness of its own. There is no doubt that this kind of animal has a remarkable capacity to interpret the intentions of its owner(s) and still the origin of the animal's behavior is not intentional but due to the development of intelligent (?) reflexes. It should be recognized as a capacity for simulation or for the formation of images of the future as acknowledged by more modern philosophers, like Charles Darwin and Jacques Monod, providing contrast to Descartes' classification of animals as automata. That birds may be less prone to learn is probably a statement that will be disputed by some ornithologists. In fact there are recent data on crows showing that they have a capacity to foresee and influence their fellow crows. A species like the parrot has a well-known capacity for imitation. The mechanisms by which many other species of birds, in connection with long distance migrations, manage to economize their energy consumption and find their way to a unique chosen tree on a different continent remain unexplained. Still a wagtail may fight its own image in a hubcap until it is exhausted.

The different levels of wakefulness can be readily identified in patients subjected to medical anesthesia. It is a well known fact that the perception of different sensory stimuli varies with the level of unconsciousness reached. Perception of sounds and tactile stimuli are among the last to disappear. This is the reason for the avoidance of disturbing sound when a patient is put to sleep. In 1942 the Swedish King Gustaf V was operated on for a large stone in the urinary bladder. The operation went well and the King expressed his gratitude to the surgeon, Professor John Hellström by bestowing on him the Royal Order of the Seraphim, but the anesthesiologist received nothing. The King said "I will not recognize you by any Order, since I heard what you said." Apparently the king had not been fully asleep when the anesthesiologist proceeded to give clearance to the surgeon to start the operation by saying "Now the old guy is asleep." Appreciation of potential remaining sensory input is particularly important at a person's death bed. A seemingly unconscious person may well retain capacity to hear and in particular to retain the tactile senses towards the very end. Careful conversation and physical contacts therefore have a particular importance in these difficult situations of saying farewell to a loved one.

Making a drastic shift in focus it can be noted that particular chemicals interrupting the function of nerves have been developed as mechanisms of mass destruction. A particular class of nerve agents was discovered accidentally in Nazi Germany in 1936 at the company IG Farben in studies of insecticides. The phosphorous — containing organic chemicals developed block the break-down of the critical neurotransmitter, acetylcholine. Small doses of the agent can kill humans on a very large scale. It is an extremely efficient chemical weapon, the production and stockpiling of which was forbidden by a 1997 UN resolution. However, it was not this kind of gas that was used for genocidal purposes during the Holocaust. Instead it was a cyanide-based compound named Zyklon B. This compound acts on cells in general by interfering with cellular respiration. The story about the rabidly nationalistic German Jewish Nobel laureate Fritz Haber, who was the inventor of this kind of compound, has already been described in my previous book on Nobel Prizes[1].

Levels of consciousness can also be identified in individuals taking psychoactive drugs. Following the cult of testing such drugs during the 1950s Aldous Huxley, the world famous novelist and half-brother of the Nobel laureate to be described in depth in Chapter 3, became a believer in mysticism during the latter part of his life. He himself tried the cactus-derived compound mescaline which has effects similar to the drug LSD (lysergic acid diethylamide). His frightening experiences of testing the psychedelic drug have been described in a book entitled *The Doors of Perception*. Regrettably his experiences inspired the so-called "hippie" generation to experiment with psychedelic drugs, in many cases with regrettable results. It can be added that LSD was also used in the treatment of certain psychiatric disorders. It exerts its effect on so-called G protein coupled receptors, which will be further described in the next chapter.

The word "robot" was introduced by the Czech writer Karel Čapek to denote a fictional humanoid. The question whether robots — various forms of mechanical devices with a computer-assisted artificial intelligence — can develop some kind of consciousness has been hotly debated. A reference made to the so-called Turing test developed as early as the 1950s[21]. The abovementioned famous mathematician Turing was an impressive code-breaker, who made a major difference to the resolution of the Second World War. His unfortunate life experiences as a result of his homosexuality, eventually leading to his suicide, represent events that are outside the scope of this book. However, he made a very interesting statement about the Turing test in an essay called *Computing Machinery and Intelligence*. In this test the computer is required to imitate its interrogators so well that they are fooled. It appears unlikely that such automata will be produced, but proponents of artificial intelligence are of another opinion.

In spite of its inherent evasiveness consciousness has been the central object of study or speculation by two of the most exceptional minds in the biological sciences, those of Crick and Edelman[1]. Crick — the discoverer (with

Watson) of the structure of DNA — together with Christof Koch was searching, fruitlessly, for a common pattern of reactivity, a functional "Cartesian theatre" in the brain. Edelman, originally a pioneer in immunology, also turned to brain studies and applied Darwinian perspectives on the maturation of *qualia* — the phenomenal properties of experience. Edelman emphasized that it is equally important to build new structures and further develop the synaptic contacts as it is to allow the death of neurons by a mechanism of pruning, and to induce a regression of surface projections. Certain surface structures, referred to as adhesion molecules, were postulated by Edelman and his colleagues to play a central role in the Darwinian phenomena of evolving brain functions, not only between species but also between individuals. The idea was that working memory subconsciously would allow information to be integrated over time and evolve into an individually defined stable interpretation of the world. The question is which neural correlates of consciousness are available for examination. Edelman emphasized, as already mentioned, that the brain is not a computer, but "a jungle." Maybe the best way of resolving the challenging situation is to use a formulation by Edward O. Wilson. He has said "To understand sensory information and the passage of time is to understand a large part of consciousness itself." To achieve a major conceptual breakthrough in our understanding of the mechanisms of memory would seem like capturing the Holy Grail and something for adventurous scientists to dream of.

Another influential scientist in the field is Antonio Damasio, who has produced several books on various aspects of consciousness. He has been particularly involved in how emotions, and their underpinnings, influence our decision making, often unconsciously, both in positive and negative directions. In this work the modern sophisticated equipment for non-invasive analysis of regional brain processes, such as functional MRI (magnetic resonance imaging) and PET (position emission tomography) scans have been extensively used. These non-invasive methods of investigation of the brain will be further discussed in the following chapter. Also some exceptional philosophers like Daniel Dennett have tried to define brain processes associated with different states of consciousness by reference to the neurophysiological or physical events in the brains — a materialistic approach. The role of our un(sub)conscious processing of thoughts and feelings was described by Sigmund Freud in his early 20th century work, but then carried to certain unrealistic extremes, not least within the field of sexuality and of dreams. The importance of the mixing of conscious and unconscious perceptions *inter alia* in our understanding of art was recently presented in a fascinating book focusing on the cultural revolution of *fin-de siècle* Vienna by the Nobel laureate Erik Kandel[22]. He has also described the recent revolutionary developments in the field of neuroscience in other books, such as *In Search of Memory*[16].

Consciousness implies a certain advanced form of being awake. An anesthesiologist is well aware of both the different steps in the process of a patient falling asleep and conversely in the waking-up process. Under normal conditions the time we spend awake and asleep is determined by our bodily circadian rhythm. Our brain does not take a complete rest when we sleep as already discussed. One theory advocated by Francis Crick is that it helps in "unlearning," to weed out materials of no use for the construction and development of both our conscious and subconscious worlds. The role of brain neuronal activities which do not lead to motor activity or some other expression was identified in interesting experiments in the late 1980s by a group of Italian researchers headed by Giacomo Rizzolatti. They found that an animal or human, who sees someone else activate a certain motor function, activates its own neurons which would have been used if it had performed the same operation. The activated neurons were referred to as *mirror neurons*. This finding has stimulated many follow-up studies of uncontrollable activations in our brain. There are many ways in which the input and output of the human brain can be influenced. The knowledge has been callously exploited in the development of mental torture. Under certain conditions our brains can be controlled from the outside without any capacity for us to intervene, as mentioned above. Professionals in advertizing have learnt to exploit our conscious as well as our unconscious world to encourage impulse buying.

Already at the end of the nineteenth century it was pointed out by William James that consciousness is not a steady state. On the contrary it is continuously changing and therefore represented by "a stream of thoughts." Similar views are expressed by many polytheistic religions. In Buddhism the term is "mindstream." Australian aborigines refer to a "dream world." Of course famous authors have for a long time appreciated the usefulness of the stream of consciousness as a narrative mode. One needs only follow the thoughts of Molly Bloom in James Joyce's *Ulysses*.

It Is All a Matter of Dance of the Molecules

Clearly there are many levels of consciousness and by an incremental evolutionary process the highest level has been reached in modern humans.

A possible distinction might be made between "perceptual awareness" and "perceptual consciousness." In particular cases the perceptual awareness can be amplified by the collective actions of specialized members of a community of animals such as ants and bees. Distinguishing instincts and learning is a never-ending challenge. To cite Derek Denton[23] "Consciousness has been honed on the anvil of natural selection …." Thus optimizing the integration of all sensory information retrieved by the complex body improves possibilities for survival. The sensory input is rich and has become enriched during evolutionary time. Sorting of molecules by taste and smell, registering the tonus of a complex mass of contractile tissue of many different functions, monitoring a richness of autonomic life sustaining systems and the advanced symbolic systems of vision and sound represent a complex orchestra. The full consciousness that allows us to plan, on a short- or long-term time frame clearly improves our capacity for survival.

There are two aspects of survival of a species. One is to find food and to manage the environmental conditions, integrating the spectrum of primordial emotions briefly mentioned above. The other is to reproduce. This is the prime driving force in the existence of both humans and animals. The fact that humans have managed to dissociate sex and reproduction has had major consequences. The unique primacy of sex remains even when culturally tuned. Cultural traits may allow it to find many kinds of outlet. The recent trend towards greater tolerance has had the effect of removing many expressions of cultural suppression — the spontaneous practice of masturbation, especially among boys, was heavily condemned, the sexuality of women was denied and suppressed and homosexuality was judged as a criminal act until recently even in advanced industrialized nations. It is thanks to the control by our brain that men do not have a constant erection. Breaking the neck by hanging causes priapism, named after the Greek fertility god Priapus, in the victim. The central role of sexuality is illustrated by the fact the French scientist Jean-Pierre Changeux, referring mostly to data by others, elaborated on the neurophysiology of orgasm in his bestselling book *Neuronal Man. The Biology of Mind*[24]. We will meet him again in Chapters 8 and 9.

A conscious subject can form a plan, make a decision to adopt it and finally execute it. Since our subject is not operating in a vacuum there is so much more that comes into the process of evolving thoughts. We are social animals and depend in particular during the period when we are growing up, on other human beings, generally the family web that surrounds us. We build our knowledge and store in our memory information retrieved from our schooling and other exposures. Our memory storage expanded by use of molecular mechanisms for knowledge retrieval which are only superficially understood. Empirical knowledge has become rapidly expanded by the addition of various means of documenting information. The earliest form of documentation was artistic expression. There are cave paintings that date as far back as 40,000 years. This is where the difference between modern man and his ancestors, and implicitly his other evolutionary forerunners, develops to become strikingly apparent. Modern humans have language, but there is no agreement as to when this emerged. This means of communication represents an unlimited form of expressing information using the principles of combinatorials. A mere twenty letters take us a long way and it serves to remind us that the genetic language, used since the origin of life, only uses four letters. In fact computer language illustrates the fact that only two are needed.

When verbal tradition was exchanged for written documentation, Socrates was worried. Our capacity to memorize would not be kept in good shape any longer. Plato was less concerned. Maybe he understood the permanent value of written documentation. Then came the printing press and for a little more than two decades the World Wide Web has been transforming the world. The essentially unlimited capacity of modern computers to store information has completely changed the knowledge landscape. Children learn to read at younger and younger ages and in the near future handwriting, not to say calligraphy, will be a forgotten means of information storage. All these developments influence the patterns of our brain processing as we move through different phases of our life. We live submerged in a world guided by the principles of digitalization — only zeros and ones are used. The possibilities for forming an interactive truly global world have never been as promising as they are today. Consciousness together with an advanced intellect also allows us to take control of our own evolution. Modern humans have developed revolutionary science and technology and therefore we can review the information content of our genomes, harvest energy, prepare food effectively by modern agriculture and communicate at the speed of light around the globe. The role of non-DNA inheritance, represented by the rapidly accrued knowledge transferred between generations, is a triumph of human culture. It serves to remind that it is this asset that has made us stewards of evolution and not the other way round. Knowledge gives power and responsibility. We will return to these major issues later in the book.

In a chapter entitled *Ambivalent embodiment* in a book published a few years ago[25] one can find many examples of the conflicts to be resolved in interpreting the time-hallowed mind-body problem. Attempts were made in this book to combine phenomenology and neurobiology, but this led to repetitive citations of reservations implying limitations in the acceptance that all kinds of mental phenomena can be explained by the natural sciences, a restriction to subscribing to "the ontological and the methodological aspects of naturalism." Although it is noted that there are a number of modern philosophers, like the abovementioned Dennett, who argue for strict naturalism, a number of reservations have been included in the text. This is illustrated by citations like "… investigations of human nature cannot and should not be a strict continuation of the natural sciences" and "… there is more to human nature than can be explained by the rigid methods that we use to expound physical nature." It seems that we want to make ourselves different from animals by allocating a qualitative uniqueness to our advanced consciousness. We find it hard to see our emotions in falling in love and developing a mature, even life-long bonding, as a consequence of complicated molecular interactions in our brains. Similarly we find it difficult to acknowledge that epiphany and religious conversions or managing beliefs and searching for a life after death should be the result of the dance of molecules.

We want to peek into a possible world of the future beyond our earthly existence. And still once we are dead all the molecules constituting our bodies eventually finish operating the way they originally did. Thus the expression body *and* mind is incorrect since they represent two aspects of the same thing. Our personality and all the rich emotions we have accumulated during our life cease to exist with us. Prior to the 18th century scientific revolution it was religion that provided an answer to the perennial existential questions. What are the explanations for the starry sky, the sun and its movement and the seasonal changes, the occurrence of rain and thunder, and not least, what happens after death? Burial rituals were already developed at the time of the Neanderthals. Polytheistic religions provided many different means of pleasing the Gods, representing different phenomena of critical importance for the survival of humans. The monotheistic religions came to have a dominating influence, since they turned out to be highly efficient instruments of power. Eventually people not only believed that there was a God, they "knew" he existed.

As science has advanced, a remarkably deep insight into different phenomena in the world around us has been provided and has also facilitated life since we have managed to harness energy and effectively use Nature for our own purposes. Undoubtedly our world view has become more materialistic but it has not restricted the full play of human emotions. The fact that they have a chemical background does not in any way detract from their central role in contributing to the richness of human existence. The transitory nature of our existence only accentuates the importance of appreciating the value of every day. As expressed by Prospero in Shakespeare's *The Tempest* "we are such stuff as dreams are made on, and our little life is rounded with a sleep." Of course temporary vibrations of emotions among close relatives and friends surviving us may remain for some time, but the only consequences of our existence that may survive for the far prospective are the genes of our offspring and possible lasting impacts that we may have made on future human culture, as discussed in the section on *Seeds and Deeds* towards the end of my previous book[1].

The Split Brain

In his later years Pavlov developed the generalization that humankind might be divided into thinkers and artists. This may appear to be an overly simplified generalization, but it was referred to in the presentation speech of the Nobel Prize in physiology or medicine in 1981. This was an important year for highlighting impressive advances in neurophysiological research. The prize was divided with one half recognizing Roger W. Sperry "for his discoveries concerning the functional specialization of the cerebral hemispheres" and the other half going jointly to Hubel and Wiesel "for their discoveries concerning information processing in the visual system" (see the next chapter). It was in reference to Sperry's work on

Roger W. Sperry (1913–1994), recipient of half a Nobel Prize in physiology or medicine in 1981. [From *Les Prix Nobel en 1981*.]

what in layman's terms has come to be called "split-brain" that the presenter David Ottoson referred to Pavlov's generalization. Since the understanding that there is a considerable asymmetry in the use of the two different cerebral hemispheres of the brain the question of consciousness and mind needed to be discussed in the light of findings made by Sperry. First, however, some

David G. R. Ottoson (1918–2001).
[Courtesy of the Karolinska Institute.]

remarks on Ottoson's background are called for.

My first contact with Ottoson was during the physiology course in the spring of 1958. In excellent lectures he presented to us the physiology of the respiratory system. This was, however, not his field of research. He was a neurophysiologist, inspired to choose this orientation of his research by Bernhard. Born in 1918 in China to missionary parents, he first studied to become a dentist but then changed orientation towards medicine and finished his studies at the Karolinska Institute in 1952. He was immediately recruited for neurophysiological studies with a particular focus on the electrophysiology of the retina and in particular also of the olfactory nerve fibers. His Ph.D. thesis presented in 1956 focused on the latter system, but his later research mostly concerned sensory reception and neurophysiologic basis of pain. He remained at the Department of Physiology II, where he finally became professor in 1973. Two years later he became an associate member of the Nobel committee for physiology or medicine, where he remained until 1984, being the chairman of the committee 1982–1984. During the ten years we worked together in the committee we became good friends.

Ottoson gave the introductory speech at the 1981 Nobel Prize ceremony celebrating the award in physiology or medicine[26]. His discussion of split brain departed from Descartes in seeking answers to the questions of the function of the mind. Growing knowledge had highlighted that although the two halves of the human brain, the hemispheres, look alike, various functions are asymmetrically located. The movements of right-handed people are controlled by centers in the left hemisphere. The left hemisphere is also the center for speech. No corresponding, apparent separate functions had been discovered in the right hemisphere and therefore this part of the brain was referred to somewhat sloppily as a "sleeping partner." Ottoson even referred to the roles of the two hemispheres as "somewhat like those of man and wife in an old-time marriage," whatever he meant by that. Neuroanatomists had long since demonstrated that the two hemispheres were connected by a massive

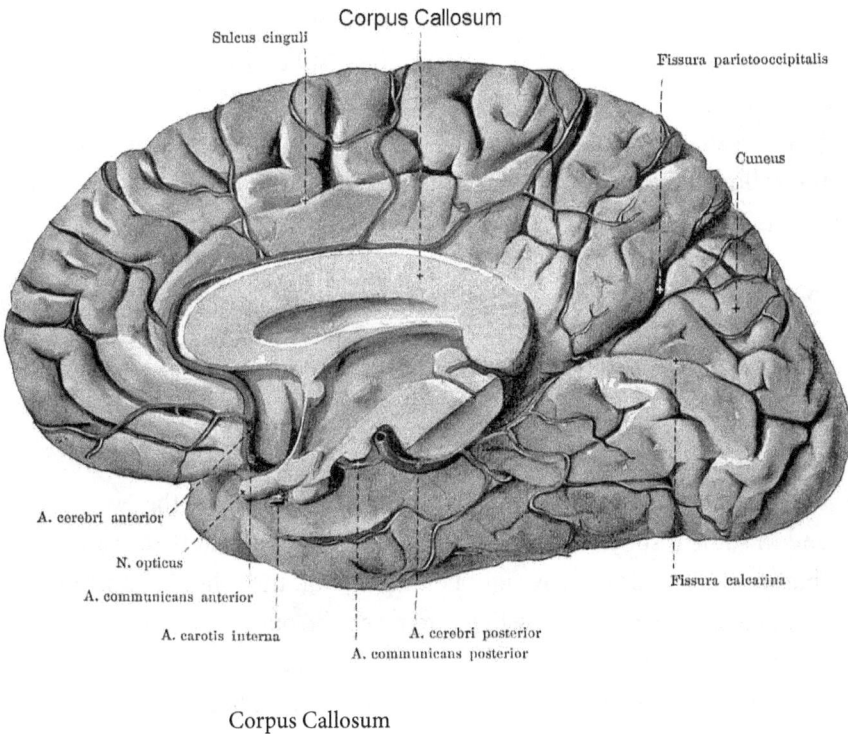

Corpus Callosum

structure the *corpus callosum*, the "tough body." This bundle of nerve fibers is the largest structure formed by white matter in the brain. In total there are about 200–250 million axons connecting the hemisphere on the one side with that on the other.

In his continued presentation Ottoson described the remarkable studies by Sperry providing unexpected insights into distinctly separate functions of the two hemispheres. These studies were made possible because of a very particular kind of operation referred to as corpus callosotomy in which the structure was cut, causing a disconnection between the two halves of the brain. The reason for this drastic surgical intervention was to dampen severe forms of epilepsy not amenable to any other treatment by available remedies. It was possible by such an intervention to alleviate the epileptic seizures of the patient. The operation represents a last resort procedure and it is only rarely performed today. When presently carried out the surgeon starts by cutting the front one-third of the structure and only continues to cut a further third if there is no effect from the operation. In the early 1960s, in tests with a group of patients who had undergone complete callosotomy Sperry made a number of very surprising discoveries. It turned out that the right hemisphere (in patients that were right-handed) was *not* a sleeping partner of the brain. On the contrary it was found that the this hemisphere, which generally is the larger one, displayed a number of unique properties. But the most remarkable finding was that the two halves of the brain in the absence of contact via the corpus callosum functioned as separate and independent units. These separate functions were described by Ottoson in the following way:

> The left brain is, as Sperry was able to show, superior to the right in abstract thinking, interpretation of symbolic relationships and in carrying out detailed analysis. It can speak, write, carry out mathematical calculations and in its general function is rather reminiscent of a computer. Furthermore, it is the leading hemisphere in the control of the motor system (in right-handed people, my remark), the executive and in some respects the aggressive brain half. It is with this brain half that we communicate. The right hemisphere on the other hand is mute and in essence lacks the possibility to reach the outside world. It cannot write and can only read and understand the meaning of simple words in noun form and does not grasp the meaning of an adjective or a verb. It lacks almost entirely the ability to count and can only carry out simple additions up to 20. It completely lacks the ability to subtract, multiply and divide. Because of its muteness, the right brain has given the impression of being inferior to the left. However, Sperry in his investigations was able to reveal that the right hemisphere in many ways is clearly superior to the left. Primarily, this concerns the capacity for concrete thinking, the apprehension and processing of spatial patterns, relations and transformations. It is superior to the left hemisphere in the perception of complex sounds and in the appreciation of music; it recognizes melodies more readily and also can accurately distinguish voices and tones. It is, too, absolutely superior to the left hemisphere in perception of nondescript patterns. It is with the right hemisphere we recognize the face of an acquaintance, the topography of a town or landscape earlier seen.

To this can be added that recent studies of "(idiot) savants," individuals with a remarkable memory capacity, suggest that they in an accentuated atypical manner shunt incoming information from their left to the more analytical right side of the brains.

Considering the distinctly different function of the two brain halves it is surprising that the "split-brain" patients in essence seemed to function normally in social contacts. They develop different strategies to evade the deficits caused by the interrupted interhemispheric transfers of information. In his Nobel lecture[27] Sperry described in more detail his different findings, but he also discussed the general implication of his discoveries under the two headings "Self Consciousness and Social Awareness" and "Progress on Mind-Brain Problem." It is not surprising that Sperry by his insight into the existence of different kinds of minds in the human brain became deeply involved in speculations on the mind-brain problem. He published several papers on the themes mind, brain and human values. He rejected dualism but instead argued for *mentalism*. He explained his philosophy in the following way in his Nobel lecture:

> The key development here is a switch from prior non-causal parallelist views to a new causal, or "interactionist" interpretation that ascribes to inner experience and integral causal control role in brain function and behavior. In effect, and without resorting to dualist views, the mental forces and properties of the conscious mind are restored in the brain of objective science from which they had long been excluded on materialistic-behavioristic principles.

And further on:

> It follows that physical science no longer perceives the world to be reducible to quantum mechanics or to any other unifying ultra element of field force. The qualitative, holistic properties at all different levels become causally real in their own form and have to be included in the causal account. Quantum theory on these terms no longer replaces or subsumes classical mechanics but rather just supplements or complements.

Denton[23] has tried to explain Sperry's thinking in the following way: "The organizational interaction within neural hierarchies generating consciousness transcends the physiological, as the latter transcends the molecular and it, in turn, the atomic and subatomic." And by another formulation of his "A veritable society of neural populations are cooperatively involved in somehow putting the ephemeral picture together in consciousness, including the apogee of the process where the cerebral processes scrutinize themselves and applaud of scorn the activities of the self."

And still it is not obvious how Sperry's thinking may resolve "the eternal chasm and irreconcilable conflict between the scientific and the traditional humanistic views of man and the world." In the light of the findings of the split-brain patients, Denton, following one of his interviews with Eccles, made an interesting and provocative footnote reflection. What he raised was a peculiar theological problem that may confront dualists in considering certain cases of these patients. If one of them had a stroke destroying the left intellectual hemisphere would this mean that they had lost their soul? And if that were the case, what about the other half of the brain which can still provide a full appreciation of esthetics and beauty?

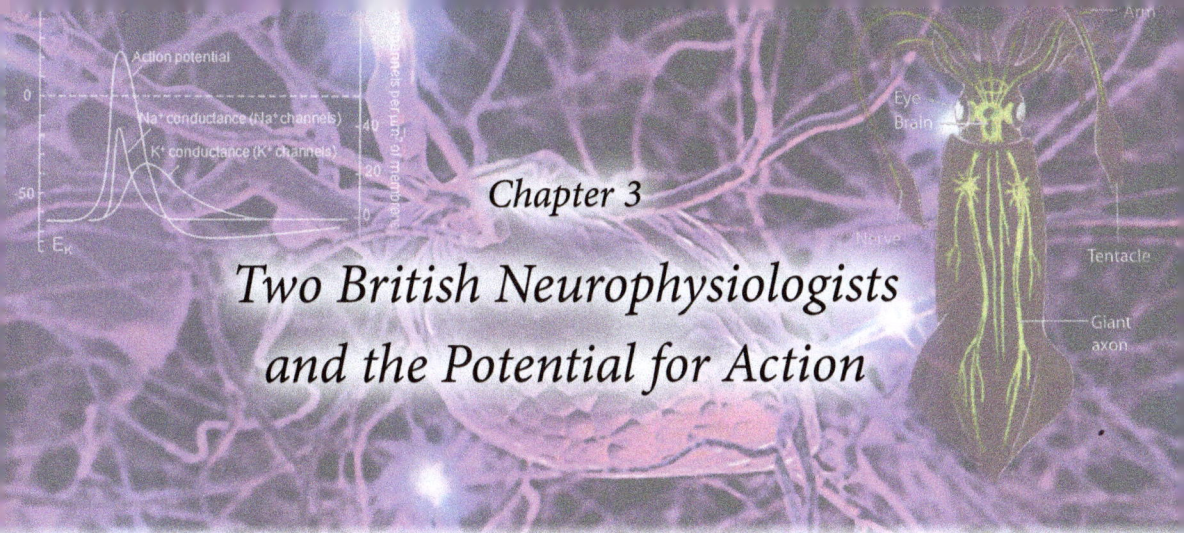

Chapter 3

Two British Neurophysiologists and the Potential for Action

THE NERVE IMPULSES

MASSIVE ION FLOW THROUGH MEMBRANES

TRANSPORTS OF SIGNALS

Essentially all cells both in animals, plants and fungi have a membrane potential. There is an electric potential difference (voltage) across their lipid-containing membrane. This phenomenon has been exploited for evolutionary purposes by the neurons that we discussed in the previous chapters. Signals can be transmitted at different speeds in their extensions, in particular their axons. The mechanisms behind the generation and transport of these signals remained an enigma for a long time but critical experiments by Hodgkin and Huxley provided a solution to the problem. The discoveries led to their receiving the Nobel Prize in physiology or medicine together with Eccles in 1963. They found that the movements of ions in and out of the cell were critical in the development of an action potential and that the conduction of such a potential was a means of rapidly transmitting signals along the axon.

It was Galvani who discovered the importance of electricity for the functioning of the nervous system, as already described, but there were two findings in the mid-1800s that shed new light on the potential mechanism. The German physiologist Emil du Bois-Reymond identified the *action potential* that propagates along the axon and soon thereafter his friend, the encyclopedic physiologist Hermann von Helmholtz, was able to measure the conduction velocity of action potentials. Some 50 years later Julius Bernstein proposed the hypothesis that there was a potential at rest — the resting membrane potential — and that a change in permeability to ions of the axonal membrane might be

121

critical in the development of an action potential. He believed that potassium ions were of primary importance in the resting potential. It had been known for some time that the inside of cells was rich in potassium and the outside rich in sodium. The mechanisms behind this separate enrichment of ions remained unknown for a long time, but progressively an extensive knowledge of active and passive channels that cross the lipid bilayer surrounding a cell has accumulated. It was Hodgkin who later, in a fruitful collaboration with Huxley, was able to clarify the nature of the resting membrane potential and the action potential and the predicted role of *voltage-gated channels* for both sodium and potassium. Such channels developed at an early stage of evolution and represent the fundamental components in the increasingly advanced electrical signaling in ever more complex nervous systems. Let us introduce these two main actors in this drama of discovery.

The Development of a Great Collaboration

Alan L. Hodgkin (1914–1998).
[From *Les Prix Nobel en 1963*.]

Alan Hodgkin was born in 1914 in Banbury, Oxfordshire into a Quaker family. He has given his own broad perspective on his life in an autobiography called *Chance and Design: Reminiscences of Science in Peace and War*[1]. The Hodgkin family has many branches including a number of prominent scholars. In medicine his father's uncle Thomas, a physician, gave his name in 1832 to Hodgkin's disease, a tumor in the lymphatic system. He was very skilled in his use of the light microscope and was the first to identify the biconcave shape of human red blood cells. Alan's cousin was Dorothy Hodgkin, the crystallographer who received the 1964 Nobel Prize in

chemistry[2] and who we will meet again in the next chapter. His father, for health reasons, had to abstain from a career as a physician and became a bank clerk. He was a conscientious objector and very consistent in his Quaker faith. Being an "absolutist", he ran a great risk of imprisonment because of the Military Service Act of 1916. An invitation to assist in investigations of the distress in Armenia gave an opportunity to contribute to much-needed humanitarian help of the wartime. His plans were to travel north from Baghdad to Baku and from there to walk over the mountains to Armenia. He left in 1918. A few weeks after his departure he died of dysentery in Baghdad, probably without having been informed that his third son had been born a few weeks earlier. The widow Mary was left with the three boys, the oldest Alan being only four years old. Being surrounded by many relatives and relieved of financial problems the three boys seem to have had a good education. Alan went to Quaker schools and at Downs preparatory School in Colwell, Worcestershire he had Fred (Frederick) Sanger, whom we will meet in Chapter 5 and Alex Bangham, who became a leading hematologist, "the father of the liposomes (Chapter 6)" as schoolmates. Alan became increasingly interested in nature and was a good ornithologist. His grandfather and his uncle were historians and he therefore vacillated between applying his rich intellect to that subject or instead to science. Eventually his interest in natural science gained the upper hand and he focused on biology and chemistry. A teacher of zoology encouraged him to widen his knowledge to also include mathematics and physics, which he did. He worked hard, won an open scholarship and became a freshman at Trinity College, Cambridge. At the College he encountered a number of important people in science, including Adrian whom we met in the first chapter. His enthusiasm for his studies can be exemplified by the following quote from his book:

> When the Physiological Society met in Cambridge, anyone keen enough was allowed to sneak into the audience and I remember a splendid debate on humoral transmission with Henry Dale, G.L.Brown and Feldberg on one side and Jack Eccles on the other. My scientific preferences were wholly on the side of acetylcholine but I thought Eccles put up a good fight. In May 1934 I was lucky enough to be present on the famous occasion when Adrian and Matthews demonstrated the effect of opening the eyes or of mental arithmetic on the Berger rhythm, using Adrian as a subject (his rhythm was unusually responsive).

During this initial phase of research Hodgkin also got to know Bernard Katz. They stimulated each other in their experimentation and this contact would turn out to be important for the post-war developments. In 1939 Hodgkin wrote to his mother from the marine biology station in Plymouth "Katz, a refugee who works on nerve, has been down here for a few days, and I have seen a good deal of him. He is going to Australia in a fortnight to work with

Eccles in Sydney. He is a very good person to talk science with." Even from an early stage Hodgkin took bold and imaginative steps in his own laboratory work. His thesis was refereed by Archibald V. Hill, the 1923 recipient of the 1922 Nobel Prize in physiology or medicine. He was impressed by the original findings made by this student, who was only 24 years old. In clever experiments using a regional freezing of the nerve Hodgkin had shown that the nerve pulse actually depolarized the axonal membrane ahead of the excited area. Already during this phase of his research development Hodgkin had learnt to work with single and not multifibre nerve preparations. His findings had consequences for the interpretation of how the spike could travel along the nerve. Hill lent a copy of the thesis to Gasser, whom we also met in the first chapter, and this led to Hodgkin receiving an invitation to work at the Rockefeller Institute in New York, which he visited during 1937–38.

The Institute provided him with a good training for his future research. In particular he learnt a lot of value for his forthcoming experimentation from Dr Toennies, the electronic expert at the Institute. Hodgkin had, already before his visit to the U.S., been in correspondence with Erlangen, whom we also met already in Chapter 1. It now became possible for him to pay a visit to his laboratory in St. Louis, Miss., for a few days. They enjoyed each other's company and Hodgkin received a lot of constructive criticism and suggestions for future experimentation from his older colleague. Most important was, however, his visit for a number of weeks to the famous Marine Biology Station at Woods

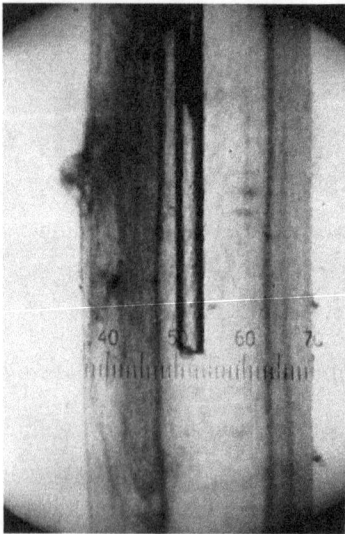

Recording electrode inside a giant axon of *Loligo forbesii*. [From Ref. 7.]

Hole. It was there that he learnt to perform experiments with the huge axons of the giant squid. As already described most vertebrate fibers are myelinated which allows a high speed transmission of the nerve impulses. Without myelination the conduction would be much slower, but one way to compensate for this is to increase size. This is the reason for the occurrence of giant axons in squids, octopuses and crustaceans, like crabs and lobsters. Their diameter is 50–100 times larger than vertebrate axons. These huge axons can be seen by the naked eye and are easily isolated and manipulated. Hodgkin learnt the particulars of the experimental design in work with Kenneth S. Cole and Howard Curtis at

Woods Hole. The size of these axons allowed an introduction of a thin glass pipette and characterization of currents by the *voltage-clamp* method introduced by Cole and Curtis. Hodgkins had already got to know these two researchers beforehand by visits to their home laboratory at Columbia University in New York. The potential use of the large nerve structures had been rediscovered in 1936 and a new era in neurophysiology was ushered in.

During his visit to the Rockefeller Institute Hodgkin met the well-known pathologist and virologist Peyton Rous. He is famous in the annals of Nobel Prizes since he had to wait almost 50 years for his prize[3]. It was in 1966 that he finally received half a prize in physiology or medicine "for his discovery of tumour-inducing viruses." The other half of this prize was awarded to Charles B. Huggins "for his discoveries concerning hormonal treatment of prostatic cancer." There will be reasons to return to this particular prize before long when the archives become available in 2017. In connection with Hodgkin's visit he also got to know Rous's daughter, Marni. They met only on a few occasions since she was away at Swarthmore College. However, they enjoyed each other's company and shared some experiences of the rich classical culture in New York at the time. When Hodgkin had left, they kept in contact by correspondence. They were to meet again during the war, an encounter that turned out to have long-term consequences, as we shall see. During his New York stay Hodgkin also had opportunities to boost his interests in nature. Together with a chemist colleague at the Institute and some other friends he canoed some 15 miles along the Ramapo River to Oakland, New Jersey. Hearing the call of the whippoorwills stayed in his memory.

Andrew F. Huxley (1917–2012). [From *Les Prix Nobel en 1963*.]

When Hodgkin returned to Cambridge the dark clouds of the impending Second World War were gathering. By means of some money from the Rockefeller Foundation he accumulated the equipment he needed in his new laboratory. In January 1939 he was ready to start experimenting. Shortly afterwards he was joined in these studies by Andrew F. Huxley. He was three years Hodgkin's junior and was at the time doing the second part of the course in Physiology. This turned out to be a critical encounter. He became interested in

Hodgkin's work on neurons and the two of them developed a close collaboration. The work was only partly performed in Cambridge. As the season for squids started it was moved to the Laboratory of the Marine Biological Association at Plymouth. Hodgkin bought a trailer and pulled all the equipment needed for the experiments along to Plymouth. At first there were no squids, but the situation soon improved and this combined with Huxley's arrival. Some very exciting experiments were initiated. Huxley measured the viscosity of the nerve interior, the *axoplasm*, and found to his surprise that under normal conditions it is solid. It became liquefied as a result of damage. He also developed a technique using two mirrors to insert electrodes into the nerve fiber without causing too much damage. Everything seemed very exciting and promising. Then Hitler marched into Poland. A brief note describing the early experimental results was submitted to *Nature* and thereafter there was a break for eight years. They must have felt some chagrin knowing that their American competitors were able to continue their work, but presumably other thoughts about the future dominated.

Hodgkin's and Huxley's ways parted but they were to meet in another very different context. It should be added that Hodgkin had lost his Quaker pacifist beliefs during a school exchange visit to Germany in the early 1930s and wholly identified himself with the fight against the enemy. Sanger (Chapter 5) as we shall see took a different attitude.

Hodgkin was recruited to the Royal Aircraft Establishment, Farnborough, where, at first, he worked on Aviation Medicine with Bryan Matthews, but very soon he was recruited to work on centimetric radar. In his biography[1] he spent more than a quarter of the book discussing this very challenging work, which demanded wholehearted commitment. He also mentioned that during this time he had an offer to join in the Manhattan project, which he declined. We have to leave that fascinating story, only to note his capacity to use his talents for many different kinds of problem solving. It might be added that Hodgkin, although generally supportive of the war efforts was very critical of certain initiatives by the military command. Thus, like a number of other scientists, Hodgkin was strongly opposed to the bombing of open cities which was advocated by Churchill and his adviser Lord Cherwell.

In 1942 one particular project concerned the introduction of an artificial cathode ray tube image of the radar target into the gun sight using an already developed spot indicator display. In the development of this project Hodgkin was aided considerably by his contacts with Huxley. He was working in London for the Admiralty Gunnery Division. In his small workshop at home in Hampstead, Huxley managed to design and develop a useful model of a gun sight. For this purpose he used a cathode ray tube image focused to infinity and reflected from a semitransparent glass surface. It would take until the last year of the war before the two scientists met again to resume their joint work in science.

In March 1939 Hodgkin hosted Dr Rous, who was visiting Cambridge. He was told that Rous's daughter Marni had received a fellowship to study in Cambridge. It must have been a great disappointment to Hodgkin when this visit had to be cancelled because of the outbreak of the war. Five years later Hodgkin suddenly found himself posted to the British Air Commission in Washington for six weeks starting in February. His task was to visit major electronic firms. The transatlantic crossing was made in the "Queen Mary." Hodgkin poetically describes this as "... when people come together for a brief period and then separated, like dead leaves eddying around one another for a moment before being blown apart in an autumn gale in the run-up to a long winter." The first thing on his mind when arriving in New York was to contact Marni Rous, with whom he had exchanged occasional letters during the war. Apparently their feelings for each other had matured and also intensified. Whereas Marni had declined his pre-war proposal she now gladly accepted. During his stay in the U.S. they managed to get married. It took a lot of effort to arrange this in the short time available and then to also try to secure a passage for Marni to England. This was not easy to organize, but contacts helped, including assistance from a liberal Supreme Court judge, Felix Frankfurter, a good friend of Marni's parents. Eventually they had to travel on separate boats, Hodgkin on the "Queen Elisabeth," jammed full of troops, and his wife on an eight-knot convoy requiring 18 days to make the passage, hence arriving eight days after her husband. She rapidly became involved in work at Telecommunications Research Establishment (TRE), the same organization where Hodgkin worked. Being an experienced editor of children's books she was allocated to the film unit. Towards the end of 1944 the young married couple were able to move to Cambridge and start a more normal life knowing that the war was coming to an end. In 1945 their first daughter was born. It was now time to get science back on track. In a letter to her parents on September 1, 1945, Marni wrote:

Alan proceeds slowly with his study ... partly because of the great inclination to return to the lab after tea. He is like a dolphin that has suddenly been released ... into the open sea. He plunges and gambols and cavorts in pure research after so long, and spends such time as he has to think of other things in the morning that he must do so much teaching when term begins again. To be sure this plunging and gamboling in the home mostly takes the form of prolonged brown studies. He may be here in body, but in mind he is far away on planes of higher mathematics from dawn to dusk and even I am sorry to say before dawn as he wakes about five thirty and then can't go back to sleep ... What is your cure Daddy? (the much delayed and unexpected 1966 Nobel prize recipient, my remark) ... Outside of worrying about his loss of sleep I am happy to see him so absorbed, and overflowing with new ideas and the opening out of new avenues to explore.

This personal description captures in a pleasant way a scientist getting prepared to make a major attack on carefully selected problems. It did not take long until Hodgkin and Huxley, eventually released from the Admiralty, were back on track. They had some very productive years ahead. As Hodgkin himself said "But by cannibalizing equipment and other expedients I managed to get my apparatus working well enough to start experiments on crab nerve fibres by the early autumn (of 1945)." A year later, Adrian received a major grant from the Rockefeller Foundation, which helped to accelerate the work of the group. Starting at the end of 1945 Hodgkin and Huxley worked closely until 1952 examining the fundamentals of the action potential using the giant axons. In this work it was critical to develop the technique for measuring the potential across the membrane by use of an intra-axonal electrode, as illustrated earlier. The methods developed were referred to as the *current clamp* technique. In their very first experiments they were taken by surprise. The true nerve action potential *exceeded* the resting potential. Hodgkin concluded: "The most important reason for making an analysis of the passive properties of a nerve fibre is that such an analysis must precede an understanding of the most complicated electrical changes which make up the nervous impulse itself." Before returning to this dynamic and successful post-war science it is time to briefly introduce Huxley.

Huxley was born into the very prominent Huxley family, which will be described in more depth below. He was endowed with a rich intellect and he displayed a particular liking for mathematics. In addition, from an early age he became interested in mechanical objects of different kinds. He was inspired to develop this talent when at the age of 12 he received a foot-operated lathe from his parents. He retained his interest in technical matters throughout his career and developed a life-long and fruitful interest in microscopy in particular. When he had graduated from Westminster School in London he won a scholarship to enter Trinity College, Cambridge. There he met Hodgkin and entered into the complex odyssey that unraveled the fundamentals of the action potential. But he also had another line of research emanating from his development of interference microscopy. He discovered the fundamental mechanisms behind muscle contraction as we shall see. Huxley had already performed some experiments with the frog sciatic nerves when he was invited by Hodgkin to join him. Their approach to the voltage clamp technique was daunting.

The Discovery

Within the six years 1946–52 Hodgkin and Huxley had resolved the fundamental problems of how nerve impulses are generated and transported in nerves. They provided a quantitative theory of the ionic events during the nerve impulse. In a way they provided an inside information on the axon transmission of nerve impulses. The two scientists made important, distinctly different contributions, with Huxley providing ideas about technical innovations and a deep mathematical insight. Hodgkin carried more of the leadership dimensions of the collaboration and also included additional highly qualified scientists in the team. Incidentally it is worth mentioning that Huxley never presented a Ph.D. thesis. Why should he, considering the genetic baggage he carried with him? In the same vein Francis Crick presented a thesis a year after the discovery of the structure of DNA, which apparently served no major purpose in his continued remarkable career. Like Huxley there are a number of other examples of Nobel Prize recipients lacking a Ph.D. academic qualification. Among laureates discussed in my previous books[2,3] can be mentioned Max Theiler, Peter B. Medawar and Gertrude B. Elion.

During a visit to the United States in 1948 Hodgkin had learnt about two newly developed important experimental techniques. To cite from a very detailed nomination in 1963 by J. Walter Woodbury and four colleagues from Seattle: "Marmont (George, my remark) and Cole (Kenneth S.) had a configuration of internal and external electrodes for use with squid giant axons which eliminated local circuit currents by reducing internal resistance and this permitted the experimenter to control membrane current and measure the resulting changes in voltage (space clamping) and to control the membrane voltage and

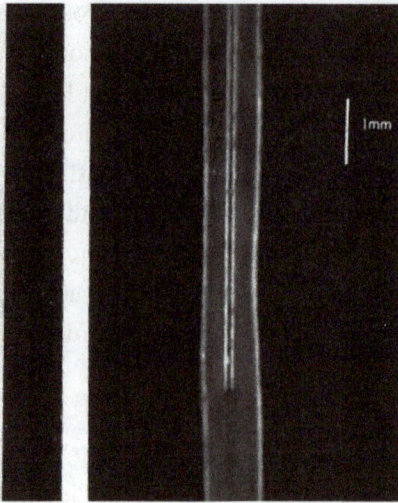

Cleaned giant axon with an 0.1 mm thick glass tube inside. [From Ref. 7.]

measure the resulting current (voltage clamping). Marmont preferred the space clamping mode of control. The other technique was transverse impalement of cell membranes with ultramicroelectrodes (having a tip diameter of less than 0.5 micra) which had just been perfected by Ling in Gerard's laboratory." Hodgkin brought the knowledge about these technical advances with him back to Cambridge. They were to be of great use in the close collaboration with Huxley.

The essence of what Hodgkin and Huxley jointly discovered was the following. By use of a cleverly designed electrode and experimenting with different ionic conditions they could outline the sequence of events. The experimental set-up was referred to as the current clamp technique as already mentioned. They demonstrated that there was a biphasic event resulting from the depolarizing current elicited by the sudden increase of conductance in a nerve axon. In the rising phase of the action potential (voltage, V, dashed line) sodium flowed from the outside into the axon (gNa, full line), the spike rose. It then turned and concomitant with a rapid movement of potassium out of the cell (gK, full line) the voltage was rapidly diminished eventually to a *lower level* than the resting state of base-line cell polarization. It was the observation of this overshoot which forced an abandonment of Bernstein's hypothesis in its original form. An explanation contingent on the occurrence of a local membrane breakdown simply could not be true. After the war Katz, whom

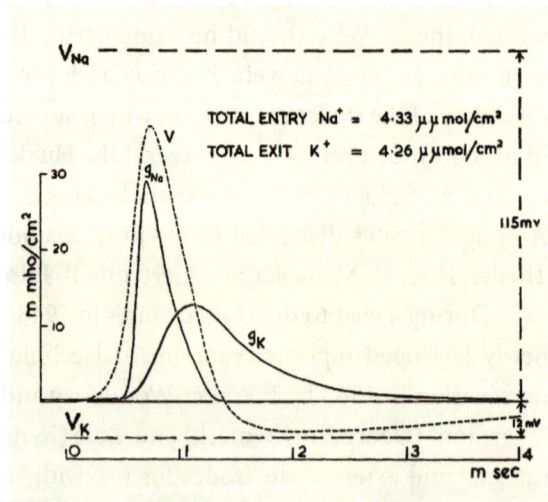

Diagram illustrating voltage changes (V) and changes in concentrations of sodium (Na) and potassium (K) inside and outside the cell membrane, respectively. [From Ref. 8.]

we will encounter again in several contexts below, discovered in work with the large axons of the *Carcinus* crab that the action potential could be abolished by a reduction of the surrounding sodium concentrations to about one tenth of the norm. This observation indicated that the overshoot could be due to sodium ions. Together Hodgkin and Katz were able to prove that this was the case by varying the sodium concentration in the squid giant axon system. What they found was that the overshoot varied depending upon the sodium concentrations in a way that suggested that the membrane must become highly and specifically permeable to sodium during the development of the potential. These findings clearly substantiated what came to be called the *ionic hypothesis*. The results were published in five landmark papers in the *Journal of Physiology* in 1952. The authors of all five papers were Hodgkin and Huxley (by tradition at the time in alphabetical order), but the first paper also included Katz as a third co-author.

The Nobel Committee Reviews Hodgkin-Huxley

The first nomination for a prize recognizing the discovery of ionic mechanisms in nerve axons was submitted in 1953 by the Nobel Prize recipient Adrian, who we have met in the first chapter (see Table 3.1, p. 132). He knew very well, of course, the details of the advances made, being from the same research environment and college as Hodgkin. The proposal only included Hodgkin and the motivation was "... his discoveries concerning the mechanism of conduction in the nerve fibre." He provided some historical perspective mentioning Bernstein's hypothesis of 1912. He then introduced the advances of Hodgkin's work, starting with the 1937 contribution of the young scientist and finishing with the series of publications in the early 1950s. He summarized these in the following way:

> Since the war Hodgkin has made a series of masterly investigations based on this finding (the critical findings by Hodgkin and Huxley in 1939, my remark) and opening up a new chapter in the biophysics of nervous conduction. He has worked in collaboration with Huxley, Keynes, Stämpfli and others but it is fair to say that *he has remained the leader* (my italics). By the use of radioactive sodium and potassium (Keynes) the movement of ions in and out of the nerve fibre has been studied quantitatively and by setting up a potential gradient across the

nerve membrane Hodgkin and Huxley have followed every stage in the permeability changes which occur when the nerve fibre is stimulated.

The final paragraph emphasized that "… it would be true to say that the problems of cell organization in the nervous system are being reduced to the molecular level."

Table 3.1. Nominations for a Nobel Prize in physiology or medicine to Alan L. Hodgkin.

Year	Nominator	Simultaneous nomination of Huxley or Eccles	Evaluator
1953	E.D. Adrian, Cambridge		Granit
1954	E.D. Adrian, Cambridge G.L. Brown, London R. Jung, Freiburg K.Kramer, Marburg		
1955	F. Bergmann, Jerusalem	Huxley	Granit
1957	F.v. Brücke, Wien J.C. Eccles, Canberra H. Herken, Berlin	Eccles Huxley	
1958	E.D. Adrian, Cambridge A.v. Muralt, Bern J.F.Donegan, Galway F.v. Brücke, Wien E. Bauereisen, Leipzig F.C. McIntosh and A.S.V. Burgen, Montreal	 Eccles Huxley	Frankenhaeuser
1959	10 nominations	3 with Huxley and one separate with Eccles	Engström
1960	10 nominations	4 with Huxley and one separate with Eccles	
1961	F. Brink Jr, New York H. Lullies, Kiel W. Wilbrandt, Bern	Huxley Huxley	Granit
1963	11 nominations	7 (5 jointly from Seattle) with Huxley	Granit

Granit, the Prime Reviewer

The committee asked Granit to do a review. It is careful and comprehensive covering 13 pages. The first paragraph takes note of Adrian's statement that Hodgkin is the leader of the group and says "There are no reasons to doubt this statement." However, Huxley was introduced early in the presentation with reference to the 1939 joint prewar experiment using a microelectrode which gave the surprising finding that an impulse not only eliminated the resting membrane potential but also caused an overshoot in the opposite direction. Granit then referred to the proposal of the ionic hypothesis in 1949 by Hodgkin, now in collaboration with Katz. The basis of this theory was that upon stimulation sodium first rapidly moved into the cell and then potassium moved in the other direction. By some hypothetical mechanisms the distribution of the two ions was restored to the original resting state. Granit then carefully reflected on the resting potential as compared to the changes resulting from the action potential. The specific movement of sodium and potassium ions demonstrated by the experiments together with Richard Darwin Keynes, incidentally a great-grandson of Charles Darwin, was also mentioned. Granit reflected at length on the action potential and how it could be explained by Hodgkin's five important publications 1952–53 with Huxley as a critical collaborator. He also discussed the priority of formulating the ionic theory and referred to the fact that as early as 1902 C. Ernest Overton had proposed that sodium was critical for the transmission of electrical signals in cells. Adrian had cited his reasoning at length in a copied section in German in his nomination.

Overton was an interesting character born in Great Britain, a student in Zürich and finally a professor of pharmacology at Lund University in southern Sweden. He married a woman who was to become the first Swedish female student to receive a Ph.D. in mathematics. Overton's main speculations concerned the role of the lipid solubility of substances and the role this had for their capacity to diffuse in and out of cells. Together with Hans H. Meyer he formulated a theory that the narcotic effects of substances correlated with the way they distributed between water and lipids. This concept has survived for a long time although as we shall see there are several means of specifically transporting ions and certain other compounds in and out of cells. Granit had such respect for Overton's proposals that he also mentioned him in his introduction at the Nobel Prize ceremony in 1963[4]. However, in his review in 1952 he wrote after his quote in German from one of Overton's publications:

Thus the original thought that sodium would penetrate (into the cell) at the impulse and be exchanged for potassium is presented, but (the way) from there to the exact proof requires a long step. Nobel prizes are not rewarded for theories but for discoveries. The only reason the theory has been mentioned here is the role it has had for their (Hodgkin and collaborators) work (true?, my remark).

The final paragraph of Granit's review has a particular importance. He wrote:

In spite of the fact that Adrian for this prize-worthy work has only mentioned Hodgkin's name it needs to be considered seriously if (by doing so) Huxley's contributions have not been passed over. Aware of Huxley's prominent research capacity, which is supported by an unusual talent for mathematics, I find it difficult without any further consideration to ignore him in this context. At least it is important to raise this discussion of principle in the Nobel committee.

The committee agreed with Granit and delayed a decision on whether or not Hodgkin was worthy of a prize.

In 1954 there were four more nominations for Hodgkin but in 1955 there was the first joint nomination of Hodgkin and Huxley (Tables 3.1 and 3.2). In the latter year Granit did another, but this time very brief review. He was worried about the fact that Adrian had pronounced that Hodgkin "... has remained the leader." He then wrote "However, since then the docent Frankenhaeuser from the neurophysiological division of the Nobel Institute has spent 8 months collaborating with Hodgkin and thus there is access to information about the local conditions (in the laboratory). *En passant* the question about Huxley's contribution to the ionic hypothesis and the working program has often been discussed, and Frankenhaeuser's view seems to be that the two scientists complement each other in a fortunate way." Since Granit had done a full review two years earlier he only gave some supplementary information. The collaboration between Hodgkin and Keynes using radioactive isotopes was referred to and the speculative proposal of the existence of a sodium pump was presented. Granit also indulged in defensive polemics concerning some of the critique that has been raised against the ionic hypothesis. He reemphasized that it was an outstanding scientific achievement. In this year Hodgkin and also Huxley for the first time are listed as worthy of a prize by the committee.

Table 3.2. Nominations for a Nobel Prize in physiology or medicine to Andrew F. Huxley.

Year	Nominator	Simultaneous nomination of Huxley or Eccles	Evaluator
1955	F. Bergmann, Jerusalem	Hodgkin	Granit
1957	H. Herken, Berlin	Hodgkin	
1958	F.C. McIntosh and A.S.V. Burgen, Montreal	Hodgkin	Frankenhaeuser
1959	J.G. Foulks, Vancouver O.E. Lowenstein, Birmingham J.D. van Nuys, Indianapolis R. Bachmann, Göttingen H.J. Deuticke, Göttingen K. Kramer, Göttingen	Hodgkin Hodgkin Hodgkin	Engström
1960	A. Fleckenstein, Freiburg B. Lueken, Halle R. Schoen, Göttingen A. Szent-Györgyi, Woods Hole M. Weatherall, London	Hodgkin	
1961	F. Brink, Jr, New York H. Lullies	Hodgkin Hodgkin	Granit
1963	7 nominations	all (5 jointly from Seattle) including Hodgkin	Granit

A Second Insightful Reviewer

After a year of no nominations there was one nomination in 1957 for Hodgkin and Huxley and two more nominations for Hodgkin, one by Eccles proposing Hodgkin alone and another which combined Eccles and Hodgkin as candidates. In 1958 there were as many as seven nominations of Hodgkin, one of which, submitted by a pair of nominators, also included Huxley. In another nomination it was again proposed to combine Hodgkin with Eccles. The Nobel committee asked Bernhard Frankenhaeuser to do a review. He was a member of Granit's research group, as mentioned (see Chapter 1, p. 66). Frankenhaeuser

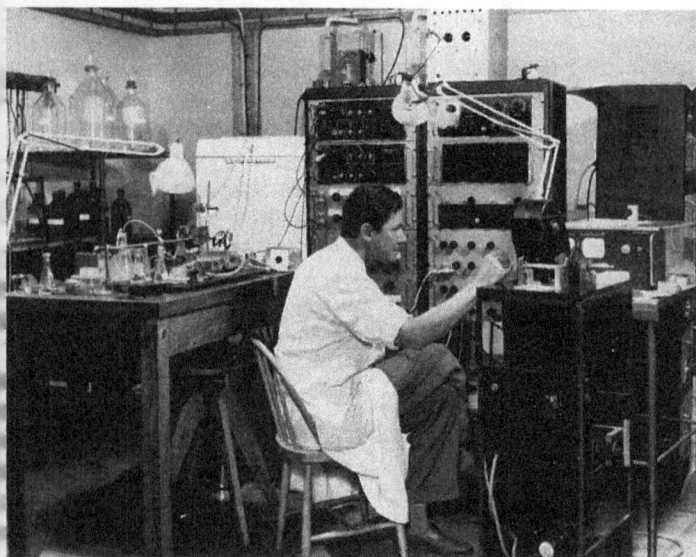

Bernhard Frankenhaeuser (1915–1994) in his laboratory. [Courtesy of the Karolinska Institute.]

was born in Finland in 1915 and received his medical education in Helsinki before moving to the Karolinska Institute. In 1946 he married a Finnish colleague, Marianne von Wright, who was later to become a very influential professor of stress research at the Karolinska Institute. She in turn was the sister of the famous Finnish philosopher Henrik von Wright, who succeeded Wittgenstein in Cambridge. After presenting his Ph.D. thesis in 1949, Frankenhaeuser remained at the Nobel Institute throughout his whole career. As mentioned he collaborated with Hodgkin in the mid-1950s and for some time worked in Cambridge. The goal of his research was to demonstrate that the ionic hypothesis introduced by Hodgkin and Huxley was valid also for myelinated nerve fibers.

Considerable technical modifications had to be introduced to achieve this goal. He and his co-workers made further refinements to the ultramicro-electrode technique introduced by Huxley and Stämpfli in 1950 to eliminate external short circuiting. A particular technical design was needed since the axons in myelinated fibers are too slender to allow the introduction of a thin pipette. Eventually it became possible to also adapt the voltage clamp technique to myelinated fibers and reveal changes in the permeability for sodium and potassium in this system too, verifying that the ionic hypothesis had a general application. This data was published in the late 1950s. Frankenhaeuser was thus a qualified reviewer with a full insight into Hodgkin's and Huxley's work.

The review was relatively brief, compared to the two previous evaluations by Granit. Frankenhaeuser recapitulated how Hodgkin and Huxley using the voltage-clamp technique could demonstrate that the main currents that the nerve membrane generates at the event of depolarization are an initial transient flow of sodium into the cell followed by an extended flow of potassium out of the cell. This movement of ions can explain the changes in current over the membrane that is the origin of the action potential (p. 130). Frankenhaeuser also summarized the complicated mathematical thinking introduced by Hodgkin and Huxley in which they had even used the first primitive computers of the time. The flow of sodium and potassium ions across the axonal membrane demonstrated by use of isotope technique in experiments by Hodgkin and Keynes was again referred to. Finally Frankenhaeuser reflected on the debate about the application of the ion hypothesis to excitable tissues in general. His conclusion was that it seemed to have a broad application although there might still be certain exceptions. In this judgment he also implicitly included his own findings of applications to the myelinated nerve fibers. The conclusion of the review was that Hodgkin and Huxley should be considered as one of the strongest pairs of candidates for a Nobel Prize in physiology or medicine, since they had made the most outstanding contribution to the field during the last half century. The two candidates remained listed as worthy of a prize.

The Temperature Increases

The number of nominations continued to increase in 1959 through 1961 (Tables 3.1 and 3.2). Nominations of Hodgkin alone dominated but there were also a fair number of proposals for him in combination with Huxley. In 1959 a third reviewer was called upon. This was the professor of medical physics Arne Engström, who came to have a major influence on the 1962 prize to Crick, James D. Watson and Maurice H.F. Wilkins[2]. His review was very brief and it did not add anything of substance. He supported the view by Granit and Frankenhaeuser that Hodgkin and Huxley were highly worthy of a prize. This was also the view of the whole committee. As time progressed the representatives of neurophysiology, Bernhard, Granit and von Euler, became increasingly eager to have a prize in their field of research. The first attempt was made in 1960, as mentioned in the previous chapter. A strong minority of the committee voted for Eccles and Magoun, but the prize this year went to the immunologists Burnet and Medawar. In 1961 it was decided to move the positions forward and the committee once again asked Granit to review the nominations of Hodgkin and Huxley this time also together with Eccles and Katz. Again Granit carried out a very thorough analysis covering 30 pages. He noted the repeated nominations of both Hodgkin and Eccles since 1953, occasionally as a combination, of Huxley since 1955, most often in combination with Hodgkin, and of Katz only since the previous year. Granit started his follow-up review with a citation from Katz in 1952. It said:

One of the objections that has often been directed against the membrane theory is that it centres around an invisible structure which cannot be demonstrated or identified by direct methods, and which, so it is claimed, has been built up mainly to satisfy the imagination of electrophysiologists. There is a good deal of truth in this allegation, but much the same might be said of such postulates as atoms and electrons, with the one distinction that these are used to satisfy the minds of a much larger population of scientists.

Granit then went on to report some work presented by Hodgkin together with P.F. Baker and T.I. Shaw at the March 1961 meeting of the Physiological Society in the U.K. using a new technique providing additional evidence for the correctness of the membrane ionic hypothesis. In this technique the giant axons had been emptied of their content and instead of their axoplasm various salt solutions containing different ions of various concentrations had been introduced. The data strongly confirmed the Hodgkin-Huxley predictions. Granit then once more reviewed the different important steps in the developments departing from Bernstein's original ion theory involving only potassium. The critical role of the five publications by Hodgkin and Huxley in the *Journal of Physiology* in 1952, also including Katz as a co-author in the first one was stressed once more, but it was also obvious from the review that Hodgkin together with a number of other collaborators had made additional important contributions consolidating the theory during the ten years that had passed. Granit unsurprisingly highlighted the important work by Frankenhaeuser in his own department. He emphasized the technical challenges that had to be met because the diameter of the myelinated vertebrate axons is 50 times less than the one of the giant axons used by Hodgkin and Huxley in their studies. Although the transmission of the current is saltatory in the myelinated nerves, as described in the previous chapter, the principles of ion exchanges are valid also for this system. These confirmatory data had been published in 1961 and the results strengthened the paradigmatic nature of the Hodgkin-Huxley 1952 discovery.

The original idea of establishing Nobel Institutes was to be able to control a discovery nominated for recognition by a Nobel Prize. Granit noted in his review:

It is comparatively rare that the Nobel institutes will work according to their statutes, which have as one basis to control (the correctness) of results proposed for recognition by a Nobel Prize. Coincidentally Frankenhaeuser has now become interested in a direction for his work, which includes this and at the same time (provides) a new development of the ionic theory as it is applied to vertebrate nerves.

Granit became almost lyrical in his writing when he stated:

Accordingly one of the big problems of physiology has also been brought a major step forward, so far that one would scarcely have believed that this generation would have been able to (take) such a step.

He summarized Hodgkin's, Huxley's and their collaborators' contribution in the following way:

... publications have led to the discovery of two systems in the nerve axon, one secretory system that can be metabolically modified and which reinstalls the "steady state" (English in the original) and is sensitive to the activators and inhibitors mentioned above, but this and metabolic processes in general only have a minor effect on the other system, to which I have paid most attention in my preceding presentation, that is, the changes of permeability which generate excitation in the nerve.

He also noted that:

... nor do we know how the development in time of the permeability changes is arranged within the few thousandths of a second that are available.

The discussion of Eccles' work that followed has already been described in the previous chapter.

Granit's presentation of Katz's impressive contributions was also very thorough and there will be reasons to return to them in a forthcoming not too distant discussion of the 1970 Nobel Prize in physiology or medicine by use of the full archival material released in 2021.

In the summary of the review it was noted that all the four candidates examined were strong contenders for a Nobel Prize. Hodgkin and Huxley seemed to represent the first pioneers of the field "... with regard to method(s), experiment and formulation of concepts." A combination with Eccles was supported by the conditions discussed in the previous chapter, whereas it was proposed that Katz be kept on the backburner since he was younger (than the other candidates) and had only been proposed for a prize since the previous year. A comment on the contestable strength of this argumentation has already been given in the preceding chapter. Granit put a major emphasis on the thematic congruency between Hodgkin's, Huxley's and Eccles' work. He wrote:

> ... all three have as a main theme (of their scientific work) the displacement of ions which indicate stimulation or inhibition of the nerve signals. Hodgkin and Huxley have focused on the axon of the nerve, Eccles on its soma (the cell body). On a previous occasion Sherrington and Adrian were combined in a Nobel Prize awarded for their studies of the physiology of neurons. In this case it was also a mainly central and primarily peripheral analysis that were combined, just as in the present case Eccles has primarily focused on the part of the nerve cell which is located in the central organ and Hodgkin and Huxley for the part that has generally been studied outside a central organ on account of the improved accessibility for study of isolated axons.

The committee was impressed by Granit's review and a majority of its member recommended Eccles, Hodgkin and Huxley for the prize. But the committee was unexpectedly overruled by the College of Teachers, which instead recognized von Békésy with the 1961 prize[2].

In his biography[1] Hodgkin mentioned the occurrence of a false alarm this year. His wife Marni received a call from a Dr Bolinder in London who said he had very good news from Stockholm and wanted to get in contact with her husband. When Hodgkin heard this he became a little suspicious. He knew that his cousin Dorothy Hodgkin had received false information about the prize in chemistry the previous year and the same year the Australian radio had announced that Eccles should receive the prize. Still Hodgkin went to the office of the Swedish journalist in London and was photographed and interviewed. At this time Huxley had become the Jodrell Professor of Physiology at University College and the two former team workers had lunch together. When they returned to Huxley's office the journalist was there with a long face. He said glumly: "Oh professor Hodgkin, we have bad news for you; it has gone to an American called Békésy for work on the ear."

A Year without Any Nomination and the Finish

In 1962 there was not a single proposal for a prize to Hodgkin and Huxley, whereas Eccles was nominated. It can be noted that Hodgkin, sometimes together with Huxley, had received ten nominations both in 1959 and in 1960, and in 1963 there were to be even more. In 1961 there were fewer nominations. However, the lack of any nomination of Hodgkin and Huxley in 1962 is very conspicuous. Eccles, Hodgkin and Huxley were prime candidates of the committee for a prize in 1961 and it would therefore have been obvious not to say obligatory that these candidates should have been available for the discussions in 1962. This was generally secured by a proposal from the secretary or any other member of the committee on the last day of nomination on January 31, when it had become obvious that no external nomination would be forthcoming. On this day in 1962 there were, surprisingly, no nominations of two of the three candidates from the invited nominators (Tables 3.1 and 3.2) and still the committee did not take any initiative to ensure that they could be discussed this year. The secretary of the committee who was von Euler, himself a neurophysiologist and obviously a proponent of a prize to Eccles, Hodgkin and Huxley, should have taken the opportunity to make a secretary nomination on the last day this was possible or have delegated this to some other member of the committee. This did not happen. The only way this can be explained is that already *before* January 31, 1962 it had been informally agreed between the chemistry committee, guided by its highly influential spokesperson Tiselius[2] — in a discreet way he seems to have been the scientist to pull the curtain — that the Royal Swedish Academy of Sciences should take care of the chemistry prize for the crystallographic studies of globular proteins and that the Karolinska Institute should give a prize in physiology or medicine for the unraveling of the structure of DNA. Implicitly this meant that the work by the committee in 1962 at the Karolinska Institute did not have *any* critical influence on the sequence of events, since in practice the prize recipients were already selected when the sorting out process started! It was not until 1963 that there was a thawing of the ice for the field of neurophysiology.

The number of nominations in 1963 compared to the lack of any nominations the preceding year is conspicuously large. One wonders whether someone with insight into the Nobel work at the institute had not inspired one or more of the nominators to provide a proposal. But on the other hand if that had been the case one would have expected a proposal including all the three neurophysiologists. Such a combination was *never* submitted by any external nominator. The proposal to combine Eccles with Hodgkin and Huxley remained a construct by the committee (Granit?).

The 1963 nominations came from leading scientists in the field. Many of the proposals were quite informative, including the repeat nomination from Eccles for a prize to Hodgkin alone. A particularly thorough nomination arrived from Seattle, signed by five nominators headed by the famous neurophysiologist J. Walter Woodbury and this material was referred to in Granit's fourth and final review. He remarked that Woodbury at about the same time and independent from Eccles and collaborators had managed to insert microelectrodes into nerve cells in the ventral part of the spinal cord. However, it was Eccles who came out first in drawing generalizing conclusions from his findings. Granit also cited a whole passage from the nomination by Woodbury and his co-nominators. It praised the discovery by Hodgkin and Huxley and read:

> It is difficult to emphasize sufficiently the uniqueness of this contribution to the physiology of nerve fibers. Prior to their work, understanding of the nature of the nerve impulse consisted of a qualitative description of its properties, e.g., all-or-nothing. In this work, on the other hand, we have a detailed quantitative description of the specific membrane properties which lead to impulse generation and conduction. Thus knowledge of nerve function was carried from the descriptive level to the level where the molecular biologist must grapple with the molecular basis for precisely specified membrane behavior. In other words, Hodgkin and Huxley carried nerve physiology from a general description to molecular biochemistry in one master stroke of genius.

The nomination by Woodbury and others also discussed at the end what had happened during the ten years that had passed since the original discovery. The main criticism had been that the glass electrodes used caused some damage to the axons leading to some degree of deterioration of the integrity of the membrane. The pipette was pushed into the large axon some 10–30 mm (see p. 124). It was argued that this could have led to some distortion of the observations, but extended follow-up experimentation in essence had confirmed the original data.

A part of Granit's final paragraph read as follows:

> I have previously included Eccles in a joint proposal with them (Hodgkin and Huxley) and would now like to recommend the same solution. Of

course Hodgkin and Huxley could be proposed for a joint prize and Eccles later for another. It would not be possible to choose another sequence of events. However, dividing the prize into three offers a solution, considering the competition between different disciplines for a Nobel Prize, since the main theme is excitation and inhibition in nervous tissue, the main methods electrical and the main results given in terms which concern ion transport through cell membranes of nerves, in the case of Hodgkin and Huxley in the axon, and in the case of Eccles the somata (cell body) of nerve cells, in particular motor neurons. On an earlier occasion the Nobel Prize was awarded to Adrian and Sherrington for their discoveries concerning the single neuron. The contribution by the former concerned the afferent message from the periphery, examined in individual nerves, and by the latter the central events analyzed by a world of concepts that defined the processes relating to the individual neuron.

In spite of this strong proposal the committee in 1963 could not draw up a unanimous proposal. There were three (Friberg, the Chairman, Hamberger — the professor of ear-nose-throat medicine, prime advocate of the 1961 prize[3] and the chemist and Nobel laureate Theorell) of the nine members who did not vote for Eccles, Hodgkin and Huxley, but instead supported the candidature of Konrad Bloch and Feodor Lynen. When the vote was taken in the College of Teachers the proposal by the majority of the committee prevailed. The trio of neurophysiologists were eventually able to receive their prize.

The Nobel Festivities of 1963

The three happy prize recipients came to wintry Stockholm to meet old friends and to make new ones. Eccles knew Granit from the early 1930s when they worked in Sherrington's laboratory and also published some articles together. Hodgkin also knew Granit and also in particular his collaborator Frankenhaeuser. Both Eccles and Hodgkin had sailed together with their Swedish colleagues in the Stockholm archipelago under the potentially more benevolent weather conditions of the Swedish summer. Granit, not surprisingly, had been selected to give the laudation presentation at the prize ceremony[4]. He attempted a popularized presentation of the work recognized by the prize, but in part it was probably too technical and specialized for the audience present. He included references to Overton's early visionary proposals and Frankenhaeuser's recent important confirmation of the ionic hypothesis in advanced studies of vertebrate myelinated fibers.

In a letter to her parents Marni Hodgkin has described in detail the events at the Prize ceremony at the Concert House (cited in Ref. 1). It is so rich in its personal presentation that it deserves to be cited in full:

> Eventually we are all ready, and are driven pompously away to the Concert Hall where the doings take place. Here we are ushered into a kind of robing room where dignitaries are getting into gowns of various kinds. All the Swedes are wearing those wonderful pleated top hats with gold buckles at the front such as one saw in the academic procession in *Wild Strawberries* (the legendary Ingmar Bergman black and white movie, my remark) and there is a dazzling profusion of medals, stars, cockades, ribbons, etc. The hall is filled with a capacity crowd, all in full evening dress, (it is now about 4.15) and all the photographers are in white tie too. Lady Eccles is behind us, with a camera and permission to snap — still more transparencies! The stage gradually fills with dignitaries and eventually the laureates shamble in and take their places. Fanfare of trumpets. Enter the Royal Family. We all stand for the National Anthem, in which the Swedes, including Princesses, actually sing the words. We're off! Procedure: piece of very light music played by orchestra in gallery at back. Lengthy citations read in Swedish (we have a translation), ending with a short citation in laureate's native tongue. Laureate is bidden to descend from platform to receive medal from King. Laureate advances, bows, descends staircase, shakes hands with King, bows to ladies of Royal household, receives medal and enormous portfolio, advances to other side of staircase, *supposedly* gives up medal and portfolio which will be on display in City Hall the same evening, ascends other side of staircase, bows to audience and returns. Well, this may sound fairly straightforward, and perhaps is with a single recipient, but with these divided prizes, three people are trundling down the stairs etc. and is the first to wait, or go on, and does he bow with the trumpets? After the trumpets? Wait for the trumpets? There is nothing like an ear-splitting fanfare to tingle the blood, but also to upset routine. Well, anyhow, nobody got it quite right. Even the Ambassadorial poet, Seferis, who spent six years at the court of St. James, forgot to bow.
>
> To do the Swedes justice, they are fond of formalities, but they appear not to mind if things don't go quite according to plan. Also there were

several things put in to make it harder. The physics people came first, and Mrs Goeppert Mayer, whose shoulder straps came right off as she sat, so she began to twitch them up in a vague way when it was her turn to rise, forgot all about returning the medal and portfolio and drifted back to her seat carrying them in a slightly absent-minded way.

Then there was Professor Natta. He could get as far as the top of the stairs but couldn't possibly descend, but it had been arranged that his son should go down the steps to receive the actual medal. Well, when this actual moment came, the King, who is 81 and well over six feet tall, gave one look and like a veritable moose he *bounded* — there is no other word for it — up the stairs and no nonsense about the boy standing deputy for the father. I found this very moving: Natta being so pathetic in his hour of glory and the King so nice. Mr Eriks(s)on, our chauffeur, confided to me that everyone, even the extreme republicans, love the King, and indeed just looking at him you can see that he is a truly lovable man.

Anyhow, after this, the routine of the procedure was rather shaken, but never mind, apart from not knowing whether to give up the portfolio and medal, our Hero acquitted himself splendidly, did remember to bow and looked younger than anyone else, so that even his critical children felt that he had covered them with reflected glory. So more music, more citations, National Anthem again and finally it is all over.

Can it be told better? But there is more to the story, because Marni's parents, who were enjoying their son-in law's hour of glory on the other side of the Atlantic, could not have had the slightest premonition that they themselves would be in the same situation three years later. But the particular story of Peyton Rous who had to wait some 50 years for his prize remains to be told in a forthcoming book. So let us go back to 1963. The three prize recipients beamed in their moment of glory.

At the banquet Eccles gave a very inspired speech addressing the students. In the first part, he said "This

Eccles received his Nobel Prize from His Majesty the King. [From Ref. 31.]

Hodgkin receives his Nobel Prize from His Majesty the King. [From Ref. 31.]

Huxley receives his Nobel Prize from His Majesty the King. [From Ref. 31.]

is the greatest day of our lives — the climax of long years of creative work. We feel a great expansion of personality. And now as I speak to you I feel elevated, as on this platform. Let me then speak to you as an old student of some 60 years and give you young students two thoughts that have come to me with special vividness in these last years." His first comment concerned space travel, much on everyone's mind at the time. Eccles believed that in general it had a negative impact and that there was no future for man outside Earth. In the light of later developments this certainly may be debated, but his romantic message was that "We and our fellow men of all countries must realize that we share this wonderful, beautiful, salubrious earth as brothers ..." His second reflection was that at the time there was a considerable underestimation of the importance of biology. Again this is unquestionable a misjudgment. Biology, not least molecular biology, has come to dominate the second half of the twentieth century. Not surprisingly he finally stated "For us the most significant questions we can ask scientifically concern the working of our nervous systems — the marvelous reception, communication and storage devices that subserve all our perceptions, our thoughts, our memories, our actions, our creative imaginations, our ideals. To the brains of our predecessors we owe all of our inheritance of civilization and culture." Wise words.

Hodgkin also gave a brief speech at the banquet[5]. He had three messages. The first was his lack of self-confidence in giving the address. By way of contrast he referred to Ernest Rutherford, the straightforward New Zealander who received the 1908 prize in chemistry. Hodgkin said "I hasten to say that my own subject is a very minor ripple compared to Rutherford's. It may seem rather like a cyclone when Sir John Eccles arrives from Australia. But in neurophysiology we have none of those vast tidal waves of discovery which shake the world to its foundations and which have such incalculable consequences for good or evil." Again this statement is debatable. Advances in the neurosciences have continued to be impressive and the more we get an insight into the uniquely complex functions of our brains, the more vulnerable we are to being manipulated. This will of course be dependent on the judgment of how to use all new knowledge. It is inherent in the dual nature of most forms of knowledge that it can be both used and misused. The balance is offered by the generation of rules for ethical behavior, an important part of the responsibility of the advancing civilization praised by Eccles. Hodgkin's second message concerned the Swedish collective sentiment. He was waiting with mixed emotions for the next Ingmar Bergman movie. He said "For myself, I have been pleasantly reminded of summer days sailing with Swedish friends among the flower-covered islands of the Archipelago. And Bergman's black medieval world reminds us that the brilliance of your summer owes something to the darkness of your long winter nights." His final remarks concerned the continuity in science, as expressed by Isaac Newton, "If I have seen further it is by standing on the shoulders of others." Hodgkin paid particular tribute to Sherrington and Adrian.

Time for Nobel Lectures

Due to the strict adherence to formal rules at this time, the Nobel lectures were given on December 11, the day *after* the festivities. A person awarded a Nobel Prize should only give a Nobel lecture after he or she has received their prize. For a number of years this rule has no longer been adhered to. As a courtesy to the prospective laureates the chosen prize recipients of today give their lecture prior to the prize ceremony, allowing them to fully enjoy the festivities. Eccles talked about "The ionic mechanism of postsynaptic inhibition"[6]. He gave a very lucid lecture introducing the concept of the synapse, paying tribute to his teacher Sherrington. Hereafter he introduced the complexity of all nerve endings which are in contact with the cell body of a neuron modulating the outgoing signal through the axon by excitatory or inhibitory influences. The cell body, the soma, was found to act as a very complex switchboard and it was revolutionary when Eccles and collaborators for the first time in 1951 managed to record the electrical activity by use of very fine electrodes within the cell body of the large motoneurons in the frontal part of the spinal cord. He also described how, in very complicated and demanding experiments, he and his colleagues tested the effect of 33 different ions in different concentrations on neuron functions. It might be noticed that in his lecture Eccles focused strictly on experimental science. No room was left for philosophizing, which he enjoyed being involved in on many other occasions. His concluding paragraph demonstrated that at the time and after a long delay he had become a spokesperson for the central role of humoral transmitter substances. In reference to a schematic picture he said:

In conclusion the figure (a schematic picture of the synaptic cleft) will serve to summarize diagrammatically the detailed events which are presumed to occur when an impulse reaches a presynaptic terminal, and which we would expect to see if electron microscopy can be developed to have sufficient resolving power. Some of the synaptic vesicles are in close contact with the membrane and one or more are caused by the impulse to eject their contained transmitter substance into the synaptic cleft. Diffusion across and along the cleft, as shown, would occur in a few microseconds for distances of a few hundred ångströms. Some of the transmitter becomes momentarily attached to the specific receptor sites on the postsynaptic membrane with the consequence that there is an opening up of fine channels across this membrane, i.e. the subsynaptic membrane momentarily assumes a sieve-like character. The ions, chloride and potassium, move across the membrane thousands of times more readily than normally: and this intense ionic flux gives the current that produces the IPSP (inhibitory postsynaptic potential, the mirror image of the excitatory postsynaptic potential, EPSP) and that counteracts the depolarizing action of excitatory synapses, so effecting inhibition.

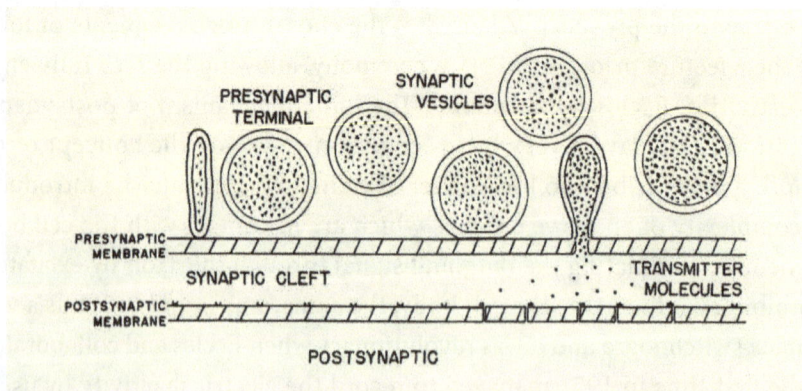

Diagrammatic visualization of the synaptic cleft with vesicles. [From Ref. 6.]

It was not easy for Hodgkin and Huxley to divide the material to be presented in their respective Nobel lectures. Their work was so intertwined. Hodgkin's lecture was appropriately named "The ionic basis of nervous conduction"[7]. He started by mentioning his early impressions of Trinity College and emphasized the particular connection this institution had with neurophysiological research. At the time when he started his studies Adrian was the dominating figure and in the previous generation Keith Lucas had played a very important role. Lucas had been a good friend of Hodgkin's father and like him had died young in the First World War as mentioned. Through his reading Hodgkin became intrigued by learning about Bernstein's membrane theory. He then referred to his early findings in the field and how he became involved in studies of the large nerve fibers of squids of the genus *Loligo*. He emphasized the critical experiment in 1939 when he together with Huxley tested Bernstein's theory and found to their surprise that the action potential was much larger than the resting potential. According to Bernstein's theory it should not exceed the resting potential. This prewar observation was the departure point for things to come when their joint research was resumed after the war. This work led to the formulation of the sodium hypothesis, giving this cation the appropriate role it should have in the development of the action potential. Hodgkin demonstrated in the last two figures that the resting potential and action potential varied with the internal concentration of K and Na in a manner which is consistent with the external effect of these ions. Improvement of the voltage clamp technique opened possibilities for important discoveries. Hodgkin said "Huxley will describe these results in more detail; here all that need be said is that by varying the external ionic concentrations it was possible to separate the ionic current flowing through the membrane into components carried by sodium and potassium, and hence to determine how the ionic permeability varied with time and with membrane potential."

Huxley started his Nobel lecture[8] in the following way:

Professor Hodgkin has told you how he was influenced as an undergraduate by the writings of four Fellows of Trinity College, Cambridge. I too was an undergraduate at Trinity, but at the time I was taking physiology seriously, in my final year in 1938–1939, there was yet another Fellow of the College who influenced me even more directly than the ones mentioned by Hodgkin, and that was Hodgkin himself.

He then went on to discuss their important joint prewar experiments and how the two researchers joined forces again after the war. After this he presented in more detail the quantitative aspects of the ionic theory. He emphasized that the measurements on which the theory was based were made by a feedback method which the researchers had named "voltage clamp." The lecture was relatively concise with a considerable amount of theory and presentation of experimental data that seemed to support the theory. The conclusions included four reservations. The first one emphasized that so far only rapid events had been studied. The two following reservations referred to some recent data that had raised certain questions. Finally he stated "Fourth, Bernhard Frankenhaeuser of the Nobel Institute here has achieved the remarkable feat of doing voltage-clamp measurements on single nodes of Ranvier in myelinated nerve fibers and has found that there are substantial differences in behavior from the squid giant axon, although the main outlines are the same. Both Hodgkin and I feel that these equations should be regarded as a first approximation which needs to be refined and extended in many ways in the search for the actual mechanism of the permeability changers on the molecular scale."

The relationship of Hodgkin's and Huxley's contributions needs to be discussed further. It is clear from the nominations, starting with the one given by Adrian that Hodgkin was the leading scientist. Also Eccles nominated Hodgkin alone. Hodgkin had close collaborations not only with Huxley, but also very productive interactions with Katz, Keynes and many others. Still it seems that there was something special about the collaboration with Huxley. There is a rule in the Nobel work that each individual candidate in a combination of two or three should be able to carry the prize on his own shoulders. It is somewhat hard to argue that this is the case for Huxley, although again the five publications in the early 1950 that cemented their ionic theory was a twin work. It could be added in addition that Huxley's equations to characterize the action potential have survived into the present time and furthermore later work demonstrated that he was also capable of making major discoveries on his own.

Hodgkin had a rich post-prize academic career. After doing some work on muscle and nerve he switched to visual research. However, he became more and more involved in administration. He was called on at short notice to take responsibility for the Presidency of the Royal Society. Originally this had been planned for Peter Medawar, but because he sadly suffered a cerebral hemorrhage in 1969 this did not come about. Hodgkin enjoyed the responsibility of the Presidency very much and he and Marni lived half the week in the apartment refurbished by the Society at Carlton House Terrace in London and the rest of the week in Cambridge. In 1972 Hodgkin was knighted and the following year he was appointed to the Order of Merit. He died in 1998, 84 years old.

A Second Important Discovery or Huxley Times Two

Huxley did not give a public address in 1963, but many years later, in 2001 he was back in Stockholm. This visit was made to participate in the celebration of the 100 years that had passed since the first Nobel Prizes were awarded. On this occasion he granted Joanna Rose of Nobel Media an interview. The interview took its starting point from a recent article with the title "Forgetfulness in science" that Huxley had written. It reflected on how not infrequently important observations were forgotten and then rediscovered. Sometimes it could be that later critical observations may have given them a completely new significance. Overton's 1902 observations pointing out the critical importance of sodium in muscle contraction and implicitly in nerve function was mentioned as a case in point. Later in the interview Huxley laconically stated "After Hodgkin and I had finished our work on nerve, we couldn't see what to do next in relation to nerve." He then referred to the rapid advances in the fields of electronics and in molecular biology and their applications to neurophysiological studies, and not feeling familiar with these he "moved into muscle."

The background to this new direction of research was his interest in developing a new kind of microscope. He was successful in improving the performance of the electron microscope existing at the time. By use of this modified instrument, Huxley, together with a visiting scientist from Germany, Rolf Niedergerke, made a number of important fundamental observations on the structure and function of muscle tissues. This led to the formulation of the "sliding filament theory" establishing a mechanism for muscle contraction. Thus Huxley truly went from brain to brawn. The same year a similar finding was made by Hugh Huxley, no relation to the historically well-known

Hugh Huxley (1924–2013.]

Huxley family, and Jean Hanson. H. Huxley had joined the MRC unit for molecular biology at the Cavendish Laboratory in Cambridge in 1948, only a year after it had started. He became a member of the crystallography group and examined wet specimens of muscle. He tried to identify the relative role of the two components myosin and actin in the muscle tissue. Hugh Huxley then learnt electron microscopy at Massachusetts Institute of Technology (MIT) in Boston. One of his fellow visitors at MIT was Jean Hanson, originally from King's College in London. Combining the observations the two of them had made they were able to formulate a new theory of muscular contraction. This finding of a fundamental new mechanism for the function of muscle fibers was in the class of discoveries which are nominated for Nobel Prizes. The first nomination for a prize to H. Huxley was submitted as early as 1959. He had just given a lecture at a congress of electron microscopy in Berlin with the title "The mechanism of contraction." A brief review was drawn up by the professor of histology Gösta Häggqvist. He concluded that it was too early to make a judgment on the importance of the discovery and no full review was done. Later on there must have been multiple nominations for this important discovery but they are hidden behind the penumbra of the veil represented by the 50-year secrecy rule. Thus it will take time until we know how close A. Huxley might have been to receiving a second Nobel Prize together with his non-relative H. Huxley.

It can be added that when H. Huxley returned to England in 1954 he first went back to the Cavendish laboratory but after a year he moved to University College London. At this location he was close to Birkbeck College and inspired by Rosalind Franklin and Aaron Klug[2] he started to examine viruses in the electron microscope using a new staining technology. He participated in the development of the important negative staining technique, publishing a seminal paper on this topic only one year after Sydney Brenner and Robert Horne presented the first paper describing the technique in 1960 that allowed high-resolution studies of virus particles, as described earlier[2], and other central biological structures. In 1962 H. Huxley returned to the MRC Laboratory of Molecular Biology, which had now moved into a new building[9]. He took advantage of his experience in using the negative staining technique on viruses and also applied it to studies of muscle tissue. The findings allowed him to consolidate his theory of muscle contraction.

To return to the interview with A. Huxley in 2001, he argued towards the end of this for more teaching of the history of science to students. The purpose would be to give a picture of how new insights are achieved and to emphasize the accumulative characteristics of natural science. Also it might be instructive to show how certain ideas are ahead of their time. It could further be emphasized how frequently one and the same discovery is made simultaneously in different laboratory settings, as exemplified by the discovery of the sliding filament theory. It would seem that not too infrequently "the time is ripe" for the making of a certain discovery. A. Huxley had a long life dying at the age of 94. He retained his incisive intellect until old age. Among his close friends he was referred to as "the gimlet," a handy tool to drill small holes. His career in many ways was analogues to Hodgkin's. He was knighted in 1974 and appointed to the Order of Merit in 1983. A year later he was elected master of Trinity succeeding his critical collaborator Hodgkin. Finally he served as the president of the Royal Society 1980–85.

Genealogy and Polymaths

The interview finally reflected on A. Huxley's ancestry and his famous family predecessors. The figure illustrates the Huxley family tree. Historically it all started with Thomas H. Huxley generally referred to as "Darwin's bulldog". He was given this nickname because he vigorously and effectively defended Darwin's theory of evolution against the church. Originally Thomas was trained as a doctor, and in this function he participated in the 1846–50 round-the-world voyage in the ship *Rattlesnake*. He wrote a book *Man's Place in Nature* in which the concept of agnosticism was introduced. He was also a paleontologist and discovered seven new species of dinosaurs. Thomas was a skilled writer

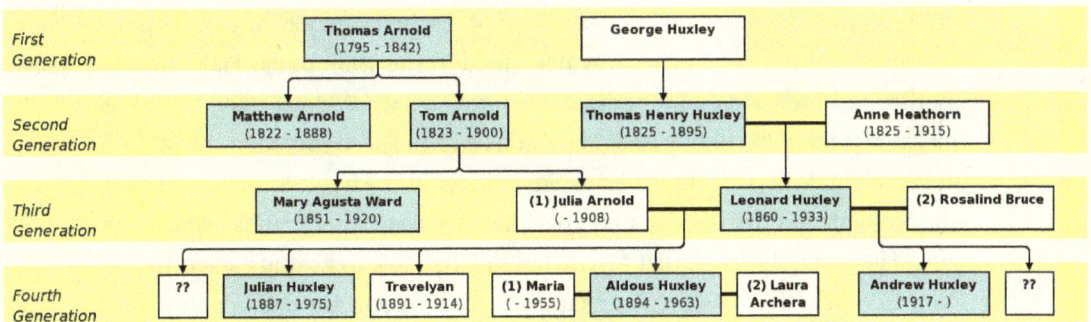

Huxley family tree

and his aphorisms are quoted into the present time. Just to give one more "The struggle for existence holds as much in the intellectual as in the physical world. A theory is a species of thinking and its right to exist is coextensive with its power of resisting extinction by its rivals." Thomas had one son, Leonard, and two daughters. Leonard became a famous English school teacher, writer and editor. He had eight children, five in his first marriage and three, including Andrew, in his second. Andrew had two particularly famous half-brothers, Julian and Aldous. Julian became a well-known biologist contributing to many fields of research, but he may be remembered most of all for his writing of popular texts. Together with two colleagues, he published in 1929–30 a popular account of the current state of biology. It was in three volumes and was called *The Science of Life*. It sold so well that he was able to give up teaching and focus only on his writing. Among his many books might be mentioned his last two. They were entitled *Eugenics in Evolutionary Perspective* (1962) and *Essays of a Humanist* (1964).

In these books he introduced important concepts like ethnic groups and transhumanism. As Secretary General of UNESCO he was able to secure the spreading of his important message on the invalidity of the concept of race and the need to change it for "ethnic group." I have his last book in my library, but I have twelve books by his famous brother Aldous, already mentioned in Chapter 2. A lot can be said about his books *Brave New World* (1932) and *Brave New World Revisited* (1958). In its conjectures, the former draws extensively on the author's deep insights into current genetics. The concept of freemartin has already been discussed in my previous book[2]. In the follow-up book Aldous acknowledged that in his first book he did not have the foresight to see that humankind could develop something as devilish and self-destructive as atomic weapons. The influential polymath and famous geneticist and evolutionary biologist John B.S. Haldane, a close friend of his, had a major influence on the description of the eugenic society described in the 1932 book. Haldane was another representative of a pedigree of intellectuals. Aldous liked to model the characters in his books on real persons and this also included the famous biologist Haldane, who appeared as Shearwater in the book *Antic Hay* (1923), "the biologist too absorbed in his experiments to notice his friends bedding his wife." However, the book that fascinated me the most as a young scientist was *Point Counter Point* (1928). The characters in this book are also in many cases modeled on different acquaintances that Aldous had. In 1963 he published a book called *Literature and Science*. This book illustrates how exceedingly well read he was. He wrote:

What is a rose? A daffodil? A lily? One set of answers to these questions may be given in the highly purified languages of bio-chemistry, cytology and genetics. "A special form of ribonucleic acid (the recently discovered messenger-RNA — see Chapter 9, my remark) carries the genetic message from the gene, which is located in the nucleus of the cell, to the surrounding cytoplasm, where many of the proteins are synthesized and so on in endless, fascinating detail. A rose is a rose is a rose, is RNA, DNA, polypeptide chains of amino acids ...

The latter association to roses is taken from a part of the 1913 poem *Sacred Emily* written by Gertrude Stein. She in her turn was inspired by Shakespeare, who wrote "a rose by any other name would smell as sweet."

Later in the same book Aldous Huxley cited from Goethe's Faust, *Grau, theurer Freund, ist alle Theorie, Und gruen des Leben goldner Baum* (Gray, dear friend is all theory. And green the golden tree of life). After this he added "For some people the contemplation of scientific theories is an experience hardly less golden than the experience of being in love or looking at the sunset." Still further on this well-informed writer stated "Science is a matter of disinterested observation, unprejudiced insight and experimentation, patient ratiocination within some system of logically correlated concepts. In real-life conflicts between reason and passion the issue is uncertain." However, in this context I believe Huxley severely underestimated the amount of prejudiced bias and emotional commitment involved in the practice of everyday science.

The aristocratic Scottish Haldane family, has deep roots in time and a lot could be said about many of its members. In the present context it can be mentioned that one sister of John Haldane was Naomi Haldane. She married a barrister G. Richard Mitchison. Originally she started a career as a scientist doing experiment in genetics together with her brother John, but she soon switched to writing. She was highly successful and became a renowned novelist and poet, acknowledged as the doyenne of Scottish literature. She was also an advisor to Tolkien in his great work being one of the proofreaders of *The Lord of the Rings*. Naomi Mitchison is famous also because James (Jim) Watson dedicated his book *The Double Helix* to her[10]. The reason for this can be seen in Chapter 15 in that book. At Christmas 1951 Watson had received an invitation from Avrion Mitchison, a famous immunologist-to-be, trained by Peter Medawar, to visit his parents' estate of Carradale on the Mull of Kintyre in Scotland. Apparently this was to become a very memorable experience and the young Jim must have been very impressed with the range of social

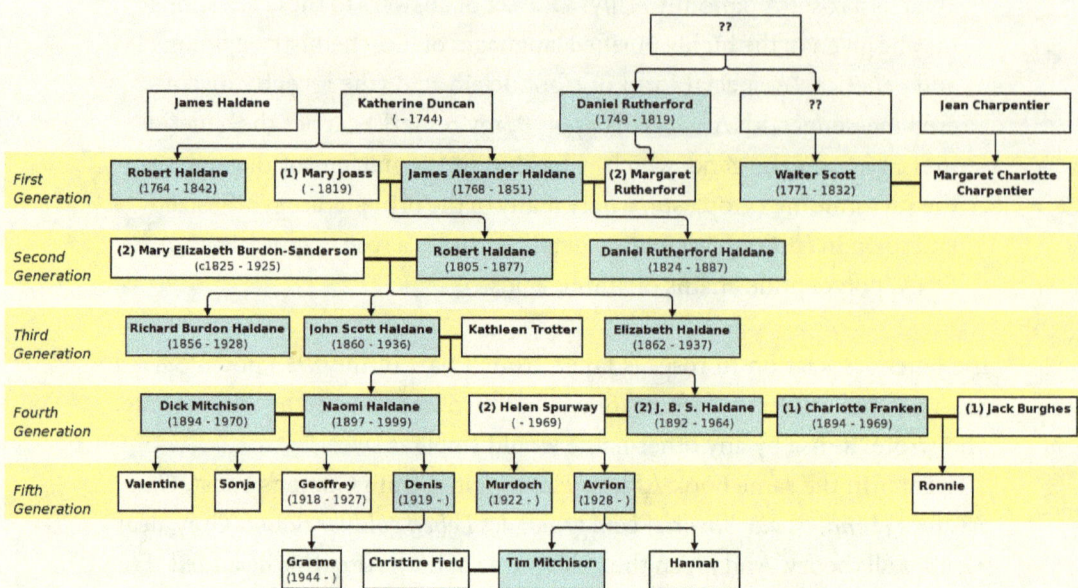

Haldane family tree

capacities displayed by Naomi. She tried to release his dormant potential for socializing by encouraging a conversation with a female author friend. It was not a success, but the author in case went on to write books that become famous. In 2007 she was to receive the Nobel Prize in literature. It was Doris Lessing.

The Sodium Pump and Local Anesthetics — A Skou Interlude

The possible existence of a pump for ions, like the one for sodium, was first proposed by Robert B. Dean at Rochester University, New York. This concept was further developed by S. August S. Krogh, the recipient of the Nobel Prize in physiology or medicine in 1920 and another famous Danish physiologist Hans Ussing. The energy-dependence of the proposed ion pumps was examined by Hodgkin and Huxley using the perfused axon system. Hodgkin summarized this in his Nobel lecture[7]:

> The molecular nature of the pumping mechanism is unknown but there is much evidence to show that in most cells it is driven by compounds containing energy-rich phosphate, such as ATP (adenosine

triphosphate, my remark) or phosphagen. Recent interest in this field has been focused by Skou on an ATP-splitting enzyme which is present in the membrane and has the interesting properties of being activated by sodium and potassium and inhibited by substances which interfere with sodium transport.

It was not until 1997 that Jens C. Skou received half a Nobel Prize in chemistry "for the first discovery of an ion-transporting enzyme, Na^+,K^+-ATPase." The history of how he discovered the sodium pump can be found in his Nobel Lecture[11] and also in a charming autobiography he published in 2013[12], regrettably so far only in Danish. He had become interested in local anesthetics. His starting-point was knowledge of the Meyer-Overton theory, already alluded to, that there is a correlation between solubility of general anesthetics in lipids and their potency of action. However, whereas general anesthetics follow this rule and are

Jens C. Skou receives his shared 1997 Nobel Prize in chemistry from the hands of His Majesty King Carl XVI Gustaf. [From Ref. 12.]

non-polar substances, local anesthetics are weak bases. In his Ph.D. thesis he examined if local anesthetics also followed the Meyer-Overton theory. For this purpose he used the intact sciatic nerve of frog legs and later nerves from which the sheath had been removed, as test objects.

Incidentally, it can be mentioned that the most effective local anesthetic for a long time was a substance with the trade name Xylocaine. This compound was discovered by Holger Erdtman, a very idiosyncratic Swedish chemist. He was trained by Hans von Euler (see front cover), the co-recipient of the 1929 Nobel Prize in chemistry. However, Erdtman's Ph.D. synthesis was not a success and he received such low marks that it should have stopped him from having a further career in science within the Swedish academic system of the time. In spite of this, von Euler employed him in the early 1930s. His task was to synthesize a compound named gramine. It failed. The tradition of the time was that new compounds were analyzed by use of all the different senses available. The color was identified, the smell was noted and in addition the taste was tested. The scientist simply put a sample into his mouth. On one occasion Erdtman noted that when he tasted one of the compounds from the failed synthesis of gramine his tongue became numb. A candidate for a local anesthetic had been serendipitously discovered and when the work on the unexpectedly discovered new group of compounds was resumed in the mid-1940s amazing progress was made.

The characterization of the new group of compounds became the topic of a Ph.D. thesis by Nils Löfgren. The compound was offered to the Pharmacia Pharmaceutical Company, but after they had rejected the proposal the offer went instead to Astra. Xylocaine was introduced into the market in 1948 and over decades it became the number one product of the company, today AstraZeneca. Over the years a number of additional indications for its use were found, like reducing the risk for arrhythmias in connection with a heart infarction. Later it was found that the compound blocks the ion channels in nerve cells and prevents sodium ions from penetrating into cells. Erdtman's rocky career ended well. He eventually became professor of natural chemistry at the Royal School of Technology in Stockholm in the mid-40s, where he developed a very successful academic activity. However, he had a very idio-syncratic personality and got into a major conflict with his graduate student Löfgren, in the process of developing and commercializing Xylocaine[13]. But that is another story although it is worth mentioning that the professor of medical chemistry Erik Jorpes, who we will meet in the next chapter, nominated Löfgren for a Nobel Prize in physiology or medicine in 1957. A preliminary review was carried out by the professor of pharmacology Börje Uvnäs. He praised the originality and importance of the discovery but concluded "However, it is not my view that Löfgren's discovery of Xylokaine is of such a fundamental importance or has led to such consequences ... that it at present can motivate a candidature to a Nobel Prize." Let us now return to Skou's work.

In his Nobel lecture[11] Skou emphasized the importance of the use of radioactive isotopes, introduced by George de Hevesy and recognized by a Nobel Prize in chemistry in 1943 (awarded in 1944)[2]. Like Hodgkin, Skou had an opportunity to visit the marine biology station at Woods Hole, an experience that changed his life. Regarding this visit in the early 1950s he said the following in his Nobel lecture:

> Scientists interested in the function of the nervous system came from
> all over the world to Woods Hole during the summer, to use the giant

axons from squids as test objects. Coming from a young University with a poor scientific milieu, this was like coming to another planet. The place was bubbling with scientific activity. I realized that science is a serious affair and not just a temporary hobby for young doctors writing a thesis in order to qualify for a clinical career. And also that it is competitive. I listened to lectures, met people whose names I knew from the textbook, and from literature, spent time in the laboratories looking on, and learning from the experiments.

When some of his experiments had indicated to Skou that there were active transports of Na⁺ and K⁺ across membranes he searched for similar evidence in published papers. He noted in his lecture "I therefore started to look into the literature to see what the substrate was for the active transport in nerves. I had limited access to the literature so there were few papers I read about active transport. The closest I could come was that A.L. Hodgkin and R.D. Keynes had shown that poisoning giant axons with dinitrophenol, cyanide or azide, decreased the active transport of sodium, suggesting that high energy phosphate esters are the substrate. And as ATP is a high energy phosphate ester, I thought it likely that it could be the substrate. Later the text read: "In 1959, after I had given my first paper on the crab nerve enzyme at the 21st International Congress for Physiology in Buenos Aires, Professor Hodgkin, who was the great name in neurophysiology and came from the famous Cambridge University, invited me for lunch to hear more about the enzyme. His interest suggested to me that the observation was of a certain importance."

He then went on to describe his approaches to characterizing the enzyme he had identified. He concluded:

It may seem disappointing that 30 years of work, since the conclusion that the membrane bound Na⁺,K⁺-ATPase is identical with the Na⁺,K⁺-pump, has not given us an understanding of the basic molecular events behind the transport. However, considering that the problem is to reveal how 1320 amino acids, inside a volume of $60 \times 60 \times 100$ Å³ can be assembled to a very efficient machine, which can convert the chemical energy from the hydrolysis of ATP into work, namely the transport of cations against their electrochemical gradient, and which can distinguish between so closely related cations as Na⁺ and K⁺, it can be no surprise that progress is slow.

Development of the Patch-Clamp Technique

It would take time before the separate ion channels for sodium and potassium postulated by Hodgkin and Huxley had been identified. One way of separating the ionic currents was to use drugs. One of these was the nerve toxin tetrodotoxin which was found to have a selective effect on the sodium channel but did not affect the potassium channel. This toxin has a particular origin. It was found in the gonads of the Japanese puffer fish. This fish, in Japanese called fugu, is a delicacy but only certified chefs who know how to dissect the fish are allowed to prepare a meal. During his circumnavigation of the world in 1774 Captain Cook nearly died from improperly prepared puffer fish.

The final conclusive proof of the existence of individual channels was obtained later. In 1991 a Nobel Prize in physiology or medicine was awarded to Erwin Neher at the Max-Planck-Institut für Biophysicalische Chemie, Göttingen and Bert Sakmann of the Max-Planck-Institute für Medizinische Forschung, Heidelberg. The prize motivation was "for their discoveries concerning the function of single ion channels in cells." Neher and Sakmann succeeded in developing a technique which made it possible to register the events in a single ion channel. Sten Grillner, professor of neurophysiology at the Karolinska Institute, gave the introductory speech at the prize ceremony[14]. He provided the following popular account of their discovery:

> This year's Laureates Erwin Neher and Bert Sakmann, succeeded in making a conclusive demonstration that ion channels exist, by developing a technique by which the miniscule currents, flowing through a single ion channel molecule, could be measured. These are currents of a thousandth of a billionth of an ampere. The technique is nevertheless, in principle very simple. A thin glass-tube filled with fluid is used as a recording electrode. The tip of the tube is pulled out to a width of only some thousandth of a millimeter. When it is brought in very close contact with the cell membrane, they form as it were, a chemical unit with each other. The ion channels, which are present in the cell membrane under the pipette opening, will then form the only connection between the interior of the cell and its outside. When one of the channels is opened a very small current will flow … .

Neher was born in southern Germany near the end of the Second World War, in 1944. He studied physics and after a year at the University of Wisconsin,

Bert Sakmann, the recipient of half the Nobel Prize in physiology or medicine in 1991. [From *Les Prix Nobel en* 1991.]

Erwin Neher, the recipient of half the Nobel Prize in physiology or medicine in 1991. [From *Les Prix Nobel en 1991.*]

Madison, he returned to Germany to earn his Master of Science degree. He then switched to biology, which led him to the Max-Planck-Institut für Psychiatrie headed by H.D. Lux. In this laboratory Neher met Sakmann, who was two years older than himself. They inspired each other in their studies of nerve functions and became good friends. Like Neher, Sakmann had also vacillated between a career in physics or biology. He learnt the voltage clamping technique applied to snails before he moved to become a visiting postdoc student at Katz's department at the University College London. Sometime after his return to Germany he received the opportunity to set up his own Max-Planck-Institute, which was eventually moved from Göttingen to Heidelberg to facilitate the development of molecular biological technique applications.

Neher started his Nobel Lecture thus[15]:

Around 1970, the fundamental signal mechanisms for communication between cells of the nervous system were known. Hodgkin and Huxley (1952 reference) had provided the basis for understanding the nerve action potential. The concept of chemical transmission at synapses had received its experimental verification by detailed studies on excitatory and inhibitory postsynaptic potentials (a reference to Katz 1966 book *Nerve, Muscle and Synapse*, McGraw-Hill Book Company, New York, was given). Hodgkin & Huxley (1952 reference) used the concept of voltage-operated gates for a formal description of conductance changes, and by 1970 the terms Na-channel and K-channel were used frequently (reference to a review by Hille).

He then went on to describe the establishment of the technique he had developed jointly with Sakmann. The latter expanded on this theme in his Nobel Lecture[16] and also described the growing knowledge about the chemical structure of channel proteins. He finished with an outlook which in part read as follows:

> Patch clamp techniques are now well established and routinely applied in combination with other techniques like recombinant DNA and fluorimetric techniques to characterize molecular details of the events underlying synaptic signaling between cells. Through the measurement of elementary currents the biophysical interpretation of the electrical signals which underlie rapid cellular communication across synapses has been simplified and can be partly understood in molecular terms. At the same time single channel conductance measurements have provided evidence for numerous isoforms of receptor channels, as well as voltage and second messenger gated channels particularly in CNS neurons. The significance of this remains to be elucidated with respect to synaptic communication in the CNS ....

The Deepening Insights into the Fine Structure of Membrane Channels

The possibility of determining the three-dimensional structure of proteins has revolutionized molecular biology. It all started with the pioneering work by Perutz and Kendrew. Their impressive findings were recognized by a Nobel Prize in chemistry in 1962 as described in detail previously[2]. Since their early contributions the field has expanded enormously, because of better equipment, better irradiation sources, better ways of marking various parts of the molecules and in particular essentially unlimited resources for accumulating and processing numerical data. Advanced crystallographic studies have completely changed the face of structural biology and erased the border between biochemistry and cellular physiology. It would be going too far to attempt a summarizing review of the extensive developments in the field in the present book. Only a few remarks will be given.

Originally, characterization of membrane-associated proteins posed particularly challenging problems because of their inherent hydrophobicity. The first membrane protein which was successfully subjected to a three-dimensional characterization was considered such an important step forward that it brought its discoverers a Nobel Prize in chemistry in 1988. This prize was given to Johann Deisenhofer, originally of German descent, but working at the University of Texas Southwestern Medical Center, and two other German scientists Robert Huber and Hartmut Michel, at the Max-Planck-Institute in Martinsried and Frankfurt/Main, respectively. The motivation for the prize was "for the determination of the three-dimensional structure of a photosynthetic reaction centre."

In his introduction[17] at the prize ceremony Bo G. Malmström of the Royal Swedish Academy of Sciences made the following statement in the middle of a paragraph:

> For a long time it has been impossible to prepare membrane-bound proteins in a form allowing the determination of the detailed structure in three dimensions. Before 1984, there were only rather fuzzy structural pictures available for a few membrane proteins. These had been derived with the aid of an electron microscopic method developed by the Englishman Aaron Klug, who was awarded the Nobel Prize in chemistry in 1982 for this achievement. But the situation had actually drastically changed in 1982, when Hartmut Michel thanks to systematic experiments succeeded in preparing highly ordered crystals of a photosynthetic reaction center from a bacterium.

I have had the pleasure of interacting with Michel, a very unpretentious and humble scientist, over many years in connection with the annual meetings of the prize committee for the Paul Ehrlich and Ludwig Darmstaedter Prize, Frankfurt.

"Humble" is a word with a built-in ambivalence. There was a Swedish Archbishop who was pleased that he was humble, but he was concerned about the fact that he was pleased to be humble and so on. In the same vein the Hungarian-born Swiss molecular biologist — cloning of interferon, molecular genetics of prion diseases — Charles Weissman has told the following story. Once there was a local rabbi who was introducing the famous visiting chief rabbi. After ten minutes of rambling panegyrics the visitor pulled his sleeve and said "Remember I am humble too!" In 2008 I had the privilege of meeting Neil Armstrong on an occasion when he gave a lecture on "Space:

The Evolving Frontier" at the Spring meeting of the American Philosophical Society. In a conversation afterwards I took the opportunity to ask him the question whether it was a blessing or a curse to be the first human being to set a foot on the moon. The answer was, not surprisingly, "both". Armstrong, like Michel, gave me the impression of being a truly humble person. Now back to the crystallography of ion channels.

Since the mid 1980s the situation has changed dramatically and a number of Nobel Prizes have been given for the characterization of the three-dimensional structure of particularly biologically relevant proteins and protein complexes, and some of these prizes have highlighted insights into the structure of membrane-bound proteins. Of particular interest is the chemistry prize in 2003, which recognized the contributions by Peter Agre, Johns Hopkins University School of Medicine, Baltimore and Roderick MacKinnon, The Rockefeller University, with the joint motivation "for discoveries concerning channels in cell membranes". The half of the prize recognizing the latter scientist had a separate additional specification "for structural and mechanistic studies of ion channels." MacKinnon's discovery concerned the structure and function of a K^+ channel.

In his Nobel lecture[18] MacKinnon first gave a general comment on the barrier function of cell membranes but then he continued:

Roderick MacKinnon, who received half a Nobel Prize in chemistry in 2003. [From *Les Prix Nobel en 2003*.]

> The modern history of ion channels began in 1952 when Hodgkin and Huxley published their seminal papers on the theory of the action potential in the squid giant axon (four references). A fundamental element of their theory was that the axon membrane undergoes changes in its permeability to Na^+ and K^+ ions. The Hodgkin-Huxley theory did not address the mechanism by which the membrane permeability changes occur: ions could potentially cross the membrane through channels by a carrier-mediated mechanism. In their words "Details of the mechanism will probably not be settled for some time". It is fair to say that the pursuit of this statement has accounted for much ion channel research over the past fifty years.

After presenting the structure and functional details of the K⁺ channel, which had been the target of MacKinnon's studies, he briefly introduced the occurrence of other kind of channels. He wrote:

A fundamentally different kind of gating domain allows certain K^+, Na^+, Ca^{2+} and nonselective cation channels to open in response to membrane voltage changes. Referred to as voltage sensors these domains are connected to the outer helices of the pore and form a structural unit within the membrane. The basic principle of operation for a voltage sensor is the movement of protein charges through the membrane electric field coupled to pore opening (references). Like transistors in an electronic device, voltage-dependent channels are electrical switches.

MacKinnon also mentioned the 1955 experiments by Hodgkin and Keynes in which they used radioactive isotopes to measure ion flow. The presentation then highlighted the explosive developments in this field. It is beyond the scope of this book to review these impressive advances. A huge family of transmembranous proteins, in total more than 140, with principally similar structure, forming six different groups, has been identified. A reference may be given to a relatively recent description of the state of the art of this field[19]. This review summarized the impressive volume of structural data in the following way:

The voltage-gated ion channel family is one of the largest of signaling proteins, following the G protein-coupled receptors (also transmembranous — TM — proteins — spanning the membrane seven times, hence named 7TM, recognized by a Nobel Prize in chemistry in 2012 to Robert J. Lefkowitz and Brian K. Kobilka, my remark) and the protein kinases in the number of family members. The family is likely to have evolved from a 2TM ancestor like the bacterial KcsAchannel. Additions of intracellular regulatory domains for ligand binding and a 4TM transmembrane domain for voltage-dependent gating have produced extraordinarily versatile signaling molecules with capacity to respond to voltage signals and intracellular effectors and to integrate information coming from these two different kinds of inputs. The resulting signaling mechanisms control most aspects of cell physiology and underlie complex integrative processes like learning and memory in the brain and coordinate movements in muscles.

The principal structural motif was introduced in the first paragraph of the same article:

> The architectures of the ion channel families consist of four variations built upon a common pore-forming structure theme. The founding members of this superfamily are the voltage-gated sodium channels. Their principal alpha subunits are composed of four homologous domains (I–IV) that form the common structural motif of this family. Each domain contains six probable transmembrane alpha-helixes (termed segments S1–S6) with a membrane-reentrant loop between the S5 and S6 segments. Structural analysis by high-resolution electron microscopy and image reconstruction show that the four homologous domains surround a central pore and suggest laterally oriented entry ports in each domain for transit of extracellular ions towards the central pore.

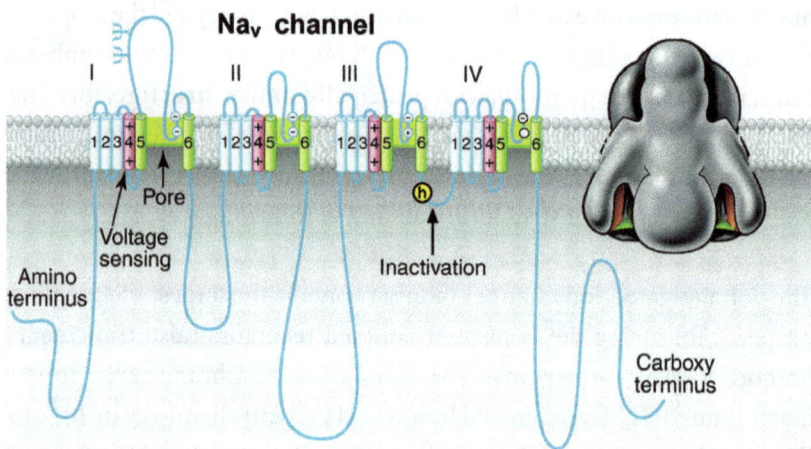

Schematic presentation of the structure of a sodium channel. [From Ref. 19.]

It is amazing how at the present time the complex functions of even large macromolecules or aggregates of them can be animated. And still the increasing knowledge leads to an astounding appreciation of the complexity of biological systems. Much remains to be learnt, in particular the mechanisms of regulation to provide the proper balance of actions in the many parts of the whole of a system.

In order to examine his crystals, MacKinnon took advantage of the irradiation resources at the National Synchroton Light Source at Brookhaven

National Laboratory (BNL) in Upton, Long Island. During his visits there he got to know a temporarily resident artist, Steve Miller, and this meeting had unexpected consequences.

Visual Art and Sciences

Miller is an artist who early in his career became fascinated by the advances in the development of computers, technology and science. He wanted to find a visual language to express all these new and dramatic developments. As he formulated it "changes in technology and changes in science allow for changes in consciousness." As further cited in the book *Colliding Worlds. How cutting-edge science is redefining contemporary art*[20] by another Miller with the initials A.I., S. Miller wanted to be a part of what the French philosopher Michel Foucault has called "epistemological breaks," which essentially is simply another term for paradigmatic discoveries. When MacKinnon met S. Miller at BNL he explained to him his work with channel proteins and how these ion channels were critical in generating the electricity that the cells of our body use for different purposes, in particular for signaling in the nervous system. Steve became excited when he was told that millions of ions per second could transverse the membrane. He wanted to learn more and visited MacKinnon in his laboratory at the Rockefeller University. What he saw made a major impression on him.

He was amazed by the unique ambience of the laboratory with equations written on the blackboards and all the hands-on busy work by the research team. Inspired by his impressions, he developed a series of art works, one of which he called *Protein#330*. As can be seen he has combined a photograph of a deduced potassium channel structure, silk-screened onto a canvas, and superimposed on this a photograph of equations from MacKinnon's blackboard. Clearly in this case the artist received inspiration for his creative work from the scientist, but the question is if it may

Protein #330. A picture of a piece of art kindly made available by the artist Steve Miller.

also work the other way round. Could it be that interaction with an artist, esthetically expressing what he experiences, could assist the scientist in the making of a discovery. This is a recurrent theme discussed in the abovementioned book on *Colliding Worlds*[20]. In most cases the answer is that scientific discoveries have not been initiated by or enhanced by meetings with artists intrigued by the culture of curiosity and obsession characterizing a productive scientific laboratory. In part this is explained by the fact that a scientist learns from nature and progressively acquires a deeper and deeper insight into its secrets, like for example the product of an incidental fortuitous choice made in a step of evolution. The situation for an artist is different. However, what a scientist and an artist readily can share is the insight into the nature of the creative process leading to the final product, a discovery or a piece of art. We will return to this theme at the end of the book.

Could it possibly be that an artist might assist in making the invisible visible — explaining to the public at large a new finding, for example the innermost nature of matter? In the figure shown above, the scientists have used one of the different available techniques to illustrate the three-dimensional structure of a protein. It is referred to as the space-filling model. This is the kind of presentation most readily grasped by our visual sense. In the Aristotelian tradition we dislike emptiness, the void —Latin *horror vacui*. Chemists, using crystallographic methods, can also use other modeling techniques that leave more space open, which is in better accordance with the true nature of the structure of macromolecules. One possibility is to use a ribbon model which emphasizes certain modules used in the construction of proteins. Such modules are the beta-pleated sheath and also the alpha-helix, discussed at length in the presentation of the Nobel Prize in chemistry in 1962 to the crystallographers Perutz and Kendrew[2] and also the 1997 Nobel Prize in physiology or medicine to Stanley B. Prusiner[3]. Aggregation of pathologically folded beta-pleated structures has turned out to be a major cause of neuropathological diseases in the brain. Yet another possibility is to use a stick-ball model, in which individual atoms are connected with a single line. The atomic distance illustrates the possibilities for reorienting individual parts of the complex structure, potentially of importance for expressing a certain function or simply to emphasize that complex biological structures are rarely rigid. In real life a considerable percentage of different kinds of protein structures completely lack a specific folding in their natural state. It is only when they interact with their specific target that they take on a partly or completely stabilized form.

Prizes in Neurophysiology after 1963

At the time of writing one additional Nobel Prize in physiology or medicine has been added to the list of prizes in neurophysiology awarded since 1963, which already included six prizes (Table 3.3). The most recent prize as can be seen from the table concerns one of the most complex of the integrated brain functions — the way we orientate ourselves in space and navigate our environment. In a popularized form it can be referred to as our individual Global Positioning System (GPS). Such a system is critical not only to us but also to all kinds of animals. Hence it is old in evolutionary terms and the

Table 3.3. Nobel Prizes in physiology or medicine between 1964–2014 awarded in the field of neurophysiology.

Year	Awardee(s)	Motivation
1967	Ragnar Granit, Haldan K. Hartline, George Wald	for their discoveries concerning the primary physiological and chemical visual processes in the eye
1970	Bernard Katz, Ulf von Euler, Julius Axelrod	for their discoveries concerning the humoral transmitters in the nerve terminals and the mechanisms for their storage, release and inactivation
1981	Roger W. Sperry (½ a prize)	or his discoveries concerning the functional specialization of the cerebral hemispheres
	David H. Hubel, Torsten N. Wiesel (½ a prize)	for their discoveries concerning information processing in the visual system
1991	Erwin Neher, Bert Sackmann	for their discoveries concerning the function of single ion channels in cells
2000	Arvid Carlsson, Paul Greengard, Eric R. Kandel	for their discoveries concerning signal transduction in the nervous system
2004	Richard Axel, Linda B. Buck	for their discoveries of odorant receptors and the organization of the olfactory system
2014	John O'Keefe (½ a prize) May-Britt Moser and Edvard I. Moser (½ a prize)	for their discoveries of cells that constitute a positioning system in the brain

critical centers are found in the older part of our brain. The Anglo-American scientist John O'Keefe, University College London, used freely moving rats with electrodes experimentally connected to their brains in his experiments. He discovered that there are certain cells in the hippocampus that kept track of the position. They were simply referred to as *place cells*. The one generation younger couple May-Britt and Edvard I. Moser, Trondheim, Norway, were interested in similar problems of navigation and they also worked for some time in O'Keefe's laboratory in London. They discovered critical structures in a neighboring part of the hippocampus, the so-called *medial entorhinal cortex.* The particular cells they identified were called *grid cells.* These cells were found to be activated according to a certain pattern when the electrode-carrying rats moved around in their box. The pattern of distribution of the grid cells was hexagonal, a topographical distribution reminiscent of the arrangement of holes in a beehive. Their recognition with half the Nobel Prize was a national cause of celebration. It was the first time a prize in physiology or medicine had ever been awarded to a Norwegian citizen.

The 1967 prize in physiology or medicine concerned neurobiological studies of the visual processes in the eye. This was briefly discussed in my previous book[2] under the general heading of *Our senses and Nobel Prizes.* Since it will only be a few years until a full review of the archives relevant to this prize will become available it seems sensible to return to it on a later occasion. This will be of particular interest since one-third of the prize was given to Granit, who played such an important role for other prizes in the field including those discussed in this and previous chapters. A central figure influencing these developments must have been Bernhard, who was also the person who gave the introductory speech at the prize ceremony. Whereas orientation in space can to some extent be compared to a tool for measuring GPS data, the eye cannot be compared to a movie camera. Bernhard expressed this in the following way in his introductory speech at the prize ceremony[21]:

> Picasso has said: "To me painting is a sum of destructions. I paint a motif, then I destroy it." The painting goes through a series of metamorphoses but "in the solution of the problem nothing has been lost. The final impression is still there in spite of all revisions." However, it is obvious to everyone that in the finished work a re-evaluation has taken place of the original elements of the motif. In some way this is a description of what happens in the visual system. An image of the outer world is formed on the retina in the same way as it is formed in the film of the camera.

The image that falls on the closely packed mosaic of light sensory cells is disintegrated, since different cell types respond to various parts and qualities of the image. The primary data are then brought together in the nerve net in which a considerable processing takes place involving not only addition but also subtraction. This characterization of the message induces an impression in which there is a re-evaluation of the image projected on the retina.

The same principle was discovered in a refined way in elegant experiments by David H. Hubel and Torsten N. Wiesel, who received half of the Nobel Prize in physiology or medicine in 1981. The half that recognized Sperry's studies of split brain has already been discussed in the previous chapter. In the presentation speech by Ottoson[22] cited in part above he also described Hubel's and Wiesel's discovery. He emphasized the importance for their work of the fact that they had been trained in Stephen Kuffler's laboratory in Baltimore.

A Master Mentor

Kuffler was one of the two most influential brain neurophysiologists during the 1950s. Together with Vernon Mountcastle he provided decisive scientific background to the critical discoveries by Hubel and Wiesel. Kuffler was born in Tap, Hungary to a bourgeois family. He studied medicine in Vienna, receiving his M.D. in 1937. Due to the impending Second World War he boarded a ship for Australia. As a postdoctoral student in pathology he met Eccles at the tennis court. As the story goes Eccles said "You play tennis too well to stay in pathology. You must come and work with me in neurophysiology." Among other things Kuffler turned out to be a very skilled surgeon. In 1939 the two were joined by Katz (see p. 92), forming a strong group for neuromuscular transmission research. It was in this research environment that the intensity of the fight between the "soupers" and "sparkers" was fought with progressively increased intensity, as already alluded to above. In 1945 two years after Eccles had left for New Zealand, Kuffler went to the United States. After a short time in Chicago he took a position at the Wilmer Institute of Ophthalmology at the Johns Hopkins Hospital in Baltimore. He built a very strong group of devoted scientists and major contributions to our understanding of vision, neural coding and neural implementations of behavior were made. It all started with a characterization of the dendrites of crayfish neurons. In 1966 he became the founder of the Department of Neurobiology at Harvard University. He brought the major part of his research group to Boston. Apparently he was a very charming unpretentious and enthusiastic leader and was given the prestigious epithet "The Father of Modern Neuroscience."

Besides Kuffler, Mountcastle, as mentioned, was another prominent figure in the field of neurophysiology in the middle of the previous century. He was born in 1918 in Shelbyville, Kentucky. Like Kuffler he developed his original work at the Johns Hopkins University. His major contribution came in 1957 when he discovered that nerve cells in the brain had a functional columnar organization. Hubel in his Nobel lecture[23] stated that the "discovery of columns in the somatosensory cortex was surely the single most important contribution to the understanding of cerebral cortex since Ramón y Cajal." Both these neurophysiologists had a major influence on the remarkable team work of Hubel and Wiesel when they started their longstanding and fruitful collaboration in 1958 in Kuffler's laboratory.

Hubel came from The Walter Reed Army Institute for Research, which he had chosen since he preferred to work in the Public Health Service instead of joining the army, and Wiesel came from Bernhard's laboratory at the Karolinska Institute. Their joint story has been told in a very personal and inspiring book also including their major shared publications[24]. The impact may not need to be elaborated beyond what they themselves have said in the present context, but a citation from the introductory speech at the Nobel prize ceremony by Ottoson[22] gives a general impression of the discoveries made. It read:

> The signal message that the eye sends to the brain can be regarded as a secret code to which only the brain possesses the key and can interpret the message. Hubel and Wiesel have succeeded in breaking the code. This they have achieved by tapping the signals from the nerve cells in the various cell layers of the brain cortex. Thus, they have been able to show how the various components of the retinal image are read out and interpreted by the cortical cells in respect to contrasts, linear patterns and the movement of the picture over the retina. The cells are arranged in columns and the analysis takes place in a strictly ordered sequence from one nerve cell to another and every nerve cell is responsible for one particular detail in the picture pattern.

To this can be added the concluding paragraph presented by Wiesel in his Nobel lecture[25]. He said:

Innate mechanisms endow the visual system with highly specific connections, but visual experience early in life is necessary for their maintenance and full development. Deprivation experiments demonstrate that neural connections can be modulated by environmental influences during a critical period of postnatal development. We have studied this process in detail in one set of functional properties of the nervous system, but it may well be that other aspects of brain function, such as language, complex perceptual tasks, learning, memory and personality, have different programs of development. Such sensitivity of the nervous system to the effects of experience may represent the fundamental mechanism by which the organism adapts to its environment during the period of growth and development.

The fact that only one Nobel Prize can be awarded each year, that there are many competing fields in physiology or medicine and that the maximum number of prize recipients is three means that a number of critical discoveries in a particular field may remain unrecognized. Kuffler and Mountcastle may represent examples of strong candidates who were never to become recognized by a prize. Choices have to be made and one example of this is a 1962 evaluation that Granit made of Kuffler, the first time he was nominated. The review is full of praise for Kuffler's scientific contributions, but the second and final part of the conclusion read (translated from Swedish):

Kuffler's research has focused on a number of problems, which have become elucidated by a wise choice of material to study, method of critical evaluation and choice of a high quality experimental approach. However, it has hitherto regularly been a question of elaborating or "fine tuning" of (a problem with) a given content of ideas. He cannot at present become included among the "discoverers", which the Nobel Prizes are assumed to recognize, even if many of his contributions should be praised, and many are admirable and executed with elegance. Kuffler, without doubt, represents one of the leaders in his field of work. The one with whom Kuffler in the first place should be compared is Katz (see the evaluation of the previous year) since both of them use (similar) techniques and have a number of interests in common in their (scientific) production. In this comparison it is obvious that Kuffler does not reach the same level as Katz when it comes to originality and importance. The case of Kuffler illustrates in an interesting way *to what a high degree one can appreciate a contribution in science without being prepared to concede that it has the particular character which is required for (recognition by) a Nobel Prize* (my italics) . At the present time Kuffler cannot be recognized as worthy of a prize, but he belongs among those whose work deserves to be followed up.

Kuffler was nominated again in 1965, this time by Feldberg and the proposal included Katz. It seems that this year the neurophysiologists in the committee were relatively passive. No additional review, neither of Kuffler nor of Katz was made and it was not even ensured that the latter remained on the list of prize-worthy candidates as he should have been. It is likely that Kuffler was nominated again on many later occasions. In the case of Katz he came back with full force five years later when he received a prize together with Ulf von Euler and Julius Axelrod (see Table 3.3, p. 169)

Another Nobel laureate, who has praised Kuffler's unique qualities as a scientist and a compassionate human being, is Kandel, who we will soon meet again in this chapter. Kandel writes in his autobiography[26] about Kuffler in the following way; "Always a first-class experimentalist, he emerged as the most admired and effective leader of the American neuroscience community." And "Until his death in 1980, he proved a friend and counselor of immeasurable strength and generosity. He took an intense interest in people, their careers, and their families." Sometimes one wonders if there should not also be a prize of a separate category which would allow recognition of those who have been found preeminently endowed to train future Nobel Prize recipients.

Still More Prizes in Neurobiology

The previously discussed presentation[2] of our senses and Nobel Prizes also included the 2004 Nobel Prize in physiology or medicine to Richard Axel and Linda B. Buck. The olfactory sense is of particular interest in relation to brain functions. Sorting out chemicals has been a basic original need for cells since the dawn of life and the sense of smell has refined the capacity to distinguish specific molecules in humans and even more so in many species of animals. The olfactory epithelium is the only place in our body where nerve cell extensions are directly exposed to the environment. The significance that this may have for a possible entrance of viruses that can spread by nerves has been repeatedly mentioned. Besides elegant molecular biological studies to explain the way a multitude of compounds of different chemical nature can be distinguished by a limited number of receptors, Axel and Buck also demonstrated the topographical imprints in the brain of different olfactory stimuli. The functional importance of the presence of a limited number of specific receptors and the combinatorial use of signals has already been discussed[2].

Among the various prizes summarized in the table it remains to also briefly introduce the 1970 and 2000 prizes. The 1970 prize focused on the humoral transmitters in the nerve terminals. This concept was introduced by Dale and Loewi, as already discussed in Chapter 1. The professor of pharmacology at the Karolinska Institute Börje Uvnäs introduced the three laureates at the prize ceremony[27]. He praised Katz's work which mostly focused on the electrical events at the motor nerve end plates, von Euler's work on noradrenaline, an original discovery that the late Hillarp (see p. 63) had also contributed to, and finally Axelrod who had clarified what happens to noradrenaline when it has been released from the nerve endings. Again it will be possible within the not too distant future to completely review the way in which these three candidates were discussed by the Nobel committees at the Karolinska Institute. Some remarks have already been made above on both Katz and von Euler.

The last but not the least prize to mention (Table 3.3, p. 169) is the year 2000 prize to Carlsson, Greengard and Kandel. They are prominent figures in the field of neurobiology and Kandel and collaborators are the authors of the most comprehensive (1709 pages!) current book in the field, *Principles of Neural Science, 5th ed.* Very superficially their extensive contributions can be summarized in a brief form by citing the last paragraph of the introductory speech at the prize ceremony[28] by the professor of physiology and pharmacology Urban Ungerstedt. He concluded:

> I am convinced that you and I will remember this Nobel ceremony for many years. This is because of the dopamine which Arvid Carlson discovered, enabling the brain to react to what we see and hear; the second messengers that Paul Greengard described, carrying the signals into the nerve cell; and the memory functions that Eric Kandel found to be due to changes in the very form and function of the synapses.

Thus Carlsson proved that dopamine was an important transmitter substance in the central nervous system. When its activity was blocked in animals, symptoms, like those in Parkinson's disease in man, developed. It was possible to restore the dopamine levels in the animals by giving the substance L-DOPA. This same compound was later demonstrated to also improve the conditions for patients with Parkinson's disease. Greengard had analyzed what happens inside a nerve cell which has been stimulated. It is a very complex series of events. Receptors on the cell surface activate enzymes in the cell wall, which in turn start the production of what are called second messengers. The discovery of a second messenger, cyclic AMP, was recognized by a Nobel Prize in physiology or medicine in 1971 to Earl W. Sutherland "for his discoveries concerning the mechanisms of the action of hormones." The cyclic AMP travels into the cell and activates a protein kinase, which leads to a binding of phosphate groups to other proteins. This cascade effect results in an altering of functions of the targeted proteins. Altogether this process involves fundamental ways of communication within a cell. The protein kinase family is the largest among protein families. The important role of proteins in this family of course also has been recognized by a Nobel Prize. This was the 1992 prize in physiology or medicine to Edmond H. Fisher and Edwin G. Krebs "for their discoveries concerning reversible protein phosphorylation as a biological regulatory mechanism." The target proteins, the G proteins, have also been recognized by a Nobel Prize. In 1994 the prize in physiology or medicine was awarded to Alfred G. Gilman and Martin Rodbell "for their discovery of G-proteins and the role of these proteins in signal transduction in cells." G stands for the high energy compound guanosine triphosphate. There are of course many additional aspects of the complex machinery of signaling inside a cell. Many questions remain to be answered, in particular the dynamic regulation through various feedback mechanisms, allowing the establishment of a state of balance — homeostasis.

Kandel's discoveries concerned the ability to form memory, as a result of activation of the mechanisms just described. He followed Crick's advice and selected a relatively simple organism for his studies, the sea slug, *Aplysia*. This organism has only 20,000 nerve cells. The exciting, still expanding research has been described by Kandel in the abovementioned, very readable autobiography *In Search of Memory. The Emergence of a New Science of Mind*[26]. As a part of some of Kandel's more recent research, in particular together with Susan Lindquist at MIT, Boston, the question has been raised if proteins of the prion kind may have a role in memory storage. Prion proteins have been discussed in a full chapter in my first book on Nobel Prizes[3]. Proteins of this family have a propensity to aggregate and form layers of beta-pleated sheets. Originally this was interpreted to be a hallmark of certain non-immunogenic, potentially infectious diseases, like Creutzfeldt-Jakob disease, but more recently it has been discovered that this kind of protein aggregation represents a much more commonly occurring origin of many diseases, in particular in the brain, like Alzheimer's disease. This rapidly expanding field was reviewed as a part of a biography by the 1997 Nobel Prize recipient Prusiner, *Madness and Memory*[29].

And the Band Played On

Our most vital organ, the brain, is well protected in our body because of the thick bone case within which it is hidden. This solid protection originally markedly reduced possibilities to examine it. Trepanation, from Greek *trypanon*, drill, viz opening up a hole through the skull bone had already been introduced by the ancient Egyptians. The original purpose most likely was to let out evil demons. The first Nobel Prize in physics in 1901 recognized Wilhelm Röntgen's discovery of X-rays. X-ray examinations of the skull was attempted, but the only changes that could be discovered at first were those that led to thinning, thickening or other modifications of the skull bones. Techniques were then developed to locally inject chemicals into the circulation giving X-ray contrast in blood vessels. In fact as mentioned in the previous chapter, the laureate Moniz made some important contributions to the early developments of this technique. The use of contrast made it possible to discover malformations or defective functions of the brain blood vessels, but in all essentials the soft matter of the brain remained inaccessible to the radiologist. This situation changed dramatically during the second half of the twentieth century. This was due to the introduction of two techniques that provided non-invasive means of studying essentially all organs in the body including the brain.

The discoveries of both these techniques have been recognized by Nobel Prizes in physiology or medicine. In 1979 Allan M. Cormack and Godfrey N. Hounsfield were recognized "for their development of computer assisted tomography" and in 2003 Paul C. Lauterbur and Peter Mansfield were awarded prizes "for their discoveries concerning magnetic resonance imaging." These techniques are generally referred to in an abbreviated form; CAT, computer-assisted tomography or CT, computed tomography, and MRI, magnetic resonance imaging. Both these awards epitomize prizes with broad clinical applications. In both the Faculty of Teachers and its successor responsible for deciding on the prize recipients, the Nobel Assembly, the question has often been raised if it would not be possible to identify a relatively larger proportion of Nobel Prizes in physiology or medicine that have a direct clinical application. Of course there are prizes that highlight advances in medicine; identification and management of infectious diseases, development and use of new remedies, understanding the metabolic background to certain diseases. However, the majority of all prizes belong in the category of physiology and highlight insights into more fundamental biological principles of potential use for the future developments in the treatment of diseases.

In my previous book[2] I have described how this longing for a prize with obvious clinical applications had dramatic consequences for the choice of recipients of the Nobel Prize in physiology or medicine in 1979. The committee had recommended a prize in immunology, but as an alternative Cormack and Hounsfield were proposed from the floor at the critical meeting of the Nobel Assembly. When the secret vote was taken there was a tie between the two proposals. Originally I wrote that the chairman had a casting vote, but Georg Klein, professor of tumor biology at the Karolinska Institute, has reminded me that this is not correct. Instead what happened was that the assembly called upon the head of administration, Eyvor Wahlbom, who said that according to the rules for the Nobel Prize the decision should be taken by lottery. This is what happened and Cormack and Hounsfield were the winners. I never use the term winner for a Nobel Prize recipient, since you get your prize because you deserve it and not by coming out first in a lottery, but in this single particular case it would be appropriate to use the term winner.

The introduction of the CAT scanning method and the MRI technique has meant a complete revolution in the examination of pathological changes in the brain. Blood clots, bleedings, the presence of tumors and many other pathological changes can be rapidly and efficiently identified. The MRI technique makes it possible to momentarily register the rapid changes in blood circulation in various parts of the brain. This allows an analysis of which parts of the brain that are activated under certain conditions of stimulation. It is possible to follow how the impression of listening to one of Beethoven's string quartets in a dynamically changing way activates various centers in the brain. In addition the technique, with certain modifications, allows registration of particular metabolic changes by use of different markers. This modified technique is referred to as *magnetic resonance spectroscopy*, MRS, also fMRI, where *f* means functional. In analogy with a radio receiver the MRS equipment can be tuned to selectively pick up signals from different chemical compounds in the brain. Furthermore, there are also particular modifications which allow a direct imaging of nerves in the body and nerve tracts in the brain. Amazing general impressions of the functional anatomy can be obtained.

All these techniques obviously have also allowed an enormous expansion of the experimental neurosciences. However, in this context there are further revolutionary techniques that have been introduced relatively recently.

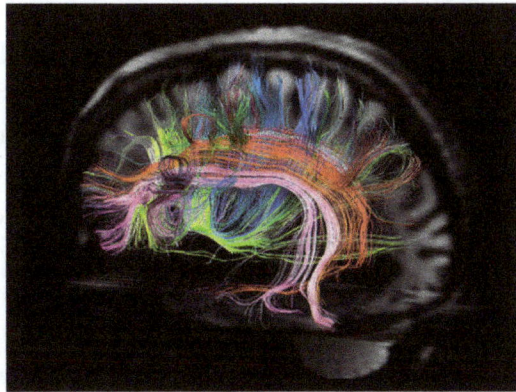

Nerve tracts in the brain demonstrated by use of different colors. [Image by Van Wedeen and L.L. Wald, Martinos Center for Biomedical Imaging.]

Since some years back the technique of *optogenetics* has taken the whole field of neurosciences by storm. As the name implies light signaling is used to selectively activate a genetically introduced foreign protein in a cell. It had already been proposed by Crick that the ideal situation would be if one could activate and turn off selected cells in the brain. This has now become possible. The basis for the technique is that a light-sensitive protein, of the kind found in the retina of the eye, as in us humans, is genetically engineered into the membrane of selected cells in the brain. In practice the so-called opsins — seven-transmembrane proteins (7TM), briefly mentioned above, that when activated by light drive different kinds of ion pumps in the cell membrane — are collected from particular hardy microorganisms found in harsh environments which only allow replication of so-called archebacteria. Incidentally it can be mentioned that the identification of the light sensitive proton pump rhodopsin and determination of its structure to be a 7TM protein in the early 1980s opened the way to the identification of a huge family of this kind of proteins, the G protein coupled receptors briefly mentioned above. Lefkowitz described in his Nobel lecture[30] what a remarkable surprise it was to him and his collaborators, when they discovered that the gene for the Beta2-adrenergic receptor they studied showed sequence homology with the gene for rhodopsin. This chance observation led to an avalanche of discoveries of a large number of G protein coupled 7TM proteins of major importance for physiology and medicine.

We have learnt that transfer of genetic material allowing the expression of heterologous functional proteins is possible across wide species barriers. However, it is not a minor feat to transfer the gene for a chosen protein into a selected cell. In inbred mice it is possible to learn about certain genes that direct the production of cells of unique characteristics forming a particular part of the brain. In mice of this kind it would be possible to associate the light-sensitive rhodopsin protein with such cells with a profiled genetic make-up. Since this approach is both difficult to use and cumbersome, researchers instead have focused mainly on the use of virus vectors. Several different vectors have been tested but at present the one mostly used is adeno-associated virus, a defective agent which under normal conditions can only replicate in a cell simultaneously infected with some type of adenovirus, the group of viruses that can cause, for example, respiratory and eye infections in us humans. However, the purpose is not to allow a full replication of the selected virus vector, but to use it as a carrier of a foreign gene into a cell in such a way that the selected gene, in the present case specifying the light-sensitive protein, becomes associated with the cell membrane allowing it to become expressed. The advantage of using a virus vector is that by carefully selecting a targeted local application of the material to be deposited, only cells in a particular limited region of the brain will become infected. The technique may also become further refined. The vector genetic construct may have a built-in limitation which determines that only nerve cells producing, for example, the transmitter substance dopamine, will become infected and express the light-sensitive protein. When the head of the transgenic experimental animal is exposed to a penetrating light of a certain wavelength, in the case of the channel protein rhodopsin-2 by exposure to blue light, the cells carrying the sensitizing protein become activated. There are also genetic constructs that allow the channels to be turned off upon exposure to light of a different wavelength. Completely new and previously unanticipated experimentation has now become possible. In essence the experimenter has access to a switch to turn selected cells on or off.

To conclude this chapter one large unsolved question will be raised and that concerns how the immensely complex signaling in the brain is structured. The signaling in motoric nerves may be straightforward, whereas that in sensory nerves is more complex. It has been mentioned that the strength of a signal may be determined by the number of spikes per time unit. The refractory time after a spike limits the intensity of spiking. Mobilization of an increasing number of neurons is another way to increase the quantity of signaling. But what about the quality of signaling? Could it be that the background wave-like patterns of electrical activity in the brain, like the alpha and beta waves discovered by Berger in some way modulate the quality of patterns of signaling. It is known that stimulus of cells responsible for recording signals directing our senses, like hearing and seeing, have a secondary indirect effect referred to as *lateralization*. Thus at the same time as a certain cell is activated, the surrounding cells are depressed in their activity — this requires of course the existence of a continuous background activity of a certain pattern. Could brain waves serve to have a similar amplifying effect? The unsolved large question concerns how all the complexity of the signaling in the brain becomes integrated and how eventually a certain perception evolves.

The complexity of the brain, with its almost 100 billion cells and of the order of 10^{14}–10^{15} synapses will keep scientists in the field of neurophysiology busy for the foreseeable future. It will take time for the human species to fulfill the call for self insight — know thyself — as specified on the pediment of the temple of Apollo at Delphi — Greek *Gnōthi seautón*, in Latin *Temet nosce.* Will we ever learn to fully understand consciousness?

Wiesel has reflected on these fundamental existential issues in the following way in the abovementioned book, published in 2005[24], describing the fruitful collaboration between him and Hubel. He concluded his introduction in the following way:

> Even now, I am struck by the complexity of the brain and how little we know about it. Can the brain really understand itself? Many of us in neuroscience assume there are no limits to what we can learn about the brain, but still we hesitate to make the claim that the richness of all human behavior and culture can be explained in biological and physical terms. In our quest, we can anticipate the benefits of new knowledge, but can we foresee the dangers that this knowledge might bring about in the future?

The Continuous Expanding Developments of the Potentials of Brain Functions

The studies by Hubel and Wiesel demonstrated the enormous importance in cats of the stimulus given to a developing organ, in their case the changes in the brain of stimulation of the visual organ, the eyes. This observation obviously has a general application. The development of our brain is dependent on the range of stimuli that it is exposed to. The limits of this potentially expanded use remain to be determined. The continuous transmutations of human culture have led to dramatic changes both in the quality and quantity of brain stimuli. At the time when we were hunters and gatherers the clusters of humans were of the order of some 200 individuals. It was groups of this kind which were able to develop an individual and collective memory. Intergenerational transfer of knowledge was important but relatively limited. Settled human civilization some 12,000 to 10,000 years ago allowed for a much broader stimulus and the collective memory increased slowly. Then came revolutionary changes first elicited by the emergence of a written language and numbers. Records could be stored and accumulated over time. The next quantum leap was when handwritten documents could be exchanged for printed material. Again the volume of stored information increased many times over.

The next phase of development was the addition of extensions of our senses. Eye glasses were an important advance allowing retention of the sense of seeing into older age, but even more important was the introduction of the light microscope. Previously hidden worlds became visible. Over the last two centuries these developments have accelerated allowing an advance of rational science that allows us to "see" even details of molecular events. Chemists happily point out that in the end everything becomes chemistry. However, physicist may argue that the resolution potentially can be increased further allowing an insight into the subatomic world.

There also has been a remarkable extension of our other senses. First we developed the means to transfer visual messages over longer distances, in most cases to warn about impending threats. Signaling using bonfires on selected peaks was used to spread a warning about an approaching enemy. Flag signaling between ships using either coded information or transfer of alphabetic messages also allowed communication over long visible distances. Via consecutively Morse signaling — sound transmitted as short or longer signals forming an alphabet; the telephone transmitting the spoken language; the radio disseminating sound messages towards the general public and finally the television allowing the full use of the sense of seeing combined with the sense of hearing, there was a revolution in the everyday stimuli of individuals. However, the true revolution in dissemination of information occurred in the early 1990s with the introduction of the World Wide Web. By this electronic means of disseminating information a rapidly accelerating process of globalization has been initiated and the final consequences of the introduction and development of this revolutionary technique still remain to be seen. A cellular phone can contain all the information needed for resolving professional as well as private issues. The specialized cells in our brains, which allow us to find our way recognized by the 2014 Nobel Prize in physiology or medicine, have become supplemented by the information supplied by the GPS we carry.

The net result of the introduction of all these technological advances is that the use of our brain has become markedly expanded. The relative influence of cultural inheritance in relation to genetic inheritance of the human species has progressively increased with time and this shift is accelerating.

The Canadian philosopher of communication theory Marshall McLuhan has made the simple statement "Technology is the extension of our nervous system." This is certainly true. The degree of stimulus of the brain of young children has progressively increased during the last century due to the series of technological developments mentioned. In the 1930s intelligence quotient (IQ) tests were introduced. The score number was averaged to 100. It did not take long until it was noticed that the average score increased when consecutive generations of children were tested. The same effect was noticed when the score was renormalized some decades after its introduction. This phenomenon has been referred to as the Flynn effect from James R. Flynn who did much to document its existence. It seems that the progressively enriched environmental stimuli lead to a wider and more effective use of our brain and its capacity for memory storage. The new electronic world also has immense consequences for the retrieval of information. Our grandchildren sometimes point out that I still look up the dictionary, but this is only partly true. During my writing of these books on Nobel Prizes I frequently retrieve material in a split second from the "web." The time saved cannot be estimated but it must be huge. Of course it remains to control the validity of the information registered. In fact modern schooling to a large extent needs to concern itself with how the information retrieved from the "web" can be transformed into knowledge. An additional challenge in schools is often that the students are more proficient than their teachers in managing the electronic media. For each individual it is important to develop his or her own mental storage of knowledge so that future information retrieval becomes effective and factually correct. In the end it is interesting to reflect on what consequences the rapidly expanding efficiency of the use of our brains will have on individual and collective creativity, the capacity for making the new discoveries that are rewarded with future Nobel Prizes. If the efficacy is increased markedly, which seems logical to predict, the committees deciding on prizes in the future may be confronted with increasingly challenging tasks.

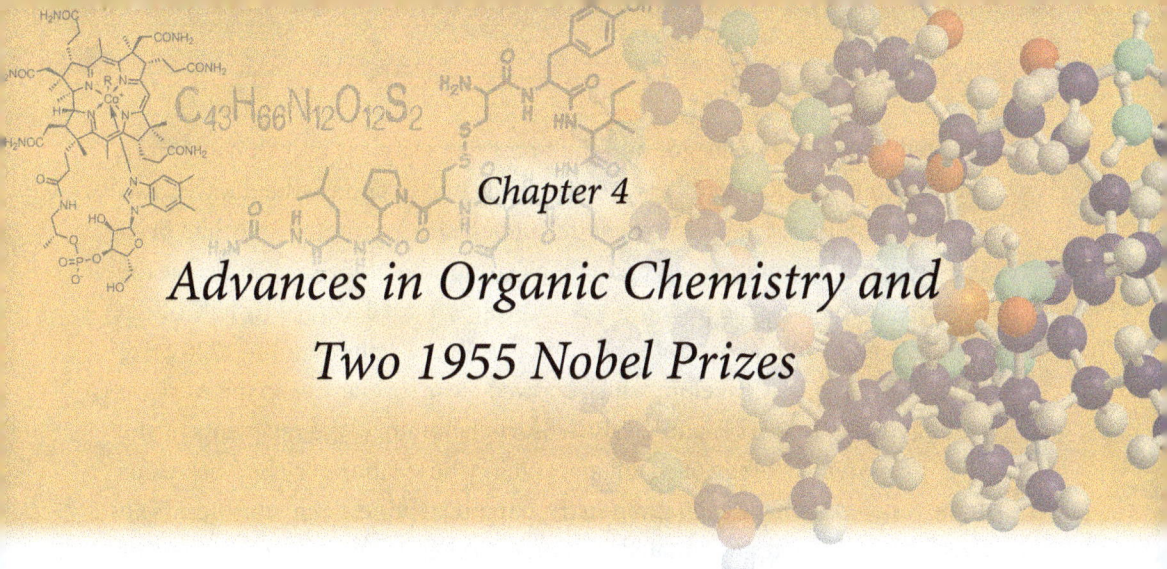

Chapter 4

Advances in Organic Chemistry and
Two 1955 Nobel Prizes

BIRTH OF BIOCHEMISTRY
A DANCE OF LARGE MOLECULES
HORMONES AND ENZYMES

In his late teenage years Alfred Nobel worked in the laboratory of the famous professor Théophile-Jules Pelouze in Paris. In this environment he met the Italian scientist Ascanio Sobrero, the discoverer of the explosive oil nitroglycerine. This experience gave a good insight into the discipline of chemistry and had a marked influence on Nobel's future successful career as an inventor and industrialist. Much later in life Nobel learnt about physiology in contacts with the visiting Swedish scientist Johansson from the Karolinska Institute, referred to in the discussion about Sherrington in Chapter 1. These later contacts made Nobel reflect on the role of chemistry in human medicine too. He speculated about new ways of destroying bacteria, in administering anesthetics and performing blood transfusions. It is therefore not surprising that Nobel in his will included both a prize in chemistry proper *and* a prize potentially also involving organic chemistry under the heading physiology or medicine. The border between awards in these two prize categories is not distinct and has shifted position over time.

The 1955 Nobel Prize in physiology or medicine recognized the Swedish chemist Hugo Theorell[1]. He was the number one candidate in that year both at the Royal Swedish Academy of Sciences and at the Karolinska Institute, a record of sorts. For unknown reasons the committees beforehand had not agreed on who should take responsibility for Theorell. Since the meeting of the College of Teachers at the Institute was held first, the assembly of course

185

selected Theorell, a colleague and the first prize recipient ever to come from the Institute. The Academy as a consequence had to choose the number two candidate on its list. It was Vincent du Vigneaud, who interestingly had been the candidate recommended by the majority of the Nobel committee at the Karolinska Institute the previous year, a proposal not accepted by its College of Teachers, which favored other candidates. In the same vein the 1970 prize in chemistry to Luis F. Leloir, which showed close similarities to the work on carbohydrates in Carl F. and Gerty T. Cori's laboratory, the recipients of the 1947 prize in physiology or medicine, and in whose laboratory he worked for some time, could have been given in the latter discipline. Each year there is an overlap in candidates nominated for a prize in chemistry and in physiology or medicine and this overlap has been increasing with time. It appears that the borders of disciplines in general become fuzzier as science advances. Many important discoveries are made in studies across borders of disciplines. The needs for interactions between the different Nobel committees over time have become increasingly important.

Chemistry in general is a wide-ranging discipline. One particular branch is organic chemistry in which carbon-containing materials are studied. The term "biochemistry" was introduced in the 1940s to specifically identify studies of chemical processes occurring in living material. The first International Congress of Biochemistry was held in Cambridge, England in 1949. Previously biochemistry had been seen as a branch of physiology. After 1949 international congresses were held every third year and the field grew very rapidly, not least when molecular biology started to develop. The International Union of Biochemistry was established in 1955. Much later "and Molecular Biology" was added to the name. There is an apparent difference in the training in chemistry at medical schools and at natural science faculties in universities. When I was a student at the Karolinska Institute medical chemistry was taught during one semester. Most of the chemistry that I later applied to my virological research during the 1960s and onwards was self-taught or picked up from other laboratories. To this should be added that virology like many other fields progressively went through dramatic changes due to the introduction of a number of new techniques. Serological (immunological)-biological studies developed into molecular investigations in my case. One may question if the limited exposure to medical chemistry during studies to become a physician might not be a limitation to future basic researchers. However, it could be that the superficial background in chemistry of medical students also might allow them to embark on experiments that someone with a much deeper insight into the discipline would have abstained from. Possibly it may in very special cases sometimes pay off to be daring, unconventional and even somewhat ignorant!

Still throughout the 1960s which is the focus of this book everyone on the Nobel Committee at the Karolinska Institute had a medical background. This situation has changed dramatically. Most students presenting their thesis at the Institute for many years past have not had training to become a physician. In 2015 the Nobel Committee at the Institute for the first time had a secretary who did not have a medical background. In the five-member committee there were two scientists, including the chairwoman, who lack a medical background. It can be discussed whether this implies some limitations, in particular in judging the value of pioneering clinical research. It is, however, not only the basic formal education of committee members that has changed with time. It seems that this may also apply to the laureates. Until 1959[1] 21 out of the total of 76 awardees in physiology or medicine had a non-medical background. In 1965, the last year examined in this book, the proportions had changed to 31 out of 90. This means a relative increase in awardees with a Ph.D. background.

During the early half of the previous century discoveries concerning hormones and vitamins had been in the focus of interest of both committees in chemistry and in physiology or medicine and several prizes had been awarded. Du Vigneaud's prize included important discoveries of the structure of two 8-amino-acids long cyclical hormones from the posterior lobe of the pituitary gland and their synthesis, as we shall see. He had also in earlier work examined the role of sulphur bonds in insulin. The discovery of insulin was recognized by the extensively discussed 1923 Nobel Prize in physiology or medicine to Frederick G. Banting and John J.R. MacLeod[1] and the hormone was the first larger protein to become available in pure form in considerable quantities. It was therefore selected by Frederick Sanger for characterization in research that, although not originally intended for that purpose, finally led to a determination of its complete amino acid sequence. His pioneering work was recognized by a prize not in physiology or medicine but in chemistry, because it was deemed to have such a major importance for the general understanding of the structure of proteins. This will be discussed in detail in the next chapter.

The three-dimensional structure of insulin was examined by the crystal-lographer Dorothy Crowfoot Hodgkin, the cousin of Alan Hodgkin who was introduced in Chapter 3. She started this work in 1934 and in 1969 she and her colleagues were able to publish the complete orientation in space of the folded polypeptide[2]. Sometimes it takes patience to do science! In the meantime Dorothy Hodgkin had received a Nobel Prize in chemistry in 1964 for her clarification of the crystalline structure of vitamin B12. Without access to archive data of the final two years this story has already been told[3]. Since Hodgkin was not included in the 1962 prize to Max F. Perutz and John C. Kendrew it was predicted that together with their own prime nominator W. Lawrence Bragg they would be active in nominating her. This is also what happened. These three laureates submitted nominations, in the case of Perutz and Kendrew, jointly, both in 1963 and 1964. In addition one nomination was made by Linus Pauling and another, including also J. Monteath Robertson as a candidate, by Robert Robinson, London in 1963. In a letter of February 14 the following year to the secretary of the committee Arne Ölander, Bragg commented on the latter nomination. In the last paragraph he praised Robertson's work, but then he stated "I have the impression, however, that he is not the genius of Professor Hodgkin". Arne Westgren made an additional evaluation in 1963, concluding that Hodgkin alone should receive the prize, but since a prize in crystallography was given in 1962 it was proposed that the committee should wait one year. In 1964 there was no repeat of the Pauling nomination but two additional Swedish nominations, one of which came from the chemist

Dorothy Hodgkin (1910–1994) escorted by His Majesty King Gustaf VI Adolf (1882–1973) at the Nobel Banquet in 1964. [From Ref. 21.]

Einar Stenhagen in Gothenburg, were submitted. He had also nominated her for a prize in physiology or medicine, but no additional reviews were done by the Nobel Committee at the Karolinska Institute, presumably because it had already been agreed that Hodgkin was the responsibility of the chemistry committee. In 1964 this committee requested two additional evaluations, one by Gunnar Hägg and the other by an exceptional external reviewer, Harry Lundin. The reviews unsurprisingly were very enthusiastic and the committee was now ready to fully support a prize for Hodgkin alone. This was to be the decision of the Academy in 1964. Her appearance as a lone female prize recipient in chemistry was exceptional, only matched by Marie Curie's prize in 1911. She and the Swedish King together led the procession entering the Blue Hall of the City Hall at the Banquet. Thirty years earlier the discovery of B12 had been recognized by a Nobel Prize in physiology or medicine to George H. Whipple, George R. Minot and William P. Murphy, the first time the prize recognized three awardees.

In addition to B12 the existence of a range of other vitamins was unraveled during the 1920s and 1930s. In 1929 a prize in physiology or medicine was given to Christiaan Eijkman "for his discovery of the antineuritic vitamin" and to Frederick G. Hopkins "for his discovery of the growth-stimulating vitamins"; in 1937 to Albert von Szent-Györgyi Nagyrápolt "for his discoveries in connection with the biological combustion processes, with special reference to vitamin C and the catalysis of fumaric acid"; and in 1943 to Henrik C.P. Dam "for his discovery of vitamin K" and Edward S.A. Doisy "for his discovery of the chemical structure of vitamin K" as mentioned earlier[3]. In chemistry two prizes were given in parallel for advances in the field of vitamins; in 1928 Adolf O.R. Windaus, who we will return to in Chapter 6, was recognized "for the services rendered through his research into the constitution of the sterols and their connection with the vitamins", and in 1937 half a prize to Paul Karrer "for his investigations on carotenoids, flavins and vitamins A and B2." The name vitamin(e) was introduced in 1911 by the Polish biochemist Casimir Funk based on his finding that rice polishing could cure polyneuritis in pigeons. It is a composite of *vita* for "life" and the chemical entity "amine," which eventually turned out not to be involved in their structure. Vitamins often play a critical role as obligatory cofactors for certain enzymes, as the examination of Theorell's Nobel prize contributions will show. We will encounter vitamins as critical nutritional components also for microorganisms in Chapter 7.

There are two central aspects to biochemistry as applied to the field of physiology or medicine. Once the chemical structure has been identified it remains to determine by which sequence of events the molecule is synthesized. This is referred to as *intermediary metabolism*, a term generally used, besides specific headings, like *hormones* and *vitamins*, in describing prizes in physiology or medicine of chemical nature[4,5]. The other critical aspect concerns the operational mechanism of the molecule. In order to understand this, the first task is to demonstrate the structure of the molecule in focus. This is determined by crystallography and the application of this technique has become impressively facilitated as already discussed due to a number of innovations and also dramatically increased possibilities for managing huge volumes of data. As a consequence the application of the technique, extended also to very large proteins and to protein complexes, has expanded markedly during the previous century[6]. The discoveries made by use of crystallography were originally recognized by prizes in physics but later in chemistry with one critical exception. This exception is the 1962 Nobel Prize in physiology or medicine to Francis H.C. Crick, James D. Watson and Maurice H.F. Wilkins "for their discoveries concerning the molecular structure of nucleic acids and its significance for information transfer in living material."[3]

The 1964 prize in physiology or medicine to Konrad Bloch and Feodor Lynen "for their discoveries concerning the mechanism and regulation of the cholesterol and fatty acid metabolism" and associated prizes in this prize category and in chemistry will be described in Chapter 6. Prior to this, this chapter will describe the expansion of the discipline of medical chemistry at the Karolinska Institute during the previous centuries. Hereafter the focus will be on protein biochemistry. A more detailed separate presentation will be given of Theorell, the first Nobel Prize recipient in his discipline at the Karolinska Institute. His toughest competitor du Vigneaud will also be presented in some depth. The following chapter will focus on the Nobel prize in chemistry in 1948 to Arne Tiselius and in 1958 to the unique two times chemistry prize recipient Frederick Sanger. His characterization of the two polypeptide chains of insulin had major implications for our principal under-standing of the structure of proteins and also for the specific functions of this important hormone. It was a true milestone in the advance of biochemistry, in particular protein chemistry.

The Development of Chemistry at the Karolinska Institute

In 1802 J. Jacob Berzelius, who was only 23 at the time was employed at a division of the School of Collegium Medicum, originally a national authority securing the competence of physicians in the Stockholm area and later more widely in Sweden. The establishment of this institute had major consequences for the development of medical education and research in the capital and also in the country as a whole. The Collegium, established as early as 1683, had the responsibility of ensuring that a student upon entering medical studies had a proper background of knowledge. The requirements include understanding of Christianity, morality and humanity, including Latin and Greek. Thus in order to qualify for medical studies a student first had to be examined by the Faculty of Theology and then by the Faculty of Philosophy. The credentials acquired were referred to as the *medicophile*. For many students it took a number of years to pass this required entry degree to medical studies. Berzelius had good previous knowledge from his high-school studies in Linköping, a city providing teaching of high quality, the students of which were referred to as "djäknar." Because of this he obtained his medicophile within six months of his studies at Uppsala University. Although still very young he was able to start his practical medical training at the Serafim

J. Jacob Berzelius (1779–1848). [From the Royal Swedish Academy of Sciences.]

Hospital in Stockholm. Furthermore he had, also in 1802, presented his thesis discussing galvanism and the potential use of electricity as a possible treatment of selected diseases. In Chapter 1 it was mentioned that Galvani's experiment with his dissected frog elicited considerable interest in the possible role of electricity in the function of the body. The early enthusiasm had waned somewhat by the beginning of the nineteenth century and in his thesis presented at Uppsala University, Berzelius critically reviewed the limitations in applying electricity for treatment of patients. He had himself tried this kind of treatment in patients with abnormal involuntary movement disorders, referred to as *chorea*, but without any positive results. The conclusion of his thesis was a healthily cautious view of the efficacy of this kind of treatment, presumably reflecting the critical attitude of an emerging scientist.

In 1806 a chemical laboratory was established in Stockholm and the following year Berzelius was appointed chair of chemistry. Four years later again the Karolinska Institute was established by Royal Decree. The early history of the Karolinska Institute has been well described by Ulf Lagerkvist, a well-qualified nucleic acid biochemist with a medical background[7]. It took time for the new Institute to move into action. Not until 1815 were its statutes finally formulated and two years later it was given its original Swedish name Karolinska Medico-Kirurgiska (Swedish for Surgical) Institutet. This was very much due to Berzelius's visionary initiatives and he became responsible for the teaching in chemistry and pharmacy. There were also two other educators at the Institute, professor Anders J. Hagströmer, in anatomy and surgery, and Erik Gadelius in medicine. Besides his central role in the establishment of the Institute, Berzelius was to become probably the foremost chemist of all time in Sweden. He established the rules of fixed proportions of reactants in the interactions and formulated a symbol language to be used to describe the chemical conversions. Furthermore he introduced new concepts like catalysis, a general term for the activities of enzymes, and isomerism, different structures of compounds with the same molecular composition. Finally, he also identified five new elements, cerium, selenium, silicon, zirconium and thorium. It was in his laboratory that Friedrich Wöhler synthesized urea from inorganic starting chemicals, results published a few years later, in 1828. As already mentioned organic chemistry devoted to carbon-containing compounds was simply a branch of chemistry at large and it took a long time for biochemistry to develop into its own dynamic and authoritative discipline. In 1808 Berzelius became a member of the Royal Swedish Academy of Sciences and served as its Permanent Secretary — in the true sense of the word — from 1819 (appointed in 1818, but absent in Paris for one year) until his death in 1848. His yearly reports were widely read internationally and seen as excellent descriptions of the advance of the discipline. In 1837 he even became one of the 18 members of the Swedish Academy (of letters), which since 1901 has chosen the recipients of the Nobel Prize in literature.

The professor of medicine in Uppsala Israel Hwasser had a view of his profession that was diametrically different from the one expressed by Berzelius. Whereas the latter was firmly convinced that physiology or medicine could only be understood by studies of the natural sciences, Hwasser believed in religious mysticism, encompassing natural philosophy and theology. He argued strongly for the medicophile requirements as the proper way to start true medical studies. It was from this perspective that he interpreted the origin of different diseases and the modes of treating them. Hwasser received solid support from his university in his attempts to prevent the Karolinska Institute from being granted the right to award degrees to physicians. His success was reflected in the fact that not until 1873 was the Institute granted the right to provide a lower degree-level qualification, the medical candidature (bachelor of medicine), and it would last as long as until 1906, until finally it was also allowed to arrange a medical doctorate examination. In this context it is appropriate to clarify some terminological issues.

After medical studies have been completed a student in Sweden becomes a licentiate of medicine, and after a period of internship he or she acquires the right to serve as a physician, to acquire an M.D. according to the terminology in English-speaking countries. In order to markedly amplify his or her medical qualifications a student in Scandinavian countries may aim for a doctorate degree. This generally involves some 4–5 years work and the thesis that is the outcome of this work is officially defended. As in the days of Luther a thesis at the Institute was literally nailed up to be publicly defended. The thesis is scrutinized by a first *opponent* selected by the faculty, today generally the only one, and earlier also by a second opponent that the candidate defending his thesis, the *respondent*, selected himself. Originally there was also an opportunity to appoint a third opponent, whose only function it was to ridicule the whole event. In each court there needs to be a jester, not least in the dense and highly focused intellectual academic environment. When I presented my thesis in 1964 we still used the formal dress of tails at the event, but this changed quickly during the late sixties. A few years later a good friend Sven Britton dressed in a jogging suit on the occasion and he also used an unusual dedication in his thesis. It read *To my surprise*. The doctorate's degree is marked by bestowing three insignia at a formal confirmation ceremony — a hat, a diploma and a ring; Latin *accipe pileum, accipe anulum, accipe diploma*. Since the work associated with developing a thesis at the Institute — or any other medical faculty in the country — is equivalent to that of obtaining a Ph.D. at a natural sciences faculty, doctoral graduates with a medical background use the combined terms M.D., Ph.D. in their curriculum vitae.

Berzelius threw a long shadow stretching throughout the nineteenth century. He truly provided a basis of natural science in the education of physicians at the Karolinska Institute. As he stated in the Foreword (translated into English) directed to King Gustav IV Adolf in his 1806 book on *Djurkemi* (Animal chemistry) "Chemistry is the most important among the sciences assisting medicine, and it will, in addition to the light it spreads over the art of medicine in general, within a short time advance some of its branches to a fulfillment, in ways that cannot be anticipated." Berzelius's mantel was much later picked up by Karl Mörner, who we met already in Chapter

Karl Mörner (1854–1917), Vice-Chanceller of the Karolinska Institute (1898–1917) and Chairman of the Nobel Committee (1901–1914). [Courtesy of the Karolinska Institute.]

1 as the first chairman of the Nobel committee for physiology or medicine, 1901–1914. He was professor of chemistry and pharmacy from 1886 until his death in 1917 and at the same time Vice-Chancellor of the Institute from 1898. Mörner discovered myoglobin in 1897 and two years later he demonstrated that cysteine is a normal constituent of proteins. Mörner was two generations ahead of me and died 20 years before I was born. Hence I could not meet him, but I did meet his grandson Karl Hampus Mörner, a count like his grandfather and by profession a civil engineer. The reason for our encounter was that in 1974 our family bought a sailing boat from him. It was a charming boat made out of mahogany and build with great craftsmanship at Storebro Shipyard at Vimmerby, a city in the county of Småland in the south-east of Sweden. The designer of the boat was Olle Enderlein. Our family of five sailed this boat through the rich archipelagoes of the east coast of Sweden and Finland for ten summers, a fantastic recreation. The name of the boat was, of course, *Karolina*. To return to the discipline chemistry at the Institute, Mörner was succeeded as professor by his student John A. Sjöqvist, who stayed at this post until 1928. During this time he was also a member of the Nobel committee. Sjöqvist was a very much appreciated teacher, but perhaps less involved in research. The very limited staff of teachers also included a junior position, established already in the very early 19th century to allow the employment of Berzelius. This position was converted to an associate professor position in 1918 and the first person to occupy this new position was Einar Hammarsten.

Einar Hammarsten (1889–1968).
[Courtesy of the Karolinska Institute.]

A Critical Leader and the Strong Advance of Organic Chemistry at the Institute

Hammarsten was a very important person in the development of chemistry at the Institute during the first half of the previous century and also in the evolving work on Nobel Prizes. We have met him repeatedly in my earlier books[1,3] as a qualified reviewer of many Nobel Prize recipients. Because of his central role in both these contexts he deserves to be described in more detail.

Blidö and the Snufkin

Hammarsten was born in 1889 as the fifth of six children. His father Fredrik was a priest in the Lutheran church. The father's attitudes were liberal for their time and he agreed to hold a wedding for a couple where the woman was already pregnant. This occurred when he was based on the island of Blidö ("the mild island") in the northern part of the uniquely island-rich (about 24,000 independent islands of very different sizes) Stockholm archipelago. The fishermen on the island appreciated this liberal attitude and allowed their priest to buy a precious piece of land on the southern tip of the island. The size of the property was seven "tunnland". Tunnland is an old Scandinavian measure of a field that took one barrel ("tunna" in Swedish) of seed to plant. It is close to 5,000 square meters in extent and hence not very different from the English acre (a little more than 4,000 square meters) which incidentally is etymologically related to the Swedish word "åker", the name for arable land. In the coming generation the land was divided between the six children and Einar got his share and in the middle of the 1930s he was able to build his own house on a high rock near the waterfront. This charming house with its multiple windows most of which include nine small panes has played an important role in the history of

Hammarsten's summer house at Blidö. [Photos by the author.]

Nobel Prizes in physiology or medicine and also chemistry. It was here that Hammarsten wrote his many reviews which are all signed Blidö and a late date in August. When he needed inspiration he would just raise his eyes and let his thoughts drift to the unique perspective-filling island-rich archipelago.

I heard about the Hammarsten family as a child in the mid and late 1940s. My father, also a priest in the Lutheran church, bought the innermost part of the bay Sandviken (the Sandy Bay) in the middle and eastern part of the relatively large island of Blidö in 1941 to be used as our summer home. In the early years when we spent our more than two-months long summer school break life was very simple. A steamboat took us to the island and a horse-driven carriage brought all our belongings to the summer house. Water was collected from the hand-made well, milk was bought from a nearby small farm, and there was no electricity or telephone. These summer school breaks offered marvelous and unforgettable times running around barefoot and living life close to nature interacting with my five brothers and our sister. There was a lot of room for development of your own unfettered imagination and fantasy. Times have indeed changed and gradually modern life was introduced; indoor water taps, a telephone, electricity in the late 1940s, cars brought to the island via two ferries in the 1950s and so on. Today a 4G connection is available allowing prompt communication from rocks, polished to a silk-like surface by the inland ice some 8,000–10,000 years ago, directly to the rest of the world.

In my recent contacts with the present generation of Hammarstens I learnt that they have been very conservative and cautious in modernizing their unique habitat. Electricity is available but no running indoor water. Drinking water is still collected from a well in the nearby village of Glyxnäs and transported by wheelbarrow. There is no road for cars into the large property. The rumors we heard about the Hammarstens in the 1940s was that they were successful academics, teachers of mathematics — my wife met one of them in this role during her high-school years — , biology and chemistry, but also that they were adventurers; mountain climbers, daring sailors and fond of powerful fireworks. The stories I have learnt to know more recently from the present generations confirm these impressions but are too many-faceted to be elaborated on in the present context.

It was through a coincidental meeting that I received an introduction to the present generations of the large Hammarsten family. What happened was that I chanced to meet Thomas Tydén, the current head of the Nobel family at a book launch arranged at Blidö by my younger brother Kaj. Thomas is the great-great grandson of Immanuel Nobel, and descendent of Ludvig, Alfred's brother. Ludvig had a granddaughter named Andriette, like his and Alfred's mother, and she married into the Tydén family. I had actually met Thomas a few years ago when I lectured to a gathering of Nobel descendants about my first book on Nobel Prizes. At the subsequent meeting with Thomas I learnt that he and his family had for some years also a summer house at Blidö and that they were near neighbors of the large Hammarsten property. With Thomas and his wife Ulla as intermediaries and with Ami Ekman and her husband as hosts a meeting was arranged at which my wife and I met some present generations of the colorful Hammarsten family, providing some of the information already cited.

As a character Einar was described as a mild and unpretentious person. He had a particular appeal to the opposite sex and was married three times. His first wife was a qualified chemist, Greta Hammarsten, also active at the Institute. He only had one child and that was from his second marriage. Regrettably there were no further descendents. Hammarsten did not like the hierarchical structure of academia at the time and in contrast to his colleagues he was certainly not a "Herr Professor" personality. These personality traits of informality most likely contributed to his success in stimulating students. He could provide good directions for their research, but perhaps he was not an ideal model of how to conduct research. He was a rather messy experimentalist, freely spreading ashes from the pipe that he eternally had in his mouth.

His habit of smoking a pipe has become documented for all time to come by Tove Jansson, an internationally highly praised Swedish-speaking Finnish illustrator and writer, the mother of the famous Moomin trolls. She died some years ago, but would have turned 100 in 2014. The well-known Moomins were presented in a series of books and cartoons, still popular at the present time. Anyone who wants to get acquainted with these charming characters can visit them at Moomin World outside Naantali (Nådendal), not far from Turku (Åbo) in Finland. Tove was the daughter of Einar's older sister and it was when she was staying with him in Stockholm and visiting him at Blidö during her university studies that she created the first Moomins. In part Einar and also one of his brothers became personified in the character Snufkin, who has a

Snufkin, one of the Moomin characters.

hat, with a broad rim encircled by a flower wreath and a pipe in his mouth. He is characterized in the following way:

> Moomin's best friend. The lonesome philosophical traveler, who likes to play the harmonica and wanders around the world with only a few things, so as not to make his life complicated. He always comes and goes as he pleases, is carefree and has lots of admirers in the Moomin valley. He is also shown to be quite fearless and calm in even the most dire situations, which has proven to be a great help to the Moomin trolls and the others when in danger.

The Moomin Valley as created by Tove Jansson found a good part of its inspiration from the unique Hammarsten property at Blidö, which I learnt to know during my visit. The property is a fascinating piece of land with heavy old round rocks rising out of the Baltic and in-between lusciously green valleys with very species-rich vegetation. It was on one of these rocks, only a few meters from the water, Einar built his abovementioned house.

The Early Developments in Nucleic Acid Studies

It is not known why Einar, when he had finished his high-school education in 1907 decided to study medicine. Maybe it was like my own situation of being the son of a priest, brought up in an environment which taught that the primary responsibility we have as individuals is to help other people, combined with a certain proclivity for mastering the natural science subjects. It seems, again like me, that Einar early on in his studies was attracted into doing science. He selected chemistry, a field in which he made a very rapid career for his time. He was appointed associate professor in this discipline as early as 1919. Only later, in 1924, did he become a certified physician and he presented his thesis the same year. Before that he had received research training abroad in Copenhagen and in Graz. The field he selected for this thesis was nucleic acids. This was a research field pioneered by his uncle Olof in Uppsala who had made some major contributions to this subject in the late nineteenth century (pp. 330–331). At the time of Hammarsten's thesis studies in the 1920s it was known that there were two kinds of nucleic acids, of thymic and yeast origin, presently referred to as DNA and RNA, but it was thought that they occurred in different kingdoms of life. However, he was able to show that "plant nucleic acid" could be found in the pancreas and somewhat later Robert Feulgen and collaborators demonstrated that "animal nucleic acid" could also be demonstrated in plants. It took another ten years before it was conclusively decided that both kinds of nucleic acids appeared in all forms of life.

Another limiting critical conceptual bias of the time was the belief that biomolecules were aggregates, colloids, of small molecules carrying some kind of important biological activity. It took a long time to acknowledge that proteins could represent large molecules and it took even longer time to acknowledge that nucleic acids in their native state had high, sometimes very high molecular weights. One obvious problem was the preparation procedures used. Due to the presence of contaminating enzymes, to be discovered in the future, the large molecules were degraded. Hammarsten used the most cautious preparation methods of the time. Further developments of the techniques allowed him to make critical observations in the thirties demonstrating that there must exist high molecular forms of nucleic acids. Very important work was performed together with his student Torbjörn Caspersson. Filtration, ultracentrifugation and flow refringence experiments all demonstrated the existence of high molecular weight nucleic acid structures. The latter studies were performed together with Rudolf Signer in Bern, Switzerland, the scientists who much later provided the nucleic acid used by crystallographers at King's College, London, Maurice Wilkins and Rosalind Franklin in their work[3].

In spite of his critical laboratory work Hammarsten was never acknowledged to have contributed to any major breakthrough. Instead he has been given credit for having two other important qualities. One was that he fully understood the importance of high quality experimental science and the need for resources to allow its full development. The other was the already mentioned great capacity to identify talented research students and giving them encouragement in their coming endeavors. He became one of the true fathers of experimental chemical research at the Karolinska Institute and established excellent contacts with the Rockefeller Foundation, which came to play a very important role for the developments. His particular personal and quite original features have been nicely described in an article in Swedish entitled *Människofiskaren* (The fisherman of humans) *Einar Hammarsten* by Ulf Lagerkvist, cited above. The latter was one of the pair of twin sons of the Swedish author Pär Lagerkvist (see book cover), who received a Nobel Prize in literature in 1951. Ulf was one of Hammarsten's pupils and developed to become an influential researcher in molecular biology. He established a collaboration with Paul Berg, the recipient of the 1980 Nobel Prize in chemistry "for his fundamental studies of the biochemistry of nucleic acids, with particular regard to recombinant-DNA." Berg and Lagerkvist together made some important studies of the specificity of the RNA adaptor molecules that carry individual amino acids to be combined in a predetermined sequence in a growing polypeptide chain[8]. After his retirement Lagerkvist became a successful author of books popularizing the history of science, frequently in relationship to the early Nobel Prizes.

The Development of an Obstinate and Strong-Willed Professor

In 1921 the pathologist Robin Fåhraeus presented a thesis that made him world renowned. He had discovered that the sedimentation rate of red blood cells from healthy and ill individuals varied and that this simple test was useful as a marker on certain conditions of disease, not least those of an infectious nature. This remarkably simple test has survived into the present time, but has now become supplemented or substituted by a test for C reactive protein. Hammarsten was the opponent on Fåhraeus' dissertation, which in fact was the first one from the department of chemistry for 26 years. Later there was an impressive sequence of theses presented. Fresh winds were blowing at the department.

In 1928 Hammarsten became full professor, succeeding Sjöqvist, and he stayed in this position until 1957. He was succeeded by Erik Jorpes as associate professor. Jorpes was a very original scientist and teacher, who I met as a student of medical chemistry in 1957. He had been brought up as the son of a fisherman on the barren rocks in Kökar, a group of islands in the south of the Åland archipelago. Because of his deep commitment to taking on societal challenges he became a Marxist. This political involvement brought him to the young Soviet Union. With only two years of medical studies as a background he became deeply involved in the health care for thousands of people in St. Petersburg and later northeast of Moscow. In 1918 he became one of the founders of the Communistic party in Finland, but soon he became disappointed by developments in the Soviet Union and in 1919 he fled to Sweden where he arrived without any means. The leader of the Swedish Social Democratic party, Hjalmar Branting, who was to receive the Nobel Peace Prize two years later, helped him and he became a pupil in Hammarsten's laboratory. Not surprisingly, Jorpes' thesis concerned nucleic acids. Theorell was the third opponent on this occasion and had decided to use fireworks to increase the excitement of the occasion. In his memoirs published in Swedish in 1977[9] he described how the rich beard of the professor emeritus in pharmacology Carl G. Santesson caught fire, but that is probably a tall story. As we shall see Theorell loved anecdotes. Jorpes later developed heparin as an important drug for anticoagulant treatment, and his students made major contributions

Erik Jorpes (1894–1973). [Courtesy of the Karolinska Institute.]

to the understanding of the complex mechanism of blood coagulation. In 1947 a special position as professor was created for him. He was less involved in the Nobel work, maybe because of the dominant influence that Hammarsten had.

By his stubborn and persistent approach Hammarsten became instrumental in bringing major resources to the Institute, part of which came from the Rockefeller Foundation. He was often ruthless and readily made both friends and foes. He had a tense relationship with Jorpes, who became chairman of the Department of Chemistry when Hammarsten had retired and the result was that he moved to another building. He also had a polarized relationship with Sune Bergström and tried to block him from becoming his successor, by ranking him second after another candidate. In the present perspective it is somewhat surprising that the Institute decided to let Hammarsten be one of the three judges for the position he himself had occupied. Theorell, whom we will soon meet in full and who was Hammarsten's favorite student, wrote in a very appreciative manner about him[9]. The text read (translated into English):

Hammarsten was the source of the development that led to the Karolinska Institute's current recognition as one of the major global centers for biochemical research. By his students in the first, second and third generation from near and far his stimulating influence has not only become of advantage to Sweden but also to the world at large.

And further on:

He was fanatically interested in research and had the rare talent to be able to encourage his students and make them believe that precisely the problem to which they devoted their interest could become of "enormous" importance. Most of the time he had the appearance of the great researcher; introvert and developing genial thoughts in his brain, sharp tongued and sarcastic. But he could also on occasion be the happiest among the happy and display pranks like a young student.

It appears that Hammarsten who might appear as a relatively gentle person, could, when a situation so required, mobilize a certain aggressiveness and obstinacy. It is said that one of his favorite expressions was an old Swedish saying "Rädda gossar får aldrig ligga med vackra flickor" — boys that are afraid will never sleep with beautiful girls. There is of course a gender analogous saying. The entertainer Mae West said "Good girls go to heaven, but bad girls go everywhere."

Jorpes, like Theorell and in spite of his animated relations with Hammarsten, was also full of praise for his contributions. In a book celebrating the 150th anniversary of the Karolinska Institute[10] he wrote *apropos* the development of the chemical discipline (translated from Swedish):

> For someone who has followed the progress (of the discipline) it can be readily testified that this development has not come by itself. It is, without doubt, as far as the Department of Chemistry is concerned, including also the Divisions for Biochemistry and Cell Research at the Nobel Institute as well as the (recently established) position as professor in medical physics, that the ascribed directions of the research activities are dependent on Einar Hammarsten, who during his time as associate and full professor at the Institute in such a devoted way — one might even allow oneself to say — unrelentingly and energetically has pursued the developments of the discipline. The personal contribution has been decisive.

In the late 1930s it was planned that the rapidly growing department of chemistry together with other preclinical departments at the Institute should move from the downtown location at Hantverkargatan 4, the building block closest to the City Hall, to the Solna campus in the periphery of the city, already mentioned above. Because of the Second World War the move was delayed. A new building for chemistry was not erected until 1949–51, although the master architect of the whole Karolinska Institute Campus in Solna, Ture Rydberg, had already finished the first complete drawings by 1937. Hammarsten and Jorpes had a major influence on the architectural design. It was a nice brick building on three floors with a central atrium space and an attached lecture hall. As a student it was easy to remember the writings on the wall of this hall. Over the entrance door it read *Panta rei* — everything flows — Herakleitos's famous saying, and on the opposite wall a quote from Goethe's *Faust* read *Wer sie nicht kennte Die Elemente, Ihre kraft Und Eigenschaft Wäre kein Meister Über die Geister.* (The one who does not know the elements, their power and properties can never become a master of their soul.)

Bringing Budding Biochemists to Flourish

I have seen Hammarsten only from a distance but never had a conversation with him. He moved together with some of his collaborators during the late 1950s from the building for the chemistry department into the new building for bacteriology at the Solna campus invited by his friend since the earlier times, professor Bernt Malmgren, introduced in one of my earlier books[3]. We received our education in microbiology in this building in the autumn of 1958.

An already mentioned major strength that Hammarsten had was to help students in building their confidence from within, making them believe that what they did was important and that they would be capable of bringing their research to a successful conclusion. He might have been a candidate for a prize for superb mentors proposed in the previous chapter. Hammarsten had a number of students that later on were to have a major influence on the work with Nobel prizes at the Institute. His star student was of course Theorell, for whom he managed to arrange the establishment of a personal chair already in 1937, as we shall see. Via Theorell Hammarsten also indirectly influenced Sune Bergström, another star scientist, but, as briefly mentioned, their personalities clashed. Bergström, the future Nobel Prize recipient will be presented more extensively in the early part of Chapter 6. Another outstanding scientist was Torbjörn Caspersson, on multiple occasions a strong near Nobel Prize candidate, repeatedly encountered in my previous books[1,3]. Other important students were Peter Reichard, whom we will meet again in Chapter 7 and also the other two following chapters, the already mentioned Lagerkvist and finally Tomas Lindahl. The latter was a classmate of mine at the Institute, whose outstanding research on nucleic acids was initially inspired by the by then retired Hammarsten. I asked Lindahl what importance Hammarsten had had for his research. He sent me the following answer (translated from Swedish):

> … I was Einar's last student and he had a major influence on me. I had Einar and Peter Reichard as role models when I took the difficult decision to interrupt the clinical part of my (medical) studies. What I in particular paid attention to was Einar's complete and artistic dedication to basic research. Even though Einar's own projects during the later years were not as impressive as his groundbreaking contributions during earlier years, together with Reichard and some others, his personality was extraordinarily impressive. If it had not been for him I would probably never have become a researcher into DNA, and I am much obliged to him for putting me on the right track.

To this can be added that a few weeks after I had received this message the Royal Swedish Academy decided to award the 2015 Nobel Prize in chemistry to Lindahl — he has been a member of the Academy since 1989 — together with Paul Modrich at Duke University, Durham, NC and Aziz Sancar at the University of North Carolina at Chapel Hill, NC for "mechanistic studies of DNA repair." The precision of DNA replication is remarkable. The error rate in insertion of nucleotides is only one in a million[3].

Most of the early development of Lindahl's earth-breaking work was done at the Karolinska Institute. He received his Ph.D. at that institute in 1967. After this he had some postdoctoral years at Princeton University and at Rockefeller University. At the latter university he did some pioneering work on ligases and endonucleases, enzymes that act on DNA, together with the Nobel laureate Edelman, referred to in the previous chapter and repeatedly mentioned in my earlier books[1,3]. Lindahl then returned to the Karolinska Institute at which time he made his first major discovery, the identification of specific enzymes for *base excision repair* in DNA. He had some very productive years at the Institute but regrettably he never received a faculty position there. He left the Institute in 1978 to become professor and chairman at the department of medical chemistry at the University of Gothenburg. After only a few years he moved to the U.K. to manage research organizations set up by the Imperial Cancer Research Fund. I think Hammarsten would have been both proud and surprised if he had known that, 37 years after his death, his last student would receive a Nobel Prize. It is a testimony to the remarkable influence he had.

The Nobel Prize Involvement

As a conclusion it is worth recapitulating Hammarsten's deep involvement in the work with Nobel Prizes. In 1929, the year after he became professor, he became a full member of the Nobel Committee at the Karolinska Institute. Apart from some odd years during the 1930s and a few years during the Second World War he remained in this position until 1947. Then he became an adjunct member for three years before he returned as an ordinary member and stayed in this position until his retirement in 1957. During the last three years he chaired the committee. He gave introductory presentations at the prize ceremony in 1931 (Otto Warburg), 1937 (Albert Szent-Györgyi), 1953 (Hans Krebs, Fritz Lipmann) and as we shall see in 1955 (Hugo Theorell).

In addition to serving on the committee for physiology or medicine Hammarsten was also an adjunct member of the committee for chemistry in the years 1948 and 1952. The double appointments in these two years serve to emphasize the interaction between the two committees, and maybe it is indicative that it was a chemist who had such dual appointments. It is a rare event and the only other example I have found was when Hugo Theorell was an adjunct member of the chemistry committee in 1950 and 1951 and simultaneously also a member of the committee for physiology or medicine in 1950, but not in 1951. We will return later to Theorell's involvement in the Nobel work in relation to his own award in 1955.

It is now time to introduce the first of the two competing Nobel laureates in chemistry and in physiology or medicine during the mid-1950s.

A Disappointed Prize Candidate Who Became Happy

When it was leaked to the press that du Vigneaud by a narrow margin had missed the Nobel Prize in physiology or medicine in 1954 because the College of Teachers supported a proposal by the minority of the Nobel committee[1], he must have been very disappointed. The fact that the same prize was denied him a second time the following year when Theorell was announced as the recipient certainly would not have made things better. Theorell has described[8] the uncomfortable situation in contacts with du Vigneaud , immediately after his own prize had been announced. Although Theorell was able to foresee that du Vigneaud would receive the prize in chemistry a few weeks later he could not, for reasons of secrecy, reveal this. It is said that when du Vigneaud finally got the message about his prize in chemistry he refused at first to believe it, saying "I have heard this twice before. So I don't believe it at all." Anyhow the prize was his, after being nominated for many years and having been on the short list for some years. Unquestionably du Vigneaud was one of the most prominent figures in chemistry at the time.

The Life of a Committed Chemist

Du Vigneaud was born in Chicago in 1901. As his name indicates he was of French ancestry. His father was an inventor and machine designer. As a freshman in Carl Schurz High School he was stimulated by two classmates to do experiments in a home chemical laboratory. Towards the end of the First World War, like other students, he assisted in farm work during the spring and

Vincent du Vigneaud (1901–1978) recipient of the 1955 Nobel Prize in chemistry. [From *Les Prix Nobel en 1955*.]

received his degree after some delay. He liked the environment and contemplated becoming a farmer. His older sister, Beatrice, managed to make him change his mind and he registered for continued studies in chemical engineering. Very soon he realized that organic chemistry was his research field *par preference*. At this time du Vigneaud was very short of money and had to find various ways of having an income on the side. The most lucrative job was to do the teaching of cavalry tactics and equitation as a reserve second lieutenant in the military, but he also made good money being a headwaiter. In the latter role he once noticed a pretty redhead. He informed a colleague that she was the woman he was going to marry and that is also what happened. Her name was Zella Zon Ford and she was an English major. However, he saw to it that she also learnt some chemistry so that after their marriage in 1924 she could become a high school chemistry teacher. They had a long rich life together.

Du Vigneaud continued his studies at the University of Illinois under Carl Shipp Marvel, also known as "Speed." He was an excellent lecturer and made du Vigneaud reflect on the relations between biochemistry and organic chemistry. Among other things he became acquainted with white rat models for metabolic investigations, which he was able to develop later in his career. He also learnt about Frederick G. Banting's and Charles Best's work on insulin. After receiving a master's degree in 1924 he worked briefly at the DuPont Company before moving to Philadelphia to study with Walter Karr at the General Hospital. Soon thereafter he accepted an invitation to come to the Department of Vital Economics at the University of Rochester, New York, where he initiated a Ph.D. project with professor John R. Murlin. In 1927 he presented a thesis with the title "The Sulfur in Insulin." Various scholarships took him to Johns Hopkins University Medical School for continued studies of insulin and then he learnt peptide synthesis abroad in Max Bergmann's laboratory at the Kaiser Wilhelm Institute in Dresden and George Barger's laboratory in Edinburgh. After this he returned to his alma mater at Urbana in Illinois but after only three years there he became the professor and chairman at the

George Washington University Medical School in Washington, D.C. Six years later he moved to the final position in his productive career. From 1938 and until his retirement he developed his science as the head of the Department of Biochemistry at Cornell Medical College in New York City. When he reached emeritus status he was invited to continue his research at Cornell University in Ithaca, New York. He died in 1978, surviving his wife by one year.

In the biographical memoir by Klaus Hofmann[11] du Vigneaud has been described as a very busy man in a highly organized laboratory. Colored slips were used at work; pink slips for proposal of new ideas; green slips for presentation of new research results and also white slips for requesting micro-analytical services. Out of these du Vigneaud preferred the green slips. When working on publication of new data he displayed a critical attitude combined with a prodigious memory for details of earlier discussions. This capacity was of great importance when an opportunity was given for an audience with the chief. Although du Vigneaud lived in the suburbs of New York he often stayed overnight in a furnished room he kept in the laboratory. This gave him opportunities to interact with his collaborators. These contacts were described as follows:

> Smoking a White Owl cigar, which he gracefully waved poised between his strong fingers, he shared a cold soft drink with us and discussed the latest research results. Speaking quietly and easily, he used words as "exciting", "surprising," "intriguing" — all suggesting great pleasure in the stepwise evolution of research. He was always highly interested in the day's results and was truly devoted to his scientific work. He felt very secure and loved his work.

And further on:

> Unquestionably, du Vigneaud was in command, and he was highly respected by his collaborators. He had a jovial manner with people, and every year he invited his entire crew to his home in Scarsdale for a picnic with softball and other entertainments. "Dee," as he was known by his colleagues over the years, associated with a great number of graduate students, postdoctoral fellows, and visiting professors. All the people who ever worked in Dee's laboratory belonged automatically to the V du V Club.

Important Contributions in Many Fields of Organic Chemistry

Insulin was crystallized by John J. Abel in 1926. Using the primitive methods available at the time du Vigneaud analyzed the composition of acid hydrolysates of the pure hormone. He identified several amino acids and concluded that the hormone was a protein. This may seem as a very trivial conclusion today, but as du Vigneaud himself expressed a number of years later "It may seem strange to speak of work establishing insulin as a protein because it is now generally accepted fact that a hormone can be a protein or that a protein can be a hormone, yet at the time (1928) there was great reluctance in accepting this viewpoint." The dominating view at the time was the concept introduced by the authority Richard M. Willstätter, who had received a Nobel Prize in chemistry in 1915 "for his research on plant pigments, especially chlorophyll." Willstätter believed that an enzyme was a combination of a small functional unit, the enzyme proper, and high molecular weight protein carrier. This was assumed to apply also to insulin. Du Vigneaud continued to make important contributions to studies of the chemistry of protein fragments, the peptides. This assisted in the development of methods to reconstitute polypeptides, advances of importance for the work later recognized by the Nobel Prize. His work in intermediary metabolism came to center on the formation of the amino acid cysteine in intact animals and the relationship to similar kinds of compounds methionine, homocysteine, cystathionine and choline. He concluded that removal of the methyl group from methionine might be a crucial step in the synthesis of cysteine.

The introduction of radioactive isotopes for studies of intermediary metabolism in the 1930s originated in George de Hevesy's discovery which was recognized by the Nobel Prize in chemistry in 1943 (awarded in 1944)[3]. The importance of the use of isotopes will be described in more detail in Chapter 6. Du Vigneaud applied this new technique in studies of the conversion of the amino acid methionine to cysteine. He was able to demonstrate that it was only the sulfur, but not the carbon chain of methionine, which was used in cysteine biosynthesis. This reaction was referred to as "transsulfuration." Other studies involved rats fed on various diets including different kinds of vitamins. The results demonstrated that choline could act as a methyl group donor for the conversion of the amino acid homocysteine to methionine. The term introduced was "transmethylation." His work also focused on substances considered to be candidate vitamins. Biotin, discovered as a yeast growth factor, turned out to be a mammalian vitamin, temporarily referred to as vitamin H, but later B7.

Oxytocin and Vasopressin

Du Vigneaud's work on these two hormones, produced in the posterior part of the pituitary gland in the brain, started as early as 1932. It was pursued in parallel with the studies described above. During the Second World War these studies were interrupted because du Vigneaud, like many other scientists, became involved in studies of and synthesis of penicillin, another sulfur-containing small compound. Important advances were made at Columbia University in the highly secret work on characterization and synthesis of this antibiotic in parallel with similar studies in England. The head of the secret project was Hans Clarke and he was surprised to hear a comment from a female scientist in du Vigneaud's regular laboratory that penicillin must contain sulfur. Out of natural curiosity he asked, "How can you know that?". She explained that the secret laboratory leaked the sulfur-containing compound benzylmercaptan into the hallway and anyone with a V du V background knew what that meant. The smell of rotten eggs meant sulfur.

Du Vigneaud's further studies of the two hormones after the war were facilitated by the development in the meantime of two new important separation techniques. One was the countercurrent distribution technique developed by Lyman C. Craig and the other the starch column separation technique developed by Stanford Moore and William H. Stein. These innovators were frequently nominated for a Nobel Prize, but only the latter two were eventually

Stanford Moore (1913–1982) and William H. Stein (1911–1980), recipients of half the 1972 Nobel Prize in chemistry. [From *Les Prix Nobel en 1972*.]

recognized by half a Nobel Prize in chemistry. This was in 1972 and the (long) motivation related to an application of the technique "for their contribution to the understanding of the connection between chemical structure and catalytic activity of the active centre of the ribonuclease molecule." After the war du Vigneaud continued his hormone studies using the new techniques. His work peaked in 1953 when he and his collaborators managed to synthesize the hormone oxytocin. Their paper concluded:

> If the synthetic product truly represents oxytocin, which it does so far as we are concerned, this would constitute the first synthesis of a polypeptide hormone. What effect slight changes in the structure of a compound of such complexity might have on chemical, physical and biological properties must be investigated.

In parallel with the work on oxytocin, the structure of vasopressin was also determined. It was found that this hormone had a similar ring structure to oxytocin but that two amino acids were of a different kind. Isoleucine was substituted by phenylalanine and leucine was replaced by arginine. There was also a variant of vasopressin with lysine in place of arginine.

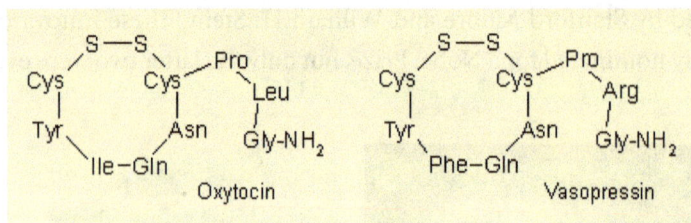

The chemical structure of oxytocin and vasopressin.

The two hormones that became the objects of du Vigneaud's important work are remarkable in their range of different activities played out in different organs of the body. They serve as an example of the fascinating *aleatoric* (Latin dice player) inventiveness of evolution. Already by the end of the nineteenth century it was demonstrated that extracts of the pituitary gland in the brain could elevate blood pressure. A decade later Dale, who we know from Chapter 1, showed that extracts from the gland also caused a contraction of the mammalian uterus. At the time a number of compounds with similar effects were already known. They were collectively referred to as *oxytocics* from Greek *oxys* and *tokos*, meaning "quick birth." A few years later it was also observed that the pituitary gland extract could stimulate milk production and also that it had an effect on urine production in the kidneys. In 1927 pharmaceutical chemists at Parke-Davis and Company were able to show that there were two different compounds in the extract, which eventually were named *oxytocin* and *vasopressin*. The two hormones were made available for research and inspired du Vigneaud's involvement. As we shall see both of them are small nine-amino-acid-long peptides which as mentioned differ only in two of the nine amino acids. A disulphide bond, which may have influenced du Vigneaud's interest in the molecule, connects two cysteins in the molecules. Later genetic studies have shown that the genes that direct the synthesis of the two hormones are located closely on the same chromosome and must have evolved by duplication of the gene material. Interestingly the two related genes run in *opposite* orientations in the genome. The separate but related genes direct the formation of the two hormones, each of which controls a number of very different functions.

Both hormones are produced in the hypothalamus of the brain. The genes direct the synthesis of a much larger mother protein (164 amino acids) from which a nine-amino-acid hormone is cleaved off. They have some direct effects in various parts of the brain, which are less well characterized, but to some extent their production is influenced by circadian rhythm phenomena. In addition they are transported and stored in the posterior pituitary gland. From there they are released into the blood stream, influencing different organs in the body. It should be added that the various effects of the hormones are critically dependent on the occurrence of specific receptors in the target organs. These receptors are decisive for the location where the activity is played out. Oxytocin as mentioned has an effect on the contraction of the uterus in connection with childbirth and also stimulates lactation. It is also sometimes referred to as the "bonding hormone" with a number of additional referred effects on sexual reproduction in both sexes. Furthermore effects on social behavior (development of fear and trust), including some possible role in autism, and wound healing have also been demonstrated. The effects of vaso-pressin are markedly distinct. When released into the blood stream it serves two functions. It retains water in the body by increasing the readsorption in the collecting duct of the nephrons in the kidneys and it also causes constriction of the peripheral blood vessels. It is the latter effect that has given it its name; Latin *vas*, container, and pressure, contract. It is presence of vasopressin that limits the production of urine as we sleep. The vasoconstriction effect leads to increased blood pressure.

Du Vigneaud was nominated for a Nobel Prize in both fields for a number of years and he was repeatedly reviewed. His first nomination for a Nobel prize in physiology or medicine was in 1943. It was Charles D. Snyder, Baltimore, MD who submitted a relatively thorough nomination emphasizing du Vigneaud's demonstration of transmethylation and his studies of biotin, both its function as a vitamin and its structure. The second half of Snyder's letter was a presentation of Conrad A. Elvehjem as a candidate. Pellagra (the name is derived from Italian *pelle*, skin and *agra*, rough; refers to a characteristic symptom of the disease) in man was demonstrated to be due to a deficient diet and Elvehjem had shown that the disease could be prevented by nicotinic acid, later in the form of niacin, vitamin B3. Theorell, an adjunct member of the committee, made an evaluation of du Vigneaud. He provided an extensive background to studies of what originally was called vitamin H by P. György, but later biotin by F. Kögl. It was György who asked de Vigneaud to help him in chemically characterizing the vitamin. This was successfully performed. Theorell was not too impressed by this contribution and wrote that du Vigneaud should not be considered for a prize at the time and that in fact Kögl's contributions were superior to those made by him. Regarding du Vigneaud's second contribution Theorell wrote (translated from Swedish):

> As concerns du Vigneaud's study on the mechanism of transmethylation it should be noted that the idea of transfer of methyl groups is not new, but has been proposed by earlier researchers. However, the experimental evidence has been provided by du Vigneaud and his collaborators, and in this work they have had decisive help from using the isotope technique developed by Schönheimer [see Chapter 6, ö becomes oe, my remarks]. The transmethylation mechanism indeed means something to a great extent new in nutrition physiology; however, its relative quantitative, theoretical and practical importance remains far from clarified, and therefore it is my opinion that the appropriate (way of acting) would be to let a possible bestowing of a prize for these studies wait for the present time.

The committee concluded that du Vigneaud's work on biotin was not worthy of a prize and that his studies of transmethylation should not be considered at the time. It would take 11 years until du Vigneaud was again in the focus of interest of the committee at the Institute. The discoveries then discussed were completely different.

There were as many as nine nominations for a prize to du Vigneaud in 1954, all by respectable scientists including Craig and the 1943 laureate Edward A. Doisy. Almost exclusively the focus of all these nominations was his studies of the chemistry of oxytocin and vasopressin. The committee was impressed and decided to commission two evaluations. One was to be made by Ulf von Euler, who was to examine the biological studies of the purified hormones and the other by Hammarsten, who would apply a more chemical perspective. Von Euler described how it was possible in 1938 to demonstrate that the two hormones had different migration properties in electrophoresis experiments. This was the first evidence that there were two different compounds. By use of more effective separation methods it was possible in the beginning of the 1950s to prepare material which almost exclusively showed only one or the other activity. Von Euler discussed the difficulty in the application of different biological tests to decide the purity of the hormone preparations. There was an inherent imprecision in the relatively crude biological tests used. The solution to the problem required an increased understanding of the biochemistry of the hormones. It therefore represented a very important advance when du Vigneaud was able to determine the different biochemical composition of the two, as it turned out, cyclical polypeptide hormones. This allowed them to be synthesized in the laboratory in 1953 and their activity examined in the biological tests available. Von Euler concluded that the chemical characterization had been pursued by use of already established methods and hence should not be presented as a discovery. To further complicate the matter von Euler brought in another candidate proposed the same year. It was Ernest B. Verney, who had discovered the existence of osmoregulators — receptors with a capacity to measure the salinity of blood — that can direct the release of vasopressin from the posterior lobe of the hypophysis. In spite of these vacillations von Euler finally supported a prize for du Vigneaud in this year.

Hammarsten's evaluation is a remarkable example of a detailed examination of a historical course of events. It extended over 19 pages and made a very fair analysis of when du Vigneaud had leaned on techniques of others and when the advances made were based on methods that he and his colleagues had developed in the laboratory. This applied both to his approaches of characterizing the molecules and to successfully managing to synthesize the two cyclical octapeptides. The final summarizing paragraphs read:

By use of electrophoresis, counter current technique and chromatography du Vigneaud has succeeded in separating and preparing in a pure form two peptide hormones which are very much alike and have a molecular weight of 1000. The purification of the peptide hormones has met with great difficulties, and there is no doubt that in his approach Du Vigneaud has shown the way for purification of these kinds of products, which in preparations in general hide among protein fractions and are easily inactivated by existing proteolytic enzymes. His choice of methods of purification and his way of applying them therefore well can be considered as a pioneering contribution of great importance. His synthesis of vasopressin and oxytocin has its basis in the determination of their structure and they are managed by a skilful use of the degradation methods of Sanger, Levy and Edman in combination with (the use of) separation methods. The contribution which may have the greater news value is the (successful) management of the synthesis, because this is the first time that octapeptides with a biological activity have been synthesized. This is without doubt a pioneering contribution.

He then concluded that du Vigneaud was worthy of a prize. The majority of the committee agreed with Hammarsten. However, no prize was awarded to him in 1954 because of the particular circumstances already described[1]. It would seem that the scientific community was upset about the way du Vigneaud was treated in this year. In 1955 there were no fewer than 24 nominations for du Vigneaud, including six from previous Nobel prize recipients. Hammarsten did a single-page review and referred to his extensive analysis the previous year. Obviously he considered du Vigneaud worthy of a prize, a view that was confirmed by the committee. However, its unanimous conclusion was that Theorell should receive the prize, as we shall soon see. The hope that remained for du Vigneaud in 1955 was recognition by a prize in chemistry, but it is very likely that he himself did not believe this to be a realistic possibility.

At the Academy du Vigneaud was progressively evaluated in the following way. He was nominated for a prize in chemistry for the first time in 1944 and then proposed yearly, except in 1946 and 1953. Reviews were carried out by Arne Fredga in 1944, 1945, 1947 and 1948 and by Tiselius in 1954. In 1955, the year of his prize one more review, now of a more general nature, was carried out by Fredga. He emphasized that there was a red thread in du Vigneaud's research and that was the chemistry of sulfur-containing amino acids. The first nomination came from two chemists at Madison, Wisconsin and emphasized that du Vigneaud already at this time was a highly regarded organic chemist. In particular his work on insulin and choline was presented as excellent. Fredga wrote a five-page review. He expressed his high regard for du Vigneaud's work on the role of sulfur in intermediary metabolism, but regarded the nomination as too general and not sufficient for serious consideration for a prize. In 1945 du Vigneaud was nominated by A. Taube from Columbia University, NY. Fredga again recommended the committee to wait. Two years later Fredga and Tiselius themselves nominated du Vigneaud indicating the respect they had for his work. But in particular they were curious about his contributions to the synthesis of penicillin. A similar curiosity was also expressed in another nomination. The review revealed that du Vigneaud's wartime work on penicillin was important but that there were other scientists who had had an even larger influence on the developments. The three nominations in 1948 did not introduce any new findings and the emphasis to a certain extent focused on du Vigneaud's impressive laboratory with a staff of skilled and qualified collaborators. This is also the spirit of an extensive nomination the following year by Robin Robinson from Oxford, the chemistry prize recipient in 1947. He wrote "The laboratory is itself an expression of his individuality and genius." No small praise, but what was du Vigneaud's discovery?

The nominations continued but the decision to wait was retained by the committee. The scene changed dramatically in 1954. Now there were four proposals of a prize for du Vigneaud and for the first time studies on the two hormones oxytocin and vasopressin were mentioned. These advances represented more of a discovery-type of research than his previous scientific work. Tiselius took care of the evaluations not only of du Vigneaud, but also of two other strong candidates. It was quite a load. Thus in addition to du Vigneaud's hormone studies he reviewed Craig for his development of the counter current technique and then Frederick Sanger for his pioneering studies of the structure of insulin, to be further described in the next chapter. He gave considerable praise to du Vigneaud's studies but noted that their success was dependent on the use of recently developed separation techniques. He finally compared the three candidates and concluded that all four would be worthy prize recipients, but since a prize in chemistry had recently been given to Archer J.P. Martin and Richard L.M. Synge (p. 282) "for their invention of partition chromatography" the committee was advised to wait. It is interesting that the committee, endorsed by the class of chemistry and the academy voting assembly in this case used the word invention. According to the rules spelled out in Nobel's will a prize in physics can be given for a discovery or an *invention* but in chemistry only for a discovery or an *improvement*. In the case of a prize in physiology or medicine only the term "discovery" applies, as emphasized earlier. In that same year of 1954 du Vigneaud was very close to getting the prize in physiology or medicine, as already repeatedly mentioned. Unsurprisingly his name therefore returned in the nominations of 1955. This year the Academy was in a unique quandary. Fredga made one more summing-up analysis of the different contributions made by du Vigneaud, integrating conclusions of earlier reviews. He emphasized that there was a connection between his different studies with regard to their focus on sulfur-containing organic compounds. In parallel Karl Myrbäck, professor of organic chemistry and biochemistry at Stockholm University, made a laudatory analysis of Theorell, which will be extensively discussed below. The committee used the following formulation in its summarizing document:

> The committee, by its evaluations and discussions, has come to the conclusion that there are two researchers who are in the lead for an award, du Vigneaud for his investigations into biologically important sulfur-containing compounds and Theorell for his discoveries in the field of oxidation ferments. It is delicate to decide which one of these (two) should be considered (to be) the most worthy of a prize. However, the committee has earlier decisively expressed that it is in favor of Theorell's candidature that it (and therefore) it recognizes that this year it (he) should be alloted the first place.

As we already learnt Theorell received the prize in physiology or medicine and as a consequence the Academy a few weeks later decided to recognize du Vigneaud, indeed a high quality second choice.

Although the synthesis of the two octapeptide hormones represented a peak in du Vigneaud's large production the Academy wanted to emphasize the continuity in his work. Therefore the prize motivation was given as "for his work on biochemically important sulphur compounds, especially for the first synthesis of a polypeptide hormone." One may wonder what the motivation would have been if instead he had received a prize in physiology or medicine. Before we join him in the enthusiasm for the prize events in Stockholm in 1955 let us get acquainted with his laureate of the same vintage, Theorell. The story about him, the first Nobel Prize recipient at the Karolinska Institute, also gives an insight into the accentuated development of the discipline medical chemistry during the middle decades of the twentieth century at the Institute.

Scientist and Musician

Hugo Theorell was a person with many talents who throughout his life enjoyed sharing with friends his enthusiasm for science and music. As in the life of any person there were of course some darker moments and his biography carries the title *Växlande vindar* (Variable winds)[9]. The book is a charming read and reveals that Theorell was a good story-teller. His full name was Axel Hugo Theodor Theorell. Among his friends he was often referred to as Theo. This is not too different from Svedberg's nickname The and like him one can say that Theorell was "The(o) main Swedish representative" of medical chemistry of the early and mid twentieth century. I have personally experienced Theorell's gift for telling the stories of his life with enthusiasm. During the 1970s and 1980s I was involved in a Referee Club, which met each month in the charming Jugend (Art Noveau) building in the very center of Stockholm owned by the Swedish Medical Association. The twenty or so members representing different medical specialties first listened to a lecture, which I arranged as the secretary of the club for many years, and then we had a nice meal together. On one occasion Bertil Josephson, a clinical chemist, and Oscar Schuberth, a surgeon, both of whom were very close friends of Theorell had invited him to share his life experiences with us. It was a very enjoyable evening and many memorable anecdotes were related.

Early Life

Theorell was born in 1903 in the town of Linköping where his father was a surgeon-major with the First Life Grenadiers regiment and also a practicing physician involved in the founding of the Linköping *lasarett* (hospital). His mother was a piano teacher and there was a lot of music in the home. At the age of three Theorell contracted a severe form of polio which at first paralyzed most of his body, but, perhaps because of his young age, there was a restoration of most muscular functions and what remained was a partial paralysis affecting both legs. This led to a tendency to fall, so to improve in particular the function of the left leg a thigh muscle transplantation was done when he had reached the age of 9. By way of compensation Theorell had very strong arms and there are a number of stories about him entering a room walking on his hands. His strong arms also helped him when he was subjected to bullying by his classmates. They quickly learnt their lesson. While convalescing from the muscle transplant he started to play the violin. He advanced quickly and began to receive violin lessons from Charles Barkel, a highly talented violinist and music teacher. He had started off as the first concert master in Norrköping a city only 40 kilometers away from Linköping. Later Barkel moved to Stockholm and in order to attend his violin lessons Theorell had to travel by train and stay overnight with relatives in the city. To cover the costs of the travel Theorell invented a complicated machine, by which he could serially produce jigsaw puzzles, which he sold at a good advance. He excelled at school and high school. Besides displaying a considerable proclivity for the natural sciences disciplines, he was also the conductor of the Music Society of the school. When it was time to plan for his university years he chose medicine. This is said to have been partly a fortuitous choice, motivated by the low matriculation costs for the Karolinska Institute. Together with three school friends he visited the Vice-Chancellor of the Institute Frithiof Lennmalm to register. Hugo could have become an engineer like his successful uncle and namesake Hugo Theorell, but this did not come about. During his medical studies he stayed with this uncle in Stockholm.

He moved swiftly through his studies and encountered a number of colorful teachers among the professors working at the Institute at the time. About the professor of physiology, Johansson, who we met in Chapter 1, he wrote[9] that he was "an in all aspects dazzling personality — erudite, kind, and full of humor. It is incomprehensible how he could live his life without getting married." The inbreeding of the academic community at the time can be illustrated by the fact that Johansson was the brother-in-law of Svante Arrhenius and the uncle of Brita Zotterman, the wife of the Swedish neurophysiologist working with Adrain, whom we also met in Chapter 1. In 1924 Theorell graduated as Bachelor of Medicine. There were three months before he could start his clinical studies. During this time he went to the Pasteur Institute in Paris to study bacteriology under the famous Albert Calmette, the inventor of a vaccine against tuberculosis. However, even before his graduation Theorell had become associated with the Department of Medical Chemistry with Hammarsten as his tutor. During his clinical studies he served as associate assistant, assistant and even as temporary associate professor. Thus he successfully made his own M.D. and Ph.D. arrangements. In 1930 he defended his thesis on May 15 and twelve days later he passed his final medical examination to become an M.D. His thesis was entitled "Studien über die Plasmalipoide des Blutes." In this work he demonstrated, as

Fåhraeus had proposed, that the sedimentation rate of red blood cells was mainly influenced by the proteins, but not to any larger extent by the lipids, in the fluid. It can be noted that until 1940 all theses at the Karolinska Institute were written in German, whereas after this the language changed to English, except of course in cases when the Swedish language was used. In his thesis work Theorell had

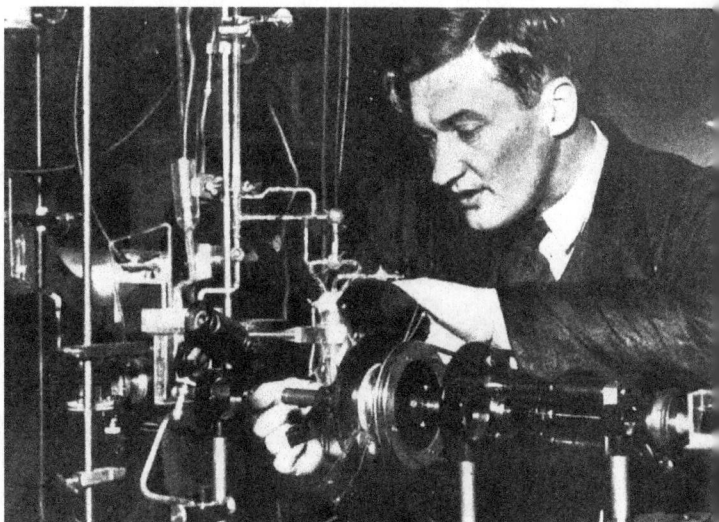

The young Theorell in the laboratory. [From Ref. 9.]

developed a new kind of apparatus for separation of molecules with different charges. It had to some extent properties similar to those of the electrophoresis apparatus developed by Tiselius, which played an important role in his work recognized by the 1948 Nobel Prize in chemistry "for his research on electrophoresis and adsorption analysis, especially for his discoveries concerning the complex nature of the serum proteins." Continuing his rocket career Theorell became a stipendiary docent at the Institute and studied the molecular weight of myoglobin by use of the ultracentrifuge together with Svedberg in Uppsala and in 1932 he was appointed associate professor in medical and physiological chemistry at Uppsala University. At this time he had already married.

Nor unexpectedly it was via his music interests that Theorell encountered his future wife. He was developing his violin playing at the Royal Swedish Academy of Music and his younger sister was a pupil at another school of music in Stockholm. She put him in contact with one of her friends at the school, Margit Alenius, who was studying to become a pianist. Music and love fitted well together and in 1931 Margit changed her family name to Theorell. For their honeymoon they went to Paris, where she took harpsichord lessons from the famous Mme. Wanda Ladowska, a world-renowned Polish pianist and teacher. Later Margit became the first teacher of this instrument in Sweden. They were to have a long and rich life together and naturally his biography[9] was dedicated to her.

Otto H. Warburg (1883–1970), recipient of the 1931 Nobel Prize in physiology or medicine. [From *Les Prix Nobel en 1931*.]

In 1933 Theorell received a Rockefeller fellowship which allowed him to work in the laboratory of the famous medical chemist Otto Warburg, who two years earlier, as already briefly mentioned, had received a Nobel Prize in physiology or medicine "for his discovery of the nature and mode of action of the respiratory enzyme." Warburg had been nominated for a prize in physiology or medicine since 1923. He had been reviewed by Sjöqvist and Hammarsten, who considered him worthy of a prize, but recommended the committee to wait. This attitude of expectation was supported also by Svedberg in 1930 and agreed to by the committee of chemistry. In 1934 there were four members of the committee of physiology or medicine. Three of them recommended a shared prize to Bernhard Zondek and Selmer Aschheim "for their discovery of the importance of the frontal lobe of the pituitary gland for the sexual functions and the pregnancy test based on this finding." The fourth member was of a different opinion. It was Hammarsten. He recommended Warburg and impressively he was successful in persuading the members of the College of Teachers to follow his recommendation. Consequently it was also he who gave the introductory speech at the prize ceremony. Warburg was a high-profile character, a bachelor and a member of the cavalry during the First World War. He retained his interest in horses and started every day with a ride. The information about his Nobel Prize reached him when returning from one morning ride. His reaction has been said to have been "Gute Zeit!" — meaning that this was due time! Warburg's continued career is unique. In contrast to all the other scientists with a Jewish ethnic background he was allowed to remain in Germany during the Second World War. Of course there were restrictions. He could not travel abroad and he could not take on leading academic administrative responsibilities. It has been rumored that his privileged treament was due to the fact that Hitler was afraid of getting cancer, a disease that was central to some of Warburg's research.

Sometime after Theorell had begun his work with Warburg, he and his wife brought their firstborn child, a daughter Eva, to live with them in Dahlem in Berlin. When the family returned to Stockholm in 1935 they soon met with tragedy. Their young daughter developed a high fever, which later turned out to be due to miliary tuberculosis. After a few days she died. This was an extremely difficult time, which their circle of close friends helped them to manage. A few months later a consolation was provided by the fact that Margit became pregnant again. During the following years three sons were born, Klas, Henning and Töres. With time they all took up the medical profession and all of them have made their own high-careers within their chosen disciplines. More recently my contacts have been with Töres, who specialized in internal medicine and became a professor of psychosocial medicine at the Karolinska Institute. He is also a good singer and violinist, like his father, and has developed a particular interest in analyzing the role of music in furthering education in general. The Sven and Dagmar Salén Foundation, in which I was involved for a number of decades, has supported this project. It has been pursued together with Fredrik Ullén, a professor of cognitive neuroscience at the Karolinska Institute and simultaneously a professional pianist, one of the foremost interpreters of György Ligeti's music.

Theorell's experience of working with Warburg completely changed and accelerated his scientific career. His focus of interest became oxidation enzymes and during his stay in Berlin he was the first scientist to produce the oxidation enzyme, "the yellow ferment," in a pure form. He was able to demonstrate that it had a *coenzyme* part (flavomononucleotide; *flavus* means yellow in Greek, later identified as vitamin B2) and a colorless protein part (the *apo*enzyme). The enzyme only showed activity when these two components were associated, a complex referred to as the *holo*enzyme. One of Theorell's favourite songs was a Danish folk song, "Too maa man vaere hvis livet skal lykkes" — there must be two if life is to succeed. He cited this in his introductory speech for Arthur Kornberg and Severo Ochoa at the 1959 prize ceremony[1]. He could have applied this citation to his own work. The finding of the critical role of reassembly of the two components for enzyme activity remained probably the most important discovery among Theorell's invariably high quality contributions to biochemistry throughout the forthcoming decades. In 1934, when Theorell had successfully purified the two components of the yellow enzyme and demonstrated that they showed no activity in the dissociated form, but could regain activity when reassociated, Warburg said: "Also, herr Thorell, da haben Sie was sehr schönes entdeckt; dafür werden Sie wohl einen Nobelpreis bekommen" — "So Mr Theorell, in this you have discovered something very beautiful. For this you may well receive a Nobel Prize." This was an early prediction of things to come twenty-one years later.

When he had returned from his stay with Warburg, Theorell worked at the Chemistry Department at the Karolinska Institute on continued leave of absence from his position in Uppsala and within a short time, in 1935, he was appointed head of the newly established Biochemical Division of the Nobel Medical Institute at the Institute. His new laboratory was in operation by 1937 and was housed in some rooms in the buildings used by the preclinical departments of the Karolinska Institute in the center of Stockholm. In the 1940s these preclinical departments progressively moved to new buildings at the Norrbacka campus in the nearby Stockholm suburb of Solna. The relocation was delayed because of the ongoing Second World War. Finally a separate building at this new location was erected for Theorell's Nobel Institute and in 1947 his research group were able to move into fresh laboratories. The new building also included a cell biology division developed to host Torbjörn Caspersson's large research group and as already mentioned in Chapter 1 an attached wing for neurobiological research. In his new building Theorell was able to expand the scope of his operations and additional laboratories were provided already in 1951 by the building of an annex, paid for by grants from the Rockefeller Foundation and from the Knut and Alice Wallenberg Foundation. The buildings were well equipped with the most up-to-date instruments for separation and characterization of large molecules. Theorell's new division developed to become an impressively productive academic laboratory and an excellent breeding ground for young Swedish scientists and also for a wide range of visiting scientists of many various nationalities.

The building for the Biochemical and Cell Research Divisions of the Nobel Medical Institute in 1948. [Courtesy of the Karolinska Institute.]

The different activities turned out to be rather wide-ranging. It all started with Theorell's crystallization of myoglobin in 1931 and his interest in oxidation enzymes emanating from his stay with

Warburg. Parenthetically it can be mentioned that Theorell, together with two collaborators, was finally successful in crystallizing his original love the yellow enzyme first in 1955, the year of his Nobel Prize. The early critical finding that the coenzyme part and the protein part, the apoenzyme, could be reversibly dissociated and showed activity, only when reassociated into a *holo*enzyme, was of importance for the future work. Theorell came to study many enzymes that serve functions for transporting oxygen and for transfer of energy in many kinds of processes inside cells. A common component in these sorts of molecular complexes — hemoglobin, myoglobin, cytochromes, oxidases, catalases, peroxidases — is the so-called heme compo-

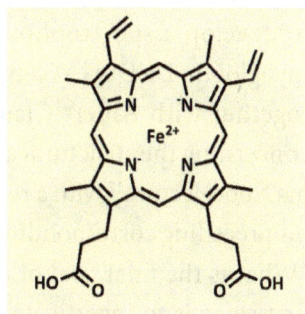

The chemical structure of a heme component.

nent. Four five-component rings with nitrogen pointing inwards allow a central binding of iron. The complex is referred to as porphyrin. The iron carries very central functions in this complex. It can occur in two forms with a charge — valence — of two or of three. In hemoglobin or myoglobin it always exists in the bivalent form allowing the binding and release of oxygen by these carrier molecules.

Cytochromes, which occur in three different forms, play a very central role in cellular respiration. Their discovery by David Keilin in 1925 was critical in the understanding of cellular respiration, as will be further discussed below. In this group of enzymes the valence of iron varies between two and three allowing transfer of electrons. Finally enzymes like catalase and peroxidase only exploit iron in its trivalent form. This makes possible a binding of peroxidase, which is then used for detoxification purposes in bacteria. Theorell and his collaborators over the years made many important contributions to our knowledge about enzyme complexes involved in energy transfer. He had already initiated collaboration with Linus Pauling before the Second World War allowing measurement of active iron in molecules. In later studies the respiration enzyme cytochrome C was characterized and towards the mid 1950s there was an emerging insight into its amino acid characteristics which allowed a prediction of the existence of an alpha-helix structure, a concept introduced as a critical form of folding in proteins, by Linus Pauling and Robert Corey[3]. Pauling received his Nobel Prize in 1954 for his earlier pioneering contributions. Collaborations between Theorell and other Nobel Prize recipients should also be mentioned.

Many Important Scientific Collaborations

He collaborated in early studies by George de Hevesy, in the use of isotopes to study intermediary metabolism, recognized by the 1943 Nobel Prize in chemistry (awarded in 1944) as already described[3]. Christian de Duve visited Theorell's laboratory for 18 months in 1946-47 and together they were able to develop a spectrophotometric quantitative determination method for myoglobin. De Duve received the 1974 Nobel Prize in physiology or medicine together with Albert Claude and George E. Palade "for their discoveries concerning the structural and functional organization of the cell." He finished his Nobel lecture[12] with a reference to the conclusions mentioned by Theorell in his preceding corresponding lecture in 1955[13]. Theorell had posed the question "What is the final goal of enzyme research?" and his answer had been "The first stage is to investigate the entire steric constitutions of all enzymes …." He then continued "In the second stage it is a matter of deciding how the enzymes are arranged in the cell-structures. This implies, as a matter of fact, the filling of the yawning gulf between biochemistry and morphology." De Duve in the last paragraph of his Nobel lecture made the following comment:

> The gulf still yawns today. But it is a particular pleasure for me to be able to tell my old friend Theo that it yawns a little less. In our efforts to narrow it, my co-workers and I have been privileged to contemplate many marvelous aspects of the structural and functional organization of living cells. In addition, we have the deep satisfaction of seeing that our findings do not simply enrich knowledge, but may also help to conquer disease.

Other forthcoming Nobel Prize recipients working in Theorell's laboratory were Christian B. Anfinsen and Paul D. Boyer. Anfinsen spent a year (1947–48) as a Senior Fellow of the American Cancer Society, whereas Boyer had a Guggenheim Fellowship that allowed him to spend a year (1955) in Sweden dividing his time between Theorell's Medical Nobel Institute, sharing in the joy of his Nobel Prize, and the ambience of the Wenner-Gren Institute for Experimental Biology collaborating with Olov Lindberg and Lars Ernster. Anfinsen in his work pioneered the advancing insights into the complex phenomena of protein folding, and the half prize he received in chemistry was motivated by "… his work on ribonuclease, especially concerning the connection between the amino acid sequence and the biologically active conformation." Boyer did not receive his shared prize in chemistry until 1997 when he was 78 years old. In his case the prize motivation was "elucidation of the enzymatic mechanism underlying the synthesis of adenosine triphosphate (ATP)." ATP is the energy-storing molecule discovered by Fritz A. Lipmann and recognized by his shared Nobel Prize in 1953. ATP production occurs in a particular cellular organelle called the mitochondrion because of its unique morphology; from Greek *mitos*, thread, and *chondrion*, grain-like. Boyer's work initially was inspired in particular by his contacts with the researchers at the Wenner-Gren Institute. He finished his 1997 Nobel Lecture[14] in the following way:

> As summarized by Ernster (the last reference given, no 122) the oxygen we use to make ATP is also a toxic substance, resulting in the production of free radicals. The mitochondrion is particularly susceptible to such damage ...

A lot could be said about the mitochondrion, the power-house of cells. Its origin is tantalizing since most likely it derives from a prokaryote that became associated with a primitive cell way back in the penumbra of early eukaryote evolution by a mechanism referred to as endosymbiosis. Thus mitochondria have their own genome and hence capacity to perform many chemical reactions. Their main role seems to be to make energy available to the cell by chemical reactions converting adenosine diphosphate (ADP) into ATP. A unique accumulation of mitochondria is found in cells in brown fat tissue. They are markedly enriched in presence of this kind of organelle and can produce of the order of 1,000 times more energy than cells in other tissues. The brown fat can be found in hibernating animals, but also in man, both in newborns and also, as recently found, in adults. It is not known what role the tissue may have in adults.

Concerning forthcoming Nobel Prize recipients associated with Theorell, Bergström finally also should be mentioned. We will learn to know him more closely in Chapter 6.

Music and Arm Wrestling

Even during the dynamic developments in his science over the four decades ending with his retirement in the 1970 playing the violin was an integral part of Theorell's life. A long chapter in his biography[9] is devoted to his rich experiences in the world of music. He carried the musical talent with him in his genes. Thus he made sure that he could keep up his professional standards and played frequently at weekly meetings of the Mazer Quartet Society in Stockholm and other groups of musicians at other events. For a long time he chaired this Society and also both the Stockholm Academic Orchestral Society, and, not least, the Stockholm Concert Society. Between 1951 and 1973 Theorell chaired the board of the latter Society. This society is responsible for the orchestra that has its home at the Concert House of Stockholm, the charming Konserthuset. This is the place where the Nobel Prize ceremonies have been arranged annually since 1926. The building was designed by the famous Swedish architect Ivar Tengbom and the style employed is referred to as the New Antique (Classic). This is apparent both from the outside appearance of the building and especially from the beautiful concert hall where the Nobel Prize event takes place. Meandering — *a la grecque* — patterns prevail in tapestries and other decorations in the grand hall.

Theorell, the violinist, among fellow musicians. [Photo from Henning Theorell.]

The Stockholm Concert hall.

At the time of writing plans are under way for a Nobel Center in the heart of Stockholm at the narrow peninsula named Blasieholmen. These are excellent plans to expose science to the public and not least schoolchildren via a modernized Nobel Museum and by staging various events. Such events may celebrate the remarkable progress of the sciences and their critical role in the advance of human culture and civilizations and also the interactions of the arts, especially literature, and the sciences. There are also some plans to include in the building a large festival hall which could house the Nobel Prize ceremony. One may have many different

The Nobel Prize ceremony in the Stockholm Concert Hall in 2014. [Courtesy of the Nobel Foundation.]

views on the need for such an additional showpiece locality used primarily for a single yearly event. My view is that bringing science to the people is better served by a different use of the available localities in the proposed building. It would seem better to continue to use the very charming, and highly appropriate for the purpose, music hall of the beautiful 1926 Concert House that has been the home for Nobel Prize ceremonies, so far for 90 years. It can be added that it is generously made available for the ceremony at no cost to the Nobel Foundation.

My uncle Johannes Norrby was the program director and then between 1950–70 executive director of the Concert House, positions he had reached by a roundabout way since he had a Master of Science in biology and planned to become a teacher. I had very close contacts with my uncle and during the 1960s and 1970s partly because he wanted to be briefed on the impressive advances in biology. When visiting his office in the Concert House we could from a nearby space, where some spotlights were located, view the scene from above. One could see the conductor, which otherwise was not possible because at this time the choir stalls that were built later did not exist. From this unique vantage point we enjoyed many concerts, not least jazz at the Philharmonic arranged by Norman Granz. All the stars of jazz at the time passed through Stockholm. In this context I may mention that I have been involved in three very different performances on the stage of the Stockholm Concert Hall. One was when our Dixieland group, *Les Saints Bleux*, mentioned in my previous book[3] were the first to perform at an evening event arranged by the entertainment company Nalen called *Black and White Show*, and the other two occasions were when I introduced Nobel Prize recipients in physiology or medicine in 1976 and 1989. I accepted the challenge of making my presentations without any manuscript in front of me. It worked like a charm the first time and with only a small memory lapse the second time.

Deep and diversified interests in music characterized my uncle. He was a superb bass singer, played a number of string instruments and was to become the conductor of the Academic Choir in Stockholm, a responsibility he held for almost thirty years starting in 1945. Theorell and my uncle became very good friends because of their joint interest in biology and music. They were able to share a deep involvement in music but also in play. As previously mentioned Theorell had very strong arms and so had my uncle. A very simple challenge was arm wrestling, first the right and then the left. In these contests they were sometimes joined by Jussi Björling, the Swedish tenor, who probably had the most outstanding voice of the twentieth century. The contests between these three men, with their well built upper trunks, often resulted in a draw. But there was even more to the meetings of music and science. As we shall see one of the important advocates and referees of Theorell's scientific contributions was Karl Myrbäck. Myrbäck was a pioneer in biochemistry and became professor of fermentation-chemistry as early as in the late 1920s and he was Hans von Euler's star pupil. Myrbäck was also very interested in music, although he did not play an instrument himself. He was for almost two decades chairman of the Academic Choir (Stockholm University Choir) where my uncle was conductor. They became close friends. But there were even more fortuitous connections. During the last polio epidemic in Sweden in 1952–53 prior to the introduction of vaccination[1] Myrbäck at the age of 53 became afflicted. Theorell, who from his own life experience knew the challenge of this disease simply said "Polio, o no!". Apparently Myrbäck managed his

Karl Myrbäck (1900–1985).

disabilities relatively well because in 1954 he became a member of the Nobel committee for chemistry and remained in this position for 20 years. Let us finish this section on music with a true story.

Theorell took every opportunities to play with colleagues. He played with both Ernest Chain, who shared the 1945 prize in physiology or medicine for the discovery of penicillin and Manfred Eigen, who shared the 1967 prize in chemistry for the characterization of extremely fast chemical reactions, both of whom were well-qualified pianists[1]. In 1956, however, Theorell had a very special encounter connected with the creation of music. He and his wife Margit were visiting Linus Pauling, with whom he had long since developed a scientific collaboration. Pauling met them at the airport in Los Angeles. His first message was that he had been approached by a musician, previously unknown to him, Willy van den Burg. He turned out to be the cello section leader of the Los Angeles Philharmonic Orchestra. He wanted to play with Theorell and his wife. Pauling therefore drove the company to a big villa in Beverly Hills. Pleasant contacts were established and music was played for some hours. On a later occasion Theorell received a letter that made him proud. Van den Burg wrote, "You are not the first Nobel Prize recipient that I have played with. Earlier I played with Einstein, but you play better than him. Both his intonation and his rhythm are as relative as his theory!" Theorell paid many visits to his friend Pauling. On one occasion my good friend Gustaf Arrhenius had taken on the responsibility for the transport of Theorell to Los Angeles airport. Time was short so Arrhenius had to speed and was caught by the police. He was perhaps a little disappointed when his senior chemistry colleague did not offer to share the fines, since he recollected this episode 50 years later.

Playing music is a great way of putting aside all other involvements, as I can testify to from my own experience. We will encounter a number of examples of successful scientists, like Bloch in Chapter 6 and Monod in Chapter 8, who were very active in performing music alongside their research activities. There is no question that performing and listening to music is a highly absorbing activity. It may potentially induce a mentally balanced state and especially in the case of listening induce an emotional state that is inducive to creative thinking. An illustration of this is the book title *Late Night Thoughts on Listening to Mahler's Ninth Symphony* by the famous essayist Lewis Thomas. It has also been documented that children who get a good share of music experience in school perform better in other subjects than those who do not. The German philosopher Arthur Schopenhauer has described music as the most metaphysical of all forms of art. However, way before his time

Pythagoras discovered the correlation between the length of a string and its tone and in addition he observed that the interval between the notes in some cases gave a clean, harmonious impression, but in other cases created a larger tension. The octave, like the perfect fifth and the perfect fourth, are unique intervals and many scientists have been curious about the potential underlying rules — neurobiological correlates (?) — of the impressions of music. Various mathematical correlates concerning both frequency and time have been defined. Without appreciating the possible influence of such correlates it is difficult to understand how "Bach could play duets with God."

Compared to other forms of art it can be noted that a tone is a part of the three-dimensional space surrounding us, as also smell may be. Listening to music may often make us forget about time, this dimension that the Nobel laureate Ilya Prigogine referred to as irreversible change. Music is also, compared to other forms of art, unique in its capacity to stir emotions, another topic of studies by neurobiologists. One need only think about the famous duet in *The Pearl Fishers* by Bizet sang for example by Jussi Björling and Robert Merrill. If you turn around after a live performance of this duet you will find that half the audience is crying! Music is a powerful mover of our emotions in many different ways. Just think of lullabies, marches that encourge young men into collective suicide in traditional war, working songs of many kinds — men pulling barges on Russian rivers, etc. Movies generally are accompanied by music that influences our emotional interpretation of what we see. Finally of course music is connected with movement in dancing, a cultural expression developed in essentially all parts of our world. A French philosopher has described music as a movement, conditionally related to passing time. This may not apply to John Cage's 4'33", a composition in which the tangents of the piano are not touched during the time specified. By way of contrast visual art, discussed in Chapter 3 and further in the concluding Chapter 9, provides impressions generally emanating from seeing an object fixed in space.

It is now high time to conclude the survey of Theorell's wide-ranging interests, which also included sailing. This interest was amplified by his collaborator as a visiting scientist on a number of occasions, the Nobel Prize candidate Britton Chance from Philadelphia. He was a pioneer in studies of fast chemical reactions, but apparently a discovery of sufficient magnitude was never identified in his huge scientific production and thus he never received a Nobel Prize. However, he was awarded an honorary doctorate at the Karolinska institute in 1960, when the Institute celebrated its 150 years of existence. Occasionally honorary doctors have interpreted the award as a consolation recognition for not receiving a Nobel Prize. Chance developed many talents during his 97 years of life and I met this energetic chemist during the last decades of his life at meetings of the American Philosophical Society in Philadelphia. This association is the oldest learned society in the U.S. established by Benjamin Franklin in 1743. Besides being a first class chemist Chance was also very skilled in electronics

Theorell sailing. [From Ref. 9.]

and in particular he was a great boat designer and sailor and won the gold medal in the 5.5 meter sailing class in the Olympic Games in Helsinki in 1952. Together with Theorell he sailed many parts of the uniquely island-rich Stockholm archipelago.

Let us now examine Theorell's road to his Nobel Prize as it is illustrated by the Nobel archive material. This has been analyzed earlier by Ragnar Björk and described in an article in Swedish[15]. He has reviewed the archives and discussed whether it could have been an advantage to Theorell that he was judged by his colleagues. However, as we shall see the committee for physiology or medicine for once used external, albeit Swedish advisors in this case. He also raised the important question of whether Theorell might have been favored in relation to other candidates in his field of research.

Early Nomination for a Nobel Prize

Theorell was nominated for a prize both in chemistry and in physiology or medicine. The first nomination was in the latter category in 1936. It was, not surprisingly, submitted by Warburg. The nomination was brief and mentioned studies of the yellow enzyme and emphasized the importance of the finding of regained activity by combining the carrier and the active group. It can be noted that at this time the protein was seen as the carrier, not the active component in itself. This interpretation came to change with time. The Nobel committee took the nomination seriously and contemplated who to choose as a reviewer. However, they had a problem, since Theorell was conducting his research at the Karolinska Institute. At the time he and all the members of his research group were formally employed by the Nobel Foundation and worked in a building which was not owned by the Karolinska Institute. This arrangement did not change until 1959, when the Medical Nobel institute was converted into a government-run Biochemical Department belonging to the Karolinska Institute. From then until 1969 Theorell was a member of the College of Teachers, later the Medical Faculty of the Karolinska Institute. The problem of evaluating Theorell in 1936 when he was an informal member of the Karolinska Institute was who to choose as a reviewer. It was the principle of the Nobel committees into the early 1960s that all reviews should be carried out by their own members. Many individual members often had a heavy load as we shall see. The committee at the Karolinska Institute in 1936 decided to make an exception and selected the Uppsala professor and member of the Nobel committee for chemistry, The Svedberg as the reviewer. This was probably the first time the committee used someone not employed at the Institute for this function. Based on his experience since 1926 of Nobel committee work Svedberg did a very careful and detailed review covering 11 pages. It was full of praise, but also contained some criticisms.

Theorell had developed a new kind of electrophoresis apparatus, by which components with different charges could be separated. In Theorell's own words it was a development of a separation technique in parallel with similar work by Tiselius, as mentioned above. Svedberg agreed that Theorell's apparatus was useful but pointed out that he had profited from a number of contributions by others. Not surprisingly he might have wanted to protect Tiselius's priority in this matter. In truth Tiselius's apparatus was much faster than Theorell's. The review appropriately gave particular emphasis to Theorell's 1934 observations that the reassembly of the active group and the colloid carrier, the protein, was required for enzymatic activity to be displayed. And still Svedberg would have liked to learn more about the background to this effect. He then discussed the dependence of the yellow ferment on other enzymes. In a particular case he criticized Theorell because in his presentation of a formula he had over-looked the fact that this had already had been deduced in 1925 by Svedberg himself. He then went on to praise Theorell's studies of myoglobin, another protein involved in respiration (oxidation), since these studies, although in their nature somewhat separate from those on the yellow enzyme, might be considered of importance in the evaluation of the potential for a prize. In the concluding paragraphs Svedberg noted that Theorell's studies of the yellow ferment, of cytochrome and of myoglobin "has brought our insights into the chemical nature and mode of actions of respiratory (oxidative) enzymes a major step forward." Still however, there remained a subdued tone of criticism. He described Theorell as a born experimentalist and yet he surprisingly said that "his way of working manually is far from exemplary." The final paragraph read (translated from Swedish):

It can hardly be doubted that Theorell in the near future will have left behind work of such importance that an award of a Nobel Prize for medicine will have become self evident. It is therefore in the opinion of the undersigned even more important that at the present time we maintain an attitude of wait and see.

This was not to Hammarsten's liking. He wanted a different evaluation of his favorite student. Thus he took the initiative to write a complementary evaluation over four pages, most of which was a critique of Svedberg's evaluation. In part of his review he had argued that in the evaluation of Theorell's contributions one needed to consider the findings made by James B. Sumner and John H. Northrop, who were studying crystalline preparations of enzymes. This work lead to them, together with Wendell M. Stanley, being awarded the 1946 Nobel Prize in chemistry[1]. Hammarsten argued that it had not been proven by Sumner's and Northrop's studies that the protein examined really carried the enzymatic activity on its own. By contrast Theorell's studies emphasized that the protein part played an essential role in the enzymatic activity displayed by the reassembled preparations. This debate highlighted the general skeptical attitude of the time concerning the importance of proteins, already alluded to in the presentation of du Vigneaud. It was not as yet accepted that it was the protein itself that could be responsible for the biological activity. This did not become fully clear until after Frederick Sanger had unraveled the complete amino acid composition of insulin, as we shall see. Hammarsten then picked on Svedberg for saying that Warburg had received his Nobel Prize for the yellow enzyme studies, which is not entirely correct since he received it for studies of metalloenzymes. He also nagged at Svedberg for making a comparison of Theorell's work and that of others without noticing that different temperatures of incubation had been used. Adding yet some information he summarized his critique of Svedberg's review. His emphasis was on the difference between physical and physiological chemistry and the conclusions he came to were as follows (translated from Swedish):

According to my opinion it (the arguments already given) is sufficient to recognize them (the contributions) to have a very high degree of prize-worthiness, if a researcher in the case of physiological processes which are difficult to access, has had a phenomenal success. ... It is my conviction that if Svedberg had been a physiologist he would not have reflected on postponing the decision. The way things are presently, it seems to me that the natural feeling of a specialist in physical chemistry of the need for completeness in the exact determination of constants to some degree has contributed to that the importance to physiology of Theorell's phenomenal success [again! my remark] in the field of experiments.

Among the results which have been submitted for evaluation of prize-worthiness, it seems to me that those presented by Theorell are the most important. It is to be expected that the value of these results certainly in the near future will become further amplified by discoveries by him and by others. I cannot see that there are *any reasons* [my italics] to postpone the decision and therefore recommend, without *any hesitation* [my italics], that associate professor Hugo Theorell should receive the year 1936 prize in physiology or medicine.

Loewi and Dale also seem to me to be to a high degree worthy of a Nobel Prize, and I am therefore of the opinion that they should be considered in the second place.

The committee accepted Hammarsten's argumentation and wrote regarding Theorell "... his discovery of the capacity of the yellow enzyme to connect and interact with the red metal ferment purified by him was worthy of a prize." In its final decision this year the committee was split. There was a majority vote for a prize to Dale and Loewi but the professors in neurology Nils Antoni and in neurosurgery Herbert Olivecrona wanted to give the prize to Harvey Cushing and Walter E. Dandy for their discovery of the technical principles of the successful surgical treatment of brain tumors; Hammarsten not surprisingly wanted to recognize Theorell, and finally the professor of physiology Hans Gertz put Dale and Loewi and Theorell in the same place, probably meaning that he wanted to see a divided prize for two separate discoveries. As already described in Chapter 1, the prize went to Loewi and Dale.

Hammarsten was not satisfied by only having put forward Theorell as a candidate to a prize in physiology or medicine. The following year,1937 he nominated Theorell in an extensive letter for a prize in chemistry. This nomination was supported by the professor of medicine at the Karolinska

Institute, H. Christian Jacobaeus, who sadly passed away later the same year. The committee at the Academy logically let Svedberg make another evaluation. One further long review was submitted. It concluded that it was questionable whether the findings were of sufficient importance to motivate a prize in chemistry. The committee agreed with this conclusion. Hammarsten then submitted another now brief, nomination for Theorell the following year, but the committee decided again to wait. It would take another 14 years before Theorell was again proposed for a prize in chemistry.

In the meantime Theorell had become a member of the Royal Swedish Academy of Sciences. This happened in 1942 and again Hammarsten had a major influence. When the class for medical sciences met in November 1942 to recommend a new member there were two candidates, the professor of practical medicine at Uppsala University, Gustaf Bergmark and Theorell. There were four members present at the meeting and including postal ballots there were 12 votes for Bergmark and two for Theorell. This did not discourage Hammarsten. At the meeting with the whole Academy a few days later he apparently argued very effectively for his candidate. In the end there were 12 votes for Bergmark and 24 for Theorell, who was thus elected at the age of only 39. Over the years Theorell came to serve different functions at the Academy and he was its President (Preses) between 1966 and 1969. There is also one physical remainder of Theorell's presidency. The original grand stairs to enter the Academy building from 1915, designed by the well-known Swedish architect Axel Anderberg, did not include any handrail. On the initiative of Theorell, who well understood the importance of a rail for handicapped persons, a central wrought iron structure with a wooden upper covering now decorates the midline of the impressive stairs.

Three years after his first nomination for a prize in physiology or medicine Theorell was again proposed for a prize, this time by H. Zangger, Zürich. Svedberg was asked to do a second review. It was short and concluded that Theorell has strengthened his candidature for a prize, but that there were reasons to wait. The committee agreed with this and proposed a postponement. This in fact meant a reversal of the conclusion in 1936 when Theorell was considered ripe for the award of a prize. In 1945 there was another recommendation of Theorell. It is a bit of a strange nomination by A. Vannotti, Lausanne, since in its major part it is a proposal for Hess. This time the committee requested the assistance of a second external reviewer, Tiselius. In contrast to Svedberg he had not as yet become connected with the work of the committee of chemistry, but that was soon to come as we shall see in the next chapter. No doubt Tiselius was a very competent reviewer. He carefully analyzed over 14 pages Theorell's contributions in full and gave a special emphasis to the more recent findings. After he had given praise in the first part of the final paragraph he introduced a reservation. He wrote "However, it seems to me difficult to bring forward any particular of these publications which has led to a discovery that is so important and so definitely secured that one without reservation can support recognition by a prize." He recommended the committee to delay awarding a prize. This view was confirmed by the committee in its concluding report. In 1945 the discovery and development of penicillin was recognized. It would take another ten years until Theorell was again nominated for a prize in physiology or medicine.

Theorell's Own Involvement in the Nobel Prize Work

Because of his appointment as professor and head of the Department of Biochemistry at the Nobel Institute of Biochemistry as early as in 1937, at the age of 34, Theorell became extensively involved in the Nobel committee work at the Karolinska Institute and also briefly at the Royal Swedish Academy of Sciences. It should be noted that this is not apparent from information about members of the committee of physiology or medicine provided by *Les Prix Nobel*. These yearly volumes give correct information about members of the committees of physics and chemistry, but until 1963, only some of the adjunct members at the Karolinska Institute are listed. Reviewing the final protocols by the committee shows that Theorell became an adjunct member of the former committee at the Institute in 1937 and remained in this position in 1938, 1940, 1941, 1943, 1944, 1947, 1949, 1950, 1952, 1953 and 1954. His absence in 1939 and 1945 can be explained by the fact that he himself was nominated for a prize, but his absence in the other four years, 1942, 1946, 1948 and 1951 remains to be explained. He was an adjunct member at the Academy in 1950 and 1951 and this could be one reason why he abstained from participating in the committee work at the Institute in the latter year. It should be noted that in 1950 he was a member of both committees, as in the other rare example of Hammarsten in 1948 and 1952. In 1950 Theorell carried out four different reviews for the chemistry committee and one for the committee of physiology or medicine. It must have been a very busy summer, but Theorell already had long experience of reviewing scientists nominated for Nobel Prizes. His son Henning has described the summer

Nobel commitments in the following way after mentioning that his father had became a committee member already in 1937:

> Membership comes with the annual summer duty of peer-reviewing possible candidates for the Nobel Prize. All our summers were spent on Ljusterö, an island in the Stockholm archipelago, where my father had bought a traditional summer house called Tomtängen from his uncle Hugo. During the first summer month he always isolated himself in a room in a little annexed house, consumed with the task of investigating the Nobel Prize candidates. The task of reading and critiquing up to 400 scientific publications and books per candidate was matched by the effort needed to transport the many crates of scientific volumes from Stockholm to Ljusterö. Before the 1950s the lack of a regular ferry meant that the crates had to be transported first by steamship from Stockholm to the local steamship bridge at Ljusterö, and then the final 800 metres of the journey to our boat bridge was completed by a cow ferry.

Later on Ljusterö was connected to the mainland by a car ferry and very soon Theorell's summer house could be conveniently reached by car. His deep involvement in the Nobel work continued. In 1958 he became a full member of the committee for physiology and medicine and Ulf von Euler was its chairman.

The following year led to a formal shift in Theorell's responsibilities. As mentioned the Government took over all responsibilities for Theorell's laboratories, which now became integrated as a department at the Karolinska Institute. The Nobel Foundation still had responsibility for the costs of maintaining the buildings. Tiselius has summarized the history of the Nobel Institutes in a review article in 1968[16]. He wrote:

> Obviously the members of the prize-awarding institutions can not fulfill the task entrusted to them if they are not themselves in close contact with the international development in the five Nobel prize fields of human endeavour. Some of them should be actively engaged in such work. This requires a high overall level of activity in these fields. The statutes of the Nobel Foundation provide the possibility of organizing Nobel Institutes for this purpose, also by investigating results which have been proposed for an award. This somewhat unrealistic idea has to my knowledge never been realized, even though the Nobel institutes indirectly have been of great importance as offering a possibility of providing especially

deserving personalities a secure position and good facilities. It would, of course, be far beyond the resources of the Nobel Foundation to assume even a partial responsibility for the high level of research in, for example, physics, chemistry, or medicine in our country. As a matter of fact, the Nobel Institute of medicine is today largely financed by the State and the Nobel Institute for Physics and Chemistry was taken over completely by the State a few years ago.

Thus in 1959 Theorell became a regular professor at the institute and implicitly also a member of the College of Teachers, and thus, for the first time had a final say in the selection of Nobel Prize recipients. In 1961 he went on to become the vice-chairman of the committee with the vice-chancellor of the Institute Sten Friberg as the new chairman. In 1964 and 1965 the positions were changed with Theorell as the chairman. Hereafter Friberg was back as the chairman of the committee for three years whereas Theorell remained as an adjunct member. He stayed in this position until 1970 when he retired and the following year his successor Peter Reichard took over as adjunct member. Over the years there were many reviews signed by Theorell. In fact if they were bound together they would probably make a book of some 400–500 pages.

Maybe historians looking at Nobel Prize reviews from the first half of the twentieth century should pause and reflect on the last line of submitted reviews. In the case of Theorell the date is preceded by Ljusterö, presumably meaning either "the light island", or perhaps it has its etymological origin in a procedure of spearing fish from a boat in ground water at night, "ljustra." As we already learnt Hammarsten has signed all his evaluations with Blidö, the mild island. The address given by Tiselius is Gröder Äng, the meadow where plants grow well, to be further discussed in the next chapter. It is not difficult to imagine the Nobel Professor hiding in his hut whereas the rest of the family is enjoying the brief light summer in the archipelago or some other charming natural scene in spacious Sweden. The problems of transporting all the material they needed for their extensive Nobel prize reviews applied to all of them, since the pristine milieus chosen were located off the main roads or waterways.

A New Wave of Nominations for a Prize

Nomination for a prize in chemistry to Theorell was not renewed until 1952. There were two proposals, one by Sumner, Cornell University, Ithaca, NY, the

1946 laureate, and the other by Jean Roche, College de France, Paris. Sumner wrote that Theorell must be very well known to the committee and expressed surprise that he had not been recognized by a prize earlier. He mentioned the work on the yellow enzyme and more recently on the iron-containing horse radish peroxidase, which also had been found to be reversibly dissociated. The committee let two reviewers, Hammarsten, an adjunct member this year, and Tiselius, review Theorell's complete production. As always the evaluations were very comprehensive. Hammarsten finished the 18-page long review with the remarks "It may not be that easy to identify *individual* (my italics) discoveries worthy of a prize in Theorell's production, the collective value of which is very important. His work on the biologically active iron-containing compounds, however, represents a fully matured contribution and it is my view that these publications have such an importance that they deserve a prize." Tiselius finished his 13-page review with an attached 6-page list of references and support for Theorell's candidature: "I do not hesitate to express as my opinion that Hugo Theorell is well worthy of a Nobel Prize in chemistry for his *collective* (my italics) chemical production." In some sentences he did, however, express certain reservations which are repeated in the summary text formulated by the committee. It noted "The committee found that Theorell's collective production on the chemistry of enzymes may well justify recognition by a Nobel Prize, but it is of the view that the results obtained in all aspects have not reached full maturation and that certain other proposals therefore seem to be stronger." The prize in chemistry in 1952 recognized Martin and Synge as mentioned earlier in this chapter.

In 1953 there was a detailed nomination by the Uppsala biochemist Ragnar Nilsson, supported by a Stockholm biochemist Christer Barthel. There was also a proposal by the Nobel laureate Hans von Euler, who supported the nomination made by Nilsson and also mentioned two additional candidates. Finally Myrbäck also submitted a proposal including a second candidate, Hermann Staudinger, who actually received the prize this year "for his discoveries in the field of macromolecular chemistry." The potential significance of large molecules was gradually accepted by the scientific community. But still there remained discussions of whether large proteins were simply carrier structures and the particular enzyme activity associated with a separate smaller molecule, an opinion dogmatically formulated by, for example, Richard W. Willstätter, the 1915 recipient of half a Nobel Prize in chemistry. Theorell's findings were of importance in this discussion, since he systematically studied different kinds of enzymes which were dependent on co-factors. He demonstrated that for a number of enzymes both components were essential and that the activity was dependent on their concomitant presence. Following the paradigmatic breakthrough made possible by Sanger's demonstration of the unique and homogenous amino acid sequence of a large protein, insulin, the field finally matured (see next chapter).

In 1954 there were seven nominations for a prize in chemistry to Theorell. A new review was carried out this time by Myrbäck, who himself had nominated Theorell the preceding year. Myrbäck must have had a very busy summer since he wrote two additional long reviews. Fortunately his attack of polio had not interfered with his brilliant intellect. Again Theorell's impressive production was summarized over some 25 pages and Myrbäck highlighted in a distilled way towards the end under separate points the studies of the yellow enzyme, the studies of iron-containing enzymes and the examination of pyridine enzymes, in particular alcohol dehydrogenase. He recommended that Theorell should be declared worthy of a prize and this was also the opinion of the committee. However, there was an unbeatable number one candidate, Linus Pauling[3].

The Critical Year

Developments reached a climax in 1955, when eventually both Theorell and du Vigneaud were finally recognized by Nobel Prizes. There were eight nominations for Theorell for a prize in chemistry, five of which came from previous Nobel laureates. One of them was George de Hevesy, who was described in a full chapter in my previous book on Nobel Prizes[3]. He introduced the isotope technique in Theorell's laboratory which provided new avenues to study intermediary metabolism. Myrbäck did an additional review focusing mainly on new publications. The synthesis made by the committee and the difficulty it had in distinguishing Theorell and du Vigneaud has already been cited. Theorell came out as number one, but in the end the chemistry prize was to recognize du Vigneaud.

At the Karolinska Institute there were, as already mentioned, a large number of nominations for du Vigneaud, but there were also three proposals for a prize to Theorell, all submitted by previous Nobel prize recipients, Warburg and the 1953 laureates in physiology or medicine, Krebs and Lipmann. The latter in fact made two nominations. One was to combine du Vigneaud with Sanger and the other to recognize Theorell. His deep respect for Theorell can be illustrated from the following quotes from his nomination:

Hugo Theorell is uppermost in my mind for the prize. He is one of the people I felt a little ashamed of meeting after I myself received the award because I had such great regard for him and felt that his stature among biochemists equaled mine or was better.

And somewhat later:

I hope to have made clear the reason for my initial remark, namely that I felt uneasy receiving the Nobel award in the presence of a man not yet so honored, whom I had admired from far back and whose contribution to biochemistry played such an important part in my own development.

These quotes highlighted Theorell's impressive standing in the field of biochemistry. Hammarsten drew up another long (15 pages) and, as it were, final review. It had a brief summary:

Theorell's position as one of the world's foremost enzyme researchers is widely recognized and renowned. No other research area may be more centrally positioned within physiology or medicine than enzymology. Investigation into the nature and mode of action of enzymes is an inevitable precondition for gaining a clear insight into the normal and abnormal functions of cells.

It is therefore my view and with reference to the review presented above, that Theorell to a very high degree is worthy of being recognized by a prize.

We may pause here for a while and reflect on Hammarsten's possible strategic thinking. Already in 1936 he had pushed Theorell forward to be identified as one of the strongest candidates for a prize in physiology or medicine and the following year he introduced him as a possible candidate for a prize in chemistry. In 1939 and 1945 he accepted that the committee at the Karolinska Institute preferred to delay awarding a prize to Theorell. During the years 1952 to 1954 he was able to follow, in the first year in person, how Theorell's candidature for a prize in chemistry grew. During these years Theorell unperturbed continued his participation in the work in the committee at the Karolinska Institute together with Hammarsten. In 1954 the two of them pushed for du Vigneaud, who received the majority of the votes of the committee members, but lost in the voting by the College of Teachers. In 1955 Hammarsten quickly swapped preferences. Du Vigneaud's worthiness for a prize was confirmed in a one-page review, but Theorell was pushed strongly. At this time Hammarsten of course knew that if the committee at the Karolinska Institute, as in the previous year had continued to recommend du Vigneaud as the number one candidate he would have received the prize. This would not have been unreasonable considering the recent breakthrough in his work on the chemistry and physiology of two important hormones and the impressive number of nominations he received for a prize. Simultaneously Theorell would have secured a prize in chemistry. But this apparently was not to Hammarsten's liking. He wanted to secure the first Nobel Prize recipient ever from the Karolinska Institute. Still one wonders why he played his cards the way he did. He could easily have secured solid nominations for a prize in physiology or medicine to Theorell during 1952–54, the time during which his credentials improved for a prize in chemistry, but this did not come about. One can of course also speculate about the possibility that Theorell himself had expressed a preference for which Nobel Prize he would like to receive. Personally I consider that relatively unlikely. The scientist who finally ended up in an uncomfortable situation was du Vigneaud, although in the end all came out well.

In the perspective of these intense involvements by Hammarsten with the two candidates a partial explanation might be given for his major omission in recognizing the greatness of another candidate, whose last chance it was to be awarded a prize these years. It was Oswald T. Avery who regrettably was never to be recognized for the pioneering discovery that DNA is the carrier of genetic information. This omission remains one of the great, if not the greatest lapse in the annals of Nobel Prizes in the natural sciences. I have previously written[1] that Theorell might be exempted as having no responsibility for this mistake, but this may not be true since he was in fact a member of the committee in the critical years 1950 and 1952–54.

The Karolinska Institute described Theorell's achievements as "discoveries concerning the nature and mode of action of oxidation enzymes." Nobel Prizes are not awarded for life contributions, but only for single well identifiable discoveries, and in chemistry also potentially for improvements of very essential techniques. These rules defined by Nobel's will apparently caused some problems in the cases of both Theorell and du Vigneaud. They were impressive in their life contributions in science and their performance in the laboratory in adapting and developing new techniques were laudable. But were they visionary scientists? The Nobel lectures are quite revealing. The titles were "A Trail of Sulfur Research: From Insulin to Oxytocin[17]" and "The Nature and Mode of Action of Oxidation Enzymes[13]."

The presentations in essence give a broad overview of their lifetime contributions in science. Theorell gave his lecture in Swedish and it included a number of practical demonstrations. In the introduction he reiterated the dominant view of the early time of his evolving science epitomized by Willstätter's statement, mentioned earlier, that "enzymes could not contain neither protein, carbohydrate nor iron and that they did not belong to any known class of chemical substances at all" as formulated in the lecture. To digress, it can be mentioned that reference to a minute undetectable amount of a proposed contaminant has a number of occasions held back important advances in science.

When Avery and his collaborators published in 1944 that genetic properties of pneumococci could be transferred by purified DNA, the scientific community was skeptical. Alfred Mirsky and many others argued that there must be some small but undetectable protein contamination. Conversely Prusiner tried to convince his fellow scientist that his prion preparations only contained protein, but no nucleic acid[1]. It took him a long time to get his voice heard[18] but eventually it became apparent that a new field of pathological protein folding had been unraveled, a field of great importance for various diseases, not least in the brain.

To return to the nature of enzymes a number of discoveries were needed to counter Willstätter's view and prove that enzymes are predominantly proteins. One important advance was the crystallization of urea by Sumner and similar work by Northrop with other protein enzymes, recognized by the 1946 Nobel prize in chemistry. Among the distinct discoveries in Theorell's work we can select his early observation of the crucial role of the combination of the protein component — the apoenzyme — and the cofactor — the coenzyme. This was discovered in the early studies of the yellow enzyme and later investigations of other enzymes. Crystallization of the coenzyme, like myoglobin, was crucial in these endeavors. He touched the right chord already early in his science and pursued this with vigor in high quality research. He was also tenacious, staying with the selected problem for a long time. For example it can be mentioned that it was in 1955, the year of his prize that Theorell finally managed to crystallize the yellow enzyme. His school building was exemplary and his laboratory was a Mecca for visiting scientists from many countries. The environment of du Vigneaud's laboratory was also extraordinary and to be a member of the V d V Club was a stamp of quality. His early work on insulin helped to turn the spotlight on proteins as carriers of fundamental biological activities. One can say that du Vigneaud's career in term of discoveries peaked in the early 1950s, very timely for his prize. It was critical when he was able to synthesize both hormones, vasopressin and oxytocin, and show that they carried the anticipated activities. From then on hormone studies became biochemical and the relevant physiological studies acquired a new precision.

David Keilin and the Possible Disadvantage of Competing with a Swedish Candidate

Except for the first Nobel Prize in physiology or medicine to a Swedish scientist, the one to Allvar Gullstrand at Uppsala University in 1911, which was just as much physics as medicine, there have been six Swedish prize recipients in this discipline until the present time. Five of them received the prize while they were professors at the Karolinska Institute — Theorell (1955), Granit (1967), Ulf von Euler (1970), Bergström and Bengt Samuelsson (1982). It has been argued that being a member of the Faculty of Medicine (before 1965 the College of Teachers) could possibly result in a more favorable treatment than that of non-Swedish candidates[15]. As an example it has been proposed that David Keilin might have suffered from being in the shadow of Theorell. Let us therefore look at Keilin's background and the way he was analyzed as a candidate for a prize in chemistry and in physiology or medicine.

Keilin was born in Moscow in 1887, but he grew up in Warsaw. The tradition of the time was for talented students to seek their higher education abroad and Keilin selected the University of Liège, where he began his studies in 1904.

David Keilin (1887–1963).

He was encouraged to study medicine, but because of poor health he decided not to do this. A year later he moved to Paris for continued studies. Once when he had left a lecture on philosophy by Henri Bergson he was caught in a heavy rain and took refuge in a shelter in a building near the Panthéon. This building housed the Laboratoire d'evolution êtres organisés, in which a lecture was being given by the well-known biologist Maurice Caullery. Keilin was attracted by the lecture and ended up as a student of Caullery. His first scientific work concerned the parasitic and free-living larval stages of Diptera (two-winged insects). He presented his thesis on this subject and was recruited to Cambridge in the U.K. as an assistant to George H. F. Nuttall, a well-known parasitologist interested in insect carriers of diseases. Keilin's curiosity about insect respiration led him to his major discovery. Using a microspectroscope he examined tissues from a horse botfly at different stages of development. In certain tissues he did not find bands indicating the presence of hemoglobin but instead he found four other bands. Their presence turned out to indicate a new compound, not related to oxidized or unoxidized hemoglobin. He concluded that the four bands represented three previously unidentified forms of haemochromogen-like compounds, which he named *cytochromes a, b* and *c*.

At the time there were two schools of thoughts on energy processing in cells, one associated with Heinrich O. Wieland, whom we will meet in Chapter 6, and the Swede Thorsten Thunberg at the University of Lund, who argued that there was an enzyme that removed hydrogen from substrates allowing it to react with oxygen; and the other by Otto Warburg, who argued that it was his *Atmungsferment* — the yellow enzyme — that catalyzed the reaction with oxygen. Without going into detail it can be summarized that Keilin's finding revealed that both schools were right. Cytochromes represented a link between a group of hydrogenases and a single oxidating enzyme. It now became possible to characterize what came to be called the *respiratory chain*. This was a major discovery. However, it needs to be added that the four bands seen by Keilin, unknown to him, already had been observed by a Scottish medical practitioner, Charles MacMunn in 1884 and he had in fact proposed that the pigment he observed might have some respiratory function. However, this was way ahead of its time and had been completely forgotten. The information had not been converted into sustainable knowledge. Keilin's (re)discovery was necessary to put it into its proper context and make it a part of established knowledge. In the perspective of its impact it was a true discovery. The description of energy transfer in cells in textbooks of today, highlights that cytochromes play a central role and they are referred to as having been discovered in 1925.

All functions of life are dependent on access to energy. One important source of this is the process of oxidation — a release of one or more electrons from certain atoms — and the coupled reduction — an uptake of electrons by other atoms. Energy is transferred and released by what in a combination is called redox reactions. Many different chemical entities can be involved in these reactions, not only oxygen. Metal ions, like iron as mentioned, often play a critical role. Expressed very simply there are two sources of energy.

One source is the carbon dioxide in the air which can be converted into sugars by use of solar energy in plants or photosynthetic bacteria. The other is energy-rich systems of inorganic chemicals, to be further discussed in the following chapter. In the former case the sugars and other organic molecules can then be oxidized by the process of respiration in living organisms with recovery of energy. In more complex organisms there are means for transport of energy by blood or lymph or by diffusion. It is easy to understand that hemoglobin in our body provides a means of transporting oxygen by blood and that myoglobin in our muscles is another means of storing readily accessible oxygen as a source of energy. The processing of energy in cells in the millions of molecular events that occur every second, driving all the different chemical reactions, the metabolism, is central to all sunshine-dependent life. Very complex chains of reactions and means of storage of energy in cells have been discovered, but the origin of this development was the identification of the family of cytochromes. They represent the central actors and prototypes of a large number of energy-transmitting compounds. One example of the central role of cytochrome c that has recently attracted considerable attention is the mechanism of apoptosis. Apoptosis — "the falling of autumn leaves" — is the critical mechanism of controlled cell death. It is the release of cytochrome c from mitochondria, structures introduced above, that elicits the proteolytic destruction of a cell. All the colored proteins mentioned above have a so-called porphyrin ring which allows the binding of iron (p. 223), as in hemoglobin and other energy-transferring enzyme systems, or magnesium as in chloroplasts.

Keilin was repeatedly nominated for a prize in physiology or medicine starting in 1932 and extending over almost three decades. In total there were more than 20 nominations. In parallel but somewhat delayed there were also nominations for a prize in chemistry starting in 1949 and extending into 1960, a total of five proposals, some in combinations with other candidates. The first nomination for a prize in physiology or medicine led to Hammarsten carrying out a preliminary evaluation concluding that no comprehensive review was needed. In fact most of the report concerned the abovementioned researcher MacMunn. He concluded "Regarding the insertion into the scheme of oxidation of MacMunn's colored substances (Keilin's cytochromes) I refer to my 1931 evaluation of O. Warburg. Thus after reviewing Keilin's contributions I have not found anything that motivates an in depth review." Five years later there was one nomination signed by four Swedish scientists and one Danish colleague, the Nobel laureate August Krogh, and another from P.E. Simola, Finland. Both nominations proposed that the prize should be split between Keilin and

Thunberg. Full evaluations were made both by Hammarsten and also by Theorell. The former retained a very critical attitude to Keilin's work, but wrote that his best work concerned another class of respiratory enzymes, the catalases. This is a special group of enzymes that serve to protect cells against potential damage by oxygen — the paradoxically potentially highly toxic substance. However, his conclusion was that this work was incomplete. Regarding cytochromes he again referred extensively to MacMunn's early observations. He raised the question of whether Keilin's contribution was a rediscovery or simply an extension of MacMunn's work. He leant towards the latter view. Theorell concluded his 11-page review in the following more respectful way:

> Keilin, without doubt, has a very prominent position among researchers studying intracellular combustion. His work on cytochromes has attracted such attention that his name is sometimes mentioned alongside Otto Warburg's; the different iron porphyrate oxidases are sometimes collectively referred to as the Warburg-Keilin system. However, it seems to me that MacMunn has a considerable share in the priority (of the discovery) in this field of research and hence one cannot allow a prize to be awarded to Keilin only for his contributions.

Theorell then referred to the fact that Keilin was a scientist in his most productive phase and recommended delaying a decision on his qualifications for a prize. The committee agreed.

In 1938 Keilin again received two nominations. One was from André Lwoff, whom we will get to know in Chapter 7. He knew the work in Keilin's laboratory well since he had been a visiting scientist there. The second nomination was a repeat of the previous Swedish-Danish nomination and included Thunberg. Theorell made some short complimentary remarks about Keilin's recent work on catalases and recommended that he should be under continued scrutiny, since the quality of his work had improved. In 1940 Keilin was nominated again and Theorell now made a more thorough evaluation. He summarized by stating that Keilin's work included both positive and negative contributions. It should be noted that Keilin and Theorell at this time were to some degree examining the same systems of energy transfer enzymes. Keilin's studies of catalases were concluded to be inferior to those made by Chance in Theorell's laboratory and his work on cytochrome was described as a "negative" contribution. Theorell had obtained different results in his research. Theorell retained his attitude that Keilin was not at the time worthy of a prize. The committee agreed. Regarding Thunberg, who had been reviewed separately, it was said that he was not worthy of a prize. In the abovementioned 1945 nomination of Theorell by Vanotti it was said that he might be combined with Keilin. The committee considered this a nomination of the latter and asked Tiselius to carry out a review. His concluding paragraph is interesting. It read:

> There is no doubt that his (Keilin's) name is inseparably associated with (the discovery of) the cytochromes. For my part I would say that I have a higher appreciation of Keilin's work than what has been expressed in the earlier evaluations by Hammarsten and Theorell. Not even Warburg was encouraged by MacMunn's work to include "the histohematins" in his work in spite of the fact that all his research at the time focused on the role of iron in cellular respiration. Keilin's contribution is, in my opinion at least, in terms of its consequences, equivalent to a rediscovery. However, the fact that his late contributions in this field have not always been incontestable and that he from a chemical perspective has to a great extent not succeeded in enlarging his results — in this case Theorell has made more important contributions — argues against his meriting a prize.

Tiselius's recommendation was not to consider Keilin worthy of a prize at the time and this was also the conclusion of the committee. As mentioned above, it likewise it did not consider Theorell worthy of a prize at this time.

After this it seems that the steam had gone out of Keilin's potential candidature. He was nominated in 1950, 1953–57 and 1959, but no more evaluations were done and he is not mentioned in the protocols by the committee.

Keilin was proposed for a prize in chemistry for the first time in 1949. It was Hans von Euler who suggested that Keilin should share a prize with Gabriel Bertrand, a respected scientist from the Pasteur Institute in Paris. Bertrand had been reviewed before and the conclusion had been that he was a potential candidate for a prize, but he did not rank among the frontline scientists. Tiselius carried out a lukewarm single-page review, noting that the two candidates did not fit together and that Keilin was a highly regarded enzymologist, and concluded that neither of the candidates needed to be further discussed that year. There was a new nomination for Keilin for a prize in chemistry the following year. This was the year that Theorell was on the chemistry committee and it was allocated to him to review Keilin. He knew Keilin well and had, as already mentioned, reviewed him a number of times for the committee at the Karolinska institute. He cited frequently from the earlier reviews at the

Institute. It was noted that if Keilin should receive a prize it should be for his early work on identification of different cytochromes. Theorell reviewed Keilin's more recent work and concluded that his contributions during the last five years did not justify a reward. Keilin should not at the time be considered for a prize. The committee agreed with this recommendation.

In 1954 there was an interesting proposal for a prize in chemistry to Keilin by two Swedish nominators, Henrik G. Lundegård, a professor of plant physiology at the City College of Agriculture and John Runnström, a very influential marine biologist and a member of the Royal Swedish Academy of Sciences. Runnström was professor at the Stockholm University (at the time referred to as the City College of Stockholm — *Stockholms Högskola*). His research pioneered the use of eggs of sea urchins for studies of embryological differentiation. In 1937 the Swedish industrialist Axel Wenner-Gren, who had accumulated considerable wealth through his company selling vacuum cleaners (Electrolux), made a major donation to the university. This donation allowed the establishment of the Wenner-Gren Institute for Experimental Biology, which initially was headed by Runnström. Wenner-Gren also made another donation at the same time for the establishment of the Wenner-Gren Foundation which has served as a major source of scholarships and resources for conferences into the present time. In 1955 he made yet another donation allowing the establishment of the Wenner-Gren Center, providing space in a tall building for research councils and other research-oriented organizations and in a semicircular building, apartments for temporary use by visiting scientists. We will return to the fate of the different donations by Wenner-Gren at the end of this chapter. Major economical mismanagement led to that essentially all the financial resources drying up and Stockholm University had to take over the running of the Institute for Experimental Biology.

Lundegård's and Runnström's proposal was rather detailed and highlighted the importance of Keilin's original discovery of a group of hematin compounds, which he named cytochromes. The proposal was reviewed by Myrbäck in a very comprehensive analysis covering 23 pages. He praised their achievements, but there was a dilemma in that Keilin by training was an entomologist, using chemical methods to study insects. However, it was his unique insight into their morphology and color and appearance in different morphological stages of differentiation that allowed him to make his epoch-making contributions. He was not a true chemist, but of course the same could be said about Theorell. In both cases the two researchers had had to learn chemistry from the outside, one as an entomologist and the other as an M.D. Myrbäck's conclusion was that Keilin's original discovery of the cytochromes could be considered for a prize but he recommended the committee to wait. The committee agreed with this. The following year Lundegård, alone, repeated his nomination, but now he proposed a combination of Keilin and Theorell, for which he presented a number of arguments. In this critical year, when Theorell finally received his prize in physiology or medicine, it was understandable if the committee of chemistry (as mentioned there were a number of proposals of Keilin for a prize in physiology or medicine during the 1950s, but no additional reviews were made) did not want to go back to the drawing board. Myrbäck made one additional thorough evaluation amplifying his earlier perspectives on Keilin and a separate review discussing the possibility to combine Keilin and Theorell. He concluded that Keilin's work was more physiologically than chemically oriented and he did not support the suggestion of combining the two scientists for a prize in chemistry. And this was the end of the story. Keilin was never declared worthy of a Nobel Prize. The final nomination of Keilin for a prize in chemistry was given in 1960, but at this time no initiatives were taken. In conclusion it could be added that Keilin became a Fellow of the Royal Society in 1926, received its Royal Medal in 1939 and its highest award, the Copley Medal, in 1951. This could have been a lesson to the committees, perhaps in particular the one for physiology or medicine.

Two Chemists Sharing the Limelight

Du Vigneaud receiving his Nobel Prize from the hands of His Majesty the King. [From Ref. 21.]

Theorell and du Vigneaud enjoyed each other's company and the attention given to them at the 1955 Nobel Prize ceremony. Du Vigneaud was introduced by Fredga[19]. He cited initially du Vigneaud's definition of true exploratory research as "the working out of a winding trail into the unknown." He then described how du Vigneaud had tenaciously and successfully followed the trail of sulfur in metabolism, but he appropriately put an emphasis on the successful synthesis of the two 8-amino-acid cyclic hormones. Hammarsten had the great pleasure of introducing[20] his favorite disciple Theorell, for whom he had worked so hard and with great success in the end. This was not the first time he carried this responsibility. As already mentioned he had previously introduced Warburg in 1931, Albert Szent-Gyorgyi in 1937 and Krebs and Lipmann in 1953. There were obvious interconnections since it was Warburg, Krebs and Lipmann who nominated Theorell for the 1955 prize. Hammarsten started with a reference to another student at the high school in Linköping, who, however, was 124 years older than Theorell. It was Berzelius, who as an antonym to *analytic* — to bring together, to integrate — introduced the world *catalytic* — to fragment, to disintegrate — which since then has been used to describe

Hammarsten addressing his favorite disciple Theorell at the Nobel Prize ceremony. [From Ref. 21.]

enzymatic activities. At the end Hammarsten could not refrain from becoming rather hyperbolical. He said (translated into English) "A fertile imagination.

Theorell in conversation with the King at the Nobel banquet. [From Ref. 21.]

An undeviating and critical accuracy. An astonishing technical skill. All scientists possess some of these attributes. Very few have all. You are one of these few. In accordance with your gifts you have chosen the most important of all tasks in biology. The purification and characterization of enzymes ...."

At the time of the 1955 prize ceremony the King and representatives of the Royal Family sat in the first row of the parquet as mentioned in Chapter 3. The prize recipients were expected to walk six steps down from the elevated podium where they were seated. It was not easy for Theorell with his handicap to manage the six stairs. The King, a highly considerate person as we already learnt, therefore took the initiative to walk up the stairs to meet Theorell on stage. They met half-way and when Theorell said "I think I can manage this" the King backed off. Theorell was a man of self-determination and self-conquest. The arrangement of meeting the prize recipients at the parquet remained until 1972. When the King, Gustaf VI Adolf had died in September 1973, his successor, the young Carl XVI Gustaf, was positioned at the stage, where since then the insignia of the prizes are handed over. 1973 also happened to be the year when for the first time I sat on the podium, as a new professor of virology and first year adjunct member of the Nobel Committee at the Karolinska Institute.

A bust of Alfred Nobel in bronze by Erik Lindberg. [Courtesy of the Karolinska Institute.]

In his banquet speech Theorell could not refrain from thanking Nobel not only for the banquet dinner but also for family meals for 19 years before the prize event. Outside his Medical Nobel Institute there was a bust in bronze by Erik Lindberg showing Alfred Nobel, which is now located inside Nobel Forum, the home of most Nobel activities at the Institute at the present time. It was to him that he directed his gratitude for a rich life, including the joy and fulfillment he derived from his science. He of course also very appropriately paid tribute to his teacher and friend Hammarsten. Du Vigneaud's after dinner speech was more conventional. He alluded to the main targets of his science and appropriately praised the contributions by his collaborators in the impressively large volume of scientific work they had jointly produced.

An Additional Note on Enzymes

Theorell's elegant work focused on a special aspect of enzymes and their mode of operation. It may be appropriate to give a wider perspective including more recent knowledge of their structure and function, to thereby put his work into context. There are a number of examples in biology where two molecules, often macromolecules, have a specific affinity to react with each other. Enzymes are one example, but also transmitter substances with which we became acquainted in the early chapters, hormones and also antibodies depend on an interaction with specific targets. All these categories react with specific chemical structures in their target molecules. Their effects vary depending upon which category they represent. Transmitter substances elicit electrical signals; hormones react with specific receptors on certain cells and trigger signals reorganizing their intermediary metabolism; antibodies can have many kinds of targets, they may neutralize the infectious property of an invading virus to give just one example. Only enzymes modify the structure of the target molecule, generally by breaking it into pieces. They operate by overcoming a specific energy barrier that keeps the different parts of the molecule together. Attempts have been made to endow antibodies with enzymatic activity, so called *abzymes*. Important work in this field has been performed by Richard Lerner and collaborators at The Scripps Research Institute in La Jolla, CA. Some of these genetically engineered products have turned out to be of practical use.

Enzymes often show a very high specific activity. In early experiments removal of most of the irrelevant material, to a large extent proteins, when successfully done left behind a considerable activity. This is the reason that it was believed for a long time that the enzyme was some small magic molecule associated with carrier substances, like protein. Eventually it became clear in the 1950s, not least because of the conceptual impact of Sanger's work that it was the protein proper that carried the biological activity. Thus enzymes have been found generally to be protein structures of varying size and conformation. Their polypeptide chain may contain all the way from 62 amino acids to in excess of 2,000 residues. In 1972 Anfinsen, one of the many early visiting scientists in Theorell's laboratory received half the Nobel Prize in chemistry for his work on protein folding. The other half recognized the abovementioned Moore and Stein.

It came as a considerable surprise when it was discovered that structures of RNA could also carry a catalytic activity. This groundbreaking finding brought Sidney Altman and Thomas R. Cech the Nobel Prize in chemistry in 1989 "for their discovery of the catalytic properties of RNA." Their observations opened completely new doors to the understanding of how life may have begun on Earth, briefly alluded to in my previous books[1,3] and discussed further in Chapter 7. In our cells catalytic RNAs have a central role in the function of our protein-synthesizing machinery, the ribosomes. To return to the protein enzymes, there are some of them, in particular those involved in energy transfer, which need the assistance of co-factors. It was Theorell's life's contribution to markedly deepen our insights into different enzymes depending on such co-factors. Let me end this section by mentioning that the term enzyme comes from a Greek word meaning "leavened" and it was in his studies of beer fermentation that Eduard Buchner discovered that the process was not dependent on intact cells, but that also the cell sap was active. This rendered him a Nobel Prize in chemistry in 1907.

A Nobel Jubilee and Theorell's Final Years

In 1975 the Nobel Foundation celebrated at the appropriate time the 75-years jubilee of the first Nobel prizes. Previous laureates who were still alive were invited to join in the festivities. On this occasion Theorell and Zotterman arranged a private dinner at the top 24/25 floors of the Wenner-Gren Center, the landmark building in Stockholm which has already been mentioned. At this party the hostess of the evening, Theorell's wife Margit, was seated between Albert Szent-Györgyi, laureate in physiology or medicine in 1937, and Pauling. This particular seating arrangement prompted Theorell in his welcome speech to remark that on her left side she had the scientist who had discovered Vitamin C and on her right the scientist who was the biggest consumer of this vitamin! It should be added that one of Pauling's major idiosyncrasies during the later part of his life was that Vitamin C in large doses could prevent infections. Several critical field trials have been conducted to see if there was some truth in his recommendation. None has been found. Vitamins as a possible food addition have a very good ring in the ears of most people. However, the major part of the extra vitamins consumed, often as a result of aggressive marketing, do not improve health. It may even be that at very high concentrations some toxic effects of vitamins may appear.

Theorell gardening. [Photo from Henning Theorell.]

Theorell's later years were a mixture of sunshine, trimming the roses at his beloved summer house and garden and enjoying a rich family life with growing grandchildren, but there were also some clouds on the horizon because of a court case relating to his alleged responsibility in inappropriate dealings in the management of the various abovementioned donations made by Wenner-Gren. Half way through the court dealings in 1975 he was exempted from the case for health reasons, but the other person involved in the case Birger Strid was found guilty and jailed for embezzling a sizeable sum of money from the foundations. In spite of this, the abovementioned donation for establishing the Wenner-Gren Foundations has survived into the present time by the addition of a considerable supplementary financial support derived from the sale of the tall building of the original Wenner-Gren Center. It continues to represent an important source of stipends for the encouragement of international exchanges of scientists and also to provide money to arrange scientific conferences. The original function of the semi-circular building has also been retained, serving as a popular place for visiting scientists to stay in. However, the Wenner-Gren Institute briefly referred to above has only survived in name as a part of a department at Stockholm University. Sadly these unfavorable events tainted Theorell's last years in life. He died in 1982, 79 years old.

In the next chapter we will meet Tiselius, whom we have already encountered repeatedly as a reviewer, primarily in chemistry but also in physiology or medicine. He grew to become a very critical arbiter in the course of Nobel prizes in these fields, but for unknown reasons he left the playing field open in 1955. He was a critical reviewer of Frederick Sanger, the unique recipient of two prizes in chemistry, who will also be presented in the following chapter.

Chapter 5

A Tale of Two Gentle Scientists Who Changed the World

PROTEIN AND NUCLEIC ACIDS
TOOLS AND INFORMATION STORAGE
CENTRAL ACTORS OF LIFE

In the discussion of Nobel prizes in chemistry and also occasionally in physiology or medicine during the 1940s to 1960s there was one particular committee member and evaluator who frequently took part. This is Arne Tiselius and he has been cited repeatedly in my two previous books on Nobel Prizes[1,2] and both of them also include pictures of him. To recapitulate, Tiselius became a member of the committee of chemistry in 1946 and was deeply involved in its work until his death in 1971. He was also a member of the Board of the Nobel Foundation, first as its vice-chairman since 1947 and then its chairman in 1960–64. He took a very active part in the expansion of science in Sweden after the Second World War, not least in the emerging field of molecular biology. He was chairman of the Swedish Natural Science Research Council, newly established by Royal charter in 1946. He stayed in this position for five years and then chaired the newly-founded Swedish Cancer Society for a further five years.

He was very knowledgeable in both physics, chemistry and medicine and he combined his broad professional insights with a very attractive personality. He has been referred to as modest, quiet and warm-hearted. These character traits were combined with a good, refined sense of humor. Furthermore he wanted to take a broader approach to problems and looked beyond the impact of scientific discoveries, considering their social, ethical and other implications. He was not a person aiming at making a career and searching for power but he was chosen to become a leader because of his capacity to find the middle way. He was an excellent arbiter and this shines through also in his advice to the Nobel committees. This has been repeatedly exemplified, not least by the way he orchestrated the 1962 Nobel Prizes in chemistry and in physiology or medicine[2]. He steered the deliberations of the committees gently but unobtrusively pointing with his whole hand. He also played a major role in the developments that led to a Nobel prize in chemistry for Frederick Sanger in 1958. This remarkable scientist who eventually received a second prize in chemistry, will be presented in full in the second half of this chapter. But first let us get to know who one of his most qualified reviewers was.

The Development of a Rich Life in Science

Arne Tiselius was born in Stockholm in 1902 into a family with a scholarly ancestry[3]. His father was a mathematician trained at Uppsala University, who was employed in an insurance company. Sadly, he died when Arne was only four years old. Arne's grandfather had also obtained a degree in mathematics from Uppsala, after which he returned to Gothenburg the city of his upbringing. He became principal of a highly regarded grammar school in that city. His surname was Johansson, but Arne's father had taken a name from his mother's family. She was born Tisell and came from a family of learned clergymen, who alternately used the name Tisell or Tiselius. Arne's own mother originated from Norway and she also came from a clergy family. He had a challenging youth not only because of the early death of his father but also because the family lost Arne's younger brother due to an infected appendix. The financial resources of the family were meager and his mother and the two remaining children, Arne and his sister, moved to Gothenburg, where the Tiselius's relatives lived. The mother attempted to improve their finances by giving piano lessons and by sewing Hardanger-seams, a special form of embroidery. Also Arne contributed to the family's income by giving private lessons to high school students. It would seem that the precarious circumstances of the family may have encouraged his early maturation. It is not uncommon in the history of successful scientists that they are the oldest among a group of children in a family that has lost its father. We already met Hodgkin in Chapter 3, who had similar challenges in his upbringing.

As Arne came of age he went to the grammar school where his grandfather had worked. At this school he was stimulated by his chemistry and biology teacher, Dr Ludwig Johansson, who had a Ph.D. in zoology from Uppsala University. In order to encourage Arne's spontaneous interest in science and in particular chemistry, this teacher provided him with a private key to the school laboratory. This allowed him to start doing his own experiments already as a student in the high school. When he had finished school and obtained his student exam he moved to Uppsala University in 1921. He chose to study chemistry, physics and mathematics. On the side he was assistant at the meteorological institute at the university, to make some extra money to live on. By good luck he became acquainted with The Svedberg, and was employed as his research assistant in 1925. A year later, Svedberg received his premature Nobel Prize, caused by a coup at the Academy[2]. This accelerated the development of his laboratory and a very creative intellectual atmosphere was established. A unique school for the emergence of new separation techniques was developed at the University. For many years it was

The Svedberg (1884–1971), recipient of the 1926 Nobel Prize in chemistry. [From Ref. 24.]

only in Uppsala that the size of macromolecules could be studied by ultra-centrifugation. In the mid-1930s the molecular weight and shapes of some 30 large proteins had been determined. The findings led to a theory that all proteins were aggregates of subunits with a molecular weight of 17,500, half a Svedberg unit. This theory, which could be interpreted to indicate that proteins only played a subsidiary role, like that argued by Willstätter, as mentioned in the previous chapter, fortunately became rather short-lived.

Svedberg had already at an early stage become a good friend of Warren Weaver, the director of the Division of Natural Sciences at the Rockefeller Foundation from 1932 to 1955. He provided important support of the separation work in Uppsala and also many other projects in Sweden, not least at the Karolinska Institute as described in the previous chapter. Weaver was very far-sighted and coined the term "molecular biology," since he believed that life should be explained by studies of molecules. He liked to quote from a textbook of the time, stating: "All that association of phenomena which we term "life" is manifested only by matter made up to a very large extent of proteins, and is never exhibited in the absence of these substances." This statement was increasingly proven to be correct by the advances in protein studies, as we shall see. Towards the end of his career Weaver also came to appreciate nucleic acids.

Tiselius's first publication together with Svedberg in the year of the prize to the latter described a new technique for electrophoretic separation of macromolecules. The moving frontiers of different proteins were identified by an ultraviolet adsorption technique exploiting so called "schlieren optics." In 1930 Tiselius presented his Ph.D. thesis entitled "The moving boundary method of studying the electrophoresis of proteins." However, Tiselius was somewhat restless and wanted to demonstrate potentially important applications of the improved technique he had developed. The idea of exploiting the fact that various kinds of macromolecules migrate differently in an electrical field was not new. Already in 1908 Landsteiner had demonstrated that the different proteins in blood serum had different migration rates and as mentioned in the previous chapter Theorell developed his own technique to separate the proteins he was studying based on their charges. However, it was Tiselius who took the first steps to make what came to be called "electrophoresis" a practically useful technique. His development as an individual and as a scientist has been summarized[3].

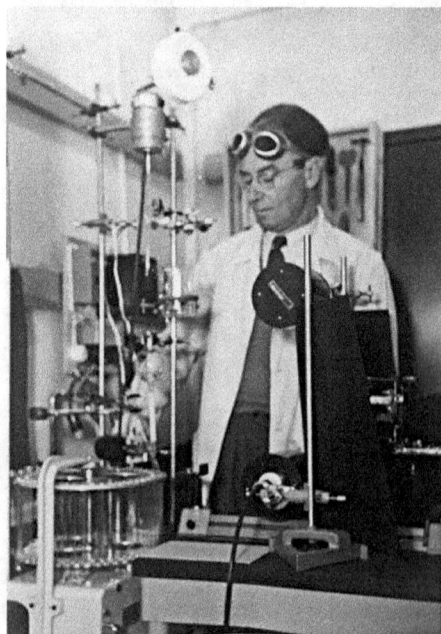

Arne Tiselius (1902–1971) as a young scientist in his laboratory.

In 1930 Tiselius found his life partner, Greta Dalén and they married the same year. The young family that soon developed brought new responsibilities, but the outlook for an academic position at the University was very gloomy. In addition, times of depression developed in Sweden after the collapse of the Kreuger companies in 1932. The only prospect was a chair in inorganic chemistry to be declared open for competition in 1936. There was a major limitation to the Swedish academic career system at the time and for many decades to come. Only a very restricted number of professorships, implicitly associated with the responsibility of chairman, were available and almost all of them were connected to particular disciplines. Professorships thus became available only when the person who held the position retired or died. Personal chairs were extremely rare. Thus Tiselius felt that he had to widen his knowledge to also include inorganic chemistry. From his extensive reading he had learnt that certain zeolite minerals had unique qualities. They could retain their crystal structure even if the trapped water was exchanged for some other fluid or completely eliminated. He developed techniques to measure accurately the diffusion of water vapor as well as other gases into the crystals. The progress of this research led to the award of a Rockefeller Fellowship and he was able to work at Princeton University at the Frick Chemical Laboratory headed by Hugh S. Taylor. His stay there turned out to be critical for his future involvement in science. At heart he still wanted to do biochemistry. During his two-year stay he established a number of important contacts, not least at laboratories managed by the Rockefeller Foundation both in Princeton and in New York. Among existing, future and near Nobel Prize scientists he got to know can be mentioned Northrop, Stanley, M.L. Anson, Landsteiner, M. Heidelberger and L. Michaelis. They knew about his important work on electrophoresis and they encouraged him to focus on experimental biology.

The appointment committee for the chair in inorganic chemistry at Uppsala University decided that Gunnar Hägg was the most deserving candidate. This was unquestionably the correct choice. He took teaching and research in this sub-discipline of chemistry to a high level. We have already met him repeatedly in surveying candidates considered by the Nobel committee in chemistry[2]. However, the appointment committee remarked that it was impressed by the progress the young Tiselius had made in his work. He might well be considered worthy of a professorship, but how would that come about? In some way Tiselius managed to return to his work on developing the electrophoresis technique and this led to the publication of some very important papers on the separation of serum proteins. A completely new phase in characterization of protein macromolecules was launched. Something needed to be done to ensure that Tiselius could stay in Uppsala. Paradoxically it was not a donor from that city, but Major Herbert Jacobsson and his wife Karin, born Broström, from a wealthy Gothenburg shipping family that provided the means for a personal professorship in biochemistry at Uppsala University. The specific focus was to be on studies of processes of importance to life by use of physical and chemical means. Laboratory space was made available to Tiselius in Svedberg's laboratory. Important support was obtained over the coming years from the Rockefeller Foundation and the Knut and Alice Wallenberg foundation.

Tiselius now had a platform to expand his creative involvements, but he lacked space. He developed fruitful contacts with the Rockefeller Institute. One of the influential scientists at the Institute, Frank Horsfall, Jr. came as a visiting scientist to Tiselius's laboratory. His name rings a bell with me, since he together with Igor Tamm, not the Nobel Prize laureate in physics, but a qualified virologist, originally from Estonia, edited the book *Viral and Rickettsial Infections of Man*, which was a bible to those of us who learnt virology in the early 1960s. When Horsfall returned to the Rockefeller Institute in New York, he brought back a Tiselius apparatus, one of the first to become established in the U.S. In 1939 Tiselius was invited to come to the Rockefeller Institute Hospital. Regrettably, his stay was limited to only two months because of the outbreak of the Second World War. After the war contacts with the Rockefeller Institute were resumed and there was an exchange of visiting scientists. Important contacts were Moore and Stein, already introduced in the previous chapter, and also the immunobiologist Henry Kunkel, who spent the year 1950/51 in Uppsala. We met Kunkel in my previous book[2].

The electrophoresis technique was further refined and eventually it was

Tiselius viewing a diagram of separated serum proteins.

also possible to use it for clinical applications. The picture illustrates the separation of human serum into the fractions albumin, alpha-1, alpha-2, beta- and gamma-globulin. The latter fraction containing immunoglobulins has been discussed at length earlier[2]. One may ask why Tiselius and his collaborators did not to a larger extent become involved in using the technique for such important applications. But as Tiselius himself has said in a review[4]. "I felt that with our particular background and our experience it would be better to concentrate upon a further improvement of the method in different directions and on applications to problems which were closer to our field of interest." It was the establishment of a technique using a filter paper matrix that paved the way for broader applications. But there were also many other modifications as the technique matured for a wider, also preparatory, use. Its development in a historical context has been summarized[5]. One of the great advantages of the electrophoresis technique was the mild way in which it treated macromolecules. This brought Tiselius a particular honorary title as head of a virtual society. The title was "Founder and President of the Society for the Prevention of Cruelty to Macromolecules."

During the 1950s Tiselius' laboratory expanded its operations and became involved in developing new forms of chromatography. The aim was to approach what Tiselius referred to as "chromatography of colourless substances." Chromatography is a term derived from Greek *chroma*, color, and *graphein*, to write. The first attempts to use an adsorption technique were made by the Russian botanist Mikhail S. Tswett in studies of the colored substance chlorophyll. The technique was rediscovered in the 1930s by Richard Kuhn at Heidelberg University in studies of carotenoids. This work was recognized by a Nobel prize in chemistry in 1939 (the prize for 1938) to Kuhn "for his work on carotenoids and vitamins." Carotenoids are pigments present in chloroplasts, the centers of photosynthesis in plants. Regrettably he was forced by the authorities in Germany at the time to decline his prize. After the Second World War, Kuhn was able to collect his diploma and the medal, but no prize money. The chromatography technique combines a stationary phase, to which different substances, originally colored products, were used as mentioned, and a mobile phase, a fluid passing through the stationary material. Various adsorption materials from which the purified components could be eluted (set free) were tested. The different profiled techniques developed were referred to as frontal, elution and displacement analyses. In parallel there were also very important developments at other laboratories, and we have already met and will again meet the names of Martin and Synge and of Moore and Stein (see previous chapter).

Until the mid 1940s Tiselius performed most of the practical aspects of the experiments himself. His manual skills were particularly good. Later on his research group grew and the practical experimentation was taken over by others. The number of his administrative responsibilities grew rapidly, as mentioned above, but he refused to accept appointment to the position of Rector Magnificus (Vice-Chancellor) of Uppsala University. He remained a humble person as reflected by his own formulation "I found to my surprise that people listened to me and often followed my advice"[4]. His research group grew and there was again a space problem. Like in the case of Svedberg, Tiselius's Nobel Prize in 1948 came in handy. The government now understood that there was a need for more laboratoratory space. Already in 1946 a separate Department of Biochemistry had been established by the Faculty of Science at Uppsala University and in 1952 Tiselius and his collaborators were finally able to move into a new building for an Institute for Biochemistry adjacent to Svedberg's old Institute of Physical Chemistry.

Tiselius's and his co-workers' research sometimes took unexpected turns. Initiating a program to develop techniques to freeze-dry human plasma for long-term storage he established collaboration with the Swedish Sugar Manufacturers Corporation. They had problems in their processing of sugar beet extracts because a slimy substance obstructed the filters used. Two of his young collaborators, Anders Grönwall and Björn Ingelman, demonstrated that the slime was composed of dextran, a branched polysaccharide containing many glucose molecules, produced by a contaminating microorganism. In order to study this dextran material further they immunized rabbits. To their surprise they found that there was no immune response, no antibodies were produced. The molecular weight of the sugar molecules was high and it turned out that a solution of them might be used to replace human serum, containing large proteins. Further studies demonstrated that it could be used as a substitute for human plasma — a plasma expander — in critical situations. The new product was a commercial success and served as a basis for the foundation of the firm Pharmacia.

In the 1950s there were also developments of a number of new kinds of separation techniques. Cross-linked dextran was used as support material in columns for electrophoretic separation. On one occasion the researchers had forgotten to switch on the electrical current. Surprisingly it was found that molecules separated simply by floating through the column. A new molecular sieving technique, exclusion chromatography, had been discovered and the key scientists in this work were Jerker Porath and Per Flodin. Still another technique, also using dextran, as well as other polymers, was the two-phase separation technique developed by Per-Åke Albertson. I have already described my collaboration with him in concentrating polio virus and how this led to my first scientific publication in Nature and also to my first encounter with a Nobel Prize recipient when I met Tiselius at Albertson's Ph.D. graduation party[1]. In the early 1960s we used many of the different separation techniques developed in Svedberg's and Tiselius's laboratories. We found, for example, that elongated molecular structures like the projections extending from the vertex capsomeres, the fibers, of the adenovirus icosahedral particles separated beautifully by exclusion chromatography.

In 1939 Tiselius became a member of the Royal Swedish Academy of Sciences and after another five years also a member of the Swedish Academy of Engineering. During the coming decades he was extensively recognized for his achievements. He became a member or honorary member of more than 30 learned societies, including the National Academy of Sciences in Washington and the Royal Society. He also received honorary doctorates from 12 universities. He had wide concerns about the use and misuse of science in society. He sympathized with Linus Pauling in his view that scientists had a responsibility to further peace. During the time of his leadership of the Nobel Foundation he initiated a series of Nobel Symposia. Their aim was to discuss the predicament of mankind in the perspective of cultural, social and scientific developments. The first cross-cultural conference was held in 1969. It was entitled *The Place of Value in a World of Facts*. Tiselius had a strict work ethic and when he had taken on a role of responsibility he committed himself fully. It was therefore difficult for him to follow the advice given him towards the end of the 1960s to be careful not to overstrain his heart. He died of a massive heart attack in 1971.

As a member of the Royal Swedish Academy of Sciences Tiselius became deeply involved in the work of the Nobel committee for chemistry. Already during his first year as a member of the committee, 1946, he was selected to review Wendell Stanley, as presented at length in my first book on Nobel Prizes[1] and also Northrop and Sumner. His conclusion was that the two latter scientists were candidates for a prize that year but that it would be wise to wait with a prize to Stanley. The committee was of a different opinion and recommended that all three scientists should be recognized. This also became the decision of the Academy and Tiselius was selected to give the laudatory address at the Prize ceremony[6]. It became the first among his many subsequent public presentations at such events. This widely-read scientist had also gained some insights into the field of virology. Towards the end of his presentation he said apropos Stanley's crystallization of the plant virus causing tobacco mosaic disease:

> It was remarkable enough when Buchner found that certain of the functions of the living cell can be separated out from it and are to be found in the expressed juice, but it appears still more remarkable that the capacity to reproduce — this unique characteristic of life — can also be exhibited by certain molecules, thus by dead substances. It must be

borne in mind, however, that, as far as we know, this capacity is only possessed by the virus molecule when it is in contact with the living cell, and that probably the latter is materially responsible for the reproduction of the virus substance.

... An extraordinarily fascinating field is hereby opened up to research workers and it is not improbable that development will lead to a closer scrutiny of the border-line between living and dead matter.

Even among scientists we sometimes hear the assumption expressed that the innermost secrets of the vital processes will always be hidden from us, that there is a wall through which we cannot penetrate. Today we do not know whether that be correct, but we know that this wall — if there is one — is considerable farther away than one had dared to believe earlier. That this is so is to an appreciable degree the result of the discoveries which have been rewarded with the 1946 Nobel Prize for Chemistry.

This is a formulation by a cautious scientist, who during his professional life would learn a lot about the dance of the molecules of life and who actively contributed to the introduction and development of molecular biology in Sweden. He probably would not have used a similar formulation ten years later when the field of molecular biology had started its rapid development leading to a completely new kind of biology. In fact there were two important conceptual revolutions. One was the full understanding of the structure and function of proteins, which is extensively discussed in relation to Sanger's insulin studies, and the other of course the identification of nucleic acids as information-carrying molecules.

At the end of 1963 Tiselius gave a provocative lecture arguing that there was a need to change the Swedish university structures with consideration taken to the new field of molecular biology[7]. After a formal review by other experts a proposal was made to move four professorships from the Museum of Natural History in Stockholm to the University of the City. This led to an intense academic feud in which Tiselius was the central figure. The museum, which in fact had been associated with the Royal Swedish Academy of Sciences until the 1950s, had a qualified division for research, but this was of the traditional kind, mostly focusing on classification of animals and plants. It obviously defended its position and presented itself as a guardian of the sacred Linnaean tradition of Sweden. There was also an intense debate in the daily newspapers and Tiselius contributed a number of articles arguing for his views. In the end he lost this fight, which was somewhat of a surprise to him and of course a great disappointment. On this occasion Linnaeus shadow stretching over more than 200 years caused a delay in the advance of Swedish molecular biological research. However, the introduction of DNA sequencing techniques as described at the end of this chapter eventually completely changed the scenery. The Linnaean classification of plants based on cleverly selected morphological criteria was finally replaced for by a direct comparison of genomes of plants, as we shall see. Empirical data were exchanged for objective experimental measurements.

To conclude this brief summary of a well-rounded personality it may be appropriate to let Tiselius himself express his view on the driving force of science and the importance of rewards recognizing the achievements. In a speech honoring the Nobel Prize recipients at the banquet in the City Hall in 1947 he said:

In science and medicine your work in searching for the truth has disclosed new laws of Nature and opened up new and vast fields for research of the utmost importance for the welfare of mankind. And, in literature, new forms of truth of a different kind have been brought to light in a way of which only art in its highest and most subtle forms is capable. When a new thought is born, or when one of the deep secrets of nature yields to the searching scientist — in this very act of creation — there is a pure and primitive happiness deeper than anything of this kind which can ever be granted to a human being to experience.

This view has been formulated in one simple sentence by the great Swedish chemist Carl Vilhelm Scheele, who some 150 years ago wrote "Det är ju sanningen vi vilja veta, och vad är det väl icke för en ljuvlighet att få tag på den!" (It is the truth we are searching for, and what a delight it is to find it!)

And further on:

In any case we do not believe that to you even the highest awards and the most wholehearted recognition can be more than a faint reflection of the deep satisfaction you must have experienced in your work.

This humble man probably had this in his own mind when he himself received the prize the following year.

Towards a Nobel Prize in Chemistry

Tiselius road to his Nobel Prize in chemistry was remarkably short and straightforward. He received his prize the second year he was nominated. However, this is not unique. During the first 50 years of awarding the prize in chemistry there are in fact five cases of prizes in chemistry awarded the first year that the scientist was nominated; 1911 Marie Curie's second Nobel Prize; 1922 Francis W. Aston; 1929 Arthur Harden (shared prize); 1934 Harold C. Urey and 1935 the married couple Frédéric Joliot and Irène Joliot-Curie. Similarly, as previously described[1] there were seven examples of recipients of a Nobel Prize in physiology or medicine who received a prize at their first year of nomination until 1960. Before being nominated for a prize in chemistry Tiselius, already several years earlier, had been proposed for a prize in physiology or medicine. In 1940 he was nominated by Burnham S. Walker, Boston, MA. Walker wrote:

> The development, in his hands, of the electrophoretic study of proteins has led to clear and definite knowledge of the number and nature of the proteins in blood and other body fluids. While his work has not perhaps borne directly upon medical problems, it has increased and clarified physiological knowledge. In this country, his methods are now being applied to the study and diagnosis of disease.

The committee decided to let Hammarsten make an evaluation. He made a brief three-page relatively critical review. He questioned whether the electrophoretic separation technique really was a mild method and discussed whether protein aggregation may not be a source of artifacts. He stated that according to him Tiselius had not made a discovery, the only criterion to be applied in the case of a prize in physiology or medicine, and concluded that at the time he was not worthy of a prize. The committee agreed with the reviewer. In 1946 Tiselius was nominated again, this time by a well respected emeritus professor from Basel, Robert Doerr. After a general philosophical introduction he recommended in the first place Tiselius and in the second place the bacteriologist René Dubos for a prize in physiology or medicine. Hammarsten provided another very short commentary, a kind of preliminary evaluation. He noted that there had been an important progress in the development of the electrophoretic separation technique but considered the field still to be under development. He concluded "To sum up I want to express as my opinion that it does not seem unlikely that Tiselius' electrophoresis and adsorption methods in the future may need to be evaluated in connection with (a search for a response to) other questions, for the solution of which it may play a role, but at present there are not sufficient reasons to initiate such an evaluation." The committee did not mention Tiselius in their summary.

A nomination of Tiselius for a Nobel Prize in chemistry was submitted for the first time in 1947 by Kurt H. Meyer, University of Geneva. Tiselius declined to be discussed as a candidate and continued to work as a member of the committee during that year. The following year there were five nominations. L. Carroll King, Northwestern University, Evanston, Illinois, Jean Roche, College de France, Paris and Bogdan Kamienski, Krakow, Poland, proposed Tiselius alone. Another nomination by Egbert Havinga, Leiden proposed both Tiselius and Rittenberg and finally a nomination by Byron Riegel, also from Northwestern University named three possible candidates, Pauling, Roger Adams and Tiselius. Tiselius contributions to both the field of electrophoresis and to adsorption analyses were mentioned and different emphasis on these two contributions were given by the different nominators. This year Tiselius withdrew from the committee work and Hammarsten was brought in as an adjunct member, as briefly mentioned above.

The committee allocated a considerable amount of work to Hammarsten. He prepared three comprehensive reviews one of which concerned Tiselius (14 pages). A separate extensive evaluation of Tiselius was made by Westgren over 12 pages. Hammarsten's review mostly emphasized the biochemical significance of the methods developed by Tiselius whereas Westgren scrutinized the work from a more methodological point of view. Attempts to separate organic macromolecules by use of an electric field had already been made at the beginning of the twentieth century. A certain degree of separation was achieved but the different components identified were not homogenous. It was also difficult to

Arne Westgren (1889–1975). [Portrait at the Royal Swedish Academy of Sciences.]

recognize the different moving frontiers. An improved technique to register the different fractions was published by Svedberg and Tiselius in 1926. The early development of a practically useful electrophoresis technique was presented by Tiselius in his thesis work and more fully developed and applied to blood proteins in 1937. In parallel, other researchers, including Theorell, as mentioned, attempted to develop alternative kinds of potentially useful techniques, but as summarized by Hammarsten "… without doubt the Tiselius method in competition with other methods and subjected to comparative evaluations by many researchers has been found, without any parallel, to be superior." He also praised Tiselius and his collaborators' work on adsorption chromatography techniques, but he noted that in this field there were a number of highly competitive developments by Martin and Synge, Moore and Stein and Craig, names we have already encountered. Hammarsten concluded that the relative value of Tiselius's work in this field as compared to that of others remained to be seen by future follow-up of the field. His final paragraph read:

> On the grounds of what has been presented above I would like to stress that the development of the electrophoresis method has been found to be of such a collective value to biochemical research and this contribution combined with the discovery, by use of the method, of properties of serum globulins and of antibodies is worthy of a prize in chemistry.

Westgren's review of Tiselius's contributions was also full of praise. He summarized:

> As has been indicated by the previous report (Hammarsten's review) the electrophoretic and analytic adsorption methods supplement each other. The former have their most important use on (in separation of) high molecular (weight) substances, the latter are mostly of value for investigations of cleavage products of proteins, amino acids and other substances with a moderate molecular weight. Tiselius's research findings are thus firmly based.
>
> It is a common feature of Tiselius's biophysical methods that they are mild. Any breakdown or denaturation of the investigated, often labile substances is not to be expected in contrast to the case with the use of the analytical methods generally employed in biochemical studies.

By the applications presented, Tiselius and his students and many other researchers have demonstrated the usefulness of his methods. A full evaluation of the results gained can only be made by a person with a full insight into biochemistry (Hammarsten, my remark). Only such a person can decide if Tiselius is worthy of becoming recognized by a Nobel prize in chemistry for his development of the electrophoretic and adsorption analytical methods and the results he has gained. However, his merits in the development of biophysical research methods are so important that an evaluation of his complete production within the field can hardly lead to any but a positive result.

The committee copied many of these concluding remarks in their relatively comprehensive overall summary. They concluded that they considered Tiselius to a high degree worthy of a Nobel prize in chemistry. However, there was one serious competitor. It was William F. Giauque. He had been nominated a number of times starting in 1936 for a prize in chemistry and in 1940 also for a prize in physics. By use of adiabatic (a thermodynamic term meaning absence of transfer of mass or energy within a system or between it and its surroundings) demagnetization he managed to reach extremely low temperatures. Under these conditions he could make high precision measurements of

William F. Giauque (1895–1982), recipient of the 1949 Nobel Prize in chemistry.

the thermodynamic properties of various substances. He also discovered the natural occurrence of two isotopes of oxygen. In addition to ^{16}O there were ^{17}O and ^{18}O. Svedberg had reviewed Giauque in 1942 and he did a new review in 1948. In the latter year nominations for Giauque had been submitted by Pauling and also by three persons well familiarized with the committee work, Hägg, Tiselius and Ölander. In the end the committee decided to let Giauque wait, but only for one year. He received the prize in chemistry in 1949 "for his contributions in the field of chemical thermodynamics, particularly concerning the behavior of substances at extremely low temperatures."

Prize Events in 1948

It was the chairman of the Nobel committee, Westgren, who had the pleasure of addressing his colleague in the committee, Tiselius, at the prize ceremony[8]. He started his presentation with a quote by the Swedish literary giant Esaias Tegnér. At the 50-year jubilee in 1836 of the Swedish Academy (of Letters) he said apropos the most outstanding chemist in Sweden during the 18th century "Scheele skedar skapelsen i härden." This is not easy to translate, but it means in essence "Scheele scoops the creation in the hearth." It is by studying the complex molecules and their fine structures involved in life processes that we can understand evolution (creation). He then described the impressive developments in biochemistry by means of Svedberg's and Tiselius's characterization of macromolecules. His conclusion read in a free translation into English (in this year *Les Prix Nobel* still only provided a French translation):

> Mr Tiselius.
> As a contrast to the majority of researchers of our time, who are specialists, each one within his own narrow field, you possess deep insights into different disciplines. You are well acquainted not only with physical chemistry and biochemistry but also with large parts of physics and medicine. It is pertinent in our time of (developing) goal targeted research organizations to ask the question if a group of scientists, designed to represent different disciplines, would have been capable of achieving the same success if they had been given the task, as a group, of attacking the problems that you have solved. I consider this as very unlikely. There is an unquestionable truth in the words Eric Linklater (the famous Scottish writer, my remark) used the previous year in a speech to students in Aberdeen. "The individual mind may, in certain activities, be more competent that the conjoint mind of a committee and the imagination that lives in a man alone has no home in a board of directors ..." "Hamlet was not written by a convocation of literary critics (English in the original)."

It was a very hapy Tiselius who humbly accepted his prize from the hands of the Crown Prince Gustaf VI Adolf. It was the first time since 1907 that King Gustaf V could not fulfill the function of presenting the Nobel Prizes. Age was taking its toll and he died in 1950. The total number of times that this King presented the prizes was 27, since there were 14 years when no ceremonies were held due primarily to the two World Wars. King Gustaf Adolf VI, the very popular King with his own academic qualifications, already mentioned in the previous chapter and in my previous books[1,2] was to present Nobel Prizes for 25 years. His grandson Carl XIV Gustaf took over in 1973 and at the time of writing he has already presented the Nobel prizes at the Stockholm Concert Hall 43 times and this number is expected to increase markedly since the Bernadotte males are generally very long-lived.

Tiselius receives his Nobel Prize from the hands of His Majesty the King. [From Ref. 24.]

In one part of Tiselius's banquet speech he said (translated from Swedish):

> Against the background of what happens in these days in certain parts of the world I may be allowed to express that I am happy and proud to be a citizen of a country where there is a freedom of research raised above political conflicts, where the understanding of the needs of the research enterprise is well understood, and where effective support, provided by the Government to science, is given in full trust. In this country there are no conditions made which might encroach on the freedom of scientists to search for the truth in the way they deem most appropriate or on their opportunities to make their results public.

In this particular comment Tiselius must have been thinking of developments at the time in the Soviet Union. The Lysenko affair will be discussed at some length in Chapter 8.

Tiselius gave his Nobel lecture three days after the prize ceremony. It was given in Swedish, but published in an English translation[9]. It did not have a very selling title; "Electrophoresis and Adsorption Analysis as Aids in Investigations of Large Molecular Weight Substances and Their Breakdown Products." After a short general introduction, highlighting the Art of Separation and the importance of isolating, defining and characterizing biological macromolecules, the rest of the lecture is mostly a matter-of-fact presentation of the developments of separation techniques. A brief historical background was presented and a credit was given to other contributors to the field. For example regarding Theorell's work described in the previous chapter Tiselius said "... but an electrophoretic analysis of enzymes was first performed by Hugo Theorell in the middle thirties with a modified transference apparatus of his own construction. This work, which among other things led to the clarification of the nature of Warburg's yellow enzyme, contributed to a large extent to making the electrophoretic method known." The most critical statement of the text, however, appeared three pages earlier when Tiselius explicitly stated "On the basis of experimental and theoretical studies on the numerous sources of error, I undertook (1936–1937) a radical reconstruction of the electrophoresis apparatus." Although a humble person, he knew his value.

Gröder Äng and Notable Nobel Reviews

The remarkable comprehension and the details provided in Nobel reviews have been repeatedly emphasized. The reviewers really had learnt to assemble all available information in the respective fields under close scrutiny. Their task was to interpret the advances, their magnitude — truly paradigmatic discoveries(?) — and impact — benefit to mankind (?). Tiselius 's reviews were exemplary and sometimes wise in their interpretations and even visionary. It might be of interest when time has passed to allow an examination of all of them together and to summarize their essence. This may allow an insight into how his understanding of the moving frontiers of science matured with time, and to attempt to grasp how he managed to develop the high degree of objectivity and fairness that characterized his judgments.

All Tiselius's Nobel evaluations are signed Gröder Äng, the date and his name. Gröder Äng is a unique lot on the island Skaftö, which also includes the nearby community of Fiskebäckskil, on the West coast of Sweden. *Äng* means meadow in Swedish, but it is not clear what the origin of the word *Gröder* is. "*Grödor*" in Swedish translates as "crops," and one may venture a guess that Gröder has its etymological origin in this word. The lot had been bought by Tiselius in 1943 during the middle of the Second World War from the family of a local fisherman who had just died. It was a relatively small lot, somewhat larger than an acre, but had a unique position. Surrounded by high rocks there was a meadow facing a small natural harbor, Gröder Hamn, and beyond that the open water with other islands at a distance. It was the ideal place to enjoy magic sunsets. In this unique setting Tiselius could withdraw from the turmoil of the world and pick his wild strawberries.

The original house on the lot was in such a bad condition that he built a completely new architect-designed house harmonizing with the magnificent surrounding environment. Tiselius got to know about this place when he spent several summers working at a nearby historical marine biology station. This station was established as early as

Tiselius in his Bohuslän summer home environment. [Photo from Per Tiselius.]

1877, making it one of the oldest stations of its kind in northern Europe. Until recently it was managed by the Royal Swedish Academy of Sciences and it was called the Kristineberg Marine Research Station. In 2008 the management was taken over by the University of Gothenburg and the center is now a part of the Sven Lovén Center for Marine Sciences. Lovén was the founder of the original station. One of the research vessels of the station was called "Arne Tiselius," but a few years ago this ship was sold. The name Tiselius, however, remains associated with the station, since his grandson Peter is a researcher at the present station. The marine biology center was used for experimental work by his grandfather during the long summers when he stayed at Gröder Äng. In the 1940s and 50s, professors at Swedish universities still had a summer break for some two to three months. Twice during the summer Tiselius took the steamer to Gothenburg and then the train to Uppsala to check on his work mail. Gröder Äng no longer belongs to the Tiselius family. It was bought by a popular television talk show host, the Norwegian Fredrik Skavlan.

Like many Swedes, influenced by the Linnaean tradition, Tiselius was very interested in natural history and he accumulated a wide knowledge of both botany and, in particular, in ornithology. In the company of his son Per he made early spring excursions to the surroundings of Uppsala where one might get a glimpse of the courting behavior of the capercaillie. His interest in birds stayed with him throughout life and in 1961 he created a private "academy," the Bäckhammar Academy of Sciences. Tiselius was its president and it included five of his special friends and enthusiastic bird watchers and their wives. The vice-president was Victor Hasselblad, the famous inventor of the camera carrying his name. The first astronauts landing on the moon used this kind of camera. Another occasionally adjunct member was the dentist and highly-regarded amateur ornithologist Per O. Swanberg. He was the national authority on cranes. Coincidentally he was my mother's cousin. The Swanberg family originated from Blekinge County in the south-eastern part of Sweden so one might conclude that the middle-class networks in the large country of Sweden at the time were indeed small. In fact there is even more to these connections. In Chapter 4 Thomas Tydén, the present head of the Nobel family, was introduced. He kindly connected me to the present generations of the Hammarsten family, as previously mentioned. Thomas is married to Ulla, born Swanberg, an offspring of another of my maternal grandmother's cousins. Hence Thomas' and Ulla's two sons carry genes both from the Nobel family and from my grandmother's Swanberg family. However, enough about the dancing genes and back to Tiselius.

In connection with the Annual Meeting of the Royal Swedish Academy of Sciences, which is generally held on March 31, the day when the Academy received its Royal insignia in 1741, the Academy used to strike a medal commemorating an important member. The medal in 2002 honored Tiselius. It carried a statement, which was a transmutation of the saying *navigare necesse est* as cited by Plutarch and translated to Latin. On the medal a diagram illustrating Tiselius' separation of serum proteins was shown and the text appropriately read *Separare necesse est,* Latin meaning "it is necessary to separate."

Medal minted by the Royal Swedish Academy of Sciences in 2002 to honor Tiselius.

A Unique Biochemist, Who Received Two Nobel Prizes

It took time until the nature of proteins, of highly variable sizes and each one with its unique amino acid sequence as it turned out, became fully understood. As was repeatedly referred in the previous chapter it was thought for a long time that they were carrier substances for other important small molecules endowed with specific biological properties, like enzymatic or hormonal activities. It was first by Sanger's methodological work leading to a full insight into the complete amino acid sequence of the two chains which were found to constitute insulin, that the relationship of protein structure and function was first highlighted.

Sanger was an exceptional scientist, hitherto the only one to receive two Nobel Prizes in chemistry. The first prize was in 1958 "for his work on the structure of proteins, especially that of insulin" and the other half a prize together with Walter Gilbert "for their contributions concerning the determination of base sequences in nucleic acids." The first biography of Sanger was recently published[10] and in November 2014 Theodore Friedmann arranged a symposium in La Jolla, CA honoring Sanger. Many of the participating scientists, including Friedmann himself, had worked for shorter or longer times with Sanger. There were many testimonies to his likable personality. I have previously written about Nobel laureates, like Thomas Weller[1] and James Watson[2] as they were still living 50 years after they had received their prizes, but this is only allowed in the case of prizes in physiology or medicine. When it comes to prizes in physics and chemistry there is, in addition to the 50 year secrecy rule, the requirement that the archive material should not be made available if the prize recipient is still alive. Each prize-awarding institution can formulate its own rules. Hence a review of the Sanger Nobel archival material could not be made until in 2014, since he did not die until 2013.

On December 11, 2001 a remarkable dinner was held at the Royal Castle in Stockholm. In addition to the Nobel laureates of the year all previous Nobel Prize recipients had been invited to celebrate the fact that it was one hundred years since the first Nobel Prizes were awarded. There were more than 160 guests and one may question if ever before in our history so much human intellect had been gathered in a single room. My wife had the privilege of being seated between the abovementioned Weller, a virologist and long-time friend, who received the 1954 Nobel Prize in physiology or medicine for discoveries that allowed the development of polio vaccines, and Sanger. This provided a natural opportunity for me to have a pleasant conversation with him after the dinner and I believe this was my only personal contact with him. It was a charming and memorable encounter. I might also have met him briefly when he was back in Stockholm to receive his second Nobel Prize in Chemistry in 1980. At this time I was heavily involved in the Nobel work at the Karolinska Institute. As a part of this work we had joint meetings between the committee for chemistry at the Royal Swedish Academy of Sciences and our committee for physiology or medicine. On one of these occasions I must have argued for a second prize to Sanger, since I recall that the chairman of our committee at the time, the chemist Reichard, already mentioned in the previous chapter, afterwards said "So you believe that Sanger should have a second Nobel Prize."

To return to the Royal table I had the pleasure to have on my right side Arthur Kornberg's third wife, a great opportunity to get some insights into the remarkable Kornberg family and of course many other matters, and two further steps to my right Joan Sanger was seated. A year or two later I invited Arthur's son Roger Kornberg to come and lecture at our Academy, which he generously accepted, and after that it did not take long until he received his own prize. Multi-generation Nobel prizes of course are rare, but they have occurred six times[1]. Thus it is even rarer for a single individual to receive two prizes. There are only four cases. Among these John Bardeen and Sanger are unique in that they received their two prizes in one and the same discipline, physics (1956 and 1972) and chemistry (1958 and 1980), respectively. However, both prizes given to Bardeen were shared whereas Sanger's prize for the structure of insulin, as mentioned, was awarded to him alone. The only laureate who might match Sanger is Marie Curie, with her shared prize in physics in 1903 and her undivided prize in chemistry in 1911.

A Quaker Mind and the Scientific Career

Sanger stands out as a unique personality and a particularly successful scientist. Words that were recurrently used in characterizing his personality at the 2014 symposium were modest, self-effacing, reserved and shy. One easily gets curious about what it was in his background and developments in his career as a scientist that made him so uniquely successful. A lot of valuable information about him can be found in the biography by George G. Brownlee[10], who was one of Sanger's important collaborators during the time of the development of RNA sequencing techniques. A major part of the book represents interviews that he had done with Sanger.

Like Anthony Hodgkin (Chapter 3) Sanger had a Quaker upbringing. For some time they even went to the same boarding school. Sanger's parents met because of the need for urgent treatment of a septic finger. The doctor, Fred's father, also named Frederick Sanger, successfully cared for his patient,

Cicely Crewdson, and in 1916 they were married. At the time they were of mature age, 40 and 36 years old respectively, and still their marriage was blessed with three children arriving in rapid succession, Theodore (Theo) born in 1917, Frederick (Fred), his father's namesake, in 1918 and a daughter Mary (May) five years later. The father had been educated at St. John's College, Cambridge and qualified as doctor in 1902. After ten years as a missionary in China he returned to England where he worked as a general practitioner in Gloucestershire. His return was caused by illness, probably tuberculosis. He had some brief contacts with science but soon became fully absorbed by his clinical involvements. Interestingly this later had some discouraging influence on the young Fred's choice of career. He was not immediately attracted to becoming a scientist, but seeing his father being busy with his patients and really "not finding time to concentrate on anything" convinced him to choose biochemistry instead of following in his father's footsteps. The mother Cicely was the youngest daughter of a wealthy family who owned a cotton mill near Manchester. The Crewson family by long tradition were Quakers and Frederick senior converted to Quakerism after the marriage. The financial resources brought to the home by the mother had importance for Fred's development during his early academic years. It helped that he was financially independent.

In the family Fred's timid personality earned him the nickname "mouse." His older brother Theo was much more outgoing. They were good friends and it was thanks to his older brother that Fred became interested in the phenomena of nature. Their home Far Leys had extensive grounds allowing wide forays into nature to search for animals or plants. Their early schooling was arranged at home by governesses. They quickly noticed that Fred absorbed and understood new knowledge more readily than his older brother. At the age of 9 Fred was sent to a Quaker boarding school and five years later he changed to another boarding school, this time Bryanston in Dorset. Gradually he seems to have adapted to the sometimes rather harsh boarding arrangements. One piece of evidence for this is that much later he sent his own children to the same boarding school. At the school he was hardworking, being stimulated in his interests in biology and chemistry by excellent teachers. During his last year of studies at Bryanston he went on a school exchange visit to Germany for some months. He stayed in a family in Salem in southern Germany. During this time he was exposed to the Hitler Youth movement, but he refused to give the salute "Heil Hitler." Still, like many others, he did not appreciate the impending threat of war and he remained a pacifist. Hodgkin (Chapter 3) had a similar experience of visiting Germany, but reacted differently. In his case it led to his losing his Quaker solidarity with pacifism, and when the war broke out he identified strongly with the war efforts as already shown.

Sanger did well at school, but he was never top of his class. However, his success at school secured his admission to St. John's College, Cambridge, where both his father and uncle had been undergraduates. In his Natural Science Tripos, biochemistry was his first choice. He was strongly influenced in this by Ernest Baldwin, one of the tutors at the College. He was a member of the Biochemistry Department at the University of Cambridge. This department had become inspired to explain biology in the terms of chemistry by its leader F. Gowland Hopkins, the 1929 recipient of the Nobel Prize in physiology or medicine "for his discovery of the growth-stimulating vitamins." It took Sanger for formal reasons an extra year to finish his College studies, but this only widened his insights into biochemistry. In order to further pursue his involvement in the field he became associated with the Department of Biochemistry in 1940. This year of the Second World War year became of particular importance to his development.

At the time Sanger still remained a committed Quaker, as just mentioned. He had signed the "Peace Pledge Union" and he was formally accepted as a conscientious objector. Prior to starting his research work he assisted for a while at a Quaker Relief Training Centre. There are pictures of him cleaning toilets. Through an anti-war group he met his future wife Margaret Joan Howe. They were married in 1940, but sadly Sanger had lost both his parents to cancer at about the same time. He started his Ph.D. work, financed by his own means, under the guidance of Norman Wingate (Bill) Pirie. We have met him before[1] because he, together with Frederick Bawden, demonstrated that purified tobacco mosaic virus not only contained protein but also about 6% RNA, a finding of major importance as would be revealed later. Sanger's contacts with Pirie, however, were short-lived because the latter moved to Rothampstead Experimental Station in Harpenden to focus on virus biology. Fortunately Albert Neuberger, a relatively young and experimentally active researcher, stepped in as Sanger's supervisor. Hopkins's guidance of the department led to the essential dietary factors becoming its main focus. Sanger's task was to study the metabolism of the amino acid lysine. This amino acid is one of nine out of the 20 amino acids that cannot be synthesized by our body, and thus referred to as *essential* amino acids. Sanger's development of his thesis was uneventful and he received his Ph.D. in 1943. To judge from later comments he did not view it as a great thesis, but there was much more to come. Still later he acknowledged that it was Neuberger who inspired him to tackle new and important scientific problems and to examine them by a careful experimental approach.

There were two events that led to Sanger initiating studies of insulin. One event was that Neuberger also left for another appointment which raised the question of the future orientation of Sanger's work and the other was the change in command of the department. Albert C. Chibnall — "Chips" — had succeeded Hopkins — "Hoppy." Chibnall offered Sanger his first salaried employment and the focus would be on studies of proteins. There were already some ongoing studies of bovine insulin in the laboratory. This hormone was the only larger protein available purified in sizable quantities and it had even been crystallized in 1926, as already mentioned in Chapter 3. Chibnall's

Frederick Sanger (1918–2013) on the right in 1947 together with Rodney R. Porter (1917–1985), forthcoming recipient of the 1972 shared Nobel Prize in physiology or medicine. [From Ref. 10.]

proposal was that Sanger should analyze the end-groups of insulin. Studies in the laboratory had already demonstrated that there were free (alpha)-amino end-groups. At the time there were a number of speculations as to whether insulin was a homogenous population of molecules or not and how many chains of amino acids might constitute the functional molecule. A lot of new methodology had to be invented and it took until about 1950 before Sanger himself for the first time saw a full determination of the sequence of the chains of amino acids as a possibility. One important step in this development was the technique developed in 1948 to separate and isolate the two protein chains of the molecule. This was done in collaboration with Sanger's first graduate student, Rodney R. Porter. Porter had been delayed in his studies, because of his involvement in the Second World War, and was in fact older than Sanger. Porter was an independent scientist who later developed an interest in immunoglobulins after having read a book by Karl Landsteiner, the Nobel laureate who had discovered human blood groups[1]. He therefore wanted to study this kind of multispecific immunoreactive proteins. Sanger told him that this was absolutely ridiculous since it was known that this kind of proteins were very heterogeneous. However, Porter was stubborn and eventually managed to reveal the basic structure of immunoglobulins. This earned him a shared Nobel Prize in physiology or medicine in 1972, a story already briefly told[2].

It took Sanger ten years to determine the position of the 51 amino acids in the two chains of insulin, named A and B, and the position of disulphide bridges by use of different techniques, many of which he had developed himself. Separation of the different polypeptide fragments was effectively achieved by use of the newly introduced methods of partition chromatography and paper chromatography by Martin and Synge, who appropriately received their Nobel prizes in chemistry six years prior to Sanger, as already mentioned. Sanger had a deep respect for these researchers working at the National Institute for Medical Research in London. According to his own saying "Nearly all of these fractionation methods were invented by them, particularly Martin. He was the great brain behind this. He was a very bright, original person and a most inspiring person to talk to. I can't say I talked to him. I listened to him, really. I think he was as near to a genius as anyone I have met."[10]. So speaks a humble scientist, who never brought to the fore his own important contributions. However, there were many. In particular deciding the positions

Archer J. P. Martin (1910–2002) and Richard L. M. Synge (1914–1994) the recipients of the shared 1952 Nobel Prize in chemistry. [From *Les Prix Nobel en 1952.*]

of the disulphide bridges was a very challenging part of his studies. It was first at the time when he managed to resolve this problem that it was eventually understood that the active molecule only had two chains and had a molecular weight of about 6000 .

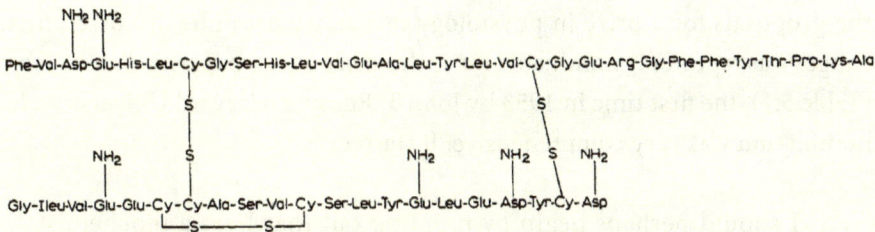

The chemical structure of insulin. [From Ref. 12.]

The impact of his work was enormous. The concept of proteins as homogenous chains of a fixed variable length with a genetically controlled stable amino acid sequence became solidified. As stated by Brownlee[10], "The scientific community was amazed at this tour de force. Moreover, this result proved beyond a shadow of doubt that proteins had unique sequences. This contradicted theories that had been advanced before, that proteins consist of periodic arrangements of amino acids, or they were heterogeneous mixtures." When Dorothy Hodgkin and collaborators finally were able to demonstrate the three-dimensional structure of insulin, as mentioned in the previous chapter, one could eventually understand the folding of the protein chain and also definitively settle the position of the disulphide bonds. Sanger's reaction at that time was "Ah, I got the disulphides right." Defining the position of the three disulfides, two between the two chains and one within one of the chains (see above figure), had in fact required a lot of ingenious experimentation. The next large protein to be fully characterized after insulin was the enzyme ribonuclease. It is 124 amino acids long. This molecule was the focus of interest in the 1972 Nobel Prize in chemistry, as also mentioned in the previous chapter.

After his great success in his protein studies crowned by his first Nobel prize in 1958 Sanger completely changed the orientation of his work. Inspired by Francis Crick he became interested in sequencing of nucleic acids, first RNA and then DNA. It took some time to reach a momentum in this new work. We will return to this phase of his professional life below.

The Nobel Committee at the Karolinska Institute Examines Sanger's Insulin Studies

Sanger of course was nominated for a prize both in chemistry and in physiology or medicine, but for reasons to be discussed it was considered appropriate at the time to recognize his achievements by a prize in the former field. However, the proposals for a prize in physiology or medicine will be discussed first. Sanger was repeatedly nominated for such a prize as can be seen from the table (Table 5.1), the first time in 1953 by John T. Edsall at Harvard University. His nomination was very comprehensive. It started:

> … I should perhaps begin by pointing out that I am a biochemist and am most familiar with contributions which are on the borderline of chemistry and biochemistry. It is, therefore, an open question whether the nomination that I wish to submit should most properly be considered by you or by the Committee which awards the Nobel Prize in Chemistry.

Table 5.1. Nominations and evaluations of Fred Sanger for a Nobel Prize in physiology or medicine.

Year	Nominator	Other nominees for a shared prize	Evaluator
1953	J.T. Edsall, Boston		H. Theorell
1955	A.B. Gutman, New York		E. Hammarsten
	G.T. Cori, C.F. Cori, St. Louis	V. du Vigneaud	
	F. Lipmann, Boston	V. du Vigneaud	
1956	A.D. Marenzi, Buenos Aires		
	L.E. Pierini, Buenos Aires		
	Ö. Özek, Istanbul		
	U. Masker, Istanbul		
	M. Clara, Istanbul		
	Z. Öktém, Istanbul		
	F.G. Young, Cambridge		
	J.H.P. Jonxis, Groningen		
	J. Gaarenstroom, Groningen		
1958	E.D. Freis, Washington		
	R. Richter, Ankara		
	J.T. Edsall, Boston		

He then introduced Sanger's achievements and stated:

> I consider that Dr Sanger's work in determining the sequence of the 30 amino acid residues in the phenylalanine chain of beef insulin and the 21 residues in the glycine chain is a really epoch-making development of our knowledge of the structure of protein molecules, which will have far-reaching effects throughout the whole domain of medical sciences.

And later he emphasized:

> ... the vistas opened up by these developments are immense.

He noted that Sanger could never have achieved his results if he had not had access to the technique of partition chromatography. Edsall also wrote:

> ... have no doubt of its profound importance, I must admit that in order to avoid any possibility of premature action it might be better to wait for another year or two in order to let further evidence on this subject accumulate.

The committee considered the nomination so important that it let Theorell make a full investigation. It covered nine pages. Theorell described the development of various labeling techniques by Sanger's clever experimenting. His first important contribution was to develop a method by which free amino groups of amino acids could be labeled. The technique introduced used dinithrophenyl and was hence referred to as the DNP method. However, according to Sanger himself the original purpose was not to solve the problem of sequencing, but to develop an end-labeling method for proteins in general[10]. As the work progressed, Sanger managed to split the disulphide bonds that kept the peptide chains together and also break each chain into pieces and label the free amino group of the end amino acid of each piece. It was then somewhat serendipitously realized that it might be possible to collect information on the full amino acid sequence of the insulin protein. Various separation techniques were used to isolate the fragments. In addition to precipitation methods Sanger together with Tiselius published an article in *Nature* in 1947 on the use of chromatographic methods for separation purposes. Sanger spent some time in Tiselius's laboratory and it seems that this should have allowed a meeting of minds and personalities. However, of greatest importance was the partition chromatography technique introduced by Martin and Synge, which, as already mentioned, was recognized by a Nobel Prize in chemistry in 1952. In 1951 Sanger and Tuppy published the sequence of the 30 amino acids in the basic phenylalanine (B) chain and a year later the first data on the sequence of the 21 amino acids of the glycine (A) chain were presented. Theorell's summarizing comments are of particular interest since they reflect the conceptual deliberations at the time about polypeptide synthesis. He wrote (translated from Swedish):

> It is obvious that Sanger's work on the composition of insulin is of major importance and interest. About ten years ago it would have been hard to dare believe that it would be possible to get a complete insight into the sequence of amino acids in a protein. Among the many interesting consequences of Sanger's results there are certain that in particular should be brought to light.
>
> The independent amino acids do not show a tendency to display any kind of periodicity, but appear to be positioned without any rule after each other, however always in a given place. There are no reversals in position of related amino acids, like for example leucin and isoleucin .... *Thus each insulin molecule should be exactly identical to all others* (my italics). Earlier, one was reluctant to believe that protein molecules should be reproducible into the smallest detail....
>
> But how would it be possible to conceptualize the synthesis of proteins? If there is no periodicity or other predetermined order (that) can be traced in the polypeptide chains that form proteins one would be forced to assume that the formation even of the simplest protein, like insulin, would require perhaps some 50 different absolutely specific enzymes (sic, my comment) to attach each individual amino acid, one at a time, until finally the chain is complete. An arrangement like this appears for many reasons to be highly unlikely, since one would expect to find a large number of only partly formed chains, which is something one does not see. Sanger believes that he can trace the occurrence of certain dipeptide sequences in a higher frequency than others .... But even if that were the case the whole protein synthesis would become exceedingly complicated in this way. It seems to me that there must be simpler and more likely mechanisms, but we can leave these aside since they do not play a role in the judgment of whether Sanger is worthy of a prize.

It would indeed have been interesting to learn what Theorell had in mind! Nucleic acid information molecules and ribosomal protein synthesis still remained discoveries of the future. In theory he could have speculated on the existence of a multipurpose enzyme managing the establishment of peptide bonds between *any* kinds of amino acids. Such a speculation would have needed a supplementary hypotheses about the possible existence of information-carrying molecules and some kind of molecular machinery that could arrange for two appropriate amino acids for the structure to be placed in juxtaposition. However, most importantly, information molecules had as yet not been conceptualized. They were very soon to enter the arena after the discovery of the structure of DNA and the discussions of a genetic language that it precipitated. Molecular biology was still in a state of gestation and remained to be delivered. In 1953 also of course remained the question of the proper establishment of the disulphide bonds.

One of Theorell's concluding paragraphs reads:

> In addition it should not be overlooked that Sanger's publications do not contain anything particularly new. They represent an application on a much more difficult material of the amino acid sequence determination performed by (a reference) on gramicidin S. This problem was much simpler than the one tackled by Sanger, since Gramicidin S only contains 5 amino acids (it is a cyclopeptide containing two identical pentapeptides, my remark). Still one cannot avoid commenting that their work is the pioneering one, which last year was recognized by a Nobel Prize in chemistry. To jump from a pentapeptide to a protein molecule is however (an advance) of such a magnitude that recognizing Sanger's contribution another year can be considered, in particular if he can demonstrate the position of the disulphide bridges.

The conclusion by the committee at the Karolinska Institute was that Sanger in 1953 was not as yet worthy of a prize.

In 1955 three new nominations for a prize to Sanger had been submitted. Gerty and Carl Cori, Nobel Prize recipients in physiology or medicine in 1947, proposed that Sanger should be combined with Vincent du Vigneaud, as mentioned in the previous chapter. The 1953 laureate Fritz Lipmann made two separate proposals in the material submitted to the committee. The first one was to recognize Theorell, as discussed in the previous chapter and the other to combine Sanger and du Vigneaud, as also proposed by the Coris. Lipmann's joint nomination of du Vigneaud and Sanger started in the following way:

> From newspaper reports in this country, it is unavoidable to recognize the fact that last year du Vigneaud had been very much in the mind of those who had to make the decision for the prize in physiology or medicine.

He referred to the above-described media leak in 1954.

There was one more nomination of Sanger for a prize in physiology or medicine in 1955. It was submitted by the professor of medicine at Columbia University, who nominated as his first candidate Homer W. Smith for his fundamental studies on kidney functions, and submitted as an alternative Sanger "for his brilliant analysis of the chemical structure of insulin and through this his contribution to our understanding of the general structure of proteins."

The committee decided that another review of Sanger should be carried out, this time by Hammarsten. As described in the previous chapter and earlier[1] he was an exceptionally influential person in the Nobel Prize work at the Institute. He did fully understand the greatness of Sanger's discovery of the structure of insulin. Hammarsten's review was very detailed and covered a total of 16 pages. It described the progression of Sanger's meticulous and imaginative research. He mentioned many of the critical challenges that Sanger had met and how he managed to resolve them, finally unravelling the complete structure of insulin. Towards the end of his review Hammarsten cited one of Sanger's central conclusions. It read:

> Also in A (the chain) it was not possible to discover any periodicity (the recurrent hypothesis, my remark) and Sanger believes it highly likely that every protein is unique in its special structure, which is suitable for its biological function.

He then described the more recent advance in which the exact positions of the disulphide bonds were determined. In addition he cited how other scientists, including du Vigneaud and Theorell, had used the Sanger approach to determine the amino acid sequence of additional biologically important proteins. Hammarsten concluded:

> Through his work to determine the amino acid sequence in proteins Sanger has made one of the greatest discoveries of this century

concerning the structure of proteins. In parallel with Sanger, when it comes to the magnitude of the discoveries made, may be mentioned Martin and Synge, Pauling and du Vigneaud (surprisingly, Theorell was not mentioned, my remark).

Sanger's analysis of the protein structure is central in the research area of physiology or medicine and it is my view that his discovery of the sequence of amino acids in proteins to a high degree is worthy of a prize.

The committee agreed with Hammarsten and declared that Sanger's contribution was worthy of a prize.

Nominations for a prize in physiology or medicine to Sanger continued; in 1956 nine nominations and in 1958 three nominations. However, surprisingly, no more reviews were done and the committee simply repeated that Sanger was worthy of a prize. One of the three nominators in 1958 was Edsall. He took notice of the impressive developments since his first nomination in 1953. The position of the disulphide bonds had been determined, the published sequence of the amino acids of the two chains of insulin had been confirmed by alternative methods, like Edman degradation, and Sanger's methods had started to be used for sequencing of other even larger polypeptide chains.

The Edman degradation technique has a certain ring to me. It allows labeling the N-terminal amino acid and detaching it from the polypeptide chain, without harming this. By this approach it was possible piece by piece to demonstrate a sequence of amino acids in a protein. Pehr V. Edman was a Swedish biochemist, trained at the Karolinska Institute with Jorpes as his mentor. After a time as visiting scientist at the Rockefeller Institute and at Princeton University in the U.S., he developed his science at Lund University and in 1957 he moved to Melbourne, Australia. It was before the mid-1950s that he developed the Edman degradation technique as it came to be called by other chemists. In the spring of 1958 I was a fourth term medical student and had just finished the course in chemistry the previous semester. I had become curious about this discipline and had obtained, together with my future wife, an apprenticeship in the department of medical chemistry. Our closest supervisors Per Wallén and Kurt Bergström, introduced me to the Edman degradation technique, but I never fully understood the principles of the technique, nor did I appreciate the purpose of the study. Since then I have learnt that it probably concerned various coagulation factors, like fibrinogen and plasminogen, studied by Birger and Margareta Blombäck. Thus I lost interest in potentially investing my time in research in biochemistry and instead, a semester later, embarked on a career in virology.

Let us now return to Edsall's nomination. His conclusion was:

> I therefore have no hesitation in repeating my earlier nomination of Dr Sanger and in emphasizing my strong conviction that his work is outstandingly worthy of a Nobel Prize. I shall not attempt to say whether he is more deserving of the prize in chemistry or of that in physiology or medicine. His work falls in the borderline area which overlaps both fields.

He also added a charming final paragraph:

> I am sorry that this letter is rather brief. I should have liked to go into more detailed analysis of Sanger's work and it significance, but I have just become Editor of the Journal of Biological Chemistry and I am so occupied at the moment with the work of that Journal that I do not have time for a more detailed presentation. I therefore offer this relatively brief statement in the hope that you will give the work of Dr. Sanger most serious considerations.

The committee certainly did this, but, although this is not documented, it seems as if already at an early stage it was agreed between the committees for physiology or medicine and for chemistry that when Sanger received his prize it should be in the latter discipline. It is not unlikely that Tiselius had a major influence on developments. The major importance of his advice, although predominantly on prizes in chemistry, but in addition also in physiology or medicine, was presented earlier in this chapter.

The Nobel Committee for Chemistry Reviews the Pioneering Protein Work

The Nobel committee for chemistry at the Royal Swedish Academy of Sciences processed the nominations for Sanger as presented in the table (Table 5.2). The first nomination for a prize in chemistry was submitted in 1952 by Irving M. Klotz, Evanston, Illinois. It was considered to be important and Tiselius was chosen to do a review. He was very familiar with Sanger's work and, as mentioned above, had collaborated with him. The review was very carefully written and initially some background data to Sanger's development as a biochemist were given. Tiselius described how Sanger and his collaborators by

Table 5.2. Nominations and evaluations of Fred Sanger for a Nobel Prize in chemistry.

Year	Nominator	Other nominees for a shared prize	Evaluator
1952	I. M. Klotz, Evanston, IL		A. Tiselius
1953	P. Denuelle, Marseille		
1954	G.T. Cori, C.F. Cori, St. Louis, Miss.	V. du Vigneaud	A. Tiselius
1955	F. Lynen	V. du Vigneaud	
1956	D.H. Campell, Pasadena, CA		
1957	M. Gruber, Groningen J.F. Arens, Groningen I.M. Klotz, Evanston, IL L.H. Reyerson, Minneapolis, Minn.		A. Tiselius
1958	C. Fromageot, Paris J.M. Sturtevant, New Haven, Conn. R. Wurmser, Paris W. Crocker, Dublin R.K. Morton, Adelaide K.H. Gustavson, Stockholm		A. Tiselius

development of an array of techniques had managed to determine the amino acid sequence of the B chain and was occupied at the time in deciphering the sequence of the A chain. He took notice that it remained to determine the position of the disulphide bonds. In spite of the fact that it was work-in-progress Tiselius concluded:

In my opinion there is no question about the fact that already at the present time Sanger's work is worthy of being recognized by a Nobel Prize in chemistry. In spite of this I would like to recommend the committee to wait and this is because of two circumstances. Sanger is apparently busy completing his work on insulin, in which it remains to determine the structure of the A chain (the committee in its September summary took notice that Sanger had presented a sequence for this chain in July in Paris at the 2nd International Congress of Biochemistry). And an award should most appropriately be given when the studies are more complete. Furthermore Sanger's work recommended to be recognized by an award has to a high degree been dependent on the developments of techniques by other researchers, among which in particular the partition and paper chromatography techniques have a prominent position.

The committee agreed with Tiselius's recommendations. Sanger was judged to be worthy of a prize in chemistry, but a decision on the time for the award was to be taken later.

In 1953 there was only one brief nomination of Sanger by P. Denuelle, Marseille. The committee did not take any action, but in 1954 when the family laureates Cori submitted a proposal for Sanger and du Vigneaud, similar to the one sent to the Karolinska Institute the following year, the committee asked Tiselius to do a follow-up evaluation. In fact he did three separate related reviews including in different combinations du Vigneaud and Craig, as already discussed above. In brief his summary recommendation to the committee again was to wait with a decision. The recent 1952 Nobel Prize in chemistry to Martin and Synge motivated a delay in recognizing in particular Craig, the father of counter current liquid chromatography. In the end Craig never was to receive a prize. Since Martin's and Synge's prize work in addition, as one of the substances examined, included a circular dimer of a pentapeptide (Gramicidin S), a remark also made by Theorell, a delay in recognizing Sanger's work was also justified. However, there was no doubt about the fact that he would be worthy of a future prize

There was a brief nomination in 1955 of Sanger together with du Vigneaud by F. Lynen, Münich, the future recipient of the 1964 Nobel Prize in physiology or medicine, to be described in the next chapter. A year later D.H. Campbell, Pasadena, CA, nominated Sanger alone, but it was first in 1957 that the committee took new initiatives regarding his candidature. In this year he received four nominations, two from the Netherlands and two from the U.S.A. Tiselius was selected to carry out yet another review. He summarized all the recent developments and emphasized that Sanger had markedly increased his strength as a prize candidate. The committee made the following summarizing statement:

Mr (this is the title used in the Academy) Tiselius, in the investigation carried out this year (attachment 16), finds that Sanger's later publications, in which among other things the sulphate bridges have been localized in the molecule, conclude an impressive scientific achievement: the complete determination of the structure of a protein molecule. By this work Sanger, without doubt has made a contribution which deserves to be recognized by a Nobel Prize.

Not least his later publications provide evidence of an extraordinary skill to overriding experimental difficulties and an intuitive capacity to find the right solution among a perplexing number of alternatives. Not only has Sanger invented important new methods — others might also be mentioned, which have had a great importance for the development of the field — he has also managed to use them to find solutions to one of the most difficult and hardest of the problems of chemistry. Even though the specific properties and biological functions of proteins most likely depend not only on the structure represented by the specific sequence of amino acids — in this situation configurations and bonds of a different kind after all appear to play an important role (the meaning of this reservation can be discussed, my remark) — evidently Sanger's contributions represent one of the most important advances in studies of the structure of proteins. The results he and his collaborators have obtained and the methods they have used in their studies of the structure of insulin have made proteins available for in-depth studies in a way that some decades ago would have been considered utopian, and an increasing number of researchers have now become involved in similar studies of other biologically relevant substances of a protein nature, enzymes, hormones and others, from which highly interesting results have now started to emerge. Sanger's later publications have strengthened his candidature to such a degree that this year he seems to be among those who in the first place should be considered for a reward by the Nobel Prize in chemistry.

It was a close-run thing but in the end the 1957 prize recognized Alexander R. Todd "for his work on nucleotides and nucleotide coenzymes."

Not surprisingly it was Sanger's turn to receive the prize in 1958. He was proposed by six different nominators, representing five different countries. Many of the nominations were quite informative, in particular the one from the Swedish protein chemist and expert on tanning Karl H. Gustavson, a member of the Academy. At the beginning of his nomination he wrote (translated from Swedish):

It is by Sanger's creative fantasy, acute perception, experimental skill, energy and persistence in his development and use of this important method, which relies on the principle of irreversible inactivation of the terminal amino group by a covalent reaction, that our knowledge about the chemical structure of proteins has been deepened to a degree which only some tens of years ago would have been considered unattainable within the nearest forthcoming decennia.

And further on:

… proteins have become included in the classical organic chemistry.

Gustavson also mentioned in his nomination the recent (1957) important finding by Vernon Ingram. Employing Sanger's technique he was able to demonstrate a difference in a single amino acid between normal hemoglobin and sickle-cell hemoglobin. This was the first experimental proof that a mutation in a gene could result in a change in a single amino acid. The discipline of molecular medicine was born, but the father of this field, Ingram, never received a Nobel Prize[2]. Ingram was proposed for a Nobel Prize in physiology or medicine in 1960 and he was reviewed by Theorell, who considered him worthy of a prize but pointed out that Harvey Itano in Linus Pauling's laboratory needed to be discussed in the same context.

The final paragraph noted that Sanger was both the pioneer and the scientist who brought to completion the analysis of proteins. The paragraph began:

Sanger's contributions to protein chemistry most likely should be considered to be the most important step forward in this vital field which has been taken since the Nobel Prize recipient Emil Fischer at the beginning of this century published his experimental proof of the polypeptide structure of proteins and created practically useful methods for synthesis of complicated polypeptides.

In spite of the fact that the outcome of the deliberations of the committee in 1958 would have been seen as relatively clear, it commissioned Tiselius to make a concluding review. Over six pages he once again highlighted Sanger's impressive contributions and provided strong support for a prize to him. The committee cited sections from Tiselius's review and wrote in the second to last paragraph in its summarizing statement:

The committee declared already last year that Sanger's work on the chemistry of proteins, in particular insulin, deserved to be recognized by a Nobel Prize, but in the choice between Todd's and Sanger's candidatures, the committee for various reasons recommended the former. The committee now is of the opinion that the time has come for Sanger to receive his award.

Naturally this became the choice when the members of the Academy gathered for the final decision on the prize.

It was probably very much thanks to Tiselius early involvement in securing a prize in chemistry for Sanger that this eventually materialized. Presumably already from the beginning he was convinced, and managed to make his fellow committee members share his view, that the mapping of the complete amino acid arrangement in insulin, a convenient and representative model protein, was so fundamental to the whole field of biochemistry that it should be a pure chemical prize. From the perspective of the Royal Swedish Academy of Sciences and the chemistry committee it was very prestigious to have Sanger included among its prize recipients and the follow-up prize in 1962 to Max F. Perutz and John C. Kendrew "for their studies of the structures of globular proteins" fitted very well into the milestone presentations of historical events in development of knowledge about protein structures. However, the 1962 deal with the Karolinska Institute implicitly meant that the Academy sacrificed inclusion of the truly revolutionary demonstration of the structure of DNA among its chemistry prizes. As a consequence it also relinquished the early prizes recognizing discoveries of the genetic code and the information processing in transcription and translation. It was much later that the Academy regained a dominant position in this area by highlighting the impressive crystallographic presentation of the molecular basis of eukaryotic transcription by the chemistry prize to Roger D. Kornberg in 2006 and three years later of the structure and function of the ribosome, the factory for protein synthesis, by the prize to Venkatraman Ramakrishnan, Thomas A. Steitz and Ada E. Yonath.

The First Visit to Stockholm

When Sanger came to Stockholm he was only 40 years old, one of the younger recipients of a prize in chemistry. In spite of his shy nature he seemed to have enjoyed the festivities. It was, not surprisingly, Tiselius who addressed him at the prize ceremony[11]. He referred to Emil Fischer's introduction of fundamental concepts of protein structure and then expanded on how Sanger's imaginative

Sanger during his visit to Stockholm in 1956. [From Ref. 24.]

and inventive methodological studies had meant a revolution of the field. He finished with a relatively long paragraph in English:

It sometimes happens that an important scientific discovery is made so to say "overnight" — if the time is ripe and the necessary background is there. Yours is not of that kind. The first successful determination of the structure of a protein is the result of many years of persistent and zealous work, in which the final solution of the problem has been approached step by step. You knew when you began to look into the structure of the insulin molecule 15 years ago that the problem was a formidable one. So did the whole scientific world. Those who knew you were confident, however, that you would ultimately succeed, and each successive publication from your laboratory strengthened our confidence. Intelligence, knowledge and skill in the mastering of the methods (were) required — we know you had them all — but in such a venture these are not enough. Without your wholehearted devotion to the task you had set for yourself, many obstacles on your way would have appeared insurmountable. Now that many years of work have been crowned with success you may look back and rejoice. You can also enjoy the satisfaction of seeing the roads you have broken and paved being used by many in their search for the building principles of the key substances of Life. However, very likely you are more apt to look ahead. It was Alfred Nobel's intention that his prizes should not only be considered awards for achievements done *but that they should also serve as encouragement for future work. We are confident that you are a worthy recipient of the Nobel Award also in this sense.* May we offer you our congratulations.

It would turn out that Tiselius's words emphasized by me in italics would be highly visionary. It was a grateful chemist who received his prize from the hands of His Majesty the King.

296 *Nobel Prizes and Notable Discoveries*

Sanger's Nobel lecture[12] departed from the notice that in 1943 fundamental knowledge about proteins included the fact that they were chains of amino acids and that there were 20 amino acids to choose from. He then went on to say:

> Practically nothing, however, was known about the relative order in which these residues were arranged in the molecules. This order seemed to be of particular importance, since although all proteins contained approximately the same amino acids they differed markedly in both physical and biological properties. It was thus concluded that these differences were dependent on the different arrangement of the amino acid residues in the molecules. Although very little was known about amino acid sequence, there was much speculation in this field. The most widely discussed theory of the time was that of Bergmann and Niemann, who suggested that the amino acids were arranged in a periodic fashion, the residues of one type of amino acid occurring at regular intervals along the chains. On the other extreme there were those who suggested

Sanger receives his 1958 Nobel Prize in chemistry from the hands of His Majesty the King. [From Ref. 24.]

that a pure protein was not a chemical individual in the classical sense but consisted of a random mixture of similar individuals.

He then referred to his teacher, Chibnall, and his early studies of insulin that opened the field for Sanger's forthcoming forays. In fact, although not mentioned in the lecture but referred to above, insulin was the only protein which could be bought in pure form at the time. It turned out to have a size that was suitable for the studies that evolved. The first step was to develop an efficient technique to label free amino groups. The introduction in 1945 of the DNP method turned out to be very important. It became the standard technique in the field which later also came to include the already mentioned Edman degradation technique.

The size of the molecule remained uncertain until well into the studies. It was only towards the end that a molecular weight of only about 6,000 was found which tallied with the size of only two polypeptide chains, the A and the B chains (p. 283). The latter chain is composed of a sequence of 30 amino acids demonstrated by the intense collaborative work with Hans Tuppy in 1949, published in 1951. Tuppy had a very productive year in Cambridge and did most of the work on the phenylalanine chain. After that he went back to Vienna and devoted his energy to administration. The A chain, also referred to as the glycine chain, with its sequence of 21 amino acids was published two years later. The main collaborator in the latter work was an Australian postdoctoral student Ted Thompson, but there was never any question about the leadership of Sanger. Eventually the position of the three disulphide bonds in the molecule was solved too and published in 1955.

Tiselius mentioned the pioneering contributions to the protein field by Fischer and he was also briefly referred to by Sanger in his lecture. Fischer was one of the giants in the field of organic chemistry. He made many important contributions. Initially his work focused on purines (nitrogen-containing cyclic compounds, later shown to have a critical role as the bases in nucleic acids) and sugars and it was for contributions in this field that he received the 1902 prize in chemistry, only the second one ever awarded. However, later in his career he was mostly concerned with studies of proteins and he made major discoveries in that field too. He identified a number of different amino acids and importantly demonstrated the way they connected to each other — the *peptide bond* — to build up long polypeptide chains. In a summarizing paragraph towards the end of his Nobel lecture Sanger wrote:

Of the various theories concerned with protein chemistry our results supported only the classical peptide hypothesis of Hofmeister and Fischer. The fact that all our results could be explained on this theory added further proof, if any were necessary, of its validity. They also showed that proteins are *definite chemical substances possessing a unique structure* (my italics) in which each position in the chain is occupied by one and only one amino acid residue.

This was a simple and revolutionary conclusion. Sanger's work became the start of expanding studies of the structure and function of proteins of increasing degrees of complexity and with varying capacity to interact with homologous proteins or with other proteins and other chemical entities. One half of the prize in chemistry in 1972 was awarded to Christian B. Anfinsen and it represented a recognition of the further developments of the field by studies of a larger protein than insulin, the enzyme ribonuclease. His Nobel lecture was entitled "Studies on the principles that govern the folding of protein chains." The introduction of crystallographic methods has provided a remarkable insight into the three-dimensional structure of proteins. The pioneering work in this field was recognized by the award of the 1962 prize in chemistry to Max. F. Perutz and John C. Kendrew described at length in my previous Nobel book[2], a discovery markedly influenced by Linus Pauling's earlier demonstration of helical structures in proteins. We have learnt a lot about protein structures up to the present time, but many questions still remain unanswered as a part of the grander inquiry into protein folding. Among these are the relative capacity for spontaneous folding of a protein or the need for additional external macromolecules to guide folding, the mechanisms for a proper insertion of disulphide bonds and also the relative importance of a distinct preformed folding versus a natural representation by a random coil. In the latter case the peptide first takes its functional structural state when interacting with its targeted molecule(s). A considerable proportion of all naturally occurring polypeptides are of this particular kind.

Developments after the First Nobel Prize — The RNA Years

As already mentioned Sanger was only 40 years old when he received his prize for the characterization of insulin. The laboratory was his second home and since he shunned teaching and administration he carried on with his innovative bench work. During 1955–64 Sanger had a more limited number of publications and he referred to this time as his "lean years." He had started to use radioactive labeling in his protein work, but the success was limited. This experience of using radioisotopes came to serve him better in his future work with nucleic acids, first focusing on RNA and then on DNA. This was the time when there was a dramatic development of molecular biology crystallizing into the central dogma — DNA gives RNA gives protein. The structure of the genetic code was on the mind of many scientists. In a review[14] Sanger presented the background to his change of field of chemistry in the following way:

> I think the move to the new laboratory (the Laboratory of Molecular Biology, see below, my remark) was probably an influence in my conversion to nucleic acid sequencing. Previously I had not had much interest in nucleic acids. I used to go to Gordon Conferences on Proteins and Nucleic Acids when the two subjects were bracketed together, and I would sit through the nucleic acid talks waiting to get back to proteins. However, with people like Francis Crick around, it was difficult to ignore nucleic acids or to fail to realize the importance of sequencing them. An even more seminal influence was John Smith, who was the nucleic acid expert in the new laboratory and who was extremely helpful to me, so that I could turn to him for advice in this new field.

Sanger's move to the new building for the Laboratory of Molecular Biology in Cambridge in 1962 was an important event. Perutz had for many years tried to convince the Medical Research Council that there was a need for a new building. Sanger had shown his interest in this project and his Nobel Prize in 1958 probably served as an important lever in the developments. In 1962 the building had been erected about a mile away from the center of Cambridge and the history of its emergence and development has been told[14]. The laboratory had three main units, the *Structural Studies* group, headed by Perutz and including also John Kendrew, Aaron Klug and Hugh Huxley, the *Molecular Genetics* group headed by Crick and including Sydney Brenner and others, and finally on the top floor the Protein Chemistry group headed by Sanger. The leading scientists can be seen in the picture. Perutz's goal was to establish an environment stripped of as much administration as possible. The emphasis should be on science and the interactions between scientists were fostered by attractive arrangements for tea and coffee breaks and for lunches. Sanger often

The Governing Board at the laboratory of Molecular Biology in Cambridge in 1967. From the left, Hugh Huxley, John Kendrew, Max Perutz, Francis Crick, Fred Sanger and Sydney Brenner. [From Ref. 14.]

joined in these events, except for the lunches, but he was probably not one of the more noticeable participants. However, he did arrange weekly informal sessions with his own group, referred to as "Bull sessions." At these meetings a selected fellow scientist presented his most recent results. Afterwards a lively discussion followed. Over the years the Sanger laboratory included a number of collaborators and its focus soon shifted from protein studies towards nucleic acids, as was just mentioned. This transition seems to have occurred without any major conflicts. In fact answering a question whether it was easier to get the second Nobel Prize than the first, Sanger said[10]:

> I think it was really. When you've got a Nobel Prize you get good facilities for research, you get good students, excellent postdocs who used to come to the lab and work with us. And another thing, having had a Nobel Prize, I didn't feel there was any obligation to get papers published. I could spend my time doing rather crazy "way-out" experiments. I think this is how you

get important advances very often, by trying the way-out experiments that are very likely not to work — crazy experiments. I did lots of experiments that didn't work. I enjoyed it more and kept at the work.

But if Sanger was correct we should have had more cases of recipients of two Nobel Prizes. The truth probably has more to do with his unique personality.

As Sanger became more and more involved in nucleic acid sequencing, he developed new and imaginative techniques. Again and again he demonstrated a unique proclivity for finding solutions to fundamental conceptual and experimental challenges. He was an impressively imaginative experimentalist. When Sanger started to work on sequencing of RNA/DNA he was confronted with a different kind of challenge. Instead of the 20 amino acids of proteins there were only four nucleotides in the nucleic acids. However, the size of the molecules generally was much larger. It seemed like an insurmountable challenge to attack DNA, which normally only occurs in the form of very large molecules. The situation developed to be somewhat more attractive when it was realized that there was a special form of small RNA molecules, originally called soluble RNA and later *transfer* RNAs, also tRNA. This kind of molecule is composed of only about 80 nucleotides. Thus it seemed like a molecule of a size that might be approached. However, one problem which immediately became apparent was that there were at least 20 forms of transfer RNA, one for each one of the different kinds of amino acids. Sanger pioneered the use of radioisotopes (^{32}P) for the nucleic acid studies. He was inspired to do this by a visit by Anfinsen. The transfer RNA selected for the studies was specific for phenylalanine. The RNA work was boosted at the time when Sanger's group had moved to the new building for The Laboratory for Molecular Biology in 1962, as mentioned.

Leslie Smith had just come back from a stay at Paul Zamecnik's laboratory at Harvard, where some of the original pioneering work on transfer RNA had been performed. He provided valuable information. Together with Brownlee and Bart G. Barrell, Sanger developed a relatively rapid small-scale method for the fractionation of oligonucleotides labeled by ^{32}P, a so called "finger-printing" method. At the time Sanger also, together with a visiting Danish scientist Kjeld Marcker, made the surprising finding of formyl-methionine, a special form of this amino acid, which was found to play a particular role for the initiation of protein synthesis in prokaryotes. The approach of partial degradation used for amino acid sequence characterization of proteins was successfully used also in the transfer RNA studies. However, Robert W. Holley's group at Cornell

University, Ithaca was even more successful in their approach. As Sanger himself expressed it "... we were beaten to that one."[10]

Hence it is not likely that Sanger might have been nominated even for a third (intermediate) Nobel Prize for his RNA studies. In the archives which will be made available during the next five years either for a prize in chemistry or in physiology or medicine it will be shown if this ever happened. Holley received a part of the 1968 Nobel Prize in physiology or medicine together with H. Gobind Khorana and Marshall W. Nirenberg "for their interpretation of the genetic code and its function in protein synthesis." Sanger had better luck in studies of another form of RNA. This was a particular kind of RNA detected in the protein-synthesizing machinery, the ribosomes, hence referred to as rRNA. This rRNA was somewhat larger than tRNA, 120 nucleotides. Its full nucleotide sequence was described by the Sanger group in 1967. A new fractionation system had to be developed. This was a kind of displacement chromatography using ion-exchange paper. This "homochromatography" technique, as it was called was later adapted for use on thin layers. It now became possible to attack even larger RNA molecules. The RNA genetic material — viruses can have either DNA or RNA as their genomes — from the bacterial virus R17 was selected. The RNA was successfully fragmented, but the full genome was never completed in Sanger's laboratory. This was achieved first in 1976 in Walter Fiers' Laboratory of Molecular Biology in Ghent, Belgium. However, Sanger and his colleagues managed to sequence a part of the RNA coding for the coat protein of the virus. Since the amino acid sequence of this protein had been determined it was possible to control the conditions of genetic coding. According to Sanger's own words:

> That was quite exciting because that was the first time that a nucleotide sequence had been determined and shown to be related by the genetic code to a known amino acid sequence in a protein. I mean one of the purposes of going into nucleic acid sequencing was to try to break the code, but in fact the code had already been broken by the time we got there. We weren't the first people to break the code but it was a good confirmation. I think it was worthwhile.

It can be added that in this phase of work on RNA Sanger often declined being a co-author on publications by his junior colleagues. Only if he himself had actively participated in the hands-on work was he willing to do that. Sanger and his colleagues were now ready to take on DNA.

Sanger receives his second shared Nobel Prize in chemistry in 1980 from the hands of His Majesty the King. [From Ref. 10.]

The Second Nobel Prize

The background to Sanger's second Nobel Prize in chemistry in 1980 will be unavailable for archival studies for many more years to come. The only direct sources available are the documentations in *Les Prix Nobel en 1980*[15,16], comments on his work by Sanger himself or by others, e.g. in Brownlee's biography[10] and the Finch book on the history of the Medical Research Council Laboratory of Molecular Biology in Cambridge[14].

The successful and practical DNA sequencing technique developed by Sanger and his colleagues dramatically changed the world of science. The impressive achievement was rapidly recognized by the shared half of the 1980 Nobel Prize in chemistry with Walter Gilbert "for their contributions concerning the determination of base sequences in nucleic acids." Publications describing both these methods had appeared in 1977. Unlike the Sanger method the Maxam and Gilbert technique could be applied to double-stranded DNA, which the original Sanger technique could not. In the initial phase the former technique therefore was preferentially used. The discovery of heat-resistant

DNA polymerases, originally from bacteria living in hot springs, changed the picture completely. The Sanger technique could now be applied also to double-stranded DNA and since his technique was much easier to automate, it rapidly and completely took over the quickly growing market of DNA sequencing.

In the early studies the major problem to tackle was the large size of most forms of DNA. However, as in the case of RNA, one could consider using the genome of a virus, which has a much more limited size than cellular DNA. The virus chosen was φX174. One particular challenge with this DNA was that it is double-stranded. An alternative was to use a genome from a bacterial virus like lambda, which is in fact single-stranded in its natural state. There were many steps in the complicated journey towards a practical method of sequencing DNA. To cut a long story short it included finding methods to synthesize a copy fragment matching strand of a single-stranded nucleic acid (a refined "primer" technique was required), to be able to stop this synthesis at a specific nucleotide, to develop separation techniques that distinguished strands only differing by one nucleotide, to find ways of separating the two strands in DNA (enzymes with such a capacity was found in certain extremo-philes, bacteria that could grow at a high temperature), to develop "chain terminator" molecules (originally described by Arthur Kornberg), etc . Many colleagues working with Sanger in the laboratory contributed to the step-wise developments of techniques. Elisabeth Blackburn, to mention only one, made important contributions to developing techniques for synthesizing a second strand of nucleic acid on a preexisting single strand using particular enzymes called polymerases. Later on she herself became a Nobel prize recipient for her pioneering studies of particular enzymes, telomerases, that control the double-strandedness of DNA at the ends of chromosomes. In 1981 Sanger and his collaborators were finally able to publish the full and correct sequence of φX174 DNA, correcting 30 errors from an early attempt (using an intermediate plus and minus strand system) to determine the exact position of the 5386 nucleotide long genome.

The viral genome was found to be even more compact than anticipated. It contained overlapping genes, a very unexpected finding. This meant that certain stretches of the DNA were used for the synthesis of more than one kind of protein. This was in a sense a blow to the classical one gene one enzyme concept defined by George W. Beadle and Edward L. Tatum[1]. They had received half a Nobel Prize in physiology or medicine in 1958. The prize motivation was "for their discovery concerning genetic recombination and the organization of the genetic material of bacteria." The use of overlapping genes provides an impressive way for Nature to compress the use of a genome, but it has the weakness that a mutation in the overlapping part of the gene can have a deleterious effect on either one or both of the proteins simultaneously coded for. Sanger's DNA studies also led to another surprising finding. As described in Chapter 4, the small energy-producing organelles, the *mitochondria*, in our cells contain a small piece of DNA that is their own. Sanger and collaborators determined its nucleotide sequence. The completely unexpected finding was that the genetic code was not categorically universal as predicted. Briefly the genetic code is based on triplets of nucleotides (a codon), as recognized by the 1968 Nobel prize in physiology or medicine which has already been mentioned. In the final code that was established the codon UGA represented a stop signal. The work by Sanger and collaborators showed that in mitochondrial DNA, surprisingly, this was not the case. As a vestigial of an old bacterial world the triplet UGA instead coded for the amino acid tryptophan.

Bo G. Malmström started his introductory address at the prize ceremony in 1980 with some more general philosophical comments[15]. He said:

> This does not mean that it is desirable, or even possible, to try to reduce all problems of biology to biochemistry. To understand our own place in the Universe we need the softer data provided by the social sciences and, not least, by literature. But the intimate relationship, seen in this century, between fundamental research in biochemistry and medical progress has demonstrated that it is not solely to satisfy the intellectual curiosity of the biochemist that we ought to try as far as possible to reach a description of life processes in chemical molecular terms."

Sanger's second Nobel lecture "Determination of nucleotide sequences in DNA"[16] — this time given two days *before* he received his prize — described the different steps in the development of the practical and effective sequencing technique. It also highlighted different interesting observation along the way; the complete sequence of φX174 DNA demonstrating overlapping genes; the heterodox coding system of mitochondria; the particular features of transfer RNA, and the distribution of protein genes, emphasizing the significance of intergenic regions. The whole lecture was very focused and avoided speculation, all in agreement with Sanger's scientific personality.

In his discussion with Hargittai[17] Sanger acknowledged that he considered "it was the dideoxy method which was the climax of my work. That did make it possible to sequence large DNA molecules." But is this really true?

The fundamental consequences of the demonstration of the amino acid composition of insulin should not be overlooked. It did not only lead to a wide application of a new technology, as in the case of DNA sequencing, it also led to a completely new way of thinking. The conceptual transformation when it was understood that each protein had its unique amino acid composition and that the components could form any random predetermined sequence, was huge. As expressed by Judson[18], "Information, for the biologist, was embodied in the extreme specificity of proteins and nucleic acids that was become evident in the early fifties. In such terms, what Chargaff and Sanger did was to reduce the chemistry of these molecules to a total dependence on information." The Nobel prize in chemistry is specified in Nobel's will to recognize a *discovery* or an *improvement*. It would seem that a comparison of the presentation of the structure of insulin with that of developing a DNA sequencing technique implicitly should be interpreted to mean a much larger relative emphasis of the discovery than on the improvement quality in the former than in the latter case. There is of course no sharp boundary between a qualified improvement and a discovery. In German there is only a two-letter difference; "erfinden" means to invent, whereas the shorter "finden" means to discover.

The Revolution of DNA Sequencing

It would be going too far to describe in full the explosive developments caused by the introduction of techniques to read nucleotide sequences of DNA. A number of milestones have been passed and the potential for the future is enormous. The first microbial genome was published by J. Craig Venter and collaborators, in particular the Nobel laureate Hamilton O. Smith, in 1995. A few years later Venter and his collaborators characterized the genome of a meningococcal group B bacterium. This had a major practical consequence. Vaccines against meningococcal group A and C have been available for a long time but now it became possible in collaboration with Novartis Vaccines for the first time to also produce a vaccine providing protection against meningococcal group B infections. The use of this vaccine will prevent the epidemics, often with a high frequency of lethal disease, among for example college youths.

The new DNA sequencing techniques have also for the first time allowed us to make an inventory of all forms of life. It is fortunate that ever since DNA emerged as a stable carrier of information, its universal language has prevailed essentially unmodified in all forms of entities involved in evolutionary events, be they cellular organisms or viruses. Previously there was a limitation in that we could examine in detail only a cellular species or a virus that we could manage to replicate in the laboratory. This limitation does not exist any longer. It is now possible to examine all DNA present in, for example, a few hundred liters of ocean water. Particles of different sizes retrieved by passing the water through filters with graded pore sizes and the final ultrafiltrate can be examined after controlled DNA fragmentation using a "shot-gun" approach. Determining the nucleotide sequence of all fragments of genes, a technique referred to as *metagenomics*, we can recognize in a remarkable way their representation as a part of a genome constituting a previously unknown form of life. In order to identify such a new species the fragments have to be lined up together via their overlapping parts in a particular form of jigsaw puzzle. The identification of microbes that cannot be propagated in the laboratory by unraveling their genome is sometimes referred to as studies of "microbial dark matter." In the end it will be possible to identify essentially all forms of life on Earth, but that will take a very long time. We will then learn, perhaps to our surprise, that the overwhelmingly largest part of these forms is invisible to us. If we truly want to study regional or global ecology we need to include these entities which are also deeply involved in the dynamics of evolution. This also applies to discussions of the extinction of species. The world of microbes and viruses dominates not only in numbers but also surprisingly in the relative representation of organic mass! The use of genomics will allow us to collect an amazing amount of new knowledge. There will be many surprising findings ahead!

The Evolution of Life

We have already had to change our view of how early life has evolved on Earth. Originally it was believed that there were two Kingdoms (domains) of organisms, the bacteria and the much more complex nuclei-containing eukaryotes. It was then shown by Carl Woese that bacteria could be divided into two major groups, *Bacteria* and *Archaea*. A distinct difference was found in the signature sequences of their 16S RNA in ribosomes. When possibilities to characterize the nucleotide sequence of the whole genome of prokaryotes became possible, pronounced differences between the DNA of Bacteria and Archaea were discovered. It was Craig Venter's group that in 1995 determined the nucleotide sequence of the whole genome of a representative of the former,

Haemophilus influenzae, as briefly mentioned and only a year later the same group in collaboration with Woese presented the corresponding sequence of an Archaea, *Methanococcus jannaschii*. Woese's revolutionary findings in biology never came to be recognized by a Nobel Prize. Probably it was difficult to fit this discovery into either a prize in chemistry or in physiology or medicine. However, in 2003 The Royal Swedish Academy of Sciences awarded Woese the Crafoord prize in Biosciences, monetary-wise half a Nobel Prize. This prize will be further presented in Chapter 9. The discovery of the two domains of microbes was very important, but it remained to identify what relative contributions they have made in the early evolution of the much more complex eukaryote cell. Its evolutionary origin still remains one of the conundrums of modern biology, although there are

Carl Woese receives the Crafoord Prize in 2003 from the hands of His Majesty the King. [Photo by Torbjörn Zadig.]

some important recent advances. Various pieces of evidence for the relationship of eukaryotes to both Bacteria and Archaea have been presented. A particular case is the energy-yielding mitochondrion introduced in the previous chapter and also mentioned above.

Some one and a half billion years ago a primitive eukaryote cell became associated with a prokaryote, an alphaproteobacterium closely related to intracellular parasites of the kind called Rickettsia. The process of merger is referred to as *endosymbiosis*. The newly acquired microorganism developed to become an integrated part of the eukaryotic cellular machinery. Its origin was revealed by the fact that it had its own DNA. This has a circular form and is replicated in an independent manner when located in the cytoplasm of the eukaryotic cell. Mitochondria have their own transcriptional and translational machinery. It is not as efficient as the machinery for replication of chromosomal DNA, and hence there is a higher frequency of mutations. Over the long period of evolution the relationship between the host and its permanent visitor has become very intimate. The mitochondrion has lost a number of its original genes, and their function has been taken over by genes of the host cell. Because mitochondria still retain a number of their own genes there are genetic diseases that occur for example in man which are due to mutations in the mitochondrion genome and not in the eukaryotic genomes. They are referred to as *mitochondrial diseases*.

The existence of extra-chromosomal DNA offers some interesting possibilities to study our ancestry. At fertilization it is the ovum that exclusively contributes the mitochondrial DNA and hence the maternal lineage can be traced back in time by characterization of nucleotide sequences of this DNA. In a corresponding way it is possible to do this on the male side by determining the nucleotide sequence of the Y chromosome which has an exclusive male origin. The dawn of modern man can be traced to the time period 150,000–200,000 years ago. We are indeed a young species or even subspecies as we shall see.

The postulate origin of the mitochondrion only provided one possible explanation for the origin of different organelles, the various membrane-enclosed structures, including the nucleus, in the eukaryote cells. Recent data strongly suggest that the eukaryote cell in its evolutionary origin is much closer to certain kinds of Archaea than to Bacteria. It was mentioned in the previous chapter that oxidation was a critical source of energy in organic life. However, there is one other source of energy that was of critical importance during the first two billion years of development of life in the absence of oxygen and probably at the present time still plays an unrecognized important role. As an alternative to oxidation many organisms can harvest energy from inorganic chemicals in the environment. This process is referred to as a *lithotropic* — feeding on rocks — as an alternative to *phototropic* — depending on sunlight. Many Archaeas are lithotropic and an interesting representative is a newly identified genetic lineage referred to as *Lokiarchaeota*, a member of the TACK superphylum. This organism has been found in deep (more than 3000 meters) marine sediments in the vicinity of an active venting site, also referred to as "a black smoker," named Loki's Castle at the Arctic Mid-Ocean Ridge. The name Loki has been taken from Nordic mythology. Loki, the son of two giants, was a complex personality, both kind and deceptive. It was he who initiated the events that led to the end of the world in what is called *Ragnarök*. Many molecular genetic characteristics of this kind of microorganism give strong evidence for similarities to signature sequences of critical eukaryote organelle systems. Taken together it has now been strongly proposed that certain Archaeas represent the main ancestry of eukaryotes.

New Basis for Classification of Plants and Organisms

On the practical side reading the books of evolution has now allowed us to substitute Linnaeus's empirical classification of plants by a systematization based on genome characteristics, as briefly mentioned above. Empirical data were exchanged for objective experimental measurements. Linnaeus introduced an artificial sexual system based on the number and arrangement of male and female organs of the plants. Departing from 5,900 identified species he defined 1,098 families which were eventually divided into 24 different classes. Using the modern technique of comparing the DNA sequences of different species of plants it has been found that all these classes, except number 15 Tetradynamia, included plants of diverse genetic origin[19]. The most commonly used system for classification of plants today is the APG (Angiosperm Phylogeny Group) system which a few years ago included 250,000 species classified into 45 orders. This Group publishes its continually updated classification data in the *Botanical Journal of the Linnean Society*! The comprehensiveness of this system grows as the DNA of the genomes of more plants is characterized. It should be emphasized that Linnaeus's sexual system has not been useless or misleading. It was a well chosen artificial system which for a long time formed a basis for much intensified research and allowed a major increase of knowledge about plants. In his hierarchical classification of everything on Earth Linnaeus used a separation into three categories — animals, plants and minerals (microbes had not as yet been identified in his time). It may seem that minerals are out of context, but unexpectedly it has been found that mineralization too is heavily influenced by the presence of microorganisms and their viruses. However, it may at the present stage be difficult to form a hierarchical classification of rocks like the one generally applied for subdivision of life forms since Linnaeus time; Domain, Kingdom, Phylum, Class, Order, Family, Genus and Species.

In contrast to Tiselius and Sanger, the genial Linneaus did not refrain from emphasizing his own contribution — *Deus creavit, Linnaeus disposit*; God created and Linneaus put into order. Still the only true interpretation of the generation of diversity (god) is to objectively interpret the genomic variation resulting from evolutionary processes. As formulated by the great biologist

Theodesius Dobzhansky "Nothing in biology makes sense except in the light of evolution." Currently we do not, as expressed by Linnaeus, "view God from the back," we look him straight in His face, although according to the Book of Exodus in the Old Testament this should not be possible. One wonders what would have happened if Linnaeus had been born some hundred years later. Would it be possible that he could have managed to separate himself from the paralyzing effect of his Lutheran orthodox monotheistic religion and have had a vision of evolution? He left the creation of species to God but allowed Nature to introduce variations and he, surprisingly, created an order of Primates (from Latin *primus*, the foremost) including man, the pinnacle of creation, *together with* the other higher apes!

Human Genomics

Naturally, a major focus has been on the sequencing of our own genomes. In 2001 the first drafts of the human genome were presented on the same occasion by the international consortium representing the Human Genome Project and the Celera group headed by Venter. Since then there have been a dramatic development with regard to speed and simplicity of sequencing and to the reduction of costs for the analyses. Because of these developments it is presently possible to characterize the full genome of a single human for less than one thousand U.S. dollars. An amazing amount of new information has been and is continuously being gathered at an accelerating speed. However, much still remains to be known before the full impact of this potentially very valuable information can be defined.

One of the original initiatives to survey a larger number of human genomes was taken by Kári Stefánsson in Iceland. He started the company deCODE Genetics already in 1996 to characterize the genomes of a large part of the population on the islands of this isolated country, totaling slightly in excess of 300,000 people, to search for genetic determinants of disease. The advantage of the Icelandic population is that it can be traced back more than one thousand years. The founder Viking male population originated in Scandinavia and a considerable part of the female founder population came from Ireland. A more recent initiative has been taken by Venter, a remarkable pioneer in genome sequencing and also in developing frontline biotechnology companies exploiting the new technologies. His latest creation is Human Longevity, Inc. (HLI) which plans to harness in full "for the first time the power of human genetics, informatics, next generation DNA sequencing technologies and stem cell advances." One goal is to improve our approach to how medicine is practiced. The target hence is to increase the quality of life and delay the process of aging. Eventually the new insights will also lead to the introduction of targeted preventive measures and to major recommendations of change of life style of the individual, partly reflecting knowledge about his genetic vulnerability.

The new tools for genomic studies have also become of great importance in forensic processes. They have allowed the exoneration of many prisoners including many death row inmates. It has also been possible to demonstrate that a large number of already executed prisoners in fact have not been guilty of the crime they were accused of. This is another and a very powerful argument against capital punishment. There are also many other applications of forensic medicine. Paternity issues can be readily resolved and for example it has been documented that Thomas Jefferson fathered Sally Hemings' six children. There remains a great question as to why this remarkable man who wrote "We hold these truths to be self-evident, that all men are created equal …" did not, like George Washington, set his slaves free after his death. He did, however, set his own four surviving offspring free and it might be added for the sake of completeness that Sally was the half-sister of his wife, who had died young. One might raise the question if it would be possible to determine who among two identical male twins is the father. Identical twins have different finger prints and eye bottom patterns, since both heredity and environment influence their development. So what about DNA? There are suggestions that its sequence characteristics in certain stem cell populations in the body may become modified to give an individually unique pattern. Thus a case of paternity might also be determined in this exceptional situation. Finally, it may be added in passing that because of the unique specificity of DNA it is also used to barcode-label expensive artifacts and art to secure identification in case they are stolen. Apparently there are no limits to the potential of using DNA in the context of legal matters. Also in general language, DNA is now used in a transferred meaning for "the heart of the matter."

Let us now return to the question of the origin of our own species as examined by genetic analysis. By a remarkable development in techniques for sequencing and identifying chromosomal DNA material from ancient bones, Svante Pääbo, whom we will meet briefly in the next chapter and his group have made it possible to look far back in time. The great challenge has been to separate out the minute amounts of extensively broken down specific DNA from all contaminating microbial DNA and even human DNA from archeologists and others persons previously handling the preparations. It is a more daunting challenge than finding a needle in a haystack! At last it has been possible to determine the relationship between the DNA of Neanderthal man, extinct since some 30,000 years ago, and to compare it to the genome of modern man[20]. Quite amazingly it has been proven that they belong to the same species and could conceive a common offspring. One wonders how many subspecies of Homo sapiens might have existed at the time when modern man developed. And when it comes to consciousness, discussed in Chapter 1, what qualitative differences may have been represented in our ancestors with a more restricted intellectual capacity than we have?

What a difference it would make if this progressively growing knowledge about the genetics of our historical forefathers became common knowledge for all the people in our world. If everyone knew, as proven by the Sanger's DNA sequencing technology, that we are all brothers and sisters and that the dark-skinned ancestors of all of us living outside Africa moved out of that continent some 60,000 years ago, our world might look very different. Hopefully it will eventually be a knowledge shared by all that there are no races and that the concept of nations is a cultural construct, ready to be erased for the good of global man. If it had been possible for him to learn about this advancing knowledge, I believe that Sanger, with his Quaker upbringing, would have enjoyed noticing this particular benefit to mankind of his sequencing technique.

It should be added that Sanger was never involved in patenting his discoveries. He had a practical altruistic view of science and simply wanted his findings to become quickly available to fellow scientists and to be used for the public good. Cesar Milstein, a fellow scientist at the Laboratory of Molecular Biology in Cambridge, who together with Georges Köhler invented the hybridoma technique[2] seems to have had a similar personality. The discovery of this technique, recognized by half a Nobel Prize in physiology or medicine in 1984, was never patented. In the aftermath one may of course speculate about what difference it would have made to the Medical Research Council in England if either one or both these two revolutionary discoveries had been patented.

There has been a dramatic increase in the speed and precision of Sanger DNA sequencing over the years. In October 2006 the new Swedish Embassy close by the Potomac River in Washington D.C. was to be inaugurated. The Swedish King and Queen honored this event with their presence. In the morning the Karolinska Institute had arranged an interesting series of lectures highlighting developments in brain research. After this the Royal couple had lunch in the White House with President George W. Bush and this was followed

The Swedish King Carl XVI Gustaf and Queen Silvia visiting the J. Craig Venter Laboratories, Rockville, MD, in 2006. [Photo by Brett Snipe.]

by a visit to the laboratories of the J. Craig Venter Institute in Rockville, MD, which I in my function as Lord Chamberlain-in-Waiting at the Royal Swedish Court had arranged. Because of the curiosity and enthusiasm showed by the Royal couple it was a very successful event. The visit was also unique because for the first and only occasion I saw the colleagues in the laboratories dressed up in ties and suits. For natural reasons the normal dress code in laboratories is very informal, to say the least.

One particular demonstration appreciated, in particular by the Queen, was the School Bus run by the Institute. This bus is used for visits to local schools and in connection with this the students can get an opportunity to try out DNA sequencing on their own. The Queen, who has a deep involvement in philanthropy with a particular focus on children and the elderly, was very enthusiastic about this particular outreach. She has by the way an honorary doctorate from the Karolinska Institute, which I in my function as Dean had the privilege to bestow on her at a promotion event in the early 1990s. However, the main reason for mentioning this visit here is the impression that the sequencing facilities gave on the Royal couple. At the time there was a separate division of the Institute employing some 60 people that managed some hundred sequence apparatuses housed in a separate building. Large volumes of new data were produced during the running of all the machines night and day. Today this whole set-up has been dismantled and substituted by a single new generation machine that can now produce in 24 hours more sequencing data than the original setup did over three years!

To Write the Books of Life

To end this section it should be added that it has now become possible not only to read the books of life but also to write them. Venter and his group have taken the studies of a genome from a microbe much further. Selecting a relatively small genome present in a mycoplasma and developing new technology it has been possible to *synthesize* the whole genome of the microbe in the laboratory as described earlier[1]. More than a million nucleotides were placed in their exact position in the nucleotide chain, a major feat. The synthetic genome molecule was introduced intact into a slightly different kind of mycoplasma with an incapacitated genome. After a few rounds of protein synthesis all the components of the cell derived from the transplanted genome. A fully synthetic new cell had been produced! Venter has described this adventurous scientific journey in a book entitled *Life at the Speed of Light*[21]. This title alludes to the fact that it will now be possible to electronically transfer the information about a genome from one laboratory to another and allow the receiver of this information to synthesize the selected genome in his own laboratory. One particular advantage of having access to a synthetic complete microbial genome is that the role of each individual gene can be assessed and furthermore that the regulatory role of non-protein coding DNA can be analyzed in more detail. This study has been concluded at the time of writing and the results are both enlightening and intriguing[22]. The mycoplasma genome has been reduced to 477 genes cutting out genes of no major importance on the rate of replication of the organism. About 15% of these genes code for proteins of unknown function. Thus a considerable amount of work remains to determine the critical role of these gene products. Most likely surprising revelations will be made in this work. When the function of essentially all the gene products has been identified it will be possible to examine phenomena of the interaction of products in the regulation of subsistence, growth and division. The phenomenon of homeostasis, the means of establishing a balance of intermediary metabolism in a single cell during different phases of life should eventually be possible to study. Later still it might be possible to also analyze the complicated regulatory homeostatic phenomena in multicellular systems under variable environmental conditions.

The Importance of Hands-on Work in Science

Sanger had a very strict work ethic, which could probably be traced back to his schooling and general upbringing as a Quaker. He was an ingenious experimentalist and worked throughout his whole career with his own hands in the laboratory. Already as a young boy he had enjoyed carpentry and welding. His daily scheme of work included starting at about nine in the morning and a lunch break for some two hours when he walked home to have a meal with his wife. He then returned to the laboratory and often stayed for the evening. One generally found him working in the laboratory, and not in his office. There is something special about the hands-on work in a laboratory. It serves as a good balance to the more speculative intellectual aspect of the commitment. The planning of experiments and the performance of them are two sides of the same coin. Manual dexterity is not given to everyone and an "all thumbs" personality may cause limitation in the practical manual work. We have already encountered examples of messy laboratory workers like Hevesy[2] and Hammarsten cited in the previous chapter. As a scientist advances in his career there is often a pattern of leaving the hands-on work to younger colleagues.

The techniques used by scientists of course vary depending on the experimental discipline. Originally pipetting was performed by mouth, but for safety reasons this was changed to the use of mechanical devices. However, the term remains in colloquial laboratory language. When I worked for a period in Michael Oldstone's laboratory at The Scripps Research Institute he blocked off time for experimental work by asking his secretary to say, in the case of incoming morning calls and spontaneous visits, that he was "pipetting." Other colleagues have made similar arrangements. The 1975 Nobel laureate in physiology or medicine Howard Temin's secretary was instructed to say to everyone who tried to contact him before noon that he was in a meeting. Another technique I learnt from Bengt Samuelsson, the star chemist and co-recipient of the 1982 Nobel Prize in physiology or medicine, briefly mentioned above. In the early 1990s when I was the Dean of the medical faculty and Bengt the Vice-Chancellor I noticed that he had such a miniature time calendar that there was only room for one fixed commitment per day! It would seem to be a general very wise rule to compartmentalize the work by some arrangement. Modern communication has been taken over by the use of email. One wonders how many of the scientists active today can wait to open their email till after lunch or later?

During my involvement in studies of viruses between the 1960s and 1990s, I could follow how the discipline became increasingly more biochemical. We continuously had to adapt new techniques. Originally it was critical to manage the technique of growing cells using various tissue culture techniques. It was then important to ensure that sterility was maintained. We used special coats, white outside the sterile room and green inside it. The colors of the wooden clogs used also changed, white outside and red or blue inside the sterile laboratory. With time progressively more efficient conditions employing laminar fluid hoods became available. The critical goal was not only to secure the absence of contaminating microorganisms in the cultures used, but also to ensure that the particular virus studied did not spread to those working with it. It was previously described[1] how a number of scientists had to sacrifice their life in order to unravel the secrets of yellow fever virus infections and develop means to prevent them. The risk of such a spread obviously varies depending upon the kind of virus studied. In modern laboratories studying Ebola virus, very strict conditions are used and the experimenter has to dress in what looks like a space suit. Chemists have long been used to working with dangerous chemicals. They were managed in specially ventilated hoods. In the 1950s there was a switch from glass to plastic material in the containers used. There was also a rapid scaling down of volumes. Milliliters became microliters. Pipettes were exchanged for the "spiral loops" invented by the Hungarian microbiologist Gyula Takátsy when serial dilutions of samples were prepared.

The introduction of the use of isotopes led to the development of completely new working practices. The amount of the invisible potentially dangerous irradiation needs to be monitored. Sanger has told a story about when the laboratory was labeling developing chickens with ^{32}P in order to isolate the egg white for studies of ovalbumin. When the chicken had laid its egg there was a "scream" from the Geiger counter measuring the dose of isotope irradiation in the room. The scientists had forgotten that an egg has a very phosphorous rich egg shell! The radiation officer was not pleased. In the 1960s we made extensive use of the separation techniques introduced by Svedberg and Tiselius. Virus components were separated according to size by rate zonal centrifugation and by density by equilibrium centrifugation. When the ultracentrifuge had stopped after the run we took out the centrifuged plastic tube, made a hole in its bottom and collected fractions by counting the drops that were slowly let out. Today this kind of "Chinese" fraction collector is no longer used. Modern science to an ever increasing degree depends on and hence requires a familiarity with relatively advanced equipment. In order to feel comfortable in their management one needs to be in regular contact with and to use the particular instrument needed for the experimental work. Gradually everything has become automated. DNA sequencing is a point in case. By now everything is automatic and made enormously more effective than the first machines for Sanger sequencing made available as exemplified above.

A beautiful experiment is one in which the controls are absolutely clean and uncontestable. My students were not always happy when I started by asking about the controls in our discussion of a particular experiment. Someone who has published what turns out to be artifacts rapidly loses his standing in the scientific community. To me it is incomprehensible that on rare occasions scientists try to forge their data. Sooner or later this will always be revealed by the large jury of colleagues that every scientist is surrounded by. If and when a scientist advancing in his career retreats to his office for an ever greater part of his time, he eventually loses contact with the experimental side of the work. This can, but need not be, a considerable disadvantage. I wrote my last scientific publication based partly on experimental work that I had done with my own hands the year that I turned 50. Theorell seems to have been comfortable participating in the development of techniques long into his career. In the case of Sanger there is no question of where he belonged. He was a laboratory bench worker throughout his whole professional life and this was a part of his great success. In one of the interviews Sanger gave[10] he said the following:

A lot of Nobel prize-winners take on big administrative or teaching jobs or do something else. I am not very good at teaching and I don't think I would be very good in administration either. I don't think I would have enjoyed any of this the way I enjoyed research. I had this opportunity, having got a Nobel Prize, to have a steady job and good facilities. It was easy to get students, particularly postgraduate students, who were well trained and this helped very much with research. I was in a position to do more or less what I liked, and that was doing research.

Sanger and Students

At the banquet following the prize ceremony in 1958 Sanger had to overcome his natural shyness and address the audience. He expressed his immense enthusiasm for his professional involvement, although he appropriately pointed out that most often nothing or very little has been achieved by the intense efforts made. In the particular part of the address directed to students he again emphasized the joy of science and underlined that it is driven by enthusiasm and new ideas. He finished the address in the following way:

> We are all student of nature and I believe the bonds of common interest between scientists and between students all over the world is doing and will do much to foster the friendship between different nations that is so much needed today.

It should be recalled that Sanger was brought up as a Quaker. Sanger appears to have been a very positive and likeable person. His science kept his spirit young. Thus when addressing the students one more time in 1980 he said:

> When I was here 22 years ago I addressed the students of 1958 as "fellow students" because although I was 40 years old I still felt that I was one of them, and I still feel the same today. I and my colleagues here have been engaged in the pursuit of knowledge. We have been learning, are still learning and I hope will continue learning.
>
> I believe that we have been doing this because we were interested in the work, enjoyed doing it and felt strongly that it was worthwhile. Scientific research is one of the most exciting and rewarding of occupations. It is like a voyage of discovery into unknown lands, seeking not for territory, but for knowledge. It should appeal to those with a good sense of adventure.
>
> When I was young my Father used to tell me that the two most worthwhile pursuits in life were the pursuit of truth and beauty and I believe that Alfred Nobel must have felt much the same when he gave these prizes for literature and the sciences. Through art and science in their broadest senses it is possible to make a permanent contribution towards the improvement and enrichment of human life and it is these pursuits that we students are engaged in.

So speaks the only two-time recipient of a Nobel Prize in chemistry. Could it be said better and more simply? The already mentioned interview that Sanger gave to Hargittai in 1997[17] and Brownlee's biography of him[10] sheds some very good light on this pleasant laureate. Sanger described how he decided not to become a medical doctor like his father, because he thought he could have a greater impact as a scientist. Sanger's upbringing as a Quaker led to his becoming a conscientious objector in the Second World War. This allowed him to get an early start in science. His contacts with science led to that him shedding his Quaker beliefs and he explained this in the following simple way:

> I'm no longer a Quaker. When I started studying science, I found it very difficult to believe something I did not know. In science you have to be so careful about truth. You are studying truth and have to prove everything. I found it very difficult to believe all the things associated with religion. I call myself an agnostic. There are things that we can't explain, like our consciousness, and I have an open mind about that.

Although Sanger may not have been a philosopher or a speculative hypothesis-driven scientist, he managed by his enormous skill and inventiveness in the laboratory and his capacity to stay with a problem to change the world of science. He tackled fundamental problems by a stubborn approach. One may ask again how important his Quaker upbringing was. He remained a pacifist and this allowed him to continue his development of scientific pursuits during the Second World War. Hodgkin, with a similar background reacted differently. He took a break from his scientific developments during the war to help his country fight Nazi Germany. For a number of years he made major contributions to the development and application of different kinds of radar systems. A good part of his biography[23] is devoted to this phase of his scientific life as described in Chapter 3. However, what Sanger and Hodgkin shared may have been an important value system, anchored in absolute truth. This search for truth obviously need not be based on some form of religious belief. Involvement in science for both of them led to a loss of previously-held beliefs. Conjectural systems of faith are difficult to reconcile with insights into the wonders of evolving life.

The first Quakers separated from the Church of England in the mid 17th century. They were barred from studying at Cambridge and Oxford by a Uniformity Act which was not repealed until 1862. Although a minor church it has had an impressive impact on the advance of human culture. One need only mention its important role in the abolition of slavery in England. Quakers generally tend to be skeptical about established authority. They only take off their hats when meeting God. Some early leading scientists were Quakers. Examples are John Dalton and Arthur S. Eddington and as discussed in this book Hodgkin and Sanger. What qualities of Quakerism could be conducive to good practice in science? The essentials referred to are truth, simplicity and integrity. To this can be added a testimony of community. The latter implies the demand for an open, uncommitted debate about problems. Thus Quakers hold that no doctrine, theory or belief can be taken as being manifestly true. It must always be subjected to debate and questioning in a constant quest for truth. This is somewhat like the Jewish tradition of openly discussing all the statements of the Torah, apparently a useful code of behavior as evidenced by the remarkable overrepresentation of scientists with a Jewish ethnic background among Nobel Prize recipients.

The Halcyon Days

Sanger the gardener. [From Ref. 10.]

At the age of 65 Sanger retired. No one was successful in trying to convince him to stay on. He had indeed achieved much more than most scientists could ever dream of. He and his wife had bought a larger home, which they named *Far Leys* like the name of Sanger's childhood home. At this new lot he could spend a good part of his leisure time gardening. Judging from pictures of his garden he was also a very skilled gardener, again a hands-on activity. Theorell also enjoyed trimming his roses. Sanger managed to give more time to his family, to his three children, none of whom contemplated a career as a scientist, and to his grandchildren. He also enjoyed marine activities, both sailing and motoring, and he built or refurbished the boats he used.

Possibly one could refer to this period of his life as his halcyon days. He entered a calm phase in his everyday activities and could enjoy his garden and the sea. According to the myth recorded by Ovid the bird Kingfisher became endowed with a capacity to calm the sea and this occurred in connection with the winter solstice, when according to the folk tale the bird was incubating her eggs. With time the word "halcyon" has come to describe blissful days and to become less connected with nesting. As expressed by Walt Whitman in *Leaves of Grass:* "Then for the teeming quietest, happiest days of all! The brooding and blissful days!"

When asked to look back at his professional career Sanger humbly referred to it having been mostly a matter of being in the right place at the right time, but that is certainly only a very minor part of the truth. His impressive achievements were widely recognized. He became a Fellow of the Royal Society as early as 1954 and received its Royal medal in 1969 and its very prestigious Copley medal in 1977. In 1967 he became a Foreign Member of the National Academy of Sciences and in 1981 he received a similar membership in the French Academy of Sciences. The number of other prizes and honorary doctorates were too many to list. On the civic side he became in 1963 Commander of the Order of the British Empire (CBE), in 1981 Companion of Honour (CH) and in 1986 he received the Order of Merit (OM). However, like Stephen Hawking, he declined being knighted. He could not see himself being addressed as Sir. His name will live on forever in the annals of science and also by the existence on the outskirts of Cambridge of the large Wellcome Trust Sanger Institute, previously The Sanger Centre, established in 1992. This institute serves as a critical source of documentation on DNA sequences and currently employs some 1,300 people.

In the following chapter we will discuss a completely different branch of biochemistry. It concerns the remarkable world of sterols and the Nobel Prize in physiology or medicine in 1964 to Konrad Bloch and Feodor Lynen.

Chapter 6

1964 — An Important Year in the World and for Nobel Prizes

BY CHOLESTEROL
FLEXIBILITY OF MEMBRANES
AND MOVEMENT OF CELLS

After the Second World War the relief at the cessation of grim violence was mixed with an appreciation of the growing tendencies towards a divided world. It was in 1945 that Churchill used for the first time the expression "the iron curtain," which was not new but highly appropriate to apply as coming events were to demonstrate. The Soviet Union together with a number of annexed and neighboring countries sealed itself off. An arms race developed leading to the build-up and stocking of nuclear arms, primarily involving the two hegemonic world powers, the USA and the Soviet Union. The rising threats to the world developed to become an important issue for international discussions. In 1963 there was an international agreement on global cessation of open air testing of nuclear weapons. One year earlier there had been the Cuban missile crisis. The world, on the brink of a nuclear war, was holding its breath for 13 days until the American political leadership, headed by President John F. Kennedy, managed to make the Soviet ships carrying nuclear arms to Cuba turn around. A year later JFK was assassinated and as a consequence an international mood of depression set in. It was a time for reconciliation and for building new connections. Also the wounds of Europe had started to heal when earlier the same year Konrad Adenauer and Charles de Gaulle had signed the Élysée agreement. In 1964 it was decided to build a tunnel under the English Channel, but also many other initiatives connecting nations were taken. Within the USA an important step in the spirit of reconciliation had been taken in 1963 when Martin Luther King gave his famous speech *I Have a Dream* protesting against racial discrimination (p. 478).

In June 1964 the leader of the Soviet Union Nikita Khrushchev visited Sweden and four months later he was removed from his position. In parallel to this Trofim Lysenko had his final fall. He had become the archetype of politically sanctioned pseudoscience and had his heyday during Stalin's time in power, but had re-emerged temporarily under Khrushchev. We will meet him again in Chapter 8. During these years the field of molecular biology was developing at a fast rate. The language of life was defined to be a redundant triplet code of nucleotides. As mentioned in the previous chapter Tiselius attempted to give this new central field of life sciences a head start in Sweden, but failed. The president of the Royal Swedish Academy of Sciences at the time, Granit, introduced in Chapter 1, wrote a letter to the Russian Academy proposing a closer collaboration between Swedish and Russian scientists. This led to an increased interaction between researchers in the two countries. In mid-October China carried out its first test of a nuclear bomb.

The Nobel prizes in 1964 reflected the changing mood in the world. In physics one half of the prize was awarded to an American and the other half was given to two Russians. The motivation for the prize to Charles H. Townes, Nicolay G. Basov and Aleksandr M. Prokhorov was "for fundamental work in the field of quantum electronics, which has led to the construction of oscillators and amplifiers based on the maser-laser principle." In chemistry the delayed prize to Dorothy Hodgkin was finally awarded, as briefly reviewed by use of supplementary complete archive data in Chapter 4. She was the third woman to receive a prize in chemistry, preceded by Marie Curie (1911) and her daughter Irène Joliot-Curie (1935, divided prize). The prize in physiology or medicine was given to two persons born in Germany, Konrad Bloch who was Jewish and Feodor Lynen, of non-Jewish background. Jean-Paul Sartre received the prize in literature "for his work which, rich in ideas and filled with the spirit of freedom and the quest for truth, has exerted a far-reaching influence on our age." However, he declined to accept it and did not collect his reward. Some years later his relatives attempted to get access to the money, but that was far too late. Then finally the peace prize was awarded to Martin Luther King, but for this particular kind of Nobel prize no motivation was given following earlier traditions. In 2010 Nobel Media AB merged with Nobel Web AB, created by an early initiative by Nils Ringertz, a respected friend and colleague for many years in the Nobel Prize work at the Karolinska Institute. The home page www.nobelprize.org is one of the most visited sites in the world and the single item most often searched for is the presentation of King's peace prize and his acceptance speech.

A Complex and Remarkable Mother Molecule

The motivation for the prize to Bloch and Lynen was "for their discoveries concerning the mechanism and regulation of the cholesterol and fatty acid metabolism." The term cholesterol has an old origin. As early as 1769 the white fatty substance in gallstones was identified by François Poulletier de la Salle and in 1815 the chemist Michel Eugéne Chavreul named the compound cholesterine, which was changed to cholesterol when it was discovered that it was an alcohol. The origin of the word is Greek; *khole* "bile" (cf. cholera, the bilious disease) and *steros* "solid, stiff" (also the origin of the word sterility). Cholesterol plays an important role in the formation of the bile secreted by the liver. It is a precursor of the bile acids it contains. In the gut the bile helps to dissolve the fats present in food. However, most discussions of cholesterol in medicine concerned its importance in coronary heart disease, the central theme of this chapter.

Cholesterol is a representative of a larger group of compounds, the steroids, which have the core structure seen in the figure. It is a relatively complex molecule containing three angularly fused, six-membered carbocyclic rings (designated as rings A, B and C) and a fourth, five-membered carbocyclic

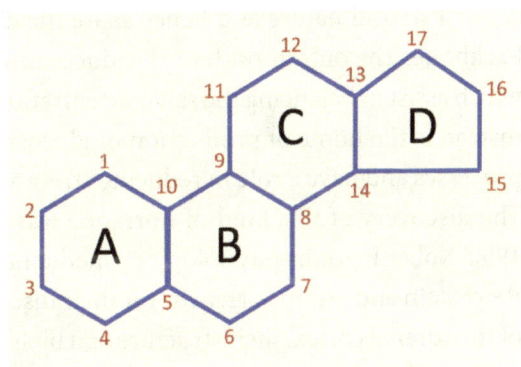

Steroid core structure.

ring (designated as ring D) fused at the edge of ring C. Apparently this carbon skeleton is very useful for the evolution of compounds with many different functions by biochemical modifications of the parent molecule. Additions of side chains and other enzymatic modifications can endow this molecule with a wide range of different unique biological functions. Attachment of an OH group at carbon 3 gives a parental molecule for a group of substances called *sterols*. The most well-known sterol, cholesterol has an additional major side chain extending from the five-carbon ring at carbon 17. This fatty substance

CH3 CH2 CH2 CH3
CH CH2 CH
CH3
CH3
CH3
HO

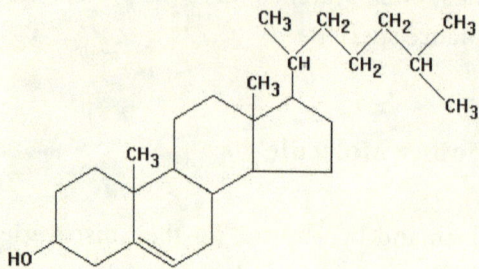

Chemical structure of cholesterol.

is a central component of the membrane of all animal cells. It provides both integrity and plasticity. It can also become enriched to serve specific functions as in the isolation layer of nerves, the myelin sheath, discussed *passim* in Chapters 1–3. Other types of steroids have other functions.

On the top of each of our two kidneys lie inconspicuous organs, the adrenal glands. In spite of their humble appearance they have a central role in balancing many critical stress challenges and metabolic controls in our body. They conduct central events using a range of hormones and in a close interplay with the frontal part of the pituitary gland in the basal part of the brain. We have met this organ already in Chapter 1, since it also produces the transmitter substance adrenaline, which derives its name from it. The glands have three compartments, referred to as the outermost, the middle and the innermost layers. Each one of them produce different hormones, which all are of a steroid nature and hence share the characteristics of the cholesterol backbone. The outermost layer produces mineralocorticosteroid hormones, which assist in balancing the salt concentrations of the blood. The middle layer instead is the home of production of glucosteroids. They regulate metabolic processes and have a role in reducing stress and also inflammatory reactions. The discovery of this kind of hormone was recognized by the award of the 1950 Nobel Prize in physiology or medicine to Edward C. Kendall, Tadeus Reichstein and Philip S. Hench "for their discoveries relating to the hormones of the adrenal cortex, their structure and biological effects" as briefly presented earlier[1]. The main hormone identified was called cortisone and it was found to have a dampening effect on the inflammatory processes leading to rheumatoid arthritis. The innermost layer of the adrenal gland produces sex hormones, estrogens in women, a critical ingredient in birth control pills, and androgens in men. The latter, often referred to as anabolic steroids, are also well-known in modern society. They are used, often illegally, to build up muscle mass. One final example of a related compound of steroids of a different tissue source is vitamin D which was the focus of the prize in chemistry in 1928, soon to be discussed. This vitamin is special in that the B ring of the steroid structure is not closed (p. 337) and the compound cannot be synthesized chemically in the body as also defined by its name. The breaking up of the B ring occurs by the exposure of precursor sterol molecules to ultraviolet light in our skin.

As we shall learn there are more than 30 chemical steps in the biosynthesis of cholesterol, which raises the immediate question of why this particular and very complex group of chemical compounds has been selected by evolution for use in a wide range of different important physiological contexts. Bloch reflected on this in the conclusion of his Nobel lecture, which we will return to.

In the evolution of life on Earth it took a long time before the complex cholesterol molecule appeared. Life emerged on Earth some 3.8 billion years ago. The first two billion years were dominated by various forms of bacteria, lacking a nucleus and with a rigid cell wall, the prokaryotes. Then cells with nuclei appeared, the eukaryotes. Oxygen started to accumulate. The eukaryotes were much larger, much more complex in structure and developed to divide by a different mechanism called mitosis. Single cell organisms, plants and mosses started their long evolutionary journey and progressively more and more complex multicellular organisms evolved. The development of animals some 640 million years ago put special demands on membrane flexibility and led to the use of cholesterol as their core constituent. The emerging capacity to synthesize cholesterol rather than collecting it from the environment was an important evolutionary step. Plants have rigid cell walls and contain certain sterols referred to as phytosterols. As an important part of the diet of animals and humans they may influence our metabolism of cholesterol, in particular if we are herbivores. Some 300 million years ago insects started to develop in parallel with more diversified plants. Insects require access to cholesterol, but have to derive it from an external source since they cannot produce it themselves. Throughout eons of time all animals that evolved used the choles-terol-rich flexible membrane. They also needed to provide the opportunity for certain cells to move around in the complex organism. Therefore cholesterol represents a core constituent of all animal life. There are in fact some bacteria, like mycoplasmas that contain cholesterol, but it is believed that they have developed a capacity to synthesize this compound by a horizontal transfer of specific genetic material from animal cells.

Steroids and the 1927 and 1928 Nobel Prizes in Chemistry

There was a great German tradition in the early development of organic chemistry. Central figures were the already mentioned giants Emil Fischer, who made groundbreaking contributions both in studies of sugars (the 1902 Nobel Prize in chemistry) and of proteins and Willstätter, who demonstrated the chemical similarities between the green pigment of plants and algae — chlorophyll — and the oxygen-carrying protein complex in blood — hemoglobin (the 1915 Nobel Prize in chemistry). The recipients of the 1927 and 1928 Nobel prizes in chemistry represent standard-bearers in this rich tradition. The Prize for 1927, awarded in 1928 recognized Heinrich O. Wieland "for his investigations of the constitution of the bile acids and related substances" and the 1928 prize was bestowed on Adolf O. R. Windaus "for the services rendered through his research into the constitution of sterols and their connection with the vitamins." These two scientists had been nominated a number of

Heinrich O. Wieland (1877–1957), recipient of the 1927 Nobel Prize in chemistry awarded in 1928.

Adolf O. R. Windaus (1876–1959), recipient of the 1928 Nobel Prize in chemistry.

times and had been reviewed in 1924 and 1926 by Olof Hammarsten, already introduced in Chapter 4. He was a very influential Swedish chemist and the uncle of Einar Hammarsten. Olof Hammarsten had concluded that the two nominees were worthy of a Nobel prize and that the strength of their respective candidature was about the same. In 1927 they were nominated again, now as a pair of candidates by the Swedish Nobel Laureate Svante Arrhenius and in another combination by a German nominator Wieland was proposed together with Olof Hammarsten. The committee noted that Olof Hammarsten had made impressive contributions during his long career, but that it was difficult to distinguish a single major discovery. He should not be considered as a candidate. Wieland and Windaus were reviewed one more time by Oskar Widman as a member of the committee. He noted the strength of their candidatures, but recommended delay in awarding a prize. This was also the view of the committee, possibly because they wanted to give one prize each to the two chemists. No prize was awarded in 1927. The following year there was only one nomination for Wieland, but five for Windaus, two of which came

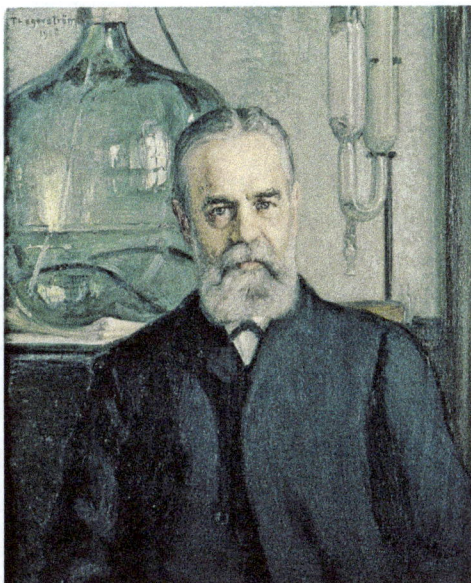

Olof Hammarsten (1841–1932). [Portrait at the Swedish Medical Association.]

from previous Nobel Prize recipients, Fritz Haber and Willstätter. Widman made two brief supplementary reviews updating the information on each

Henrik Söderbaum (1862–1933), permanent secretary of the Royal Swedish Academy of Sciences (1923–1933) and chairman of the Nobel Prize committee for chemistry (1927–1933).

of the two candidates. The committee took note of "the exhaustiveness and exactness (of the work), the perspicacity (in the approach) and the unusually great experimental skills (applied to the work)." It was concluded that there was no need to wait and in fact it would not be appropriate to delay any longer. Thus they received their separate prizes a hundred years after Friedrich Wöhler, a student of Berzelius had made the discovery that a substance of importance in the living organism, urea, could be synthesized in the laboratory.

The Prizes to Wieland and Windaus were introduced at the prize ceremony by Henrik G. Söderbaum, the permanent secretary of the Academy and Chairman of the Nobel Committee for Chemistry[2].

His remarks were relatively general. After having expressed concerns about the threats of increasing specialization in the pursuit of science (one wonders what he would have said about present-day science) he commented on the sterols studied by Wieland. He first referred to the gall acids, the main constituent of bile, and then to substances toxic to the heart muscle. One of the substances mentioned was bufothaline, from toads as indicated by its name; *Bufo* is Latin for toad. Other substances with effects on the heart had long been known to be present in plants, like *Digitalis lanata,* woolly foxglove. He noted that in addition there was a broad representation of many other sterols in plants, the phytosterols, which of course were ingested by mammals. Further he mentioned that the substance was found "not only in bile but also in the brain, in nerve substance, in the egg, in blood, and presumably in all cells." In one later paragraph he wrote:

> Finally we come to a group of compounds, which indeed have been known only for a comparative short time, but during this short time has attracted all the more interest, not only from the professionals but also from the public at large. Who today does not know about the vitamins, these enigmatic substances, which have such a pervasive importance for life itself, *vita,* and which because of this justifiably have received their name. But the difficulties which in this case have confronted the scientist, were, in comparison to those already mentioned, many times more accentuated and in most cases one has had to restrict oneself to characterizing these substances by (measuring) their physiological effects.

He then mentioned in general terms that Wieland had made major progress in identifying chemical structures of sterols and proven that the bile acids and cholesterol were related. Regarding Windaus he emphasized the studies demonstrating that the cardiac-effecting glucosides, the already mentioned foxglove product, were related to cholesterol and bile acids and further that *ergosterol* (a sterol found in fungi and protozoa) was an interesting possible source of vitamin D. Irradiating 5 mg of this compound with ultraviolet light gave material with an efficacy corresponding to 1 liter of cod-liver oil. Intakes of this kind of oil were the recommended means of substituting a vitamin D deficient diet and hence preventing rickets. For young people brought up during the Second World War the taste of this oil was not one that they in general remember with enthusiasm. For some reason I did not mind the taste of it.

In their Nobel lectures[3,4] the two prize recipients provided detailed presentations of the impressive advances they had made in their chemical studies. Wieland had done most of his academic work in Munich, presenting his doctoral thesis in 1901 at the Ludwig Maximilian University and becoming a professor at the Technical University of the city. After a brief period in Berlin he returned to succeed Willstätter as the Chair of Chemistry. He also had close connections with industry. His cousin Helene married the founder of Boehringer Ingelheim Company, Albert Boehringer. Over time this company became very successful and its first research building was named the Heinrich Wieland building in 1938.

In his lecture[3] Wieland paid tribute to studies of bile by the two Swedish chemists, Berzelius and Olof Hammarsten. His starting point was knowledge, from the work of others, about the relative occurrence of carbon, hydrogen and oxygen atoms in sterols; $C_{24}H_{40}O_{1-5}$. He formulated his scientific journey in the following words (translated from German):

> The problem is not very attractive from the experimental viewpoint. There is no nitrogen, which adds interest and variety to the treatment of alkaloids (nitrogen-containing amines, my remark). Only carbon, hydrogen and a little oxygen, all in the traditional combination, which does not lead us to expect any surprising results. The task would appear to be a long and unspeakably wearisome trek through an arid desert of structure. True, the wanderer in this apparently so unattractive region finds friendly landscapes at all stages of his journey, and the large quantity of substances bringing him nearer his goal, accumulates around him like dear companions, although, clothed in the plain garment of colourlessness, they do not stand out in their appearance or in their properties. But the driving force, which steels the perseverance, lies in the problem itself.

Colors were important in the early developments of chemistry in Germany. A company IG (Interessen Gemeinschaft — joint venture) Farben was an amalgamate of different dye-making corporations, some with a long history. When it was formed in 1925 it represented the largest chemical industry in the world. Chromatography (color separation) techniques were mentioned already in the previous chapter, which also described Tiselius's attempts at examining the separation of colorless substances.

Without going into details it can be noted that Wieland in his work progressively came closer to the sterol structure as we know it today. Towards the end of his presentation (translated from German) he discussed the structure of the molecule and said:

> This, then, characterizes in rough terms the status of our knowledge of the chemical constitution of the bile-acid molecule. Despite the unsuccessful attempts of physiologists at producing experimentally the transformation of cholesterol into cholic acid in the animal organism, it must be assumed that the vegetable sterols provided by food are transformed by the cell according to the system of the decomposition reaction carried out by Windaus on pseudocholestane. When we attempt to get a picture of the mechanism by which cholesterol (the correct chemical formula was given; $C_{27}H_{46}O$ shows that its molecule consists of three isoprene groups and a residue of twelve rectilinearly-bound carbon atoms.

Later we will return to the source of the cholesterol we have in our bodies. Wieland's final paragraph noted that once the constitutional problem had been solved it would be time to synthesize the compound.

Wieland was a very important mentor to Lynen, as we shall see, but not only that, his youngest daughter Eva married his top research student very soon after he had presented his Ph.D. thesis. As repeatedly emphasized, forthcoming Nobel prize recipients often have experience of working with previous Nobel laureates, but the kind of family connections exemplified by this case must be rare. Lynen received his prize 36 years after his father-in-law and also in a different discipline, physiology or medicine. Alan Hodgkin, whom we met in Chapter 3 also married the daughter of a Nobel laureate. Peyton Rous, however, represented a very different branch of medicine and due to the unique circumstances of his very much delayed prize, Hodgkin received his prize three years *ahead* of his father-in-law.

Windaus had his background in Berlin where he also began his studies of medicine. He became fascinated by Emil Fischer's lectures and started to learn chemistry in Freiburg in parallel with his other studies. He then returned to Berlin to work with Fischer. During these years he established a close friendship with Otto Diels, who was born in the same year. Diels also received a Nobel prize in chemistry, but he had to wait 22 years longer than Windaus since this happened first in 1950 when he was 74 year of age (Table 6.1, p. 373). Windaus returned to Freiburg in 1901 and presented his thesis *On Cholesterol* two years later. He demonstrated that bile acids and sterols were closely related. Besides his work on sterols he made many other important contributions to organic chemistry. In his Nobel lecture called simply *Nobel-Vortrag*[4] and naturally presented in German he discussed cholesterol at large and tried to deduce its physiological importance. He could of course have noted that it was the main constituent of cellular membranes, but instead of doing this he focused on attempts to understand how it was metabolized in the body and what other kinds of more general physiological roles it might have. Departing from its defined chemical formula he reviewed its occurrence in the animal and vegetable kingdom and also in fungi. Like Wieland he raised the question of the origin of cholesterol in the body. He said:

> If the human and animal organism were really not in a position to produce cholesterol from other constituents, then the cholesterol of the higher animals, the carnivores and herbivores, would necessarily originate in their food. But as the herbivore does not receive cholesterol but phytosterol in its food, it must possess the ability to absorb the phytosterol of its vegetable nourishment and to convert it to cholesterol.
>
> To test this assumption (Rudolph) Schoenheimer (see below) carried out experiments on rabbits in the Institute of Pathology in Freiburg. He found 96.9% of the ingested sitosterol (0.2 g per day) in the excrement, and concluded from this that sitosterol is not absorbed by the intestinal canal of the rabbit. Therefore there can be no question of the lack of capacity of a herbivore to use the phytosterol in its food to form choles-terol of its body substance. On the contrary, it must be in a position to form cholesterol from other components of its nourishment.

We will meet Schoenheimer again very soon but it could be added that it was after he had spent the academic year of 1930–31 at the University of Chicago that, upon his return to Freiburg, he initiated very critical follow-up experiments. Schoenheimer had first tried the experiments in patients but got rather frustrating results, not allowing clear-cut conclusions. However, the studies of mice allowed the important demonstration that the animals only synthesized large amounts of cholesterol when the substance was not provided in the diet. It would take time before the importance of this early observation of a feedback mechanism controlling the synthesis of cholesterol was fully understood[5]. The observation was an early premonition of the elucidation of the occurrence of this kind of mechanism in a very different biological system, in bacteria, to be demonstrated by François Jacob and Jacques Monod two decades later, discussed in depth in Chapters 8 and 9.

Windaus returned to this question later in his lecture:

If small doses of cholesterol are fed to rabbits deposits occur at first only in the intima of the blood vessel, which is the organ most sensitive to cholesterol in the rabbit. This observation is of great interest, for according to (Ludwig) Aschoff (see below) a disease pattern arises at this point which is identical with that of atherosclerosis in man. The hypothesis of a genetic connection between ingested cholesterol and atherosclerosis in man thus cannot be dismissed. Since phytosterols, which cannot be absorbed, bring about no such symptoms, it should be investigated whether or not true atherosclerosis presents itself in societies which are truly vegetarian (abstaining also from milk and eggs).

He concluded that the human body must be capable of synthesizing cholesterol and emphasized that "the synthesizing capabilities of the animal have been gravely underestimated here as in many other cases."

It is very surprising that Windaus in his lecture did not mention that he as a student of Aschoff in 1910 was the first researcher to link cholesterol to atherosclerotic lesions. He had done some straightforward experiments comparing the amount of cholesterol in normal and atheromatous aortas. He demonstrated that the amount of so-called free cholesterol and bound cholesterol was increased by factors of 7 and 26, respectively, in the diseased blood vessels. This observation probably inspired Russian physiologists in St. Petersburg to conduct the important feeding experiments discussed later. At the time Aschoff was one of the world's most influential pathologists. He had a particular interest in the function of the heart, but also in atherosclerosis. The latter word derives from Greek and combines *atheros*, meaning gruel, and *sclerosis*, meaning hard. Already in the 19th century German pathologists had identified the plaques of atherosclerotic changes on the inner surface of the aorta and shown that they contained a cheesy substance. In 1906 Aschoff had proposed that they contained cholesterol as mentioned by Windaus in his Nobel lecture.

In a summarizing part of his lecture Windaus speculated:

What then, is the biological role of cholesterol itself? Many attempts have been made to solve this question. A number of scientists place in the foreground the physical — especially the colloid-chemical — properties of cholesterol, and they point to the ability of cholesterol to emulsify fat, and to its importance for the permeability of the cell. The conditions prevailing here still badly need explanation.

The final third of his lecture discussed the recent findings of a connection between sterols and the anti-rickets vitamin (later called D). As was mentioned before, this vitamin is related structurally to steroids. The critical difference is that the B six carbon ring is not closed. Hence it cannot be synthesized

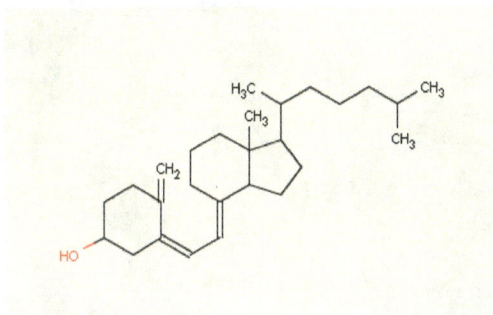

Chemical structure of vitamin D.

by the body, the definition of a vitamin, but needs to be supplied in its final form or a form that can be converted to the active substance under certain conditions. As already mentioned the unsaponifiable portion of fish-liver oils had been found to be an effective remedy of rickets. Windaus discussed the various means of improving the content of the vitamin by different physical and chemical treatments. He particularly emphasized the efficacy of ultra-violet irradiation. The effect remained unexplained since there was no difference between the chemical composition — $C_{27}H_{48}O$ of non-irradiated and irradiated material. Later it was demonstrated that irradiation breaks open the B rig.

A central question that remained to be answered was how the complex molecules of sterins were synthesized. In order to solve this problem there was a need for access to new technology. One important technique was the possibility of labelling specific atoms and following how foodstuffs were processed by the living body. In the laboratory one wanted to demonstrate how a complex molecule could be built up from smaller components. This field was pioneered by George de Hevesy, described at length in my preceding book[6], but there were also important contribution by other scientists.

The Use of Radioactive Isotopes for Studies of Intermediary Metabolism

Rudolph Schoenheimer (1898–1941).

Besides Hevesy another scientist of Jewish background played a central role in the critical developments of the techniques for the use of radioactive isotopes in the late 1930s. It was Rudolph Schoenheimer. He was born in 1898 in Berlin and studied medicine at the Friedrich Wilhelm University in that city. At the University of Leipzig and later in Freiburg he received a solid training in pathology and organic chemistry. He was inspired by contacts with the already briefly mentioned Aschoff. Schoenheimer later became the head of physiological chemistry in Aschoff's Institute of Pathology in Freiburg. At this institute he met Hans Krebs, a productive encounter although Schoenheimer described him as a loner. Schoenheimer correctly predicted that Krebs would eventually receive a Nobel Prize, which he did in physiology or medicine more than twenty years later. Schoenheimer spent the academic year of 1930–31 at the University of Chicago. After his return to Freiburg he published the important cholesterol feeding experiments mentioned above. Very soon afterwards he was forced to leave Germany because of his Jewish background. Thus he has been included among the many scientists presented in the book *Hitler's Gift: The True Story of the Scientists Expelled by the Nazi Regime*[7].

Schoenheimer settled at Colleges of Physicians and Surgeons, Columbia University, New York, where he built a strong research group. This group included David Rittenberg, who had a radiochemistry training from Harold C. Urey's laboratory, and later also Bloch, another German Jewish refugee and one of the two central figures of this chapter. The new approaches to studying metabolic steps by use of isotopes were made possible by Urey's identification of isotopes of hydrogen, recognized by the Nobel Prize in chemistry in 1934, and by the discovery of techniques of producing other new radioactive elements, by Frédéric Joliot and his wife Irène Joliot-Curie, recognized by the same kind of prize the following year. Using isotope tagging techniques and applying advanced mass spectrograph technology, Schoenheimer and his collaborators could establish, among many other findings, that cholesterol was a risk factor in atherosclerosis. He developed a theory of a dynamic steady state of cholesterol metabolism. His work was very much admired and it led to that his being invited to give the Dunham Lectures at Harvard in 1941. Sadly his lectures had to be read for him since he had committed suicide at the time. He had been fighting his tendencies towards manic depression throughout his life. Although aged only 43 at his death he had already been nominated for Nobel Prizes both in Chemistry and in Physiology or Medicine. Jean Medawar and David Pyke speculated freely in their book "If he had lived he would have been a likely winner of a Nobel Prize." A number of modern historians agree with this statement and view Schoenheimer as the single individual most responsible for the birth of molecular biochemistry.

In fact Schoenheimer was nominated together with Hevesy for a prize in chemistry in 1941. The nominator was none other than Urey, who through working at the Department of Chemistry, Columbia University, knew the activities of the nearby departments well. He carefully introduced the two candidates over two pages and in a separate letter he commented on the absence of five reprints by Schoenheimer among those that he had sent separately by surface mail. His shipment of documents, surprisingly, considering the war conditions, arrived. It represented the only set of submitted attached material for a prize in chemistry in 1941, a year when there were 23 nominations of 18 candidates. In total there were more than 80 reprints, most of them representing articles published in 1938 and later. Schoenheimer's department must have been an impressive powerhouse. Urey remarked that Hevesy already had the idea of using isotopes during the 1920s but that the field did not open up until in the 1930s when new useful isotopes could be produced. He then stated "The two men were first to take up these new studies, particularly as they applied to biology." Hevesy used radioactive phosphorus in classic experiments and Urey noted that he deserved to receive a Nobel prize. Regarding the second candidate he wrote:

> Professor Schoenheimer, on the other hand, made use first of deuterium and of stable nitrogen 15, and is now using stable carbon 13 to make studies on the intermediary metabolism using these tracer atoms. His work again is a classic, and I believe for this work he deserves the award of the Nobel prize.

Urey then made some general comments on the important developments in studies using isotopes and finally wrote "… only these two men have made their studies with a care and with a precision that makes their results definite and final. Much of the other work, I fear, has muddied the waters rather than clarifying them."

The committee did not subject Schoenheimer to a review. They made the following, perhaps somewhat surprising remarks, in their summarizing comments (translated from Swedish):

> Schoenheimer has been proposed by Urey to share this year's prize with Hevesy. Last year the committee proposed to the Academy that "in the event that the current circumstances are deemed to be suitable (acceptable?) for awarding Nobel prizes, the prize in chemistry should be awarded to Hevesy for his studies of the use of isotopes as indicators in studies of chemical processes. At the time Schoenheimer was not proposed for a prize, but the committee is of the view that Schoenheimer's publications in addition to those produced by the founder of the method, Hevesy, are not of such importance that it is justified to divide the prize."

When seeing the material Urey submitted as evidence of Schoenheimer's impressive contributions, it may be pondered if the decision not to investigate him was correct. It seems that he should have been considered for a review. Possibly it was of importance that the chairman of the committee at the time, Svedberg, had already carried out repeated reviews of Hevesy and may have considered the case closed. However, a review of Schoenheimer was done at the Karolinska Institute in 1942. He had been nominated by Theorell. The committee asked Hammarsten to do a review. In line with his other reviews it provided a careful analysis and it covered 16 pages. Hammarsten emphasized that almost all the work concerning labeling of compounds by isotopes had taken advantage of naturally occurring stable isotopes, deuterium, nitrogen 15 and carbon 13. In nature no difference was made between the mother substance and the naturally occurring stable isotopes. Hammarsten preferred to talk about the Schoenheimer School and he pointed out the advantage provided by the presence of Urey's laboratories at the same university. Sometimes he made a contrast between contributions in the U.S. and in Europe — he often referred to the studies by Krogh and Ussing (see Chapter 3), indicating a tendency of the former to overrate their own relative contributions. The meaning of the "school" — terminology can be deduced from the following underlined paragraph on page 5 of the review (translated from Swedish):

It should be emphasized that the synthesis of isotope-containing derivatives for use in animal experimentation and to an even higher degree the fractionation from the organs of fatty acids and amino acids in a purified state or in a form that allows isotope analysis, represent, without comparison, the most time-consuming part of the investigations and require a high degree of skill. One natural consequence of this is that the Schoenheimer school has had (included) a very large number of collaborators, with whom his contributions should be shared. Most likely one should view Rittenberg to be on an equal footing with him.

Hammarsten reviewed advances in Schoenheimer's and his collaborators' work on lipid metabolism, in particular cholesterol, and on protein metabolism. He pointed out studies in which he judged Hevesy to have priority. In the case of protein turnover new concepts had been introduced by the Schoenheimer school. There appeared to be a previously unrecognized turnover with a continuous breakdown and synthesis of proteins. It would take long into the 1950s before a full interpretation of these observations was possible, as discussed in the previous chapter. The isotope techniques also offered new insights into the transfer of methyl and sulfur groups in proteins, work developed in some cases, in collaboration with du Vigneaud (Chapter 4), also based at Columbia University. The critical formation of creatinine in the liver and its accentuation in certain diseases was another field where Schoenheimer's group had made critical contributions. Hammarsten finally concluded "that Schoenheimer has not made any discovery worthy of a prize," a conclusion supported by the committee. Hammarsten also noted that he had not found any publications by Schoenheimer from 1942. Apparently he did not know the sad truth that Schoenheimer had died. It should be emphasized again that the conditions of the Second World War implied major restrictions on the international exchange of information and that presentation of Nobel prizes for obvious reasons had been put on hold.

Another Swedish Star Chemist

At the Karolinska Institute, Sune Bergström, mentioned briefly at the beginning of Chapter 4, played a central role in the award of the 1964 prize to Bloch and Lynen. Bergström was born in 1916 and grew up in an academic home in Stockholm. The biographical information about him is relatively sparse.

In *Les Prix Nobel en 1982* he only listed years of Degrees, Appointments, Memberships and Honorary Memberships and Honors and Awards, an impressive list, but of limited value when the aim is to identify his personality. By accident I got to know about his heritage on his mother's side.

In 1972 when I became professor of virology and chairman at the Karolinska Institute, Bergström was the Vice-Chancellor, having full control of all major developments at the Institute. In that same year I received the encouragement of becoming selected as the best teacher of the Institute, the "Mäster." At the student feast when the award was presented I took the opportunity of reading some selected paragraphs out of an 1851 second edition of a handbook about medical self-care in the home (*Handbok i Husmedicinen*) written by August T. Wistrand. Bergström who was present at the event told me that his maternal grandfather, a physician, was named Wistrand and that the author of the book I had cited was a brother of his great grandfather. August Wistrand was a physician with very broad-ranging interests. He was the author of many books and his qualifications have been summarized in the following way (translated from Swedish): "(He was) quick in perception, sharp in his judgments and (with) a prodigious memory (Swedish *häst* — horse — *minne* — memory), which, combined with an outstanding capacity for work made W. one of the foremost people within the rapidly growing administration of the health care system in Sweden." However, Bergström's great grandfather on his maternal side, named Alfred, was also a physician. For health reasons he did not practice medicine, but became a highly respected official in the field of medicine and served as extraordinary professor of forensic medicine at the Karolinska Institute. Like his brother he was a productive author and one of his books, a handbook of forensic medicine for physicians and lawyers, he wrote together with him. Sune's father Sverker Bergström was a very talented mathematician, who sadly died of the Spanish flu when his son was only two years old. Thus like Alan Hodgkin and Tiselius, Bergström presumably developed early being dependent on his mother as a single parent.

The other more personal indirect insight into his personality concerns his own genetic offspring. In the previous chapter the impressive DNA studies on archeological materials by Svante Pääbo were briefly described. On the personal side Bergström married in 1943 and in a separate relationship became the father of Svante in 1955. I have had the pleasure of meeting Svante on a number of occasions and have been impressed by him both as a scientist and as a human being. His book *The Neanderthal Man*[8] is not only a description of his impressive advances in science but also provides some candid remarks about his personal development. He can write unsentimentally about the different kinds of love of humans in their rich and diversified dimensions. After this musing about the dance of genes in generations preceding and following Bergström let us return to the more mundane matters of his curriculum vitae.

After graduation from the high school Bergström began his studies at the Karolinska Institute in 1933. At a very early stage he decided to get involved in science outside the regular curriculum. The introduction to his Nobel Prize lecture[9] is helpful in presenting the course of events. He started with Jorpes (see Chapter 4) and contributed to his pioneering work on heparin. Jorpes was of the view that there was a major limitation that no one in Sweden worked on lipids and steroids. He therefore sent Bergstöm to England in 1938, where he worked with G. A. D. Haslewood, an authority on bile acids at Hammersmith Postgraduate Medical School. Plans were then made for Bergström to work in Edinburgh, but the war broke out and by use of a Swedish-American Fellowship instead of the one he had received from the British Council he was able to work for eighteen months during 1940–42 in the U.S. at Columbia University and at the respected Squibb Institute in New York with Oskar Wintersteiner.

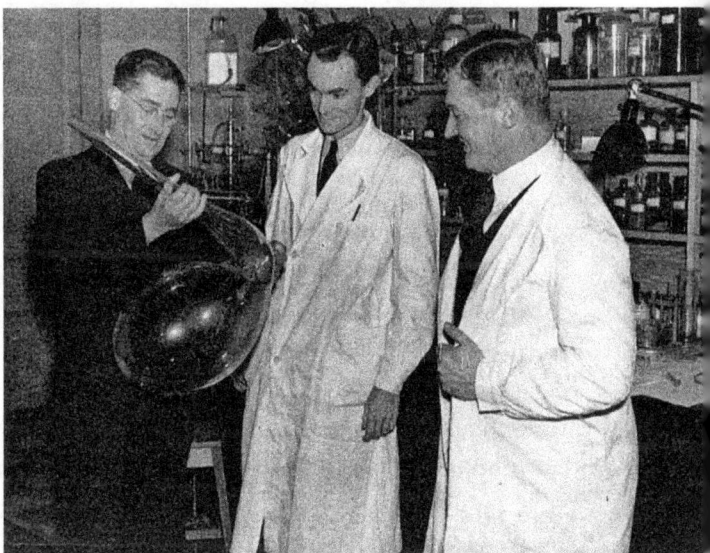

From left: Howard W. Florey (1898–1968), recipient of a shared Nobel Prize in physiology or medicine in 1945, admiring one of Berzelius' retorts, together with Bergström and Theorell. [Courtesy of Karolinska Institute.]

Problems of cholesterol autooxidation were examined. When he had returned home he initiated studies of autooxidation of linoleic acid and was appointed research assistant to Theorell at The Medical Nobel Institute. Bergström presented his thesis in 1944. The title was "On the oxidation of cholesterol and other unsaturated sterols in colloidal aqueous solution by molecular oxygen." Together with Theorell he worked on the soy bean lipoxygenase enzyme.

Bergström's introduction to and subsequent deep involvement in work on the prostaglandins can be identified with precision[9]. On October 19, 1945 Bergström gave a presentation at a meeting of the Physiological

Society of the Karolinska Institute. Theorell was the chairman, Zotterman the secretary and Ulf von Euler signed the minutes of the meeting. It was after the meeting that von Euler asked Bergström if he might be interested in studying the small amounts of his lipid extracts of sheep vesicular glands that he had stored since before the war. Collaboration began and by use of Craig's counter current extraction method a high degree of purification of major components in the material that von Euler in the mid 1930s had called *prostaglandin* was achieved.

Bergström's road towards his joint 1982 Nobel Prize in physiology or medicine together with Samuelsson and John R. Vane "for their discoveries concerning prostaglandins and related biologically active substances" will be something for future science historians to review. However, it can be noted that already in 1965 he was nominated by Lynen to receive a prize together with Ulf von Euler. The committee asked an external reviewer Myrbäck to make an analysis of Bergström's contributions. Myrbäck, whom we have got to know as a very meticulous reviewer, analyzed Bergström's impressive advances in his studies of prostaglandins. He emphasized the importance of his collaborators and mentioned in particular Samuelsson. In his summary Myrbäck noted the important effects of prostaglandins in stimulating smooth muscles with a potential consequence on the blood pressure and also their effects on fat metabolism. He praised the extensive structural analyses made and also commented that it was Bergström's and D. A. van Dorp's research groups who had independently elucidated the biosynthesis of the different compounds and examined their metabolism. Myrbäck concluded "… that Bergström's work on the chemistry and biochemistry of prostaglandins is a contribution that well deserves to be recognized by a Nobel Prize." This was also the conclusion of the committee, but Bergström would have to wait 17 years until his shared prize.

In 1947 Bergström became professor and chairman of the Department of Physiological Chemistry at Lund University. He built up a very strong group of researchers, some of whom, including Samuelsson, already mentioned above, followed him when he returned to the Karolinska Institute in 1958 to become professor of chemistry there. His further career followed a straight line; Dean of the Medical Faculty (1963–66), Vice-Chancellor of the Institute (1969–77), Chairman of the Board of Directors, The Nobel Foundation (1975–87). Like von Euler, Bergström held this position in the year of 1982 when he received his Nobel prize. The Deputy Chairman of the Board, Tore Browald, a warm-hearted director of a major Swedish bank, took over the responsibility of giving the opening address at the prize ceremony. During the later phase of his life Bergström had a deep involvement with the WHO Global Advisory Committee on Medical Research in Geneva. He died in 2004 at the age of 88 years.

Developments Towards the 1964 Nobel Prize in Physiology or Medicine

It is now time to introduce the two main actors of this chapter, the laureates in physiology or medicine in 1964, Bloch and Lynen. As we shall see they were rather contrasting personalities. One who followed the rich German traditions of high quality chemistry research even enduring the challenges of the Second World War, and the other who was forced to emigrate to the United States, and initiated an impressive career contributing to the post-war hegemony of that country in science. They were nominated for a Nobel Prize both in chemistry and in physiology or medicine.

A Renaissance Scientist

Lynen was born in 1911 in Munich and stayed in that city throughout his life. He was the seventh of eight children. One of his older brothers stimulated his interest in chemistry. In 1930 he began his university studies. Wieland, whom we have already met, was his mentor both in his undergraduate and graduate studies. Lynen's thesis, presented in 1937, was entitled "On the Toxic Substances in Amanita." Amanita is a genus of mushrooms, some of which are very toxic. Lynen continued his postdoctoral work in Munich. He managed to keep up his academic activities during the war. He was a very daring alpine skier and broke bones of his legs on a number of occasions. One of the early accidents apparently caused

Feodor Lynen (1911–1979), recipient of one-half of the Nobel Prize in physiology or medicine in 1964. [From Ref. 22.]

him to have a bit of a limp, which exempted him from military service. Still it must have been hard times, not least seeing colleagues of Jewish background being forced to leave the country. Although he was never a member of the National Socialist German workers' party he was able to advance in his career and become a Reader in Chemistry in 1942. There was a certain hesitation in the grading of this thesis; it included a qualification "only with concern." A former graduate student Hildegard Hamm-Bruchner has recalled her mentor in the following remarks[10]:

> He neither endangered his life by conspiratorial activities nor by death-defying conduct, but he has always proven himself as an example of integrity and civil courage and endeavored to preserve the chemical institute as an oasis of decency.

After the war German scientists were not well regarded by their colleagues in other countries and at the First International Congress of Biochemistry in 1949 in Cambridge only four German chemists were invited. One of them was Lynen and he was probably an excellent ambassador. "Fitzy" as he was called by his friends has been described as having a cheery nature with a good sense of humor and a fondness for partying. One of his many visiting scientists M. Daniel Lane, wrote:

> By this point in my sabbatical in Lynen's Institute, I began to recognize certain habits of "the Chief." For example, he had the habit of working in his office until late in the afternoon. Then, around dusk, i.e. 6.00–6.30 p.m., he would emerge to make "rounds" in the Institute, moving from one bench to the next to survey the day's progress or lack of it. Of course not one of the about 30 investigators would consider leaving until after he had passed through. He ran a "tight ship"! Fitzi had an uncanny memory and could recall details of experiments done weeks earlier.

And later on he said:

> Although Fitzi Lynen was a hard driving biochemist, he did like to socialize over a beer or a martini. On Friday afternoons Harland (Wood) would often bring a half-gallon bottle of Gilbey's gin to the institute and prepare martinis in the second floor laboratory.

Like Lane, a very large number of scientists could have testified to Lynen's decisive influence on their academic careers. He died at the age of 68 a few weeks after an aneurysm operation.

The Value of a New Homeland

Bloch's life history is different, because he had a Jewish background. He was born in 1921 in Neisse, Upper Silesia, at the time a part of Germany. The extended Jewish family provided a rich intellectual and cultural milieu. At his bar mitzvah his uncle offered him a choice between a cello or a canoe as a gift. He himself leaned towards the latter but his parents convinced him that he should choose the cello. This turned out to be of decisive importance later in life. In 1930 he went to Germany to study chemistry at the Technische Hochschule in Munich. During this time he was exposed to the German chemistry

Konrad Bloch (1912–2000), recipient of one-half of the Nobel Prize in physiology or medicine in 1964. [From Ref. 22.]

giants of the time, Hans Fischer, Windaus, Wieland and Willstätter. In 1934 he had acquired the degree of Diplom-Ingenieur in Chemistry, but now he could no longer stay in Germany because of the rapidly rising anti-Semitism. Thanks to a generous recommendation by Fischer he was appointed research assistant at the Schweizerisches Höhenforschung Institute in Davos, Switzerland. This city and the sanatorium located there at the time was the setting for the famous novel *The Magic Mountain* by Thomas Mann. Thus the target of his research became the tubercle bacillus and under the direction of Frederic Roulet he examined its lipids and came to the conclusion that it did not contain cholesterol. This was in contrast to earlier findings by the authority in the field, Rudolph J. Anderson, at Yale University. Bloch, trained in the German tradition wrote to Anderson and asked him what might have gone wrong in his experiments. To his great surprise he received a response saying that Bloch's bacillus material probably was purer than the one used by the research group at Yale University.

Bloch had been permitted only a limited time to work in Switzerland and this was coming to an end. In this very worrying situation Bloch wrote to Anderson and asked for support to emigrate to the United States. He received two letters, one that said that he had been appointed assistant in biological chemistry and the other that there was no salary attached to this. Using only the first letter during a visit to the American consul in Frankfurt he received a — lifesaving? — visa for immigration. When he reached the United States he found his way to New Haven to visit Anderson. He, however, recommended that Bloch contact Hans Clarke at the Department of Biochemistry at Columbia University in New York. An informal interview with Clarke was arranged and towards the end of this Clark asked Bloch "Do you play any instrument?" Bloch dutifully answered that he played the cello and since Clark needed an enlargement of his chamber music group an agreement was sealed. This may not, however, be the whole story, since there were also very strong letters of recommendation from Willstätter and Fischer. Thus by very fortunate circumstances he could enter the Department of Biochemistry at Columbia University in New York. As we have already seen, this was a hothouse for studies of intermediary metabolism employing the new isotope technologies. After Bloch had received his Ph.D. under Clark he was invited to work under Schoenheimer's guidance and also to collaborate with Rittenberg. This accelerated his maturation as a scientist and together with the latter he initiated his very important and durable involvements in studies of the biological synthesis of cholesterol.

In 1944 he became a naturalized U.S. citizen. Thus he was to belong to the large group, about 30%, of all U.S. Nobel Prize recipients who were foreign-born. Two years later he moved to the University of Chicago. The chairman of the Department of Biochemistry at that university was Earl Evans, another product of Clarke's department at Yale University. Bloch built a group of many enthusiastic and capable students and major progress in understanding the metabolism of cholesterol was made. He spent a year in 1953 at ETH (Eidgenössische Technische Hochschule) in Zurich before he could establish his own laboratory at Harvard University. In 1954 he was appointed the Higgins professor of Biochemistry and became the Chairman of the Department of Biochemistry in 1968. Bloch has been described as a very modest person — kind, generous and soft-spoken, who outside his deep involvement in science found time for skiing, tennis and music. He died 88 years old. In the obituaries he was referred to as having provided "an outline for the chemistry of life" and to have been "a marvelously perceptive biochemist and genius." It is now time to describe how Bloch and Lynen were evaluated for a prize by the committees.

The Committee of Chemistry Examines Bloch and Lynen

In 1959 a nomination was submitted to the Academy by P. S. Sarma, Madras, India, to give the prize in chemistry to David E. Green working on the citric acid cycle (for which Krebs had received the 1953 prize in physiology or medicine) and to Lynen. The proposal was reviewed by Myrbäck over 12 pages. His conclusion was that the two candidates did not fit together and that Lynen's work on fat metabolism and acetyl coenzyme A was in such a dynamic phase that there were reasons to delay awarding a prize to him. The following year there were two nominations for Lynen from Germany, one from Otto Hahn, who received the 1944 Nobel prize in chemistry. Myrbäck did another extensive review and again he proposed that the committee should wait. One strong argument was that there were also other important contributors to the field of cholesterol synthesis, in particular Bloch. As in the previous year the committee agreed with this conclusion. Lynen was proposed again in 1961, both alone and also together with Bloch by the recent (1959) physiology or medicine prize recipient Arthur Kornberg. There was also a third nomination from M. Visconti, Zurich, including one additional candidate, Karl Folkers, Merck, Sharp & Dohme Research Laboratories, Rahway, New Jersey. We will soon return to his particular candidacy.

In 1962 there was a repeated nomination of Bloch and Lynen by Kornberg and three more nominations separately for Lynen. No review was carried out. Kornberg persisted and repeated his nomination of the combination of Bloch and Lynen in the following two years. In parallel there were again separate nominations for Lynen alone in both these years. New reviews by Myrbäck were commissioned. In 1963 he asked whether a separate prize could possibly be given to Lynen for another set of important contributions, the role of the substance biotin at enzymatic carboxylation, but in 1964 he was primarily following the line of recommending a joint prize to Bloch and Lynen. The committee agreed with this proposal, but noted that probably the contributions discussed had also been recognized by the committee for physiology or medicine. Because of contacts between members of the two committees the chemists at the Academy should have known that Bloch and Lynen were the prime candidates at the Karolinska Institute in 1964.

A Highly Respected Chemist in the Pharmaceutical Industry

As already mentioned there was a 1959 nomination for a prize in chemistry to Bloch and Lynen, which also included Karl Folkers. He was a high caliber industrial chemist who had been nominated for a prize both in chemistry and in physiology or medicine as early as 1949. His frontline contributions involved studies of many medically important substances like streptomycin, B12 and penicillin. Tiselius did an early investigation and came to the conclusion that it would be best to wait with a prize in chemistry and to follow the coming developments. No obvious major discoveries could be identified. There were also many other contributors to the forays into research areas selected by Folkers. In 1954 he was nominated again, but no further action was taken for a prize. In 1958 there was another nomination for a prize in chemistry to Folkers and this time Myrbäck was selected as the reviewer. He agreed with Tiselius that Folkers and collaborators had made many important contributions, but that still there were reasons to delay a decision on prize-worthiness. The recent discovery of an "acetone replacing factor" in cholesterol synthesis, *mevalonic acid*, was commented on. This factor had been identified in studies of microorganism growth factors, using a technology introduced by others.

A discussion of the later finding was an important issue in the already mentioned 1961 nomination which also included Lynen. Myrbäck, this time together with Fredga, contributed another long and detailed evaluation. The discovery of mevalonic acid by Folkers and his collaborators was critical to the full understanding of the biosynthesis of cholesterol, but there was some uncertainty about his personal contribution to the very first observation of the substance. Hence it was concluded that Folkers was not strong enough to fit into a combined proposal, but that Bloch and Lynen formed a powerful and a natural combination. The prize in chemistry for this year went to Melvin Calvin for his studies of carbon dioxide assimilation by plants. The final nomination of Folkers for a prize in chemistry was submitted in 1962 and this time the nominator was none other than Lynen. Among Folkers' different contributions he mentioned in particular the discovery of mevalonic acid and its critical role in the synthesis of so-called isoprenes. These compounds have an important role in the synthesis of many different biological substances, including squalene, which as we shall see is central in the synthesis of cholesterol. In spite of the authority of the nominator the committee was not impressed by the proposal. Folkers' contribution was not considered to match that of Bloch and Lynen. The committee referred back to the investigation the previous year.

In parallel with nominations for a prize in chemistry there were also nominations of Folkers for a prize in physiology or medicine. The first one, submitted in 1949 by Jorpes, recommended a prize jointly to Folkers and Selman A. Waksman. Folkers and colleagues were the first to produce crystalline streptomycin. The professor of pediatrics Adolf Lichtenstein did a comprehensive review of Waksman, but Folkers was only briefly mentioned in the text. The committee did not consider Waksman ready for a prize at the time and Folkers was not mentioned in the concluding protocol of the year. In 1951 there were two new nominations of Folkers now together with E. Lester Smith from Glaxo laboratories, England, but they both concerned his contribution to the elucidation of the chemical structure of vitamin B12. The committee let Theorell make a full investigation. He noted that the mere chemical characterization of a vitamin, like Folkers' characterization of vitamin B6, pyridoxine, was not at the time sufficient to justify a Nobel prize, but there was something special about the elucidation of the chemical structure of B12. Theorell argued that the low concentrations of this particular vitamin under physiological conditions made purification particularly difficult. A surprising finding was that the vitamin contained cobalt. Theorell concluded that Folkers and Smith were worthy of a prize, but that they could wait since the field was still in an important phase of development. The majority, but not the whole committee, agreed that they were deserving of a prize. The next year the same duo was nominated again, this time by none other than the committee member Theorell. He referred to the reviews he had carried out in the two previous years. Interestingly, Bergström in parallel submitted a nomination of candidates for the discovery of streptomycin, including Waksman, Folkers and Winterstein the latter of whom he had worked with in the U.S.

The committee let Bergström review Theorell's proposal and Einar Hammarsten took care of the nomination of Waksman and others. Bergström's review was relatively brief and gave support to Theorell's conclusion the previous year. Folkers and Smith were considered worthy of a prize. The committee supported this view. However, when it came to Folkers' involvement in examining the chemistry of streptomycin the situation was different.

Hammarsten's summarizing view was that Folkers, Winterstein and two other scientists he had also reviewed were not worthy of a prize for their characterization of streptomycin. Only Waksman was a worthy candidate. Surprisingly the committee was hesitant and proposed a delay in recognizing Waksman's discovery. Instead the committee recommended unanimously that Hans O. Krebs should receive the prize and that Theorell should give the introductory speech at the prize ceremony. This is not what finally happened.

The College of Teachers overruled the committee and selected its own candidate, who was Waksman. It would take too long to describe in this context the history of the nominations of Waksman. Briefly he had been nominated every year since 1946 and he had been repeatedly reviewed by Liljestrand, Anders Kristenson (a specialist in tuberculosis), Bernt Malmgren, Einar Hammarsten, Sven Gard, Arne Wallgren, John Hellström and Jan Strömbeck. The professor of pediatrics Wallgren, who had provided one of the early reviews, was selected to introduce Waksman at the prize ceremony. One can read between the lines that there must have been considerable crises in the prize work at the Institute this year. To this can be added that the prize to Waksman has been extensively debated and is the main theme of the book *Prize Fight: The Race and Rivalry to be the First in Science*[10]. It has been argued that Waksman did not give due credit to his younger collaborator Albert Schatz. Schatz and another collaborator, the technician Elisabeth Bugie were in fact nominated in 1952 and reviewed by Hammarsten. He concluded that they, like Folkers and Winterstein, were not worthy of a prize for their studies of streptomycin. A lawsuit settled out of court gave credit to Schatz's critical contribution towards the discovery of streptomycin and Rutgers University also emphasized his contribution by bestowing a medal on him. However, his academic career never took off.

Let us now return to Folkers' and Smith's work on vitamin B12. The importance of this work was emphasized by further Swedish nominations; 1954 by Bergström, 1959 by Einar Stenhagen and 1960 by Hans Lagerlöf. No more reviews were carried out and the two candidates remained deserving of a prize in the protocols of the committees. Arne Engström, professor of medical physics at the Institute, did a review of Dorothy Crowfoot Hodgkin in 1960. The final paragraph read (translated from Swedish):

I would like to present as my view that solving the crystal structure of B12 is a contribution worthy of a prize. If a prize is to be given in this field, the discovery and purification of B12 should however also be considered, since the chemical and crystallographic studies have developed hand-in-hand. Among the group of researchers that have been active within the sphere of crystallography, D. Hodgkin takes a prominent position and if the Nobel committee after its discussions decides to present the prize for this year for the discovery and structural determination of vitamin B12, Dorothy Hodgkin's name *should be included* (my italics).

The committee concluded that Dorothy Hodgkin, Folkers and Smith were worthy of a prize for their discovery of vitamin B12 and the characterization of its structure. This is as close as Folkers came to a Nobel prize. As we already saw the discovery of the structure of vitamin B12 was recognized by the prize in chemistry to Dorothy Hodgkin alone in 1964 (p. 188).

Finally it should also be added that Folkers was never nominated for a prize in physiology or medicine for his critical contribution to the understanding of cholesterol synthesis. However, Bergström gave due credit to the discovery by him and his colleagues in his examinations of Bloch. For example he wrote in his 1959 review (translated from Swedish):

As regards the intermediary categories of molecules in the formation of squalene from the (two-carbon) acetate this field (of research) offered a rather confusing picture with a number of different proposals for the use of different labelled C_4, C_5 and C_6 acids in the synthesis of cholesterol. Unexpectedly this whole field became clarified when Karl Folkers and his collaborators (1956) were able to isolate and identify (β, δ)-dihydroxy- (β)- ethyl valerian acid (reference to an article by Tschen and Bloch in 1957) as a growth factor for a microorganism. It was then found that this acid, which was given the name mevalonic acid, was also used to a higher degree than any other chemical compound for the in vivo synthesis of cholesterol. However, during this process the carboxyl part of the mevalonic acid was removed.

Although Folkers had made this important contribution, he was not mentioned in the summary of Bergström's reviews, but he was appropriately referred to in Bergström's introductory speech at the prize ceremony as we shall see. Folkers who has been described as one of the most imaginative and productive organic chemists of the twentieth century was never to receive a Nobel prize. It is worth repeating that the prize is given for a clearly demarcated discovery and not for high quality life-long contributions. The most outstanding of Folkers' contributions was probably his identification of the structure of vitamin B12. This was praised in the many awards he was to receive during his long life.

Karl Folkers (1906–1997).

Reviews for a Prize to Bloch and Lynen at the Karolinska Institute

Already in 1953 there was a nomination of Lynen by Lipmann, the prize recipient in physiology or medicine that same year. Lipmann received his prize for the discovery of coenzyme A and its role in the intermediary metabolism. He was very impressed by Lynen's recent findings of an acetylated energized form of coenzyme A. It was interpreted to be critical for the way the molecule operated. Hammarsten did a two-page review showing that he was less excited than Lipmann. Hammarsten wrote (translated from Swedish):

> It was well known to Lynen that Lipmann had shown that activation by acetic acid of Co-A was achieved by a binding of acetyl to Co-A, which after Lipmann had also been demonstrated by Ochoa. However, before Lynen's discovery it was not known to which part of Co-A the acetyl and pyrophosphate were attached, but that — disregarding which kind of binding that may exists — it should be one providing a readily accessible energy. It is important to recognize the starting conditions of Lynen's discovery, because he himself quite apparently tries to minimize the contributions by Lipmann and Ochoa.

Not very generous words. Hammarsten concluded that Lynen's discovery was not of the magnitude that motivated a prize. The committee agreed that at the time Lynen should not be considered for a prize. As we shall see Theorell carried out the final review of Lynen in 1964. In that review he mentioned that there would have been a discussion already in 1953 about including Lynen in the prize to Krebs and Lipmann, but that seems to be highly unlikely.

There was a lot of interest in acetyl-CoA in the early 1950s. This had its source in the early work of Lipmann on the role of phosphates in energy metabolism for which he was recognized by a Nobel Prize in physiology or medicine together with Krebs in 1953. The "active acetate" was found to be used not only in what is called Krebs cycle but also in the metabolism of fats as shown by Lynen. The substance was used in both prokaryotes and in eukaryotes and it was thought to be synthesized using two reversible enzyme reactions. At the time it was believed that different mechanisms were used in the two groups of organisms. Lipmann and Lynen with their colleagues were interested in this problem and believed that in higher organisms instead of the single step analysis favored at the time there should be a three-step reaction. Paul Berg, who at this time was a young scientist[10] was also interested in this problem. He had just moved to Arthur Kornberg's new laboratory at Washington University, St. Louis. The reason for his curiosity was the potential use of the reaction in an even wider context as proposed towards the end of Lipmann's Nobel lecture. He had suggested a possible role in both nucleic acid and protein synthesis. Since Berg was interested in nucleic acid synthesis, the leading theme of the laboratory he was joining he wanted to further develop the data published by Lipmann and Lynen. However, he was in for a surprise.

Following the rule guiding the work in the laboratory that was his new home he used purified enzymes. A famous saying allocated to Arthur Kornberg was "Don't waste clean thoughts on dirty enzymes," but he himself has stated that the origin of this maxim was the American biochemist Efraim Racker. The result of Berg's work was that in contrast to the conclusions reached by Lipmann and Lynen, the reaction in focus was directed by a single enzyme responsible for a two-step processing. The energy rich adenosine (a building stone of nucleic acids) triphosphate reacted with the two-carbon acetate which eventually in the presence of added CoA led to the appearance of acetyl-CoA and adenosine monophosphate. Berg had to conclude that the two famous scientists had erred and he could present his new data in Atlantic City, with both of them in the audience. They accepted their mistake and Lynen even told Berg, "You are the only person to have proved me wrong in an experiment." A series of important papers, mostly with Berg as the single author, were published the same year in the most prestigious journals of the time for biochemists, *Journal of Biological Chemistry* and *Nature*. Interestingly, in a later interview with Istvan Hargittai[12] Berg argued that he considered this contribution to be of an even higher value than the development of the recombinant DNA technique for which he received his half Nobel Prize in 1980. The latter achievement he considered to be more predictable. After this digression it is time to return to the evaluation of Lynen for his Nobel Prize.

It would take six years until a nomination more directly concerned with discoveries of relevance to the metabolism of cholesterol was again brought to the table at the Karolinska Institute. It was in the form of a single nomination for a prize to Bloch by Leopold Ružička from Graz. Ružička had received the Nobel prize in chemistry in 1939 (Table 6.1, p. 373) for work that had relevance to the early insights into cholesterol metabolism. The nomination also included two other candidates in different fields. Called upon by the committee, Bergström now entered the picture. He produced a very lucid and succinct evaluation. The review is accompanied by hand-drawn figures. We will return

Bergström's hand-drawn formulas accompanying his Nobel Prize reviews.

to certain details below but as an introductory overview (to be used together with the figures) it is worth translating (from Swedish) the first paragraph of his summary. It reads:

> The simple (structure) acetic acid forms mevalonic acid containing six carbon atoms, after which three such molecules are brought together to become an alcohol containing, after a loss of three carbon atoms, 15 such

atoms. Two such (molecules) are condensed to the long and branched hydrocarbon *squalene* [my italics] (a chain) which contains 30 carbon atoms. In a single simultaneous reaction this (molecule) is cyclized into a steroid containing four rings (lanosterol) still containing 30 carbon atoms. The sequence of events and the final reshuffling of the double bonds to the finished structure of cholesterol (see p. 328) have also in principle been clarified.

This is a highly simplified description of a sequence of events that in reality includes as many as 36 individual steps. It was a major feat in particular by Bloch and to a certain extent Lynen to map this complex synthesis. Of course there were also many other scientists contributing particular pieces of critical information. Bergström mentioned especially the importance of studies by Folkers and his group, which have already been discussed above.

Bergström's conclusion was that Bloch was worthy of a prize and this was confirmed by the committee. However, another five years would pass before a prize was eventually awarded and during this time the magnitude of the discoveries had become clarified and amplified, partly because Lynen had become a part of the package. In the year that followed, only Lynen but not Bloch was nominated. Bergström was asked to make a preliminary investigation. He took notice of the quality of the nominations and mentioned that he essentially agreed with their content. He therefore logically recommended that a full investigation should be carried out. In the end this was postponed to the following year because Bergström became responsible for another evaluation of a different strong candidate, Luis F. Leloir from Argentine. This scientist was recognized two decades later by a Nobel prize in chemistry "for his discovery of sugar nucleotides and their role in the biosynthesis of carbohydrates." Lynen was not mentioned in the final 1960 protocol by the committee at the Institute.

In 1961 Bergström's successor as professor of chemistry at Lund University, Bengt Borgström made a joint nomination of Bloch and Lynen and also included an additional strong candidate in the field George Popják. This year Bergström carried out another extensive review including all three candidates. The evaluation overlapped in parts both in the text and in the accompanying figures used with the review he had presented two years earlier. He noted that Bloch following the previous evaluation had already been declared worthy of a prize, but that the other two candidates had not been examined earlier. For unknown reasons he did not refer to the 1953 evaluation of Lynen by Hammarsten, which led to his being declared not worthy of a prize at the time. It was also remarked in the introduction that Lynen had been nominated not only for his cholesterol work but also for his contributions to the understanding of the mechanism of action of biotin (vitamin B7). The latter contribution was not included in the review, which focused on the development of insights during more than two decades into the synthesis of cholesterol. The Nobel Prize in chemistry to Windaus and Wieland in 1928 was mentioned and it was noted that after the introduction of certain modifications the correct structural formula for cholesterol had been presented in 1932 in a publication by Otto Rosenheim and O. King.

The important contributions by Rittenberger and Schoenheimer in 1937 were described. These were followed up in the early 1940s and onwards by Bloch and Rittenberger. This early work was done using the isotope techniques in intact animals, but later organ culture techniques to be managed in the laboratory were developed. In addition the range of isotopes used was expanded to not only include naturally occurring rare isotopes, like ^{14}C, but also artificial radioactive isotopes. The progressively interpreted pathways of synthesis were analyzed stepwise. The critical role of activated acetic acid as an energy source, later discovered by Lynen, as mentioned above, was emphasized. Bloch proposed at an early stage that the molecule squalene represented an important intermediate form of molecule in the complicated sequence of molecular steps in the synthesis. This critical insight was confirmed by other researchers. Several scientists, including Popják (together with John W. Cornforth), contributed to the understanding of how squalene was processed to form the carbon tetracyclic structure, *lanosterol*. The understanding of the last step by which three methyl groups were lost, converting lanosterol to cholesterol, was described. Altogether the findings were referred to as one of the most elegant structural chemical pieces of research in the late 1950s. It was emphasized that the major contributions had been made by Bloch but that Lynen's contribution of the activated acetic acid was critical.

Bloch also played the dominant role in emphasizing the importance of the intermediate molecule, mevalonic acid, serving as a precursor to squalene. However, as already mentioned it was Folkers and his collaborators that identified the existence of mevalonic acid and its importance in the synthesis of cholesterol. The competition for priority in defining who was first in the intense race to understand the complete synthesis of cholesterol was emphasized. In many cases it was a matter of days or weeks in the time manuscripts for publication were submitted to scientific journals by different research groups. Bergström mentioned in particular Popják's contributions with respect and was ambivalent as to whether he should be considered worthy of a prize. He emphasized that all his important contributions had been made in collaboration with Cornforth, adding that the latter was one of the foremost organic chemists at the time and mentioning on the side that he was severely handicapped by total deafness since his childhood. Cornforth did in fact receive half a Nobel prize in chemistry in 1975 "for his work on the stereochemistry of enzyme-catalyzed reactions" (Table 6.1, p. 373). Regarding Popják Bergström concluded that his contributions did not match those by Bloch and Lynen and in the end he was never nominated again. The final recommendation obviously was that both Bloch and Lynen should be declared worthy of the prize. This was also the conclusion of the committee. In 1962 Bloch and Lynen were proposed by Theorell, now vice-chairman of the committee. Bergström was asked to update his review from the previous year. He briefly noted that during the year that had passed the two candidates had been involved in other kinds of studies; Bloch had analyzed the formation of unsaturated fatty acids and Lynen had examined the mechanism of action of vitamin B12. He emphasized that they remained very strong candidates for a prize, an opinion endorsed by the committee. However, as mentioned in Chapter 3, this was the year when the Nobel prize in physiology or medicine had been informally agreed on already before the work of sorting out candidates had started.

The following year there were as many as six nominations for a prize to Lynen, some of them relatively comprehensive. One of them included a parallel nomination of Gerhard Schramm, the discoverer of infectious virus RNA, discussed at length previously[6]. Bloch was nominated in a particular proposal by a group of Russian scientists. It was the retired professor of pathology at the Karolinska Institute, the 82-year-old Folke Henschen, whom we met in Chapter 1, who served as an intermediary in submitting the nomination. The four Russian scientists proposed a combined prize to Nikolay Anitschkow and Bloch. This is not just a spurious nomination to assist in easing cold war tensions, but Anitschkow, at the time of nomination 78 years old, had indeed contributed some classic work on coronary heart disease in the early part of the century. In fact he had already been nominated for a prize in physiology or medicine earlier. The first time was in 1937 when he was proposed by P. Nolf, Brussels. On this occasion Henschen made a brief note to the protocol, but Anitschkow's candidature was not discussed further. In 1961 a new nomination of Anitschkow was submitted by J. Hamilton, Toronto. The committee now took some action. Åke Wilton, the successor to Folke Henschen as professor of pathology, was commissioned to make a preliminary review of Anitschkow (also spelled Anichkov in the English literature). This analysis was relatively extensive to be a short preliminary evaluation. It covered two and a half pages. Among pioneers in studies of the role of diet for formation of atherosclerotic plaques Wilton first mentioned Alexander I. Ignatowski. In 1907 he fed experimental animals varying diets including beef meat, chicken eggs, also separately, the egg yolk and egg white, and milk. This work was followed up by other fellow scientists in St. Petersburg. It was found that brain substance could also induce the changes. The exact chemical cause of the changes of course could not be defined. Animal proteins and lipids were referred to. The critical experiment was made five years later when Anitschkow and Semen Chalatow (also spelt with a v) fed animals with pure cholesterol.

For a number of years prior to the First World War Anitschkow had worked in the laboratory in Freiburg headed by the great German pathologist Aschoff. As mentioned Windaus in Aschoff's laboratory had demonstrated in 1910 that atherosclerotic plaques contained markedly increased concentrations of cholesterol. It was a few years earlier that Anitschow had been a student in Aschoff's world famous pathology department. This provided a natural inspiration for his later cholesterol-feeding experiments evaluating the possibilities that this substance could cause formation of atherosclerotic plaques. Wilton was somewhat concerned about the fact that the first results of feeding animals pure cholesterol were presented by Chalatow. He judged that this might infringe on Anitschkow's claim of priority. The conclusion of his preliminary review was that Anitschkow should not be considered for the prize. The discovery proposed for recognition was too limited and had been made too far back in time to be recognized. Since the review was only preliminary Anitschkow was not mentioned in the summary protocol by the committee. However, the truth of the matter was that Chalatow was a graduate student mentored by Anitschow. Whereas the latter spent the rest of his long professional life developing his cholesterol theory, Chalatow pursued other fields of medicine. Thus Anitschkow had priority in the discovery[13], but it was far too late to recognize this in the early 1960s. However, it can be added that his cholesterol theory was included in a 1998 book on *Medicine's 10 Greatest Discoveries*[14].

In 1963 the committee did not make any further analysis of Anitschkow's work and in fact no further analysis of Bloch and Lynen was performed. Although no new review had been carried out, three members of the committee, including its chairman Friberg and vice-chairman Theorell did not agree with the majority of the committee that a prize should be given in the field of neurobiology, but recommended Bloch and Lynen for the prize, as briefly mentioned in Chapter 3. A year later, one of the earlier Russian nominators repeated his previous joint nomination of Anitschkow and Bloch. The letter began as follows:

> Thinking that fruitful cooperation of Russian and American scientists in the same field of research is a striking example of the importance of peaceful collaboration of scientists of different countries for the welfare of humanity, I desire to put forward my colleague, the eminent pathologist N.N. Anitschkow of the USSR Academy of Sciences, together with Dr. K. Bloch (USA) for the 1964 Nobel Prize for physiology and medicine as authors of remarkable investigations in cholesterol pathogeny of atherosclerosis and cholesterol metabolism.

In other nominations, Borgström again proposed the combination of Bloch and Lynen whereas Lynen alone was nominated by two German scientists, including Warburg, the Nobel laureate. The committee asked Bergström to review Bloch once more and Theorell to make an investigation of his good friend Lynen. Anitschkow was not reviewed and he did in fact die on December 7, 1964. Bergström's evaluation naturally to a large extent was a repeat of his previous three reviews. In an early paragraph he wrote (translated from Swedish):

> The clarification of the biosynthesis of cholesterol has with justification been considered one of the foremost achievements in biochemical research during the 1940s and 1950s. In this work Bloch and his collaborators have been responsible for the most important contributions. Already in this context it should be mentioned that two other research groups, Lynen and his collaborators in Germany and Popják and Cornforth and their collaborators in England, on a number of points have contributed to our understanding of this process. Regarding the first step in the biosynthesis of all kinds of lipids, the so-called two-coal fragment metabolism, Lynen has made fundamental contributions, which have been mentioned in previous investigations and which are discussed in Professor Theorell's investigation (this year) regarding Lynen.

Bergström then again discussed the different major consecutive separate steps in the biosynthesis of cholesterol, already outlined above. In the end repeating his praise of Bloch's many important contributions he concluded:

Bloch's contributions have given us a rational basis for further studies of lipid metabolism under normal and pathological conditions. The relationships between diseases of blood vessels and the dietary supply and turnover in the body of cholesterol and fats have during later years become ever more apparent. Knowledge of the mechanisms for the formation and metabolism of these compounds in the body are in a longer perspective indispensable to attack the problem of blood vessel diseases from a medical and therapeutic point of view.

Theorell's review focused on recent contributions by Lynen. Important advances had been made in three areas. These were (a) the elucidation of the constitution of Warburg's respiration ferment in 1953, (b) important work on the biosynthesis of fatty acids and (c) the function and biosynthesis of biotin. His conclusion was that by the recent contributions in different fields Lynen had added further strength to his candidacy. He noted that the fact that a single scientist can make more than one important contribution, as in the case of Einstein (?!, my addition), only made a recognition of the contributions even more urgent. However, since a prize is given for a single discovery, also in the case of Einstein (however, the prize awarded was not for his most important discovery), the arguments brought forward by Theorell should not have been seen as a strengthening of Lynen's case, which in fact was not needed. Nobel prizes are not given for the accumulated scientific contributions throughout a long career, as has been repeatedly emphasized here.

The committee under Theorell's chairmanship was unanimous in proposing Bloch and Lynen for the 1964 prize. This proposal was endorsed by the College of Teachers. The two prize recipients happily came to Stockholm to receive their prizes.

The Nobel Events in Stockholm in 1964

In the absence of Sartre the natural scientists dominated the podium of the Concert Hall in Stockholm. Naturally it was Bergström who gave the laudation address to Bloch and Lynen at the prize ceremony[15]. He started by pointing out that important events had occurred at the Karolinska Institute. It had been decided by the Government that the following year it should become a university. This meant that in the coming year it would be the newly established Faculty of Medicine that would take the decision on the Nobel

Prize. Thus one of the last tasks of the College of Teachers, existing at the time, had been to award the 1964 prize to Bloch and Lynen. Bergström then described in popular terms the discoveries of the critical steps in cholesterol synthesis. He emphasized the importance of the use of isotopes for evaluation of the metabolic steps, mentioning by name Hevesy and Schoenheimer. He also emphasized that other groups had made important contributions to the accumulating knowledge, in particular Pópjak, Cornforth and Folkers. In a few phrases in German Bergström emphasized that the recipients represented an extension of Nobel laureates in chemistry of Munich origin — von Baeyer, Hans Fischer, Wieland, Emil Fischer and Willstätter. He then switched to English and asked the prize recipients to step forward one at a time to receive their prizes from the hands of His Majesty the King. Bergström belonged to the generation whose second language in school was German. However, he also had a good command of English, after his two research sojourns in England and in United States. It can be noted that he presented his thesis in 1944 in English and not in German as was the practice at the time.

After the Second World War the relative influence of the United States in natural sciences increased rapidly and as emphasized this also led to certain important changes in the culture of science. It became more democratic and as a consequence it mattered less where the scientist was positioned in the academic hierarchy. It was what one said and not who one was that should decide the impact of a given remark. In a speech in Swedish as Chairman of the Nobel Foundation at the prize ceremony more than ten years later, 1976, Bergström referred to this "democratization of science," as mentioned earlier [1]. During the 1950s and 1960s English gradually became the language of science. I wrote my thesis in English in 1964 and all the seminars in our department were held in English.

The Nobel Prize ceremony is a Swedish event and therefore the native language of all the speakers has been used until the early 1990s. Since the prize recipients and members of the diplomatic corps present in general cannot understand Swedish, printed translations of all speeches are always provided. In 1992 Samuelsson who had just become the Chairman of the board of the Nobel Foundation decided to give his introductory speech in English. All other speeches were given in Swedish. It should be added in passing that reviews of the different yearly volumes of *Les Prix Nobel* provide somewhat misleading information. From Lars Gyllensten's second year as Chairman of the Board in 1988 the text was printed in English and referred to as a translation from the Swedish text. This remains the correct information until his last presentation in 1992. However, by an editorial mistake this remark remained attached to the heading of opening addresses until 2006. Thus all presentations by the Chairman of the Board since 1993 have been given in English. Like all Swedish scientists of his generation Samuelsson is highly proficient in English, but it was not his native language. The motive for switching to English can readily be understood. It was simply a matter of courtesy since all those present at the ceremony understood this language. The new practice naturally raised the question if the introductory speeches to the different prizes also should be given in English. Not surprisingly the Swedish Academy reacted very strongly against this. The prize in literature highlights the great value of diverse languages and the global richness of many indigenous cultures. It was simply not possible to introduce a prize in literature by an author writing in a language other than English by use of that language.

But there were also problems in general of Swedes speaking English if they have not been brought up with this language. A full command of another language is possible only if there is an extensive exposure to this language before the age of 11–12 years. The format of a text of a presentation by a speaker using a non-indigenous language of course can be controlled by a skilled translator to be grammatically correct, but the problem of pronunciation remains. The Swedish language is characterized by an equal stress of the syllables. And there is also finally in addition the problem of developing a refined rhetorical presentation including alliteration and other means to capture the audience when a non-indigenous language is used. This can only be developed to the full if one is using one's native tongue(s). In my view therefore only the Swedish language should be used throughout the Nobel Prize ceremony. Anyone who does not understand Swedish should be provided with printed translations of the speeches. When it comes to transmission by television and various other digital media simultaneous translation provides a simple solution. Of course as a matter of courtesy it is natural for the different speakers to add a paragraph in a different language to open or to conclude their presentations. After Samuelsson there have been two consecutive chairmen of the Board of the Nobel Foundation who have both chosen to speak in English, but the rest of the ceremony has been conducted in Swedish. There have of course been natural exceptions. The female chairperson of the Nobel committee at the Karolinska Institute at the time of writing, Juleen Zierath, is English-speaking by birth and hence logically at the 2013 ceremony she gave her introductory presentation in English. The same applied to another member of the present committee at the Institute, Ole Kiehn, whose mother tongue is Danish. As a compromise he also used English in his 2014 introductory address. A final remark in this context concerns Nobel's name. In 1976 when I gave my first introduction of Nobel prize recipients Bergström at the rehearsal emphasized to me the importance of presenting the name with equal stress. It is a prize created by Nóbél and not a "noble" prize, which often confuses even naturally Swedish-speaking persons, not least when they use English.

After this long digression let us now return to the 1964 events. After the introduction by Bergström, Bloch and Lynen received their insignia from the hands of His Majesty the King. At the banquet the Vice-Chancellor of the

Bloch and Lynen receive their Nobel Prizes from the hands of His Majesty the King.

Karolinska Institute, Sten Friberg said: "I believe that I may be permitted the indiscretion of revealing that the only problem relating to your prizes was to decide whether they should be awarded in medicine or chemistry." This may not be a fully accurate comment. The archival materials reveal that Dorothy Hodgkin some years back was considered to be the responsibility of the Academy (see Chapter 5) and that Bloch and Lynen fitted best into the field of physiology or medicine.

Both prize recipients gave a speech at the banquet. Bloch acknowledged his most important mentor by saying "More, however, than anyone else it was the late Rudolf Schoenheimer, a brilliant scholar and a man of infectious enthusiasm, who introduced me to the wonders of biochemistry." He then cited the French-born American scholar Jacques Barzun who had said "Science is, in the best and strictest sense, glorious entertainment." Throughout his long life — he became 104 years old — Barzun in his very prolific writings formulated a huge number of expressions that later turned into quotes. I came to know about him through his grandson, Matthew Barzun, who was a very popular Ambassador of the United States in Sweden. He was interested in my first book on Nobel Prizes, which he wanted to use as gifts to the American Nobel prize recipients of the year. I handed copies of the book to him at a private meeting on October 27, 2010. On this occasion he generously gave me a copy of his grandfather's book *From Dawn to Decadence: 500 Years of Western Cultural Life, 1500 to the Present*. This book, a fantastic read, was finished by Barzun when he was 93 years old. A very generous dedication was added to the book I received. It read: "To Erling, In thanks for your gift of your great book and in hopes that this inspires new chapters and new sparks of discovery and serendipity. Regards, Matt Barzun, grandson of author." Regrettably the Ambassador had to return to the United States prematurely in order to orchestrate the financial campaign to secure the re-election of President Obama. After the successful conclusion of this campaign he became the United States Ambassador at the Court of St. James to the United Kingdom, the most prestigious position in the United States Foreign Service.

In Lynen's speech in German at the banquet he took inspiration in a quote from Richard Wagner's *Die Meistersinger* (The Master Singers). Hans Sachs, the shoemaker and poet, in the Third Act sings "Euch macht ihr's leicht, mir macht ihr's schwer gebt ihr mir Armen zu viel Ehr (You take it lightly, but for me you make it hard; you do me, poor man, too much honour)." Interestingly he then in particular addressed Tiselius and said "... I would like to thank him for the fact that it has happened also to me to become included among the trusted personalities (the Nobel laureates), an entrustment that makes me bow in wonder: for the objectivity, carefulness and altruism that you apply when you direct and put in order (the Nobel prizes). It is thanks to you (and your colleagues) that the Nobel prize today has achieved such an excellent reputation in the whole world."

Bloch's Nobel lecture[16] in its printed form covered more than 20 pages and again recapitulated the *tour de force* of deciphering the many steps in the synthesis of cholesterol. He gave due credit to his mentors and to other groups of scientists making essential contributions during the more than 20 years of expanding knowledge. The exploration of many different fields, allowing unexpected connections were alluded to; the use of mutants of yeast, learning from attempts at rubber synthesis, experimenting with intact sharks or with the exposed lipid-rich shark liver (acknowledged to be unsuccessful), etc. The following paragraph from the *Perspective* with which he finished the lecture, deserves to be cited:

Along with the interest in chemical and enzymatic aspects of choles-terol biosynthesis there has been a growing appreciation of the role which sterols might play as fundamental cell constituents. The known metabolic transformation of cholesterol, the conversions to steroid hormones and bile acids, surely serve a specialized function since they take place only in vertebrate species. However, in many tissues and cells the function of cholesterol is clearly not metabolic. In organisms which do not metabolize sterols but nevertheless produce or require it — and this is true for all but the most primitive forms of life — sterols must play some role as structural elements of the cell. Comparative biochemistry suggests what this function might be. Sterols have not been found in any bacteria or in the blue-green algae, i.e. in primitively organized cells which lack the various membrane-bound intracellular organelles. The elaboration of membrane-enclosed structures devoted to specialized functions is now viewed as a landmark in evolutionary diversification and it would appear that the parallel development of the biosynthetic pathway to sterols is one of the biochemical expressions of these morphological events. The sterol molecule is not distributed at random inside the differentiated cell but appears to be mainly associated with the cytoplasmic membrane and its endoplasmic reticulum. We do not yet know why and for what specific purpose the sterol molecule was selected during the evolution of organisms. One may speculate, however, that the rigidity, the planarity and the hydrophobic nature of the molecule provide a combination of features that is uniquely suitable for strengthening of the otherwise fragile membrane of the more highly developed cell.

Lynen started his lecture, given in German[17], by describing how he got excited about enzymes in the 1930s. He described the stage set by previous contri-butions by Berzelius, Pasteur — the paradoxical phenomenon that oxygen inhibited fermentation — Warburg, Theorell and not least his main mentor Wieland. He then naturally first highlighted his own major contribution to the understanding of the complicated process of cholesterol synthesis. It was the discovery of "activated acetic acid" a very critical extension of Lipmann's co-enzyme A findings. The rest of the lecture gave a panoramic view of the range of very important research problems pursued in Lynen's dynamic laboratory. It is an impressive *tour de force* covering some 40 pages. However, most of the work presented did not in a critical sense have a direct bearing on the central theme of biosynthesis of cholesterol. Still they could be excused since the prize motivation in its final form also included fatty acid metabolism and that was what Lynen mostly discussed. His subtitles translated to English were *Fatty Acid Cycle, Biochemical Action Mechanism of Biotin, The Biosynthesis of Terpenes, The Multi-enzyme Complex of the Fatty Acid Synthesis* and *Biological Regulation of the Fatty Acid Synthesis*. The general character of the lecture was mostly hands-on chemistry and less philosophizing about the impact of the findings made. As in the case of Theorell it was difficult to clearly distinguish the critical single discovery. These kinds of "renaissance" biochemists running large and very qualified laboratories with many high class collaborators covered broad grounds and almost paralyzed the committees by their many impressive contributions. Sometimes it might have been easier if just a single major breakthrough could come to the fore. In the end the critical question, finally also raised by Lynen, was if the knowledge gained would allow physicians "to control fatty acid medicinally." In order to get a perspective on this, a fast-forward move to the 1985 Nobel Prize in physiology or medicine is in order.

A Remarkable Duet of Clinician Scientists

During my more than 20-year involvement in the work of the Nobel committee there is one prize I remember for particular reasons. This is the prize in 1985 to Michael S. Brown and Joseph L. Goldstein "for their discoveries concerning the regulation of cholesterol metabolism." As our yearly work progressed we were asked frequently by our clinically active colleagues if we could not identify a discovery with an obvious immediate clinical relevance. It was in this perspective that the prize to Brown and Goldstein was perfect. Here were two physicians performing their clinical duties and in parallel pursuing high quality in-depth experimental research using the full arsenal of molecular and cell biological techniques available. In a clever way they learnt from the mistakes of nature and could deduce an understanding of fundamental metabolic processes. They also had the joint talent to choose a very important medical problem, namely the influence of the level of cholesterol circulating in blood on the development of blood vessel disease. It had been known for a long time that high levels of cholesterol lead to the formation of plaques in blood vessels which are then the source of coronary infarcts and of stroke. The history of these findings was presented above. What was also finally attractive about their joint prize was that it illustrated a unique example of teamwork between two scientists, an exceptionally constructive *folie à deux*. They presented individual Nobel lectures cutting the huge material into two halves, but the printed version of their lectures appears in one piece[18]. So, what were the highlights in their series of discoveries?

Michael S. Brown in the back and Joseph L. Goldstein, sitting. [Courtesy of Goldstein.]

Let us put the answer to that question on hold and provide a framework to the coming presentation by noting what they said in their brief banquet speeches. Brown commented that even when they had become deeply involved in experimental work they remained clinically active physicians, but he also emphasized that in order to make progress in the experimental work two attributes were required — basic training and technical courage. One very critical aspect of basic training was to choose one's mentor and Brown as well as Goldstein were fortunate in this regard. Both of them benefitted from the time they spent training at the National Institutes of Health, Bethesda, MD. Brown received his introduction to experimental work from Earl R. Stadtman, a revered biochemist with deep insights into metabolisms of cells and their energy requirement, and Goldstein was associated with Marshall Nirenberg, a highly original researcher, who shared the 1968 Nobel Prize in physiology or medicine with Robert W. Holley and H. Gobind Khorana "for their discoveries concerning the interpretation of the genetic code and its function in protein synthesis." Goldstein also had some training in the laboratory of Arno Motulsky, the father of pharmacogenetics (the genetic variation in response to drugs) together with whom he started his work on the genetic basis of hyperlipidemia in coronary heart disease.

When discussing boldness in application of techniques Brown emphasized the need to be capable of shifting from one technique with which one may have become well acquainted to a completely new one. To be daring and bold are important qualities. He hoped that he and Goldstein might provide encouragement to the new generation of creators and innovators in medical research. Goldstein in his banquet talk emphasized that his and Brown's efforts were not additive but that there was a multiplying effect. He then highlighted that medical problems were identified at the bedside and that one of the best ways of learning about physiological phenomena and their mirror image, pathological conditions, was to examine mistakes by nature. Critical among such mistakes are the genetic errors, the consequence of harmful mutations occurring in a single (heterozygotic) or double (homozygotic) form. Goldstein in his speech emphasized the significance of the name that Nobel had given the prize they had just received:

> Of all the Prizes endowed by Alfred Nobel, only one has an ambiguous name — the Prize for physiology or medicine. Nobel believed that physiology was an experimental science like physics and chemistry. On the other hand, medicine was an empirical art that would rarely merit a scientific prize. To the contrary, however, many of the advances in biology during the subsequent 85 years were made by people trained in medicine who attempted to solve medical problems.

Goldstein then briefly mentioned the father of studies of genetic diseases in humans, Archibald Garrod. His original observation was simple. It concerned rare patients with urine that darkened upon exposure to air. The patients were diagnosed with the exceptional and not particularly harmful disease that was to be called alkaptonuria. He had evidence that in order to develop the illness a person needed to receive genes from both the father and the mother, it was an autosomal recessive genetic disease. The name alkapton was coined to identify a class of substances with affinity for alkali. It turned out from later studies that the problem was an incomplete oxidation of the amino acids tyrosine and phenylalanine. This led to the formation of a particular metabolite, predominantly homogentisic acid, which was responsible for the black urine. Garrod was able to identify three other similar kinds of genetic diseases and thereby defined the concept *inborn errors of metabolism,* a very visionary concept at the time. It should be emphasized that this observation was made more than 50 years before the nature of the genetic material was defined. Using the DNA sequencing technique discussed in Chapter 5, we can define today precisely where in the human genome the critical gene is located and what the decisive nucleotide change is. The first time it could be documented that a change in a single nucleotide in DNA could result in the insertion of the incorrect amino acid in a protein was the studies in the late 1950s by Ingram of hemoglobin in normal individuals and in patients with sickle cell anemia, as briefly mentioned above and in my previous book[6]. Ingram was the father of molecular medicine who was never recognized by a Nobel prize.

In his brief allusion to genetics Goldstein also mentioned Avery and interestingly also Peyton Rous, who demonstrated that viral genes can have an influence on the development of cancer. As mentioned the archival material describing Rous's long and rocky road to his prize will soon be available in full, but it will be two decades until the full story of how Brown and Goldstein became Nobel prize recipients can be written. Since I was involved in the process I have some insights that I cannot use in the writing of this book, but Viktor Mutt's introductory speech[19] and in particular Brown's and Goldstein's comprehensive joint Nobel lecture[18] provide a good basis for describing the developments. Mutt was a very successful biochemist, who pioneered purification of a large number of bioactive peptides from natural sources. I first met him as a teacher in medical chemistry. No one could write long carbon chains at the blackboard as fast as he did! We then became colleagues for a number of years in the Nobel committee. He was an exceptionally meticulous reviewer and a preliminary investigation which normally only covered a single page, in his case most often extended over four to five pages. Cholesterol metabolism was not a field within which he worked, but he gave an excellent introduction of the field to a wider audience[19].

Viktor Mutt (1923–1998). [Courtesy of the Karolinska Institute.]

A Receptor-Mediated Pathway for Cholesterol Homeostasis

This heading was the title of Brown's and Goldstein's joint Nobel lecture. They started their collaboration in 1972 and chose to examine a particular human genetic disease, familial hypercholesterolemia, FH. In patients with this disease the concentration of cholesterol in the blood is increased to levels much higher than the normal. They contract heart infarctions early in life. In order to study molecular events in these patients, fibroblast cells originating from them were cultivated in the laboratory. A number of critical observations were made in these studies, but some additional background needs to be provided to understand their significance. As we have learnt there are two sources of cholesterol in the body, preformed material present in the food we eat and synthesis from a simple two-carbon substrate in cells in our body. The main source of the latter endogenous cholesterol is the liver. Already the early studies by Schoenheimer indicated that there was some kind of equilibrium between the amounts of cholesterol synthesized within a cell and those externally provided ready-made molecules, originating from the diet. Although many important insights were provided by Brown's and Goldstein's pioneering early and follow-up studies into how this homeostasis was established, certain questions remain even into the present time.

The fact that cholesterol is such a central molecule in life, being a predominant constituent in the membranes of all our cells and also serving as the father molecule of a plethora of biologically very essential molecules like for example bile acids, anti-inflammatory as well as sex hormones and the related fat-soluble vitamin D, has made it a very attractive target for scientific studies. As many as eight Nobel prizes in chemistry (Table 6.1) have centered on the structure of the molecule or contributed methods critical for examination of the structure and synthesis of cholesterol. In addition two prizes have been given in physiology or medicine, the one to Bloch and Lynen, the central theme of this chapter and then the 1985 prize to Brown and Goldstein.

In order to understand how cholesterol is processed in the body the first problem to consider is how this fatty substance is transported by the blood. It is completely insoluble in water. Therefore multicellular organisms have developed certain transport vehicles for cholesterol. Oil droplets containing a large number of cholesterol molecules covalently connected to long-chain fatty acids, referred to as esters, form the core and to make this aggregate soluble in water it is surrounded by cholesterol and phospholipid. The complex also

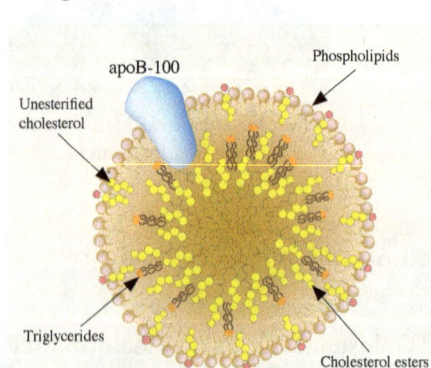

Apoprotein B-100.

Table 6.1. Nobel Prizes in Chemistry concerned wholly or in parts with studies of cholesterol.

Year	Name(s)	Prize motivation
1927	Wieland, H. O.	for his investigations of the constitution of the bile acids and related substances
1928	Windaus, A. O. R.	for the services rendered through his research into the constitution of the sterols and their connection with the vitamins
1939	Ružička, L. (shared)	for his work on polymethylenes and higher terpenes
1947	Robinson, R	for his investigations on plant products of biological importance, especially the alkaloids
1950	Diels, O. P. H. (shared)	for their discovery and development of the diene synthesis
1965	Woodward, R. B.	for his outstanding achievements in the art of organic synthesis
1969	Barton, D. H. R. Hassel, O	for their contributions to the development of the concept of conformation and its application in chemistry
1975	Cornforth, J. W. (shared)	for his work on the stereochemistry of enzyme-catalyzed reactions

includes a single copy of a large protein, Apoprotein B-100. It has been found that there are four major classes of blood plasma lipoproteins; very low density lipoprotein (VLDL), intermediate density lipoprotein (IDL), low density lipoprotein (LDL) and high density lipoprotein (HDL). They have different levels of importance in the context of the development of atherosclerosis. Once it was found how cholesterol could be distributed by the circulation in the body the next question was how the cholesterol in these circulating packages could enter the cell. This was clarified by the introduction of a *receptor concept* by Brown and Goldstein.

There is a critical rate-determining enzyme in cholesterol synthesis in cells. It is HMG-CoA (3-hydroxy-3-methylglutaryl coenzyme A) reductase. We will encounter this central enzyme repeatedly below. Using their cell cultures Brown and Goldstein could measure the enzyme activities in cells exposed to lipoproteins. They found that LDL efficiently suppressed the reductase activity, but this was not observed in experiments when the genetically defective cultured cells from patients with FH were used. Isotope-labeling experiments verified the existence of a receptor for LDL in the normal individuals. In further experiments it could be demonstrated that the lipoprotein complex attached to the cells was taken in by a cellular transport system, the lysosomes, originally identified by de Duve. Together with Claude and Palade he received the 1974 Nobel prize in physiology or medicine "for their discoveries concerning the structural and functional organization of the cell." These scientists were impressive personalities, whom I had the privilege of meeting during my second year on the Nobel committee. The LDL attached to receptors at the surface of cells was engulfed and transported to their interior by the membrane-enclosed sacs of lysosomes. Inside these vesicles cholesterol was released and this eventually led to suppression of the HMG CoA reductase activity by intracellular signaling. There are many such systems within cells, generally referred to as *second messengers*.

Further analysis showed that receptors were recycled and could return to the cell surface. The receptor was successfully purified and its chemical nature defined. It was a protein containing more than eight hundred amino acids. Five separate domains were distinguished; the part binding LDL, a second part displaying similarities to a growth factor (epidermal growth factor, EGF), a third part to which sugar chains attached, a fourth part that spanned the cellular membrane and finally a short tail in the cellular cytoplasm. The following year, 1986, the Nobel prize in physiology or medicine recognized the discovery of growth factors including EGF. The prize recipients were Stanley Cohen and Rita Levi-Montalcini. The latter Laureate was the *grand dame* of Italian science who died in 2012, 104 years old.

Brown and Goldstein and their collaborators as a next step mapped the genetic representation of the LDL receptor using techniques for sequencing of DNA, discussed at length in the previous chapter. They discovered a very complex architecture including coding and intervening non-coding regions. Against the background of this knowledge it was possible to deduce different kinds of mutations that might lead to functional defects in the receptor. As generally has been experienced when molecular genetics dissections are undertaken, highly complex relationships were found. It is easy to generalize one gene-one enzyme, one mutation — one dysfunction, but it is important to realize that proteins function as a whole and that they often have intrinsically complex structures. The complicated field of regulation of gene expression will be further discussed in Chapters 8 and 9. The final chapter in the beautiful story told by Brown and Goldstein concerned how the receptor functions in the body. It was demonstrated that the LDL receptor has dual functions. The first one was that the LDL production was reduced by removal of a precursor, IDL (intermediate density lipoprotein) from the circulation. The other mechanism was the already mentioned enhanced intracellular degradation. Thus in summary a deficiency of LDL receptors leads to an accumulation of LDL due both to over production and to delayed removal. This insight led to a discussion of how receptor functions could be regulated and what the perspectives were for development of drugs.

Before noticing the summary of their Nobel lectures, one particular problem that relates to the FH patients that were so critical to Brown's and Goldstein's studies deserves to be recognized. The homozygotes that do not have any LDL receptors cannot be treated by drugs — *statins* — that interfere with cholesterol biosynthesis. In the Nobel lecture a six-year-old girl with sky high cholesterol levels who had encountered repeated episodes of myocardial infarctions was mentioned. The only way forward in this case was a heart-liver transplant performed by Startzl, whom we met in Chapter 2. The operation was successful. The transplanted liver did not only lower the plasma cholesterol levels, but it also for the first time endowed the patient with receptiveness to statins. This was a beautiful illustration of the critical importance of the cholesterol receptors on the liver cells. In their concluding remarks Brown and Goldstein departed from Konrad Bloch's summary of his brilliant studies of the biological synthesis of cholesterol. They then said as their own concluding remarks:

> Finally we have learned that regulation of this receptor through drugs and diet can profoundly change the LDL-cholesterol level and that saturation and suppression of receptors may contribute to the high incidence of hypercholesterolemia in industrialized society. It is hoped that these insights will lead to a deeper understanding of the biology of cells and thereby to more effective forms of treatment for diseases such as familial hypercholesterolemia.

It may also be appropriate to cite in this context one of Mutt's concluding paragraphs in English at the prize ceremony[19]. He said:

> This knowledge forms a rational basis for development of methods for the treatment and prevention of the widespread disabling diseases known to be a consequence of variations in plasma cholesterol concentrations.

You have also demonstrated something else: how successful co-operation can be, a principle that should perhaps be more widely applied, both in science and in other areas of human endeavour.

Two Scientists Who Did Not Rest on Their Laurels

It has often been discussed to what extent a Nobel Prize may have a dampening effect on the continued scientific involvement of a recipient. It serves to remind us that Nobel's original goal was to provide conditions for high quality research, but since the prizes often have come to mark particular discoveries recognized in full often only some 10–20 years after they have been made this generally has not been the case. There are of course exceptions, in particular when a recipient is still at his or regrettably more rarely, her dynamic middle years at the time of the prize. We met a great example in the previous chapter, Sanger, and Brown and Goldstein provide us with another impressive example. They had come a long way at the time when they received their joint prize in 1985, but there was much more to come. All along the way they have applied the latest technologies, fostered critical international collaborations and managed to ask the proper crucial questions. Their exemplary pursuit has had a huge impact on modern medicine. Nobel's formulation "to the benefit of mankind" has rarely been as appropriately applied as in the case of cholesterol and coronary-heart disease and the possibilities of introducing prophylactic preventive measures. This has become possible because of the unique new molecular insights into the ways by which the level of cholesterol in our cell membranes is controlled.

It is worth pausing for a moment and contemplating on how this very critical homeostatic mechanism has evolved. The almost one hundred trillion cells of our body serve many different functions. Differentiated tissues of some 200 kinds have been recognized. In some way the central organ, the liver, manages to monitor the conditions of membranes of all cells throughout our body and to find a balance of synthesis for the whole very complex organism that is appropriate for the conditions at any given time. It would take too long to review in detail all the findings made by Brown and Goldstein after they had received their Nobel prizes. I almost wrote prize in the singular because never before have I encountered such a unique sharing of responsibilities and of progress made. They have themselves written exemplary reviews of the development of the field. The delicate regulatory mechanisms discovered have been summarized in a 2009 review with the title "Cholesterol feedback: From Schoenheimer's bottle to Scap's MELADL[5]." The first half of the title may be understood from the presentation above, but what does the latter part mean? The review describes the molecular details of the delicately balanced regulatory mechanisms. The molecular feedback has been found to involve membrane-bound transcription (control of information flow from DNA to RNA) factors which have been named sterol regulatory element-binding proteins (SREBPs). They in turn have a partner with a sterol-sensing role, a protein called Scap. A critical part of Scap is a peptide containing six amino acids giving it the name MELADL (one letter for each amino acid). Together this insight will explain the latter two words of the title of the review, although we need to know a good bit more to understand the nature of the relevant mechanisms. Maybe it would suffice to note that the outcome of the action of these different mechanisms is that the end products of the biosynthetic pathways inhibit their own synthesis.

In order to get a broader understanding of the consequences of the advancing insights into the mechanisms of balancing the cholesterol levels in our bodies and their importance for disease a recent 2015 review can be recommended for reading. The title is "A century of cholesterol and coronaries: From plaques to genes to statins[20]." All the studies made converge on LDL as the primary cause of atherosclerosis. The accumulated evidence included not only experimental studies but also population based genetic, epidemiological and therapeutic studies. In the future we will take a more refined approach to who to treat and when. The importance of examining our individual genomes, discussed in the previous chapter, will take on an increasing importance.

The Discovery of a "Penicillin" for Cholesterol

The story of the development of statin drugs has its own fascination and includes many twists. It has been well covered in another review by Brown and Goldstein, a tribute to the discoverer of the first inhibitor of cholesterol synthesis[21]. Again only some brief comments will be given. Since the liver is the main site in the body where LDL receptors are expressed it would seem to represent the prime target for influencing the concentrations of circulating cholesterol. One possibility would be to prevent readsorption of bile acids, but a much more appealing approach has turned out to be to attempt interfering with cholesterol synthesis at a very early stage. At the time of the

Nobel lectures in 1985 the very first important steps to specifically block cholesterol synthesis had been taken. The original discovery was made by a Japanese biochemist, Akira Endo, at the Sankyo Drug Company. He worked as a researcher at this company between 1957 and 1978. During this time he spent two years at Albert Einstein College of Medicine, Bronx, New York studying cholesterol metabolism. When he had returned to Japan he activated an interest he had had for a long time, to study fungal metabolites. His hypothesis was that fungi might use certain chemical compounds to protect against parasitic attacks. Since synthesis of cholesterol was necessary for production of ergosterol, a central component of the fungal membrane, a blockage of this synthesis might be a mechanism to exploit. Out of about 6000 compounds studied there were three metabolites originating from *Penicillium citrinum* that showed the effect searched for. The first effective inhibitor found, later called *statin,* was referred to as ML-236B. There were a number of characteristics that made this compound attractive for further studies. It worked at low concentrations and showed capacity to compete with the critical substrate HMG-CoA. The fact that ML-236B had a structure resembling mevalonate could explain its function. The compound was therefore later also referred to as mevastatin.

Akiro Endo

As described above the critical target enzyme HMG-CoA reductase is responsible for the formation of mevalonate from the activated two-carbon acetyl-CoA discovered by Lynen. Mevalonate by several metabolic steps then leads to formation of squalene, the central metabolite in the synthesis of cholesterol. Endo's first publications were presented in 1976. The original one was in Japanese in the national journal *Seikagaku (Journal of the Japanese Biochemical Society)* and was accidentally brought to Brown's and Goldstein's attention. They were very impressed by the presentation of the critical inhibitor and established contact with Endo. He promptly sent them a sample of the compound. They also planned to meet, taking the opportunity when Endo was visiting Philadelphia to participate in a symposium on drugs that affected lipid metabolism. It was a disheartened Endo who came to Dallas a few days after the meeting. There had been almost no listeners at his talk!

However, collaboration developed and the first article on statins outside Japan was published jointly between the Japanese and Dallas group in 1978.

The work on statins and on understanding their mechanisms of action developed rapidly, but it was a long and rocky road before the pharmaceutical industry caught wind of the developments. Brown and Goldstein served as consultants to the Merck & Co. and they frequently took the initiative in encouraging developments. Still advances were slow in coming. The Japanese company Sankyo had stopped its trial in 1980, because they deemed that there was an increased risk of tumors in the dogs used for the animal testing. Endo, who had left the company in 1978, judged that what had been observed did not represent true tumors, but morphological change in intestinal cells including convoluted endoplasmic reticuli. These changes were, according to Endo, due to the high doses administered. In further collaborations between Brown and Goldstein and Merck & Co. it became apparent in the early 1980 that the statin drugs acted by up-regulating the presence of receptors for cholesterol on liver cells and that this resulted in a reduction of the level of LDL in the circulation. And still the management of Merck & Co. continued to vacillate. However, due to initiatives by Roy Vagelos and Ed Scolnick it was finally possible to accelerate the developments. In 1987 the first statin for human therapy eventually appeared on the market. It was Lovastatin, put into production by the Merck & Co. The rest is modern drug history. A multi-billion dollar industry was born.

To the Benefit of Mankind

The introduction of the first effective statin to interfere with cholesterol synthesis by blocking the enzyme HMG-CoA reductase led to enormous activity in the pharmaceutical industry. The debate about the proper use of statins has involved many specialists and very diverse opinions have been expressed. There is of course no question about the fact that blood cholesterol levels influence the risks for coronary-heart diseases. Since there are no shortcuts to general health it should be emphasized that lifestyle choices are fundamental adjuncts to possibilities for long-term high quality living. The primary needs for physical exercise and for the use of a moderate and balanced diet need not be emphasized. In addition the use of statins in cases when the endogenous level of LDL is increased should be seriously considered. I have personal experience of using this kind of preventive medicine since more than

20 years. There are by now studies on large clinical materials that document a beneficial effect. Clearly there are side effects — indications of liver dysfunction, muscle problems and an increased risk of developing diabetes. Other potential side effects have been discussed too. The long-term preventive treatment obviously needs to be controlled by a physician. The arguments from both sides, those who advocate the use of statins and those who do not, have been rather clamorous and reference is often made to "statin wars." This book is not the place to discuss the arguments presented. Because the potential income of a preventive lifelong treatment is of considerable, possibly even decisive value to the pharmaceutical industry, it is of particular importance that the level of ethics in designing/evaluating clinical trials and in marketing a product registered for general use should be very high.

The 2015 review by Goldstein and Brown[20], already referred to, besides giving a lucid state of the art stands out in two ways. The first one is the description of the impressive contributions made by these two inseparable scientists after they had received their Nobel prize(s) in 1985. They have, in the spirit that was the foundation of Nobel's will, and similarly to the case of Sanger presented in the previous chapter, taken command of future developments. The insights provided by Brown's and Goldstein's post-1985 work into the very complicated mechanism of the function of cholesterol receptors on liver cells are very impressive. In addition they have pushed the industry towards developments that they, in a visionary way, saw as a unique possibility to manage the very important public health disorder atherosclerosis. By doing this they have served the ultimate aim of Nobel's will, expressed in the formulation "shall have conferred the greatest benefit on mankind."

Considering the enormous impact of the development and use of statins in medicine it is natural that the original discovery of the first active compound should be recognized by prizes. In 2006 Endo received the prestigious Japan Prize and two years later he was recognized by the Lasker-DeBakey Clinical Medical Research Award. At the time of writing a Nobel Prize is still pending.

Chapter 7

Magnifiques, Nos 3 Nobel
(Our Great 3 Nobels)

The headline cited in the title above was run by *Paris-Presse* across the whole front page on October 15, 1965. The reason for the euphoria was the telegram that had arrived the previous day from the Karolinska Institute announcing that François Jacob, André Lwoff and Jacques Monod were to receive the

Nobel prize in physiology or medicine "for their discoveries concerning genetic control of enzyme and virus synthesis." The work by the three scientists was interconnected but also distinctly profiled as the somewhat convoluted prize motivation indicated. The three scientists were each about half a generation apart, with Lwoff being the senior researcher and Jacob the youngest. Lwoff had initiated the developments by his imaginative studies of the interaction between bacteria and their viruses, the *bacteriophages*. These studies led to some major

Three happy Pasteur scientists — *from the left*: Lwoff, Monod and Jacob — at the time of the announcement of their Nobel Prize in physiology or medicine. [© Institut Pasteur]

discoveries and Lwoff went on to become a statesman of virology as mentioned previously[1], he inspired Monod in his early work on the influence of selected sugars on their metabolism in bacteria, which led to the formulation of completely new concepts of activation and control of gene expression in microorganisms, the topic of the next chapter. Their first contacts dated from the dawn of the Second World War, but the critical acceleration of Monod's work occurred first during the later parts of the 1940s. A few years later Jacob, after some hesitation, was accepted to work in Lwoff's laboratory and started his pioneering studies of the genetic basis of lysogeny. After still some more years a remarkably productive intellectual intercourse involving Monod and Jacob took off. In a way Jacob came to serve as a middleman between the research on phage pursued by Lwoff and the bacteria physiology problems examined by Monod. At the time of the final discussions for a prize a number of new concepts concerning the expression of genes critical to the rapidly evolving insights into the fundamental mechanisms of molecular biology had been presented by Monod and Jacob. The committee therefore had an option to recommend only these two scientists for the prize but according to George Klein it was thanks to his and Gard's argumentation that Lwoff was wisely included. Klein has also told me about an encounter in the early 1960s between President de Gaulle and the long-serving Swedish Prime Minister Tage Erlander. De Gaulle complained about what he interpreted to be an animus towards French scientists by Nobel committees and asked Erlander to intervene in this matter. The obvious answer that politicians had no influence on the selection process managed by the prize-giving institutions and that any attempt to interfere in the process most likely would be contra productive surprised the President very much.

The enthusiasm of the press had two causes. The first one was that a long time had passed since a French researcher had been recognized by a Nobel Prize in the natural sciences. André F. Cournand, who received the 1956 Nobel Prize in physiology or medicine together with Werner Forssmann and Dickinson W. Richards, Jr. "for their discoveries concerning heart catheterization and pathological changes in the circulatory system" was born and educated in Paris, but he moved to the U.S. and became a citizen of that country as early as 1941. His Nobel Prize work was performed together with Richards at Columbia University in New York. One might also have referred to Bovet, who was described in Chapter 1. Most of his Nobel Prize work was done at the Pasteur Institute, but he was claimed as a Swiss laureate, since he was born in that country and as an Italian laureate, because he worked in Rome when he received the prize in 1957. Thus one needs to go back as far as 30 years to 1935 to find a true French natural scientist recognized by a Nobel Prize. It was in this year that Irène Joliot-Curie and her husband Frédérick Joliot had received the prize in chemistry, as already mentioned in the previous chapter and in one of my earlier books on Nobel Prizes[2].

The second reason for the enthusiasm of the press was the attractive visibility of the three prize recipients. As the already cited newspaper wrote, "The beautiful and hard adventure of three Frenchmen who are a team in life as in their work." However, the press did not only praise the scientists for their discoveries but also for their important patriotic contributions during the Second World War. Lwoff and Monod had both played important roles in the Resistance in Paris and Jacob had made courageous sacrifices as a member of Charles de Gaulle's liberation army. The latter was an exceptional war hero. In fact it was his war afflictions that forced him to change career from becoming a surgeon to become a scientist, as we shall see. Everything the press could ask for was there — bright scientists, courageous patriots and in addition personal involvements in music and in the arts on the side. The three intrepid and chivalrous Frenchmen rapidly became public figures. It was when these three scientists were crammed together into the small attic at the Pasteur Institute originally inhabited by Lwoff that amazing things started to happen in the 1950s.

A Unique French Institution

Institut Pasteur, the Pasteur Institute, is a unique French non-profit private foundation created as its namesake by the great Louis Pasteur. It has been described as an "anarchic autocracy"[3]. Pasteur had broad scientific interests with an emphasis on biochemistry and microbiology. At the non-profit institute he created in 1887 he wanted to provide opportunities for basic research as well as the development of products based on the discoveries made. Over the years he had made impressive contributions such as removing microbial contamination of milk and wine by heating — pasteurization — developing a cure for a silk-worm disease saving the French silk industry, and he had developed a vaccine against anthrax. However, it was the successful prevention of rabies in a young boy bitten by a rabid dog that led to the initiative to establish the institute.

This success stirred the media and the public. Pasteur together with his close collaborator Pierre Paul Émile Roux decided to establish a private foundation by public subscription. It would be independent of governmental support and unattached to the university system, a unique situation in France. By a special law the Institute became identified as a public utility. Income to support the researchers was to come from production of vaccines and sera for passive immune protection. Some decades later affiliated companies had been established in Brussels and in the former French colonies in Africa and Indochina. The structure of the newly-established institute was very forward-looking for its time. One of the first courses in microbiological techniques in the world was given at the young institute.

The Institute today occupies two large blocks in Montparnasse on the left bank of the Seine. The history of the development of the institution has been briefly summarized by Jacob[4]. Originally Pasteur bought a piece of land close to what was then called rue Dutot. The first building on this property was referred to as the "Microbie" laboratory. Then another piece of land on the other side of the street could be bought and after Pasteur's death a building for chemistry was erected. Piece by piece more land was acquired and new buildings for new functions were added. A hospital for contagious diseases with attached pavilions was built and there was a special pavilion for tropical diseases and later a building especially for tuberculosis. The present address of the Institute appropriately is 25–28 Rue du Dr Roux, commemorating

The original building of the Pasteur Institute.

the co-founder of the institute. The original building remains and has been modified in parts with Pasteur's original apartment converted into a museum. The cellular section of the apartment was converted by Pasteur's son, Jean-Baptiste, into a vault which was transformed into a crypt, highly decorated in the Art Nouveau style. When Pasteur died in 1895 the Government offered his family the option of having him buried in the Panthéon. However,

Pasteur's crypt. [© Institut Pasteur]

his widow was not agreeable to this and after temporary burial in a cathedral, Pasteur's ashes were transported to the newly established crypt. Once a year, on the anniversary of Pasteur's death on September 28th, the crypt is open to all employees at the institute, who have an opportunity to pay tribute to its founder. The name of the young boy whose life Pasteur saved by his rabies vaccine was Joseph Meister. He later became a caretaker at the institute and remained in this position until his death in 1940, when he committed suicide ten days after the German invasion. The story that he did this to not have to let German soldiers into Pasteur's crypt is apparently not true. Most likely it was concern for his own family, whom he had sent away and thought might have been killed, that elicited the act.

There have been major ups and downs in the financial conditions of the institute over its long existence, but it has remained a home of high-quality science. Over the years a total of eight Pasteur Institute scientists have been recognized by Nobel Prizes. Roux did not receive a Nobel Prize in physiology or medicine although his invention of serum for passive immune protection against the effect of the diphtheria toxin was an important contribution. However, the principle of passive immune protection had been pioneered by Emil A. Behring, who received the first prize in this category in 1901. Roux became a foreign member of the Royal Swedish Academy of Sciences in the same year after a joint nomination by the Vice-Chancellor of the Karolinska Institute, Mörner, who we already met in Chapters 1 and 4, and others. Roux presented a Ph.D. thesis that described studies on rabies virus infections and he was a co-worker with Pasteur in the development of the vaccine against this virus. Other important contributions from the Institute were the identification by Alexandre Yersin of the causative agent of bubonic plague, *Yersinia pestis*, and later the development by Albert Calmette and Camille Guérin of a means to immunize against tuberculosis by use of the BCG vaccine. The first Nobel Prize awarded to a member of the Institute was the recognition, in 1907, of C. L. Alphonse Laveran, the discoverer of "the role played by protozoa in causing disease," especially the malaria parasite. The field of parasitic diseases has only rarely been the focus of interest in the medical Nobel work, but it came back in full force in 2015. In that year the prize was divided, with one half going to William C. Campbell and Satoshi Omura "for their discoveries concerning a novel therapy against infections caused by roundworm parasites" and the other half to Tu Youyou "for her discoveries concerning a novel therapy against malaria." In the meantime a prize in 1927 (the prize for 1926) recognized a claim that Spiroptera, a parasitic nematome worm, was a cause of carcinomas, representing probably the major stain on the bright shield of the Karolinska Institute in the context of Nobel Prizes and in 1946 the use of DDT as a contact poison against several arthropods, uncontroversial at the time, as discussed in my first book on Nobel Prizes[1], was the motivation for a prize.

The next prize to a Pasteur scientist was awarded the year after the prize to Laveran in 1908 to Ilyich Mechnikov, a prize shared with Paul Ehrlich as also discussed earlier[2]. The same book also described briefly the 1919 prize to Jules Bordet for his studies of antibodies and complement. In 1928 there was one more prize, this time to Charles J. H. Nicolle "for his work on typhus." It would then be 37 years before any scientists at the Pasteur Institute were recognized again, by the 1965 prize, the subject of this and the following two chapters. The last two Nobel Laureates among the eight are Luc Montagnier and Françoise Barré-Sinoussi, who were honored in 2008 for their identification of a virus — later named HIV (human immunodeficiency virus) — as the cause of AIDS. As mentioned already in Chapter 1 Bovet who received the 1957 prize, did most of his prize work at the Pasteur institute. It could be added in relation to the 1963 prize on ionic mechanisms of nerve transmission that Jean-Pierre Changeux (p. 460) in 1970 isolated the first receptor of a neurotransmitter, acetylcholine, at the Pasteur Institute. In spite of this and his many other notable additional scientific achievements, such as contributing to the early formulation of the concept of allostery, a phenomenon to be presented in the following chapter, he has not been recognized by a Nobel Prize.

I have had the privilege to pay many visits to the Pasteur institute, since it has always hosted scientists of major importance in virology. In the 1980s many of the conferences concerned HIV and AIDS. Besides Barré-Sinoussi and Montagnier, I befriended Simon Wain-Hobson and others. Together with Montagnier and a respected British virologist, David Tyrrell, I was for a number of years a member of a committee — "the three wise men (?!)" — that gave a unit under the European Union advice on how to support a unique chimpanzee colony kept in the Netherlands and to allow the use of the animals for research. This was a delicate matter, since without using chimpanzees as experimental animals, the remarkable protein infectious agents, later called prions[1] and the identification of hepatitis B viruses would not have been possible. On the other hand their position as our closest relatives in an evolutionary perspective necessitates very careful judgments of their potential use in scientific research. Today only a few highly selective experiments are performed using this kind of animal.

The focus of the 1965 prize in part had its origin in another branch of research pioneered at the Pasteur Institute. It concerned studies of bacterial viruses, the *bacteriophages*. Felix d'Herelle at the Institute was one of the discoverers of these kinds of infectious agents and gave them their name. This important contribution came close to being recognized by a Nobel Prize, but in the end it did not make it[1]. Important studies of bacteriophages were made during the 1930s by the married couple Eugène and Elisabeth Wollman, of Russian Jewish origin, and followed up by their son Élie L. Wollman who joined Lwoff's laboratory in 1945 just a few weeks before Monod. It was Élie who knew the history of the Pasteur Institute well and told it to Jacob. The occupation of France by the Germans in 1940 did not lead to interference with the, at the time, very much reduced research activities at the Pasteur Institute. No attempt was made to gather information. The Germans relied on their own authority in the research fields of infections and immunity. The relative freedom the Institute had led to it becoming the home of certain resistance activities, partly on the initiative of Louis-Pasteur Vallery-Radot, a physician named after his grandfather. We will return to these activities in the next chapter, but it should be mentioned that Lwoff himself was involved in the Resistance movement as a member of two networks, the Cohors-Asturies and Shellbum.

On one occasion the Germans became suspicious when there was a typhoid epidemic among Wehrmacht troops near Paris in 1943. The source of the epidemic was later found to have its origin at the Institute. A sample of the bacterium causing the disease had been retrieved by an employee and by collaboration with an accomplice a batch of butter used by the German soldiers had been infected. As a consequence the German authorities requested an inventory list of dangerous infectious agents at the Institute. In addition they demanded lists of the names of employees. The two older generation Wollmans and three additional laboratory technicians were incarcerated and sent to the Auschwitz concentration camp where they died. The Wollmans contributed a lot to pioneering work widening the understanding of the relationship between phages and bacteria and we will return to this in discussing Lwoff's revolutionary work on the phenomenon called lysogeny, which he initiated in 1949. Already before that he had gained some major insights into studies of bacteria and parasites, so let us follow his early life and his development as a scientist.

The Young Scientist

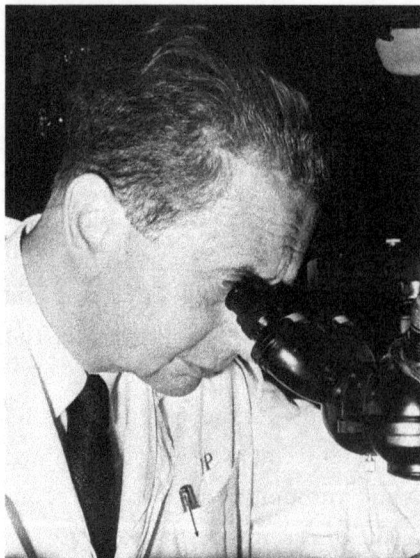

Lwoff at his microscope. [From Ref. 23.]

André Lwoff was born in 1902 in Ainay-le-Château in Allier in central France to parents of Russian descent. His father was a psychiatrist and had a firm belief in the importance of science, whereas his mother was an artist and sculptor. Metchnikoff was one of his father's friends and these contacts stirred André's interest in experimental biological research. It is said that André saw his first microbe in the light microscope at the age of 13. However, his father considered it important for André's future that he had a formal education. Thus he enrolled in the vocational training to become a physician, but in addition he took different courses to learn research techniques. In fact he probably enjoyed his science more than his medical studies. His early work was carried out during summers at the Marine Biological Laboratory at Roscoff in Brittany under the supervision of Edouard Chatton. He was a famous protozoologist (someone studying single cell eukaryotic organisms of animal of plant origin) who had made major contributions to distinguishing nucleus-carrying cells — eukaryotes — and cells without such a central structure — prokaryotes, like bacteria. His main interest was pathogenic protozoa, like trypanosomes. The original mentor-student relationship developed into a rich friendship, lasting until Chatton's death in 1947. Chatton opened the doors for Lwoff to further develop his experimental studies in 1922 under Félix Mesnil at the Pasteur Institute. He became Lwoff's second mentor. Mésnil had been Pasteur's secretary for some time and then became an authority on microbial infections, in particular parasitic diseases. He was an important collaborator of Laveran's in the studies demonstrating the parasite responsible for visceral leishmaniasis, also referred to as Kala-azar in India. Later in his career he pioneered experimental infection of chimpanzees with the parasite *Plasmodium vivax*, causing malaria.

In these environments Lwoff's first work naturally concerned parasites. Many of those he studied were of medical importance, like trypanosomides. He examined their life cycle and in particular their nutritional requirements. A major part of his interests concerned protozoans with hair-like organelles at their surface, the *ciliates*. These organelles serve many purposes — movement, attachment, feeding and collecting sensory signals. The corresponding structure in eukaryotic cells is called flagellas. He retained an interest throughout his life in ciliates and his last two scientific publications, at the age of 88, concerned these organisms. In 1927 he received his M.D. and five years later his Ph.D. His thesis concerned the role of different growth factors in the life cycle of certain protozoa and the change in requirements when the microorganism adapted to various degrees of parasitism. This was important work highlighting the role of such growth factors. Very early on Lwoff adopted an evolutionary perspective on the systems he was examining.

He had an open mind and deduced that developments could include *both* gains and losses of functions, a heretical view at the time when evolution was interpreted as signifying progressive advances based on modified or acquired new characteristics. Later it was found that an important part of evolutionary events are *regressive*. Already as a result of his early research he had been recognized by a number of prizes from the French Academy of Sciences. In 1929 he was appointed head of his own laboratory at the Pasteur Institute. Lwoff has said the following about this early phase of his scientific career "It was Mesnil who initiated me into the Pasteurian disciplines. When I entered the Pasteur Institute, many representatives of the heroic period of microbiology were still there. At the twilight of the Golden Age, the vibrations of the Pasteurian epic had not yet subsided and were still able to awaken echoes in the enthusiastic soul of a young student. It is with passion that I devoted myself to the study of ciliates."

Already in the mid-1920s Lwoff met his future wife, Marguerite. They married in 1925 and remained inseparable companions in private life and at the laboratory bench until the early 1970s and in their private life for a few more years until she died in 1979. André liked to work with his own hands and preferably only in a small group, including his wife and a technician. The couple had no children. With the help of a scholarship from the Rockefeller Foundation the two of them were able to spend a year in Otto Meyerhof's laboratory. Meyerhof had received the 1922 Nobel Prize in physiology or medicine (bestowed in 1923) for his studies of muscular metabolism. Lwoff examined the effect of protohaematin, a form of the oxygen-carrying hemoglobin, on the growth of flagellates, a group of parasites with whip-like organelles on their surface. In this work he discovered the first growth factor ever identified to be required by this kind of organism. In 1936 the couple went abroad again on another Rockefeller scholarship, this time to David Keilin in Cambridge, U.K., whom we already got to know in Chapter 4. In this intellectually inspiring environment they investigated the role of additional growth factors for an even simpler biological entity, the bacterium, *Haemophilus influenzae*. During this time various vitamins had been found to be essential for the growth and well-being of animals and man. Many of these vitamins were shown to serve in the metabolism of cells, not to provide energy or offering a catalytic function of their own, but as co-factors to specific enzymes, critical biomolecules we already met in Chapter 4. The "V factor" studied by the Lwoffs turned out to be the same substance as the co-enzyme purified by Warburg. Lwoff was able to show that the different vitamins identified also played a critical role in the life-cycle of even the simplest organisms. He discovered several new growth factors and developed the first synthetic medium that allowed propagation in the laboratory of a free-living ciliate protozoon, *Tetrahymena pyriformis*, which has served as an important model organism in later studies.

All these new findings had major consequences since it dawned on the scientific community that bacteria and cells of higher organisms were similar in many regards when it came to metabolic processes. Hence the more readily managed microorganisms potentially could be used as a relatively simple model system in the laboratory allowing general conclusions applying to all forms of life. Lwoff summarized his work in a book written, during the German occupation of Paris, in French with the lead title *L'evolution physiologique* and published in 1944. Already at this time and to a large extent unknown to Lwoff, because of the war, chemists like Lipmann had began to use bacteria to examine energy transport in cells, George Beadle and Edward Tatum had initiated genetic studies in microorganisms demonstrating that genes could direct the production of critical enzymes and finally Max Delbrück and

Salvador Luria had presented early studies that showed the role of genes in both host bacterial cells and in bacteriophages. Developments in all these three areas eventually led to Nobel Prizes in physiology or medicine, in 1953, 1958 and 1969, respectively.

In 1938 Lwoff was appointed head of a microbial physiological laboratory and moved to the famous "Le Grenier" — the attic — in the E. Duclaux building at the Institute. This setting has been described in the following way by Horace Judson[3]:

> Lwoff's fabled attic was three flights up, at the back corner, and consisted of a corridor about twelve yards long, high-ceilinged but windowless and crowded with equipment, with five small laboratories opening off it and tucked beneath the slope of the mansard roof. Lwoff's lab was at one end, Monod's (who joined him after the liberation of Paris in the Second World War, my remark) diagonally across at the opposite end. There were also a tiny secretarial office, a kitchen space around the corner of the stairs, and a cold room not bigger than a broom closet. The elevator was small, slow, grumbling, often out of order.

It was in this setting that the intellectual high tensions feeding a unique creativity developed. These developments, as we shall see in this and the following two chapters, were critical to the maturation of the field of molecular biology

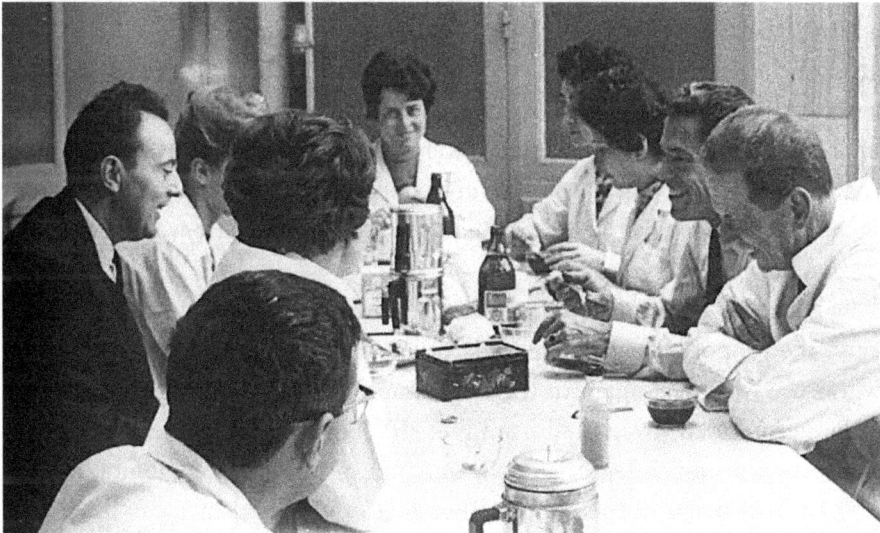

Lunch gathering in the attic hosting Lwoff's laboratory. Jacob is seen to the left and Lwoff and behind him Monod to the right. [© Institut Pasteur]

which dates from the mid-1950s. However, it took some time to create the momentum, in large part due to the Second World War. During the first year of the war Lwoff was called upon temporarily to do military service and for the rest of the war the laboratory, as already mentioned, served the Resistance. After the war there was a need to catch up on scientific developments, in particular in the U.S.

A symposium on *Heredity and Variation in Microorganisms* was arranged in 1946 at Cold Spring Harbor, NY. This was attended by Lwoff and Monod. One ambition of the meeting was to reopen contacts between European and American scientists in the field. Lwoff was known because of his pioneering work on nutrition of microorganisms and Monod had already in the 1930s spent some time at the California Institute of Technology in Pasadena, CA (Chapter 8). A number of very important contacts between the Pasteur scientists and American colleagues were established. The meeting included several highlights. Joshua Lederberg and Edward Tatum for the first time announced results of experiments demonstrating a successful exchange of genetic properties, *mating*, between different strains of bacteria. The "phage school" led by Delbrück, described earlier[1] had achieved spectacular results in their studies of fundamental genetics using this model system. Delbrück and Alfred D. Hershey had furthermore shown that homologous phages could show *genetic recombination*. Many years later they became the recipients, together with Salvadore E. Luria of the 1969 Nobel Prize in physiology or medicine "for their discoveries concerning the replication mechanism and the genetic structure of viruses." However, at the time Delbrück and others in the group believed that bacteriophages always caused a destructive replication, a *lysis*, in the bacteria. They did not trust a phenomenon of quiet persistence of phages in the host cells, as proposed by earlier scientists. It is said that this challenged Lwoff to take up studies of bacteriophages, since he had a different opinion influenced by the older generation of Wollmans. He wanted to prove Delbrück wrong.

Lwoff was well updated on the developments in early molecular genetics. With the support of a Rockefeller Foundation grant he and Boris Ephrussi, whom we will meet again in the next chapter, in 1948 arranged a meeting in France entitled (translated from French) "Biological Units Endowed with Genetic Continuity." Regrettably the outcome of this meeting did not become well-known because the proceedings were published in French. At this meeting André Boivin from Strasbourg together with a fellow bacteriologist presented data from studies of *E. coli* supporting Avery's 1943 discovery of the transforming principle of DNA from pneumococci. In addition he and his colleagues had studied the DNA content of animal cells in both their haploid — with half the number of chromosomes as represented in germ cells — and diploid state, the normal set-up of chromosomes in somatic cells. The diploid cells were found to contain twice the amount of DNA per cell as compared to haploid cells. A part of Lwoff's summary — formulated as early as 1948, as mentioned — translated into English read as follows:

> The transforming principle of pneumococcus is deprived of proteins and appears to consist exclusively of deoxyribonucleic acid. This is probably the case also for *E. coli.* The importance of DNA is indicated by the fact that diploid nuclei have twice the amount of DNA as haploid nuclei. The study of the transforming principle of pneumococcus has led to the conclusion that the purine and pyrimidine bases are not present in equimolar proportions. This gives an inkling of a possible explanation for the specificity of nucleic acids.

These poorly disseminated comments were way ahead of their time, but clearly Lwoff was prepared to tackle crucial microbial genetic phenomena, in his case lysogeny.

Lwoff and Lysogeny

In 1949 Lwoff phased out his studies of microbial metabolism and turned to bacteriophages, in particular their capacity to cause persistent quiet infections. There was a long tradition at the Pasteur Institute of studies of these kinds of viruses, as already mentioned. Lwoff was of course well acquainted with the phage work by the Wollmans of the older generation. Already in 1936 he contributed a commentary to their work in an appendix published in *Annales de l'Institut Pasteur*. It is apparent from this early publication that Lwoff had a very broad insight into microbiology at the time and also that he had a great talent for combining knowledge into new insights and conjectural hypotheses. He was well anchored in the French cultural tradition, allowing advanced intellectual synthesis and also philosophical reflections. At the same time he had good knowledge of European cultures outside France, having worked abroad, and added to this he also at an early date built up contacts in the United States appreciating the rapid advances of natural sciences in that country during and after the Second World War. His theoretical analysis of the phenomenon of virus persistence in 1936, the *lysogeny*, etymologically, with a potential to cause disruption of cells, foreshadowed much of his experimental work on phages which, however, he did not start until some years after the Second World War. It is apparent from his thinking in the mid-1930s that he was a qualified intellectual synthesizer and it is not surprising that he was the first scientist to give a solid definition of a virus in 1957, as referred to in my first book on Nobel Prizes[1] and discussed below. In his 1936 reflections he departed from the concept of viruses as "modified genes" and mentioned the pioneering findings by Rous, cited in Chapter 3, that virus in chickens could have the capacity to cause tumors. He referred to the Rous sarcoma virus as a "transmissible mutagen." He also reflected on the fact that plant viruses as well as bacteriophages could appear both in an active *and* in a dormant, *latent* state. In a very visionary statement, made already in 1936, he said in a text cited in translation in[5]:

> In summary, the action of genes can be explained by the hypothesis that genes act in cells not directly as catalysts, but as inducers that are activated in a cyclic manner. The gene appears to exist in two states, active and inactive. The reactivation of the gene appears to be linked to division. This hypothesis agrees perfectly with the properties of the lysogenic principle and the mosaic virus, both of which are activated as a result of cell division. This common relationship of the lysogenic principle and the mosaic virus to cell division is not just due to convergence … It appears to us to be an expression of the fundamental similarity between the "virus-gene" and the "normal gene …"

It should be noted that these remarks were made before the one gene-one protein concept had been identified by Beadle and Tatum and way in advance of the identification of the chemical nature of the genetic material and the launching of the field of molecular biology. He foresaw the emergence of the field of biochemical genetics. Before moving on to this topic let us reflect on the virus concept.

Viruses are cellular parasites. Provided that there are particular receptors on the surface of the target cell — normally serving various critical physiological functions — there are possibilities for the virus to enter the cell, which occurs by different mechanisms. There are various kinds of viruses which can infect all forms of life; bacteria, other single cell organisms, plants, insects and vertebrates. Viruses are truly *ubiquitous* and wherever there is life there are viruses. As we now know there are many forms of relationships between viruses and cells that they exploit as their hosts. The genes of a virus invading a cell can take control of the cellular machinery and direct it to produce the components needed to assemble new viruses in large numbers which are released either momentarily by a disruption of the cell or progressively over a shorter time usually leading to cell death. However, there are many other possible consequences of a virus attack on a cell. A chronic infection may become established. The virus remaining in the cell may rest in a quiescent form and possibly be activated later or it may influence the gene expression of the host cell changing its properties. This applies to all monocellular organisms; the prokaryotes — *Bacteria* and *Archaea* — and eukaryotes, like algae, protozoae, certain fungi — and the large world of multicellular organisms. In the latter the presence of certain viruses may lead to the development of a lack of capacity of cells to respond to the critical cell division controlling mechanisms. In this case the infected cell turns into a tumor cell, which displays varying degrees of unregulated cell multiplication behavior. All these different forms of virus-cell interactions have now been and continue to be analyzed by use of molecular biological techniques. However, it all started with studies of the interaction between certain bacteria and their viruses, the phages.

It was the abovementioned d'Herelle who at the Pasteur Institute developed his studies of ultrafiltrable infectious agents in bacteria for which he in 1917 coined the word *bacteriophage*, as already introduced. His theory of their independent infectious properties was not accepted by the famous Belgian immunologist Bordet, mentioned earlier, who was soon to receive a Nobel Prize in physiology or medicine. He did not believe like d'Herelle that the phage was an independent infectious entity but that it had a cellular origin. In a way both turned out to be right. D'Herelle was studying what later was called *virulent* phage that caused the destruction, *lysis*, of the bacteria, whereas Bordet examined persistently infected, what he called *lysogenic* strains of bacteria. The term later introduced by Lwoff for the latter kind of phages was *temperate*. D'Herelle believed that there was only one kind of phage that could attack all bacteria, but this was shown by Macfarlane Burnet, in some important early work, to be wrong. Burnet, the recipient of the 1960 Nobel Prize in physiology or medicine, was described at length in one of my previous books[2]. He demonstrated that there were many different kinds of phages, distinguishing them by use of immunological methods. He was also able to show that the specific infection by a phage was contingent on the presence of certain structures on the surface of the bacteria, one of the first notifications of the conjectural existence of a *receptor*, a concept that developed to become very fundamental in biology. Finally he was able to distinguish the two different kinds of interactions between the virus and the bacteria, the lytic interaction caused by virulent phages and the lysogenic state established by temperate phages. The stage was then set for the important developments in the 1940s when work on phages was resumed, using them as a means of studying the action of genes.

When Lwoff initiated his studies of lysogeny in 1949, he chose to study a large bacterium *Bacillus megaterium.* This had previously been used by other scientists, including the older generation Wollmans. It is said that Lwoff disliked statistical analyses and therefore he avoided examination of populations of cells and instead researched the events in individual cells. At the time it was known that a lysogenic culture sometimes produced small amounts of phage, but it was not known if this was due to activation of the dormant infection in some cells leading to a full cycle of replication with release of phages or if there was a slow release from some cells of small amounts of virus. Using a micromanipulator Lwoff and his collaborator Alan Gutmann examined the fate of individual bacteria isolated in micro drops. They found that occasionally there was a complete lysis of the cell with release of a large amount of virus, but in most cases no virus was found. It was concluded that the relationship between the phage and the cell was an *all or none* phenomenon. Either the virus remained quiescent in the cell, and in this state it was referred to as a *prophage*, or there was a complete cycle of phage replication with destruction of the host cell. These findings convinced the scientific community that lysogeny was a true phenomenon. It remained to define the relationship between the bacterium and the prophage. In order to examine this some method for activating the dormant virus was needed. Many different treatments of the lysogenic bacteria were tried in experiments performed together with his collaborators Louis Siminovitch and Niels Kjeldgaard but all in vain. Then one day they struck lucky. Monod was using a UV lamp to induce mutations in his *E. coli* cultures and Lwoff tried the same technique to activate the prophage. It was a great success. As expressed by Monod, who had great interest in music "The phenomenon was reproducible with the regularity of a metronome." Much later in 1965 when Lwoff delivered the First Keilin memorial lecture[6] himself, he said "As far as I can remember, this was the greatest thrill of my scientific career — for the first time in my life, I had a feeling of having discovered something."

The phenomenon of induction had a major impact on the work on different kinds of viruses. The role of latent animal viruses and cancer development was one field of great interest and another concerned latent viruses and their role in plant infections. It then remained to interpret what effect the UV irradiation might have and this took some time and included the future important work by Jacob (Chapter 9). However, the discovery coincided timeously with the demonstration of the critical role of the phage nucleic acid in the infectious process in studies by Martha Chase and Alfred Hershey using isotopes separately labeling nucleic acids and proteins and the discovery of the structure of DNA by James Watson and Francis Crick[1,2].

In 1953 Lwoff summarized the important advances hitherto made in a review[7]. His clarification of the nature of lysogeny made this a popular topic for studies by other scientists. This was not to Lwoff's liking. He sincerely disliked competition and therefore decided to switch to studies of animal viruses. There had of course already been some competition in the field of studies of lysogeny. One of the competitors was Giuseppe Bertani and he has summarized the

Milislav Demerec and Lwoff at Cold Spring Harbor in 1953. [From Ref. 3.]

developments in a late life review[8]. Bertani was born in Italy, but his studies brought him to the U.S. where in 1949 he became a research fellow in Milislav Demerec's laboratory at Cold Spring Harbor. He had no experiences of working with bacteria, but was guided to examine spontaneous and induced mutations in streptomycin-dependent *E. coli* strains. Once he observed a colony that looked sectorially nibbled. Having no idea what this meant he went to the neighboring laboratory. The fellow scientists suggested the obvious explanation that it was all about phage contamination. Then one of them said

Giuseppe Bertani (1923–2015).

"There is also something called lysogeny ...." This put Bertani on the track of new studies of his *E. coli* cultures. Later the same year he moved to work with Luria at Indiana University in Bloomington. Luria supported further work on lysogeny although personally he was only moderately enthusiastic about this. But there were other problems. They had chosen to work on the *E. coli* system and the phage lambda discovered by Lederberg's wife at the time, Esther. However, since at this time the Lederbergs had already started successful genetic mapping of the genome of this organism Bertani's choice of system was not to their liking. Hence another system needed to be looked for.

At the time Luria, Watson, as a final-year graduate school student in Bloomington, and Bertani were all sharing a smallish laboratory. Finally the potentially pathogenic bacterium *Shigella*, the etiological agent of dysentery, was chosen,

Joshua and Esther Lederberg at Cold Spring Harbor in 1953. [Photo from Karl Maramorosch.]

although there were wild protests from Watson, who was scared of pathogenic organisms. Progress was made in studies of the kinetics of release of lysogenic phages in the bacterium. Plaques of different characteristics were found and it turned out that the system included three different kinds of phages, P1, P2 and P3. It was at this time that Lwoff published his great breakthrough using UV light to activate lysogenic phages and cause cellular lysis. This was then tried in many different systems, like *E. coli* and lambda, and worked a treat, but to Bertani's chagrin, the lysogenic phages in *Shigella* could not be induced! This eventually allowed Bertani to unravel some particular and important properties unique to the *Shigella*-P phages system, a story that is too comprehensive to be elaborated in this context. However, in order to carry out his work he needed to develop a new medium in which he could grow his bacteria. The medium was called LB broth, which incorrectly has been cited as Luria-Bertani medium although in truth it means *lysogeny broth*.

There is a special additional reason to mention Bertani and his research in the present context. The field of bacteriophage research was of course carefully followed at the Karolinska Institute. In 1960 Bertani was recruited from the University of Southern California to a professorship at the Institute. He established his activity at the newly built Wallenberg Lab 60 where he formed a research group. Over the years he would train many bacterial and phage geneticists, mostly of Swedish background, and I have myself taken part in a course he gave to budding molecular biologists. We also had joint seminars between Sven Gard's department of Virus Research and Bertani's department. Bertani became a pioneer of microbial and molecular genetics in Sweden, but he never became directly involved in the work on Nobel Prizes. He returned to the U.S. first after 21 years to the Jet Propulsion Laboratory in Pasadena, CA. His American wife Elisabeth was herself a qualified researcher. She was the only competitor I had for the position of successor to Gard, which I was finally appointed to in 1972. The most qualified virologist in Sweden at the time was Lennart Philipson. He did not enter the competition for Gard's chair, since he preferred to retain the chair of microbiology that had recently been arranged for him at Uppsala University. However, if he had decided to join the competition, my professional life would have developed very differently and it is doubtful if the present Nobel books would ever have been written. However, that is counter-factual history.

Animal Viruses and the Evolution of a Statesman

In 1954 Lwoff had been invited to the U.S. by the National Foundation for Infantile Paralysis. The foundation promised major support for Lwoff's work if he shifted to studies of poliovirus, which he did. At this time there were rapid developments in the poliovirus research field. As previously described[1] John F. Enders and his collaborators Frederick C. Robbins and Thomas H. Weller had managed to grow the virus in regular tissue cultures of fibroblasts or epithelial cells, crushing the long-lived dogma that the virus could only grow in nerve cells. This discovery which was recognized by the 1954 Nobel Prize in physiology and medicine paved the way for laboratory studies of the virus and in particular for the rapid development of a vaccine against the disease caused by it. The newly developed tissue culture techniques seemed to Lwoff to offer opportunities to do genetic studies with poliovirus. In order to prepare himself for his new adventures, he and Marguerite made a long journey through the U.S. visiting many laboratories. At the National Institutes of Health in Bethesda they met Wilton Early, in New Haven Joseph Melnick, in Boston John Enders, in Toronto Raymond Parker, in Pittsburgh Jonas Salk, in Minneapolis Jerome Syverton and finally in Pasadena Renato Dulbecco and Marguerite Vogt, the latter already mentioned in Chapter 1. They stayed for a longer time in Pasadena to learn techniques for handling animal viruses, which are completely different from growing bacteria to examining their phages.

Back home at the Pasteur Institute they embarked on studies of poliovirus. In some experiments Lwoff employed the previously successfully used technique of single cell/drop infection. One aim was to map the genetic properties of the virus. Interesting findings were made in examining the capacity of different strains to grow at various temperatures. In this context it was discussed what importance the fever associated with an infection could have on the rate of virus replication. It was possible to change the development of a poliovirus infection in animal models using different mutants of the virus, but this never came to result in any means to develop a live attenuated virus for vaccination purposes. In fact the rapidly developing immunizations, originally using inactivated virus eliminated almost all infections in industrialized countries. Certain mutants sensitive to selected chemicals like guanidine were also found and Lwoff described his work on such mutants in detail in the first Keilin lecture in September of 1964[6]. This was arranged by The Biochemical Society in the U.K. to honor the great chemist discussed at length in Chapter 4. However, in the end Lwoff's genetic studies with poliovirus had no lasting benefits. Later studies have shown that poliovirus is far from ideal for genetic studies. It was much later before it was eventually demonstrate that there could be an exchange of genetic material, recombination, between strains of this kind of virus. Its genome is single-stranded RNA with what is called positive polarity, which means that it can serve directly as messenger-RNA (see Chapter 9). This kind of genome cannot associate with the cellular genetic material and therefore the virus is always virulent and inevitably causes cell death by lysis.

As the techniques for sequencing RNA and DNA developed as described in Chapter 5, they were quickly applied to different kinds of viruses, because of the attractive limited size of their genomes. The RNAs of poliovirus of all three types were characterized. But this was not all. As the techniques not only to read, but also to write — synthesize — nucleic acids developed, genomes of selected viruses were produced in the test tube. This was also done with poliovirus and the RNA produced was shown to be infectious on its own. In fact Eckard Wimmer at Stony Brook, New York and colleagues first synthesized the DNA version of the genome. This was then used as a mold for virus RNA. The fact that the laboratory-derived poliovirus RNA was infectious caused considerate debate. Was it appropriate to synthesize infectious agents, to create life? We will return to this question below. Obviously, possibilities to synthesize the virus genome provided means to perform any kind of genetic studies. It would have been impossible for Lwoff to foresee these remarkable developments at the time when he initiated his poliovirus studies.

Although Lwoff's research after 1953 may not have led to any major discoveries he was to have a major impact in the virus field at large by his statesman-like actions. There were two major contributions that led to the virus community regarding him with awe. The first was fundamental since it concerned the definition of a virus. This occurred in 1957 in an article entitled "The Concept of a Virus[9]." In his presentation Lwoff used an interesting preamble:

> My ambition is to show that the word *virus* has a meaning ...
> "Frenchmen" said Paul Valéry, "deem possible ... that a prodigiously
> diverse ensemble of highly complex phenomena can be and must be
> condensed and finally reduced into a few plain formulae, at the same
> time necessary and sufficient." Belonging to a hyperlogical extrovert
> nation, I have coined numerous definitions as if I had really penetrated
> the essence of things. (So,) I shall defend a paradoxical viewpoint, namely
> that *viruses are viruses*.

However, he took the issue further and did in fact offer a definition. It read:

> ... infectious, potentially pathogenic, nucleoproteinic entities possessing
> only one type of nucleic acid, which are reproduced from their genetic
> material, are unable to grow and to undergo binary fission, and are
> devoid of a Lipmann system.

These were fundamental and useful definitions that have stood the test of time. The concluding part of the definition implies that viruses cannot generate the energy required for their replication. This is provided by the host cell as well as many other metabolic tools required for the synthesis of the virus-specific products.

Understanding the nature of viruses opened the way for dividing them into groups, a system of classification. These developments took a major step forward in 1962 and again Lwoff played a central role. In this year another symposium was arranged at Cold Spring Harbor and several of the publications discussed means to identify the order of viruses. In my two previous Nobel books[1, 2] discussions of the symmetry structures used by viruses in building the shell that protects their nucleic acid were described and their use for classification purposes was mentioned. At the 1962 symposium Lwoff together with Robert Horne and Paul Tournier proposed a system that has remained central for development of a taxonomic system for viruses[10]. It highlighted four decisive characteristics; type of *nucleic acid* — DNA or RNA; symmetry of the *capsid* (a word created by Lwoff for the protein covering the nucleic acid giving it protection from the environment) — helical or cubical; presence or absence (referred to as a naked virus) of an outer membrane structure, the *envelope*. All these three criteria are categorical in that a virus has either one or the other of the potential characteristics. The fourth criterion was more qualitative and concerned the dimension of the whole virus particle, the *virion* — another word created by Lwoff — be it a non-enveloped or enveloped virus. Viruses vary enormously in the number of genes they have represented in their genomes and also in the size of their virions. A lot has happened in the world of virus classification after the pioneering contributions in 1962, but the fundamental criteria used for separating them into groups have remained.

A Rich Personality

Let us pause for a moment and reflect on the personality of this intellectually gifted and visionary person by citing some remarks made by Jacob in his famous biography[4] and in the biographical memoir of Lwoff that he and Marc Girard together wrote for the Royal Society[11]. He presented Lwoff as a tall, straight-backed and slender man of refinement. This refinement was apparent in every aspect of his personality: in the way he dressed, the wine and food he enjoyed, his command of language and handwriting, etc. In those whom he selected to be his friends he had complete trust, even to a degree of credulity. For example when Monod was involved in helping Hungarian scientists to leave their home country illegally after the 1956 Russian invasion, to be discussed in the following chapter, Lwoff provided him with a considerable sum of money, without of course any security. Lwoff created a special warm, open and lively imaginative atmosphere around himself in groups with others. As Jacob formulated it[11] "Around André an exceptional atmosphere prevailed, mixing enthusiasm, clear-mindedness, non-conformism, humor and friendship. Anyone who had not lived and worked in the attic could not appreciate the warmth, the wealth and the generosity of a personality who often hid himself behind a kind of haughty reserve." Lwoff had a natural paternal authority in his relations and Jacob always addresses him as "monsieur." But Lwoff also kept track of individuals who he disliked and expressed his attitude by irony and sarcastic remarks.

Jacob praised Lwoff's remarkable intuition. As Gunther Stent wrote in a brief biographical memoir for the American Philosophical Society "I tried to style my scientific persona as a hybrid of Lwoff's intuitive-optimistic-synthetic and Delbrück's logic-pessimistic-analytic approach to research." Lwoff combined a huge knowledge of biology with a rich humanism. He was also a man with an enormous demand for precision of thought and language. This left major imprints on his collaborators, Monod and Jacob who, however, resolved this in somewhat different ways as we shall see. The demand for clarity sometimes led to the creation of new words. Many words of importance in virology and molecular biology originated in Lwoff's laboratory; prophage as we have learnt; virion and capsid (as well as its components, the *capsomers*), which were also mentioned above. Even in intellectual reasoning and in his science Lwoff acted like an artist, and on the side he was in fact a good painter as we shall see. Lwoff had a general love for painting, sculpture, music and "those things that awaken the spirit."

My first distant encounter with Lwoff must have been when I participated in my first congress of microbiology in Montreal in 1962, but I have no memory of this. What I do recollect is the dinner in Sven Gard's home in connection with Lwoff's visit to Stockholm to receive his Nobel Prize. Since I had presented my thesis in 1964 I was in my early post-doc period and felt more professionally at ease than when I met Macfarlane Burnet in the same situation five years earlier. Like Burnet, Lwoff was one of the very prominent figures in the field of virology and left an unforgettable imprint on those who were privileged to meet him. In fact I also met Lwoff on one more occasion. As already mentioned he was heavily involved in the work on classification of viruses. He was therefore invited to be the honorary speaker at one of a series of International Congresses on Comparative Virology, arranged in the Laurentian Mountains outside Montreal in the late 1960s by Eduard Kurstak, a previous graduate student of Lwoff and Karl Maramorosch. I had the pleasure to be present at this congress speaking about recent result from our studies of the immune-biology and structure of adenovirus particle components. The meeting allowed a gathering of the most influential virologists of the time and the picture arranged by Maramorosch tells this story. In the front row Lwoff is sitting to the right and on his right side are the two organizers, with Maramorosch humbly in the center. In the row behind many of those who came to lead the field of virology through the rest of the century can be found; Purnell Choppin, David Baltimore — the Nobel Prize recipient in 1975 — Norton Zinder, Jordi Casals, an unidentified colleague, Neville Stanley and Sam Dales. I cannot refrain from citing from a communication from Maramorosch in response to an email from me thanking him for photograph and congratulating him on his 101st birthday! He wrote "Thank you for the birthday wishes. I delayed my reply because I attended a conference this week in Las Vegas with my daughter. I won $4 at the casino." The rest of the email contained information about the congress in Canada including data on virologists in the picture and information about his recently published memoirs *The Thorny Road to Success*. As cited earlier, scientific endeavors can give a happy and long life with vitality until an advancing age and apparently also a possibility of late wins at the casino!

Lwoff at the lower right together with the organizers — Kurstak left and Maramorosch in the middle — of a conference on Comparative Virology in the Laurentian Mountains outside Montreal, Canada, in the late 1960s. The virologists in the back row are identified from left to right in the text. [Photo from Karl Maramorosch.]

The Reviews of Lwoff's Work by the Nobel Committee

In my first book on Nobel Prizes there was a chapter entitled *Nobel Prizes and the Emerging Virus Concept*, which included a section on *Bacteriophage replication and the studies of genes*. At the time of writing archive material until 1959 was available for examination. In this section a sketchy historical description of the development of the field and in particular background information on the father figure of phage genetics, Delbrück was presented. The first nomination of the leading figures in the field occurred in 1955. Four U.S. phage virologists, Seymor S. Cohen, Delbrück, Hershey, and Luria had been nominated by C.W. Jungebluth, New York. The committee decided to let the professor of bacteriology, Berndt Malmgren, and not the frequently used reviewer, the virologist Sven Gard make an evaluation of the candidates and the field. Malmgren made a very comprehensive examination noting the development of knowledge that led to bacteriophages being accepted as true viruses. He emphasized in particular the studies of so-called T-even phages, the tadpole-shaped, DNA-containing phages, infecting the common gut bacterium *E. coli*. This kind of phages had been demonstrated to be excellent tools for studies of the genetic interactions between different strains, *recombination*, and the interaction between the phage genome and the genome of the host, *transduction*. Malmgren's conclusion was that the pioneering studies of bacteriophages had given results of such major importance for the discipline of virology in general, for fundamental cell research and for basic genetic studies that it was relevant to discuss them in relationship to a Nobel Prize in physiology or medicine. Hence he recommended that the candidates should be considered worthy of a prize. This was also the conclusion of the committee which, however, noted that a prize could not be given to all four at the same time.

Four years later Hershey was again nominated, this time together with Gerhard Schramm and Heinz Fraenkel-Conrat, the discoverers of the infectious property of isolated tobacco mosaic virus RNA. A very comprehensive review was carried out by Georg Klein, as already described briefly[1]. Klein had become professor of tumor biology in 1957 and was already involved as an adjunct member of the committee the year after, 1958. In his 1959 review he considered both the discovery of infectious RNA and also, in support of Malmgren's recommendation, Hershey's demonstration of the central role of DNA in phage infection, worthy of a prize. It would, however, take until 1969 before the founding fathers of phage genetics were recognized by a prize, as already referred to above.

Lwoff was nominated for a Nobel Prize in physiology or medicine for the first time in 1957. He was included in a list of seven potential candidates representing four different categories provided by the recipient of this prize in 1952, Selman Waksman. He offered to provide more information if the committee had a particular interest in any of the proposed candidates. Lwoff was mentioned together with Delbrück. The latter had been reviewed two years earlier, but the committee decided to let Gard review the new candidate Lwoff. Gard had a broad experience of managing reviews for Nobel Prizes[1,2]. Over 16 pages he presented the full picture of the historical background to the concept lysogeny, the development of this concept and in particular the discovery of the induction phenomenon. Gard first discussed a situation in which a bacterial culture survived an attack of a certain phage, but then intermittently released small amounts of phages. This carrier state could be cured by treatment by specific antibodies. Thus it differed from the true lysogenic state on which antibodies had no curative effect. During the mid-1940s there were the beginnings of an insight into different phases in the replication of viruses.

After a virus particle had attached to the receptors presumed to exist at the cell surface and by some, at the time unknown, mechanism had become internalized in the cell there followed a short phase when no infectious virus could be discovered. This was named the *eclipse* phase. For a while it was speculated that the lysogenic state might represent an extended eclipse phase. Lwoff's one-cell culture analyses demonstrated that this was not the case, but that the lysogenic state had some hereditary characteristics. The hereditary traits could be demonstrated by the UV induction experiments. X-rays were found to have a similar effect. Furthermore certain chemicals like peroxidases also could be inductive. The common denominator in all these cases was the exposure to mutagenic or carcinogenic agents, which stimulated a lot of work in a related area, namely carcinogenesis.

The remaining parts of Gard's full review focused on the follow-up experiments by Jacob on the conditions of establishment of a lysogenic state. Lwoff speculated that the genome of the phage most likely became associated with the chromosome of the bacterium. However, it remained to demonstrate what the effect of UV irradiation might be, expressed in molecular terms. This took some time to clarify and required the accumulation of additional knowledge about the interaction of nucleic acids and blocking protein substances, so-called *suppressors*. These mechanisms will be discussed at length in the following two chapters. Suffice to note that in the end the mechanism was not clarified by use of Lwoff's *Bacillus megaterium* system, but by use of the special lysogenic state of the phage lambda in *E. coli* K-12, the popular workhorse among bacteriologists and geneticists. This bacterium was introduced by Esther Lederberg, as mentioned. The Lederbergs demonstrated that the hereditary property of lysogeny could be transferred from one bacterium to another by recombination of hereditary material achieved by genetic interaction between different strains of *E. coli* K-12. Lwoff demonstrated that the metabolic conditions of the host cell influenced these events. Starving of cells delayed the activation of the phage genome by induction.

At the end of his review, Gard made some interesting comparisons with current understanding of the field of animal viruses. In August 1957 when he submitted his review this authority on animal viruses appears to have been convinced that all of them had RNA as their genetic material and thus were substantively different from the DNA-containing phages. This is a somewhat surprising statement because already in the early 1940s it had been demonstrated that poxvirus contains DNA. It is possible that this finding had been relegated to oblivion and rediscovered first in the mid-1950s when DNA was demonstrated to be the genetic material of many phages. After this identification of the presence of DNA in many animal viruses followed suit. In spite of the caveat introduced, Gard discussed the potential importance of knowledge about lysogeny for the understanding of potentially similar phenomena caused by animal viruses, "virogeni" (?). He discussed cancers caused by animal viruses and also dormant (*latent*) infection with herpes viruses, like the repeated local occurrence of blisters caused by herpes simplex virus and speculatively mentioned the persistent infection caused by hepatitis B virus a long time after exposure to the agent.

His conclusion was that studies of bacteriophages had given results of such principal importance for the discipline of virology in general, for fundamental cell research and for basic genetic studies that it was relevant to discuss them in relationship to a Nobel Prize in physiology or medicine. Gard fully agreed with Malmgren's statement in 1955 and emphasized that study of phage genetics was of relevance to human medicine. He then highlighted Lwoff as the leading scientist in his group, which also included a number of collaborators. The conclusion was that Lwoff's discovery of the induction phenomenon and his studies of the mechanisms and replication of viruses in bacteria was worthy of a Nobel Prize. The committee supported Gard's view.

In 1958 there was a nomination for three animal virologists who had pioneered the discovery of influenza virus, as described earlier[1]. One of these virologists Richard E. Shope, besides his crucial contributions to the discovery of influenza virus had made pioneering studies of virus persistence in papillomas (benign warts) in rabbits. In his review of this nomination Gard discussed the pertinence of possibly giving a prize for the discovery of viral persistence. He noted that the most important work in this area had been done with phages and therefore proposed that the best candidates would be Lwoff and Lederberg. The latter scientist had been nominated for the first time in the same year, 1958. He had been included in a very thoughtful nomination of five candidates by George P. Berry at Harvard University, which has been discussed at length earlier[1].

Lwoff was included in the group together with Beadle, Delbrück, Hershey and the first-time nominee Lederberg. There was no further reviewing of Lwoff but the newcomer on the committee Klein was chosen to examine Lederberg. Klein was very enthusiastic about his discoveries and considered him a strong candidate. Thus Lederberg had pioneered genetic studies using bacteria and demonstrated that under certain conditions strains of bacteria could exchange selected genetic properties. In addition he had found that certain viruses could be used to carry host genetic material from one bacterium to another. This phenomenon was called *transduction* and turned out to have a major importance since it offered a different mechanism than mutation to provide bacteria with new properties. A classic example is the capacity of the potentially disease-causing agent *Corynebacterium diphtheria*. Under normal conditions this is a harmless microorganism, but if it is lysogenized by a phage carrying the gene for toxin production it becomes a very dangerous pathogen. Since the gene that provides the capacity to produce the toxin in the diphtheria bacterium is an inherent part of the phage genome a separate term *lysogenic conversion* was introduced.

Klein's proposal was to combine Lederberg with Lwoff as already suggested by Gard. This was an attractive combination highlighting various forms of *horizontal* transfer or genetic material between bacteria. However, as we shall see it would be even more appropriate to combine Lwoff with Jacob, since the latter pioneered the molecular understanding of genetic mechanisms of lysogeny and also fundamental mechanisms of genetic control (Chapter 9). The bacterial geneticists George W. Beatle and Edward L. Tatum were already at the time considered strong candidates by the committee. When Klein, apparently at Bergström's suggestion brought forward Lederberg as an additional leading figure in bacterial (including phage) genetics the committee decided to include him although he had been nominated for the prize for the first time the same year[1]. There was only one dissenting voice in the committee. It was Gard who favored the combination of Lederberg and Lwoff, as in Klein's original proposal, but Lwoff eventually had to remain on the list of candidates worthy of a prize. It can be added that when Lederberg received his prize he was only 33 years old, until now the second youngest recipient of the prize in physiology or medicine. Frederick G. Banting was 32 and Watson 34 years old when they received their prizes. The motivation for Lederberg's half prize was "for his discoveries concerning genetic recombination and the organization of the genetic material of bacteria."

In 1959 there was no nomination of Lwoff and thus he is not mentioned in the summarizing protocol by the committee but in 1960 he is nominated again and this time together with Monod. The nominator in 1960 was the immunologist Melvin Cohn, an American with a pronounced liking for Paris working at the Pasteur Institute for five years starting in 1948 and during briefer stays thereafter. He was a close collaborator of Monod and thus knew the work in the laboratory very well. We will meet him again in the following chapter. The nomination was very comprehensive and included not only a description of the main achievements in their work but also attached reprints and a detailed *curriculum vitae*. Part of the curriculum was retrieved from material prepared in connection with Lwoff's appointment as professor of microbiology at the Sorbonne, a position he held between 1959 and 1968, still retaining his laboratory at the Pasteur Institute. The first five pages described Monod's contributions and the following four Lwoff's work. The description of the latter first discussed the discovery of the induction of lysogenic virus by different treatments and hereafter more recent findings in studies of poliovirus. Cohen's description of Lwoff's examination of the induction phenomenon recaptured what has already been described above. He reiterated that the virus remained in bacteria not as a particle but as a genetic element, the prophage, integrated into the host cell chromosome.

Cohen also finally mentioned Lwoff's important early work on protozoa and in particular his discovery of the role of certain metabolites which could be demonstrated to be essential not only in eukaryote cells but also in the microorganisms. He wrote "… the most important consequence is that these studies contributed in a decisive way to the demonstration of the unity of biochemistry in the living world." Bacteria and eukaryotes to a large extent shared central metabolic mechanisms. The committee did not do a follow-up review of Lwoff, but relied on Gard's 1957 evaluation. It let Malmgren, professor of bacteriology, whom we met in Chapter 4 and earlier in this chapter, make a review of Monod, to be discussed in the next chapter. Lwoff remained worthy of a prize, but the prize this year went to Burnet and Peter B. Medawar, which kept Gard very busy[2].

In 1961 the nominees from the Pasteur Institute had been expanded to three. A nomination by Jean Boyer, Paris, included Jacob in addition to Lwoff and Monod. In an attachment in French covering 19 pages Boyer presented the contributions by all the three candidates. This nomination described in some detail all Lwoff's contributions in science, giving an emphasis to his studies of lysogeny. There was also a separate nomination of Lwoff by another Paris professor, Robert Fasquelle. The committee which had not met Jacob before let Gard do a preliminary review in which he promptly emphasized that it was appropriate to include Jacob in a comprehensive review to follow. Hence it was decided to let Gard do another review including all three candidates and unconventionally the committee also let a professor of biochemistry from Uppsala, previously associated with the Karolinska Institutet, Gunnar Ågren, carry out a separate review of Monod. Gard's analysis of Monod and Jacob and Ågren's evaluation of Monod will be reviewed in the following two chapters. Regarding Lwoff Gard recapitulated the previously-mentioned findings of lysogeny as a persistence of the genetic material of a (temperate) phage in a bacterium, hypothesized to be in association with the bacterial host genome. He then discussed Lederberg's recent findings about genetic recombination and transduction in bacteria already referred to above. Gard referred to these developments in understanding the evolution and expression of genetic traits and in particular the role of horizontal gene transfer in nature as a means of acquiring new properties. This general finding gave an extra importance to the discovery of lysogeny, which provided the first insight into the possibilities of such a gene transfer by a virus. The general challenging theme of gene transfers vertically by asexual or sexual reproduction *and* horizontally by the action of viruses and evolution is a theme we will return to below. Lwoff was considered by the committee to be a qualified candidate for the prize, but this year he was joined by Monod and Jacob, as we shall see.

The following year, 1962, the three candidates were again nominated, this time by four different recent Nobel Prize recipients. The developments in Monod's and Jacob's work apparently had markedly strengthened their collective candidacy for a prize as will be further discussed in the following two chapters. Cournand, the already-mentioned 1956 recipient of the Nobel Prize in physiology or medicine, had taken his responsibility to nominate candidates to the prize seriously. His four-page nomination letter was accompanied by a

17-page attachment describing in some detail the contributions of the three candidates. The co-recipient of the prize to Cournand, Dickinson W. Richards, supported the nomination of the three candidates in a shorter letter. Two other recent Nobel laureates, Lipmann and Tatum also submitted strong nominations, but in Lipmann's case Jacob was not included. The committee decided that an additional review should be done, but this was not carried out. Possibly the committee decided to wait since the outcome of the deliberations in 1962 had already been given at an early stage of their work. The three candidates remained listed as worthy of a prize in the September protocol by the committee.

In 1963 there were new nominations for all the three candidates. Étienne Wolff, a highly respected French embryologist, member of the Collège de France and a foreign member of the Royal Swedish Academy of Sciences proposed Jacob and Monod. Gard noticed on the last day of nomination that there was no proposal for Lwoff and hence he submitted a nomination for all three candidates referring to the earlier reviews that had been commissioned by the committee. It was decided to let Peter Reichard, whom we already met as a student of Hammarsten and as the successor of Theorell in 1971, do a review of all the three candidates.

Peter Reichard. [Photo by the family.]

At the time of this review Reichard was 38 years old and temporarily, for a period of two years professor of medical chemistry at Uppsala University. In 1963 he returned to the Karolinska Institute as professor of medical chemistry. He is a first-class biochemist having made important contributions to our understanding of ribonucleosides, the building-blocks of nucleic acids. Over a number of years we worked together on the Nobel committee at the Institute and he was a very effective chairman for three years. Reichard noted the four reviews that had previously been carried out and therefore emphasized in his review in particular Monod's and Jacob's pioneering contributions to the emerging field of molecular biology, especially their discovery of the genetic repressor system and what came to be called messenger-RNA. Thus he only briefly summarized Lwoff's contributions from the early 1950s, not adding anything substantial to the reviews done earlier. In his conclusions he noted the critical importance of Lwoff's identification of the existence of prophages and their interaction with the genetic material of bacteria. The insights into the function of temperate phages were noted to have a particular medical relevance because of their potential importance for the development of cancer research at the time. Reichard and later the whole committee supported Gard's view that all three candidates were deserving of a prize.

In 1964 there were new nominations for the three French candidates. Professor J. N. R. Grainger from Dublin proposed Jacob and Monod and the 1953 Nobel laureate Lipmann reiterated his 1962 recommendation for Lwoff and Monod. There is an interesting second half to Lipmann's nomination. It has the heading "Proposal to include Erwin Chargaff with Monod and Lwoff." As previously discussed[2] Chargaff had made a very important contribution to the insights into the chemical structure of DNA by his identification that pairwise the purines and pyrimidines adenine-thymine, guanine-cytosine occurred in equimolar amounts and further that the proportion between these base pairs varied between hereditary material from different forms of life. The first observation was explained when Watson and Crick unraveled the double-helical structure of DNA, but neither Chargaff himself used his data to make a prediction of the proper structure, nor did Watson's and Crick's resolution of the structure seem to have been critically dependent on access to Chargaff's data, although they were well aware of their existence. The archives indicate that a new review of the three candidates should be done, but in the end no such review was carried out. The three candidates remained prize-worthy in the summarizing report by the committee.

In 1965 there were 13 nominations for a prize to Jacob and Monod, but only one of these also included Lwoff. The latter proposal was submitted by Arne Engström, professor of medical physics and central in the 1962 prize for the discovery of the structure of DNA. However, he was not a member of the committee in 1965 and furthermore it was not a matter of a last-minute proposal since it was submitted in mid-December 1964. It might be understandable if the committee at large was impressed by the emphasis of the contributions, some of them relatively recent, by the duo Jacob and Monod. However, Gard was given one more chance to keep the trio intact since he was requested to do one additional review. He noticed to start with that three of the nominations came from Paris, seven from London, four of which, one headed by Peter Medawar, were from the National Institute for Medical Research, and three from Gent. The succinct review over eight pages recapitulated the different individual and joint discoveries and emphasized the central role played by Jacob in linking the different findings together. The summary paragraph read:

> It seems to me not to be justified to emphasize the contributions by either one of the scientists above than that by any other. The contributions made by each one of them have been necessary to achieve the final results and they have complemented each other in an exceptionally fortunate away. I therefore want to propose that the 1965 Nobel Prize in physiology or medicine should be awarded jointly to Jacob, Lwoff and Monod for their discoveries of mechanisms for the transmission of genetic information and for genetic regulation of the synthesis of viruses and enzymes.

This was also the conclusion by the committee although one can see that there must have been extensive discussions as to whether Lwoff should be included or not. Gard needed the full support of Klein to convince his colleagues that the trio should be kept intact. The meeting of the full committee on August 26th did not lead to a final decision and the protocol only noted that this would be given later. It was not until October 14 that the committee finally announced that it had agreed to recommend to the College of Teachers to give a prize jointly to all three candidates.

A Culturally Enlightened Nobel Lecture

In Chapter 9 the Nobel Prize events in 1965 will be described, but it may be appropriate to mention already in the present context the content of Lwoff's Nobel lecture[12]. President Charles de Gaulle had decided that Frenchmen in official functions abroad should use their native language — *la langue de culture*. Since English already at this time had become the language of science the three laureates made a compromise. Lwoff spoke in French, Jacob gave half his lecture in French and the other half in English and Monod, finally, only spoke English. All lectures were published originally in French, but English translations are readily available via the official website of the Nobel Foundation. I had the pleasure to listen to all three lectures, but regrettably my acquaintance with French did not suffice to allow me to enjoy the precision and beauty of the language that Lwoff and to some extent Jacob used. I have had to rely on the English translations.

As a preamble Lwoff cited Plato, who had said that all things arise out of their opposite. He then compared the biological order of the molecular society of an organism to social order. This order was contrasted with revolution and anarchy. Viruses were introduced as "... have not failed to follow the general law. They are strict parasites which, born of disorder, have created a very remarkable new order to ensure their own perpetuation." It took a long time, however, before it was accepted that viruses represented their own genetic elements. For some decades it was believed that they were some kind of undefined simple form of bacteria and there were even some diffuse ideas, advocated by Bordet among others, that heredity was not linked to structure. Burnet in the late 1930s as mentioned above put the lysogeny concept on a firmer foundation and brought evidence of the role of independent virus genetic material.

Lwoff in his lecture then introduced the father figure in the use of phages to examine the function of the genetic material, Delbrück, and referred to Hershey's 1952 experiment identifying the critical role of the virus nucleic acid in replication. The importance of understanding the complete growth cycle of a virus in cells was emphasized, including the earlier mentioned quiet eclipse phase, with complete absence of infectious particles. Because of the existence of the eclipse phase viruses could not be minibacteria, replicating by binary division, which was believed well into the 1920s. Virulent phages were described as completely destroying their host once there was a massive production of new virus particles. But then Lwoff noted that the relationship between the virus and bacterium did not always have this dramatic character. The authorities of the phage genetic work at the time did not believe in lysogeny as mentioned and this challenged Lwoff. He embarked on the studies that led to his own pioneering discovery of temperate phages causing lysogeny. It was Lwoff's experiments with single cells in isolated droplets that proved that lysogenic bacteria could be passed through 19 generations without the release of any infectious virus. His own explanation of his choice of experimental approach was simply "I have already said that I do not have a statistical soul, that my mind tends to the concrete, and that I like to observe because I like to see." The painstaking experiments leading to the discovery of the induction by physical and chemical agents that shared the property of disturbing metabolism of nucleic acids was naturally highlighted.

In 1954 when Lwoff abandoned the studies of temperate phages and lysogeny, the field had already been taken over by Jacob and Wollman, and we will return to their pioneering work in the last chapter. Lwoff finished his lecture by referring to some of the advances that they had made in elucidating mechanisms of gene regulation, the identification of regulatory and operator genes and their products. He then concluded his lecture by reflecting on what his studies of lysogeny in bacteria could mean for the understanding of infections by animal viruses. In particular he discussed the potential application of the new knowledge to understand persistent, latent infections with for example herpesviruses, like herpes simplex virus blisters, and also for different viruses, which at the time had been discovered to play a role in the development of cancers.

The Aftermath

Lwoff was 63 years old when he received his Nobel Prize and had already at this time left experimental science and instead grown into an exceptional statesman of virological science and microbiological sciences at large. Already before he had received his Nobel Prize in 1965 he had received many honors. In addition to several prizes in the field of science he received the Médaille de la Résistance in 1964 and became Commandeur of the Légion d'honneur in 1966. He was a foreign member of both the Royal Society and the National Academy of Sciences in the U.S. and also academies in several other countries. Surprisingly he did not become a member of the Académie des Sciences until 1976, an academy that before that had bestowed on him a number of prizes. Jacob has given an explanation for this fact[13]. The reason was that the Pasteur group was rather critical of what they thought was an antiquated Academy. One of the rules was that a scientist had to personally announce his candidacy and of course Lwoff and his co-laureates could not dream of doing this. Finally both Lwoff and Jacob were elected as members, but surprisingly Monod never became a member of the Academy. Lwoff had many honorary doctorates from different famous universities.

Already in 1962 he had become Chairman of the International Association of Microbiological Sciences, a responsibility he carried until 1970. He was a member of the World Health Organization advisory board 1967–71. In 1968 he left the Pasteur Institute and became Director of the Cancer Institute in Villejuif (CNRS) where he stayed until 1972. During 1977–1989 he was appropriately the President of the Association for Development of the Institut Pasteur, an institution on which he as a scientist and academic leader had really put his hallmark. This may have been more of an honorary position, because after 1972 he lived a mostly private life, focusing with considerable success on his painting and writing. Since his mother was a professional artist he received some early inspiration to paint, but he did not start this himself until the age of 57. He was inspired to do this by Marchelle Wahl, a good friend of his who was both a physician and a painter. After he had produced his first still life in her studio, painting remained an obsession. The colorful painting shown was done by Lwoff in 1975 and is owned by my virology friend Wain-Hobson. Ullmann has told a charming story[14] of when

Painting by Lwoff from 1975. [Courtesy of Simon Wain-Hobson.]

she visited an exhibition of his paintings in a gallery. The return of the sale at the exhibition was to be used for the aid of Soviet dissidents and to support the Pasteur-Weizmann Council. There was a particular gouache that Ullmann liked a lot, but when she asked to buy it Lwoff said that it was probably too expensive for her. She bought it anyhow and with particular satisfaction. It might serve as an expression of gratitude to Lwoff for the already mentioned sizeable sum of money that he provided to Monod to cover the expenses for the arrangements to get Agnes Ullmann and her husband out of Hungary illegally. Lwoff had never acknowledged to her that such a transfer of money had ever taken place.

Lwoff's rich personality led to many involvements outside the field of science. He was a staunch defender of human rights and opponent of capital punishment. Throughout his life he stood up for justice, peace, freedom and the universality of science. A very special issue was raised immediately after he and his two colleagues had been informed about the Nobel Prize they were to receive. The three of them were asked to accept the presidency of the Honorary Committee of the French Movement for Family Planning (MFPF — le Mouvement Français pour le Planning Familial). The contraceptive pill introduced in the U.S. in 1961 was not available for legal use in France. After the First World War in fact all forms of contraceptives were banned in France to boost the growth of the population, a policy obviously supported by the Roman Catholic Church. This regulation, surprisingly, was retained way into the 1960s. This had led to a large number of — naturally, illegal — abortions in the country. The three scientists accepted the invitation to the presidency and in fact it only took until 1967 before the ban on contraceptives was dropped.

Lwoff had a Jewish background from either one or both of his parents. This is generally not mentioned in his biographies, but it becomes highlighted by his involvement as a staunch defender of the state of Israel. For example he acted with vigor when the Arab countries in an unholy alliance with countries of the Soviet empire tried to ooze Israel out of UNESCO. He was very active in assisting the developments at the Weizmann Institute and together with French philosopher and political activist Simone Weil he established the Pasteur-Weizmann Council, which he chaired for a number of years. His last book, published in 1981[15], was a collection of speeches he had given on the death penalty, torture, anti-Semitism, racism, etc. Intellectually he remained a young man throughout his long life. He died aged 92 years in 1994.

Drivers of Evolution

Before ending this chapter it may be worth reflecting on the impact that the discovery of the concept of lysogeny has had on our understanding of factors influencing the relation of genetics and evolution. This is a large topic to be discussed here only in a selected and summarized form. There will be a particular emphasis on the origin and evolution of viruses. Insights into the mechanism of lysogeny taught us that a foreign complex of genes can invade a cell and remain intact for a long time in some kind of association with the host cell genome. This introduced a new dimension in our interpretation of genetics. It had already been realized that mutations could be due to spontaneous or induced changes in the genome of the host but not that foreign genetic elements could also have an important role in the process. Thus a view of genetics as a *vertical* process of information handling was supplemented by the appreciation that there could also be a *horizontal* spread of genetic information, in particular in the world of microorganisms and viruses. In the eukaryotic world the emergence of sexual replication some billion years ago meant a revolution for evolutionary developments. Doubling the sets of genes in cells and dividing special germ cells into two kinds, male and female, allowed dramatic new possibilities for gene representation and the appearance of combinatorial variations. Chromosomes from the two parents containing different homologous sets of genes could be reshuffled and in addition there were many possibilities for a random genetic recombination between homologous parts of chromosomes. The latter is in fact another, but qualitatively very different, example of horizontal gene transfer.

From Traditional Genetics to the Central Dogma

Examination of genetic events started with Gregor Mendel who, studying observable phenomena like wrinkled and smooth peas, defined the simple rules for vertical inheritance. Once the gene concept had been consolidated in the early 1900, experiments using mutagenic substances could be used to show the connection between genetic changes and emergence of new properties. These advances were recognized by the Nobel Prizes in physiology or medicine in 1933 to Thomas H. Morgan for his identification of the role of chromosomes and in 1946 to Hermann J. Muller for his discovery of mutagenesis by X-rays. We will meet Muller again in the next chapter. It remained to identify the chemical nature of genes. Thus the true revolution(s) was set in motion when this was achieved. Identifying the critical role of the two forms of nucleic acids and the development of tools allowing examination of their properties in detail set the stage for remarkable advances continuing into the present time. No longer was it only a question of identifying spontaneous changes but it became possible to bring about mutational changes with a precision in time and space. A new discipline, molecular medicine, was born. It also became possible to identify the source of specific genetic changes. The emergence of the central dogma, mostly thanks to Crick, forever changed the nature of genetics; by *transcription* and *translation*, respectively, the information flow implicitly meant that DNA was copied into RNA and with this RNA as a messenger the amino acid sequence of a protein was specified. Furthermore as will be discussed in the coming two chapters complex systems for the control of gene expression have been unraveled. But how did all this become established by irrevocable steps of evolution?

Cellular life emerged on Earth some 3.8 billion years ago. It was a critical step in the development of a self-replicating system when the molecular events that formed the basis for growth and division of the chemical system became enclosed within a sac formed by a lipid-containing membrane. The shift from using RNA as the information-storage molecule to the much more stable double-stranded DNA was another critical step but it is highly conjectural how it came about and when it happened. The early cells, the prokaryotes, had no nuclei and the processes of formation of messenger-RNA and synthesis of proteins were coupled in them, a situation that remains in these cells into the present-day. The cells lived in an atmosphere without oxygen for the first two billion years and they used alternative energy sources as mentioned in Chapter 5. Gradually the concentration of oxygen in the atmosphere increased and the emerging organisms learnt to exploit this for generation of energy but they also needed to develop means to control its toxic effects. The many times more complex and much larger nucleated cells, the eukaryotes, emerged by a development over a long time. DNA was allocated to the nucleus. Thus in this kind of cells there is a separation of the transcription of the messenger RNA, which occurs in the nucleus, and the translation into proteins which takes place on the ribosomes in the cytoplasm. In addition eukaryotic DNA is associated with the basic histone proteins. This relationship adds an extra, complicated and only partly known, layer of control of gene expression in eukaryotic cells.

Horizontal Gene Transfer and Recombinant DNA Technology

Lwoff pioneered the critical insight that viruses could be virulent and replicate in a way that eventually caused the death of the host cell, but that they occasionally also could be temperate and remain in a physiologically normal, or as we shall see, altered cell. As already mentioned the young Lederberg in parallel managed to demonstrate in 1946 in collaboration with Tatum transfer of genetic properties between strains of *E. coli*. Also bacteria were shown to have sex. However, this statement needs to be qualified. Sex as we know it from eukaryote cells is based on the occurrence of a duplicate set of genetic material.

The cells are *diploid*. The only exception, as mentioned, is the gamete cells, the eggs and the spermatozoa, which have only one set of genetic material. They are *haploid*. Upon fertilization the diploid state is reestablished. By way of contrast bacteria only have a haploid set of genetic material. Upon division the DNA is duplicated and one set ends up in each one of the daughter cells. The "sex" that Lederberg discovered is an example of horizontal transfer of genetic material from one bacterium to the other, leading to the appearance of new properties in the latter.

With time it has become apparent that there is a very complex world of interactions between horizontally introduced genetic material and host cells. Besides viral DNA which is packaged into a protein shell for its transport there are various forms of non-chromosomal DNA in bacteria, referred to as *plasmids*. They replicate independently from the chromosome and can endow bacteria with different properties, like antibiotic resistance. There are means by which they can be transferred from one bacterium to another. Another variant of non-chromosomal DNA structures are *episomes*, which can appear both in a free form and also integrated into the bacterial genome. The term episome was introduced by Jacob and Élie Wollman, and hence we will encounter it again in the last chapter.

With the advancing knowledge it became possible to develop laboratory techniques to modify genetic material and to combine genetic material between different organisms in a way that cannot happen under natural evolutionary conditions. Although the data generated by scientific endeavors are value-free, as emphasized by Lwoff the new approaches raised ethical questions. Recombinant-DNA technology was already mentioned briefly above in Chapter 6 in the discussion of Berg's contributions. Using engineered plasmids it has been possible to introduce genes of eukaryotic origin into bacteria to use them as a factory for production of selected foreign proteins. The development of this technique and its industrial applications have been discussed in the recent biography of Berg[16], in a book describing emergence of biotechnology at Stanford University[17] and also by Stanley N. Cohen one of the pioneers in the field[18]. Briefly the technique has been developed to allow a safe and highly useful production of human proteins of value to medicine, like the human growth hormone, insulin, erythropoietin, etc.

The application of gene technology to plants has been more controversial, with a more conservative, cautious approach within the European Union as compared to many other regions in the world. Genetically modified organisms (GMOs) remain a matter of debate although there is no experimental proof of negative effects on health in connection with their use. In discussions of GMAs it has struck me that there is the lack of a wider perspective of biomedical ethicists on the problem of gene conservation and modification. A very critical dimension in this context is the existence of a *gene pool* — the totality of all genes and their aptness to be operative — and this is determined by *both* the environment and by indigenous modulations of genomes. It is mainly the latter that is in the focus of interest of ethicists. However, when humankind more than 10,000 years ago started to intervene successfully in Nature and select monogenetic seeds of grains or as another example non-toxic nuts this of course had as a major consequence that in a not foreseeable way the total gene pool changed. It can be noticed in passing that two Nobel Peace Prizes have been given for improvements in agricultural procedures by selecting spontaneous variants with a better reproducing capacity; one to John Boyd Orr in 1949 and one to Norman Borlaug in 1970.

A recent call by the Nobel laureate Richard J. Roberts, mentioned also in the following chapter, has raised the serious question of the consequences of the resistance towards the use of the long-available "golden" rice. This form of rice has become endowed with a capacity to produce its own vitamin A, which native rice does not contain. The World Health Organization has calculated that about 250 million people suffer from vitamin A deficiency. This leads to blindness in hundred of thousands of children and also to premature death in many of them. Roberts has raised the obvious question — whether the failure to make available this genetically modified rice, the use of which, as in the case of other GMOs, has not been demonstrated to carry with it any defined risks, is not morally unacceptable. And still this is only a part of a much larger problem, namely how to supply the projected more than 10–12 billion people that will live on Earth at the time of 2050 with food without overexploiting potentially available land.

Stewards of Earth

So what about evolution at the present time? Because of the enormous impact of the rapidly growing human population after the Second World War we have developed to become the dominant species in the world. The present era is therefore appropriately referred to as *anthropocene*. We are now the stewards of Earth and the "footprints" we leave will have an impact for all forthcoming generations to come. But we are not only severely impacting the environment;

we have also developed tools to control genetic events in their smallest detail. The most recent advance is the adaptation of the use of CRISPR — clustered regularly interspaced short palindromic (reads the same in both directions) repeats — Cas adapted for precise genetic modifications in human cells. Often in biology we can learn amazing thing from studying Nature. A system has been discovered in prokaryotes that provides an acquired immunity to foreign genetic elements like the plasmids and phages discussed above. In an ingenious way the CRISPR spacers can recognize specific locations in foreign DNA and cut them into pieces. The breakdown of the identified foreign DNA is achieved by an associated Cas9 enzyme with appropriate nucleic acid cutting activity. There is a partly corresponding mechanism in eukaryotes, the RNAi. The discovery of the latter system in eukaryotes was recognized by a Nobel Prize in physiology or medicine in 2006 to Andrew Z. Fire and Craig C. Mello "for their discovery of RNA interference — gene silencing by double-stranded RNA."

The bacterial CRISPR system is unique in its adaptability to use for gene editing — adding, disrupting or changing sequence of selectively targeted genes. In 2012 it was shown that the technique could be used also in animal and human cells. Until then it had been successfully tried for genetic engineering in industrially important bacteria. By use of the further developed technique it was experimentally demonstrated that it was also possible to prevent hereditary genetic disease due to a single gene defect in experimental animals by introduction of targeted changes of DNA in germ cells. Potentially therefore the technique could also be used in humans for targeted repair of dysfunctional DNA. Initial targets could be hereditary diseases caused by single base-pair mutations, as in cystic fibrosis and sickle-cell anemia. There has been a sound reluctance among some scientists to start applying the new technique for germ line modifications in humans. At present discussions about applications of the technique continue although there is no formal moratorium on its use. Further deliberations will decide when and for what purpose it will be applied. However, we may hesitate for a while to consider the consequences of the application of the new technique in humans in a wider cultural perspective.

We need, as discussed above in the case of gene modified plants, to consider the concept of the total gene pool. A major change in this pool has developed as a consequence of the introduction of modern medicine. The new forms of treatment that have been made available has allowed the survival of patients with life-threatening infections, endocrinology diseases, cancers, etc. In many cases individuals surviving such challenges later could reproduce causing a major change in the human gene pool. These matters have not been the subject of discussion by ethicists, but when instead there is a discussion of making a specific change in the genetic material of stem cells or even germ cells this immediately elicits an intense debate concerning ethical principles. This of course in itself is of great value and a reflection of the responsibility that the introduction of new tools inherently means. However, there is a possibility that potential dangers are overrated in the light of what is referred to as the precautionary principle. In addition there is a risk that, unintentionally and unconsciously, cultural remains of respect for ingrained belief systems influence the logic of thinking. This is not a denial of the fact that various such belief systems may assist humans in the course of their life, providing moral guidance and emotionally resolving certain existential dilemmas. However, there is no designer of DNA. As already mentioned earlier in this chapter there is an absolute need for an impenetrable divide between the world of reproducible and controllable facts generated by science and the belief systems of different kinds generated by human culture to provide advice about how to manage life well — to nurse dignity. As we shall see Monod emphasized that it is all a matter of chance and necessity.

However, it is likely that as in earlier cases even when a temporary moratorium might have been introduced, a way forward will be found and then it will be possible to lift any restrictive regulations. Once a technique is available it is likely to be used sometime in the future. A critical step will then have been taken by the human species to selectively manage the evolution of our own genome. At that time we will have become conductors of *both* the world of human genes and of the environment on Earth. This is a major responsibility. Let us hope that it is not true, as Isaac Asimov said "The saddest aspect of this (science) is that it gathers knowledge faster than wisdom"?

Classification of Viruses and the Species Concept

The basis for discussion of evolutionary development is the identification of a species, as mentioned in Chapter 5. In general terms a species in the eukaryotic world is demarcated by the capacity of its members to produce a vital offspring upon sexual interaction. However, this kind of concept cannot be applied to microorganisms, and, in particular, not to viruses. Bacteria for example are grouped into categories referred to as species by comparing similarities of homologous properties — *polythetic* classification — but this is different in the case of viruses. Lwoff and his colleagues pioneered the introduction of the use of properties of the virus particle, the virion, as a primary divider in practical taxonomic considerations, as discussed above and earlier[2]. Different virus families do not show genetic relationships and hence their genomes cannot be used to draw a tree illustrating genetic interrelationships. In the case of Bacteria and Archaea classification is also based on like properties, the polythetic criteria, but in this case partial trees of life can be constructed. However, the picture is blurred in part because of the frequent occurrence of horizontal transfer of smaller or larger pieces of genetic material. By way of contrast, comparison of eukaryote genetic characteristics allows the construction of a tree with many bifurcations reflecting the continuous evolution over long periods of new species and higher orders of classification. At its root is found the most primitive single cell eukaryotes and further back relations can be traced to selected properties of genes found in Bacteria and, as emphasized in Chapter 5, in particular Archaea.

The control of the nucleic acid replication in viruses has an almost 1,000 times lower fidelity than the corresponding doubling of DNA in eukaryotic cells. Hence, in an evolutionary perspective a virus is much more adaptable than various forms of cellular life. A practical consequence is that the virus population released from an infected cell is relatively heterogeneous. It is therefore referred to as a *quasispecies*. This has practical consequences for example when the virus is exposed to a site-specific (monoclonal) antibody or to a specific antiviral as a means of drug treatment of an infection. The virus readily spurns antibody-resistant or drug-resistant variants during its replication. Because of the latter, an efficient treatment of a virus infection optimally requires the use of more than one kind of drug, each one displaying capacity for a distinct mechanistic interference. It has only been possible to develop a successful treatment of HIV infections, since drugs have been discovered which attack a number of different critical steps in the virus infection.

The fact that propagation of a virus leads to the appearance of a genetically heterogeneous population of virus particles raises the question of the use of DNA or RNA nucleotide sequence determination for the identification and characterization of a newly encountered virus. In 2003 there was an outbreak of severe acute respiratory syndrome (SARS) in Asia with a limited spread also to other parts of the world. Characterization of the nucleic acid of this newly encountered virus demonstrated that it contained RNA and that the architecture of the genome showed it to be a new form of Coronavirus, a kind of virus previously known to give upper respiratory infections in man and many diverse animal diseases. SARS represented a relatively severe form of disease but fortunately its continued spread locally and internationally was possible to prevent by introduction of appropriate epidemiological interventions. More recently there has been an extensive spread of Ebola virus during 2014 and 2015 in West Africa involving the three countries of Guinea, Liberia and Sierra Leone. The origin(s) of this exceptionally large epidemic of the disease is not known and it took time to bring the epidemic under control. Major international efforts coordinated by the World Health Organization were required. The virus responsible for this disease was originally identified in 1976 in Zaire. At that time it was not possible to use nucleotide sequencing of the virus genome for characterization and identification of the agent. Electron microscopic examination by European scientists revealed that the virus had an unusual elongated form. A virus with a similar morphology had previously been encountered, the Marburg virus, also capable of causing very severe and potentially lethal disease. It took immunological studies by scientists at the Communicable Disease Center, Atlanta, GA to reveal that Ebola and Marburg viruses were immunologically distinct. A new virus of potentially major importance for humans had been discovered. Because of the particular form of the virus particles the family, including the two kinds of viruses, was named *Filoviridae*, from Latin *filus*, long thread.

The four criteria for classification of viruses originally introduced by Lwoff and his colleagues as mentioned above of course were markedly expanded when it became possible not only to determine the type of nucleic acid in the virus but also to characterize its nucleotide sequence. The representation of the different genes revealed the family to which a newly encountered virus belongs, provided of course that a similar virus had been seen before. Because of the unique resolution of the nucleic acid sequencing technique it is of great value for tracing the epidemiological spread of strains of agents like, for example, measles virus or HIV. In spite of the detailed knowledge about the virus that molecular genetics techniques allow the identification of, immunological techniques remain important to apply. From a human or animal perspective it is important to determine if the virus encountered belongs to the same (sero)type as any previously characterized virus. This has important practical consequences. The virus present in, for example, a live measles vaccine has markedly different genetic characteristics from wild measles virus strains but it gives an immunity that protects against any kind of wild strains of the virus. It would seem logical therefore to equate serotype with the species of a virus. However, the more detailed information provided by genome sequencing has raised questions as to whether additional criteria should be used. The high resolution of the genome nucleotide sequencing technique thus has led to intense discussions between "lumpers" and "splitters" in taxonomic discussions. It would take too long to further develop this discussion in the present context. Suffice to notice again the practical importance of characterizing two related viruses by determining whether they give cross-protective immunity or not. That is if they represent the same type of virus or are distinct.

The problem of lumping or splitting information has a general importance. For example the way our brain functions, discussed at length in Chapter 3, is the question of lumping or splitting information we collect by our multitudes of perceptions. This applies both to physical objects and to abstract concepts. Our brain excels in helping us decide, in a split second, if the impression is congruous with an object we have learnt to know earlier — is it a rose or some other flower — or is the concept proposed harmonious in most aspects with something we have learnt earlier or are there some certifiable nuances that allow it to be characterized separately. All these are matters of seminal importance for survival which could be discussed at length, but it is now time to return to the more direct genetic aspects of survival and evolution.

Interaction between Biological Entities

Darwinian evolution is frequently referred to as a brutal process with survival only of the fittest. Tennyson referred to it as "red in tooth and claw." However, it has been learnt that collaboration is an equally — or often even more — important aspect of evolution — it is "green in mergers and acquisitions." A very good example of this is our own body as already cited in my previous books[1, 2]. The human body contains 10-100 trillion cells but it also includes ten times more microorganisms, our *microbiome*. This includes many kinds of organism *and* their viruses and their role in food processing, protection against different pathogenic agents on mucosal surfaces and skin, etc., is presently the target of intense studies in laboratories around the world.

An interesting historic example of close collaboration between biological entities is, as already mentioned, the eukaryotic organelle, the energy-giving mitochondrion. This organelle was introduced in Chapter 4 and further decribed in Chapter 5. It must have happened on many occasions during the immense eons of time that prokaryotes have become temporarily associated with emerging eukaryotes. Many consequences of this can be seen at the present time. There are a number of bacteria that can only replicate as parasites in eukaryotic cells and there are special forms of such intracellular parasites, like Rickettsia (mitochondria-related), the source of for example epidemic typhus and also other diseases. It was speculated for a long time that the more complex DNA viruses could have emerged by a backwards — retrograde — evolution from such intracellular prokaryotes. However, their mode of replication argues against this, since there are no traces of a binary division mechanism. Instead the viruses, by remarkable skills and effectiveness, reprogramme the functions of cells to produce separately the virus genetic material and the proteins controlled by the virus genome. Additional events exploiting cellular mechanisms lead to an assembly of virus particles to be released for transmission of the infection to other cells. The large animal DNA viruses have genes that are absolutely essential to allow a production of new virus particles. These are referred to as *essential* genes. But there are also many other genes, referred to as *non-essential*, which are not required for the replication of the virus in the laboratory. Instead they direct the synthesis of proteins which influence the conditions of virus-host organism interactions. Some of the latter genes have been found to have their origin in the host cell genome. This highlights that there are undefined mechanisms, facilitating rare events that allow a shuttle of cellular genes into viral genomes and often, with certain kinds of viruses, also in the other direction.

The commonly encountered forms of nucleic acids in nature are the double-stranded DNA and variants of single-stranded RNA. However, in various kinds of viruses all different potential forms of nucleic acids can be encountered, but DNA and RNA do *not* occur simultaneously, as repeatedly emphasized. Thus there are viruses containing single or double stranded versions of either kind of nucleic acids. The single-stranded versions can be of two kinds, a plus strand form which in the case of RNA can serve directly as messenger or a minus-strand form that needs to be copied to allow control of protein synthesis. Further the nucleic acid molecules may be linear or circular. All in all it seems that virus infectious agents in their evolution have used the complete nucleic acid tool box of molecular variants, arguing for their evolution also as an integrated part also of the early phase of emergence of life, *the RNA world*. The genomes of some of the smallest viruses have such a limited size that they can be readily synthesized in the laboratory. Thus scientists, as already mentioned, have successfully synthesized the poliovirus genome and the genome of the influenza virus that caused the worst virus pandemic ever, the Spanish flu of 1918. It could be documented that the artificially made genomes were fully functional. When these data were about to be published the question was raised if it would be dangerous to present this information, since it theoretically might be misused for example by bioterrorists. Fortunately it was decided that it was more important to make the new information available to fellow scientific colleagues than to restrict its access.

The 1975 Nobel laureate Baltimore devised a scheme illustrating that viruses have exploited efficiently all different potential combinations of copying of RNA and DNA of all possible forms[19]. Thus not only the kind of nucleic acid but also its anatomy is critical in virus classification. Baltimore was recognized as one of the co-discoverers of a special enzyme *reverse transcriptase*, which can copy RNA into DNA. This is a critical enzyme in a large group of viruses, the *Retroviruses*. Thus although retroviruses have an RNA genome they operate like DNA viruses. When their genome has been transcribed into DNA they can associate with the cellular genome. This can have many consequences. Members of this family have been demonstrated to be capable of causing cancer and also to cause persistent infection eroding the immune system, exemplified by HIV, the virus causing AIDS. Already in the second decade of the twentieth century Rous, mentioned in Chapter 3 as the father-in-law of Hodgkin, found that the virus given his name could cause leukemia in chickens. This discovery was recognized by a Nobel Prize in physiology or medicine almost 50 years later, in 1966, a record of sorts. There will be reasons to return to this prize once the archives are released for analysis in the near future, in 1967. This leads us back to questions about how temperate viruses, to use the term applied to phages, can influence the operation of the genetic material of their host cells.

Cancer is a disease based on a derangement of the normal regulation of cell proliferation due to an accumulated number of genetic changes. Infections with certain viruses represent one source of such genetic changes. In a forthcoming review of Rous's discovery there will be reasons also to examine the 1989 prize in physiology or medicine to Michael J. Bishop and Harold E. Varmus "for their discovery of the cellular origin of retroviral oncogenes." They made the critical observation that retroviruses as they became associated with the host cell genome during the infection could hijack cellular genes. In their case it was Rous's virus that had picked up a cellular gene that could cause sarcoma, hence named *src*. When this gene, belonging to a group later referred to as *oncogenes*, was transferred to normal cells, leukemia could ensue. In later studies the oncogenic effects of retrovirus infections has been demonstrated to not necessarily imply the hijacking of a gene. Depending on where the virus is inserted in the host genome many different changes of growth behavior can be observed. The harmful effect can be alternative on genes that promote or suppress the development of cancer cells. It obviously will also be imperative to review in this context the half prize given in 2008 to Harald zur Hausen "for his discovery of human papilloma viruses causing cervical cancer." His contribution was discussed as an example of persistent virus infections in my preceding book[2].

In addition to retroviruses, members of the herpes family, which contain DNA, can remain dormant in cells in a latent state as also mentioned in passing and described previously[2]. Upon activation simplex virus can cause blisters for example around the mouth and varicella (chicken pox) virus can spread by the nerves over a segment of the body, a dermatome, leading to zoster. Because of their capacity to persist in a harmless form, herpesviruses, like retroviruses, have been companions to vertebrates throughout long periods of evolution. It can be deduced that their DNA genome contains genes that are essential for their effective reproduction in a cell or for establishing a state of latency, but also other genes, many of which have been picked up from the host during a lengthy evolutionary chess game. These "non-essential" genes referred to above are used by the virus to achieve a balanced infection in the host, the proper pace of the acute infection facilitating the spread of virus to other susceptible hosts.

In the perspective of all these various consequences of the effects of virus infections let us once again reflect on the nature of viruses. The central question that needs to be answered concern how viruses shall be classified biologically.

A Virocentric View of Biology

The question of the relationship of viruses to different cellular forms of life was discussed at length in my first book on Nobel Prizes[1] and the conclusion was that the inquiry into whether viruses were live or dead was inappropriately framed. The problem in part was the difficulty of defining what we mean by life. Scientists interested in searching for extraterrestrial life of course need to have access to an agreed definition. The one generally used is simple. It states: "Life is a self-replicating chemical system, with a capacity to make mistakes." Personally I like the "mistake" part, and of course it is a truism that without spontaneous or induced mutagenic changes there will be no evolution. Viruses, which are parasitic sets of genes exploiting the machinery of living cells clearly do not have the independence needed to call them live. But let us instead ask the appropriate question, namely if viruses play a role in evolution? The answer then is a resounding yes. The emergence of life remains very speculative but many scientists support the notion that it started with an RNA world, as mentioned. The reason is that it is only this "clever" molecule that can serve both as an information carrier and also function as an enzyme. The latter fact was highlighted by one of the conceptually most important Nobel Prizes ever, the one in chemistry in 1989. It recognized Sidney Altman and Thomas R. Cech, "for their discovery of the catalytic properties of RNA."

Viruses contain either DNA or RNA as their genetic material, never both on the same time as repeatedly mentioned. Since they are cellular parasites it is tempting to conclude that they could not have evolved until the first cells had appeared on Earth some 3.8 billion years ago. This turns out to be a gross misinterpretation of facts. One critical such fact, namely that many of them have an RNA genome strongly suggests that they have their early ancestors in the pre-cellular RNA world. There is in addition one particularly puzzling aspect of the genomes of viruses. When genes for various proteins not previously encountered are compared in different forms of cellular life it is generally possible to identify the existence of evolutionary relationships with some previously characterized proteins. Ancestors of newly encountered gene products are most often found. This is not the situation in the case of virus proteins. About 80% of the coding sequences for these products have *not* been encountered in previous studies to deduce the amino acid sequence of proteins of other viruses or cells by nucleic acid characterization. Inspired by terms used by astrophysicists to manage inexplicable findings the term "virus dark matter" has been applied to this surprising and challenging finding. The more one will examine in the future the genomes of hitherto not encountered viruses the larger the number of genes representing this dark matter will become. It is anyone's guess how large the population of newly discovered genes may eventually be found to be.

Viruses, like any single or multiple cell organism, being actors in the evolutionary drama, want to secure their survival. Hence a virus that is too effective in killing its host will cut off the branch it is sitting on. Thus they *coevolve* with their hosts and in these interactions there is an emphasis on balance in the evolutionary process. Effective competition and constructive synergism are equally important phenomena in evolution, as mentioned. Viruses are ubiquitous and wherever there are cells there are viruses, as repeatedly emphasized. This applies even to cells adapted to survive under the most extreme conditions. There are reasons to believe that this has been the case ever since the first cellular organisms appeared on Earth. Examination of genomes of different representations in for example water on Earth reveals that each milliliter contains about one million microorganisms and at least ten million virus particles. This means that all water on Earth contains a minimum of 10^{24} virus particles, ten times more than all the stars in the universe! What do these viruses do and how do they survive?

Parasitism does not end with viruses potentially causing cell destruction on their own. There are in fact viruses that are parasitical on viruses, like for example adeno-associated viruses, mentioned in Chapter 3, and the delta agent in individuals chronically infected by hepatitis B virus. The delta agent is a small defective RNA virus which has to borrow the hepatitis B virus envelope to transmit between cells. Because of its appearance in this context it was named hepatitis D virus. The double infection with the two viruses B and D has been found to lead to markedly more severe forms of hepatitis. And there are examples of even more defective agents, like the *viroids* in plants. These are the smallest infectious agents known. They consist solely of circular, single-stranded RNA containing a few hundred nucleotides. They cannot direct the synthesis of any new protein. Still they can facilitate the production of copies of themselves. The RNA of some viroids has catalytic effects, which allows self-cleavage and ligation to a circular form. It serves to remind again that in evolution the critical factor is to find means of survival. Parasitic genes may do this without influencing the physiology of their host and hence be considered harmless, but they may also cause disease.

These advancing insight into the genetic particulars of viruses has led to the formulation of a *virocentric* concept of the evolution of life[20, 21]. This concept which becomes progressively strengthened is based on a number of crucial observations. To repeat, viruses parasite on all cellular forms of life; viruses represent the physically most abundant and genetically most diverse entities of biology; viruses use all conceivable strategies in the replication and expression of their genomes. In summary viruses represent a unique and all prevalent category of biological entities. In the coming decades we will witness a major revolution in our view on the role of viruses in the emergence of life and the evolution of progressively more complex forms of life. There will be reasons to come back to these fascinating developments in the future.

In summary, it seems that in the world of extra-chromosomal genetic elements with direct or indirect effects on the host cell genome, a phenomenon originally observed by Lwoff, anything can happen. The result may sometimes be a development of a disease by destruction or change of behavior of cells, but there may also be many events that potentially promote evolution by widening the number of variants from which selection can be made. There are reminiscences of shuttling of genomes in and out of the eukaryote genome in particular in the case of retroviruses. Our own genomes contain a large number of defective retroviruses that have remained, as it seems, without playing any demonstrable role, positive or negative, at the present time. Viruses have been described as "packages of bad news" and they are generally looked upon as critical pathogens to humans, animals and plants. What we may have over-looked in such an interpretation of their central role in Nature is that these truly ubiquitous agents may also play many different very positive roles and further physiological processes. This has been discussed in a report entitled *Viruses Throughout Life & Time. Friends, Foes, Change Agents*[22]. Possibly they still may remain important in the present phase of evolving life, but in particular they certainly have had a critical role in accelerated evolution way back in time. It would seem that the progress towards complex life has been to a large extent influenced by horizontal gene transfer. Thus it is not an unrealistic proposition to say that without viruses and other forms of horizontally transmitted genes, evolution *would not* have taken its course towards the complexity of vertebrates with an advanced form of consciousness that presently exists in us humans. So, maybe, in spite of the negative impact of certain disease-causing viruses, it is in fact *thanks to* other ancestral viruses inferred to be important drives of evolution, that we, the human species, have been granted the privilege to make our, in geological terms, brief visit to this planet.

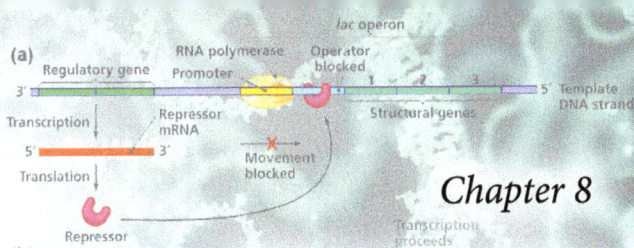

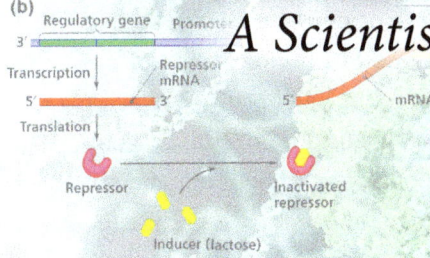

(a)
Regulatory gene   RNA polymerase   Operator
Promoter   blocked

3′

Transcription

Repressor
mRNA

5′   3′

Translation

Movement
blocked

Structural genes

5′ Template
DNA strand

Transcription
proceeds

Repressor

(b)
Regulatory gene   Promoter

3′

Transcription

Repressor
mRNA

5′   3′

Translation

5′   mRNA

Repressor

Inactivated
repressor

Inducer (lactose)

Chapter 8

A Scientist of Many Talents

A BRILLIANT SCIENTIST
MIND OF RICH CREATIVITY
BUT WITH A JANUS FACE

Nature is not democratic. Sometimes it bestows a plethora of talents on a single individual. This is well exemplified in the case of Jacques Monod. In addition to a brilliant analytical intellect he had a broad cultural background and became deeply involved in music, in particular directing choirs and playing his cello. His timeless book *Chance and Necessity — Le Hasard et la Nécessité*[1] originally published in French in 1970 discussed the ethical and philosophical implications of the molecular biology that developed during the 1960s. In this development Monod and Jacob were central figures. But as we shall learn Monod also had deep involvements in human rights issues and was highly committed to the Resistance movement in France at the time of the German occupation during the Second World War. He also had strong left-wing political commitments which, however, were modified by the passing events.

Monod did not write his autobiography but there is a copious material about him and his life in Horace Judson's book *The Eighth Day of Creation*[2] and in the recent book *Brave Genius: A Scientist, a Philosopher and Their Daring Adventures from the French Resistance to the Nobel Prize*[3]. Judson was careful to formulate his impressions of Monod by repeated personal contacts. However, one may ask if it is possible to give a general and timeless presentation of an individual based on a few encounters. One is always dependent on the nature of the beholder, the person interacting with the individual in focus and the circumstances of this interaction. It seems that

Monod could be described in many ways. Agnes Ullmann who was his friend and collaborator has written excellent biographical notes on him[4]. I cannot add much personally to the interpretation of Monod's personality since I only met him once. At the time of the Nobel festivities in 1965 I was invited to a private dinner in Gard's home where the guest of honor was Lwoff, as mentioned. When it was time for coffee we went to the neighboring red brick chain house at Solna Kyrkväg where Malmgren lived to join Monod, Jacob and the other guests. My impressions remain of a very vivacious person, but nothing more. Someone blessed with Monod's many talents would happily succumb to a panegyric presentation, but others whom he might have hurt deeply and possibly even for life by an acerbic remark would of course give a very different impression. What interests us in this book is why he decided to become a scientist and how his different unique qualities allowed him to pioneer the early conceptual advances in the, at the time very young, field of molecular biology. As always in the case of career developments there were special circumstances that favored his maturation as a scientist; in fact, in his case, two particular encounters. As we shall see he was slow to enter the field of research. This did not happen until the mid-1940s when the ten-year-older Lwoff put him on the right track by mentioning to him the concept of enzyme induction of which he had no prior knowledge. The second important encounter occurred ten years later, when the ten years younger Jacob received the inspiring insight that there could be parallels between the genetic control in the lysogenic phage system he was studying and enzyme control system in microbes which was the target of Monod's studies. This led to a remarkably fruitful collaboration in the coming decade. Their joint contributions to the emerging field of molecular biology were critical.

In our studies of Nobel Prize recipients we have repeatedly encountered the importance the ethnic-cultural background of an individual may have on his success as a scientist. The most striking examples are the scientists with a Jewish background who, remarkably, are overrepresented by a factor of about hundred among the prize recipients in the fields of natural sciences. Both in Chapters 3 and 5 we met prize recipients who had a Quaker upbringing, Hodgkin and Sanger. Could it possibly be that groups of religious dissenters potentially forced into a diaspora could develop unique cultural patterns to contribute to social cohesion facilitating survival during times of oppression? And may such patterns foster the development of successful scientists?

Monod's father had a Huguenot background. It was during the 16th century that Protestant movements confronted the dominant Catholic Church associated with the worldly power in a theocracy. This happened in Germany during the 17th century leading to intense fighting over thirty years which was not resolved until after the Peace of Westphalia, in which for the first time the national authority emerged as a consolidated concept. One of the national actors heavily involved in this war was the small northern country of Sweden which temporarily developed into a major power around the Baltic Sea. A very stable and influential Lutheran Church was established in the country and it was not until the year 2000 that full religious freedom was established and a newborn child did not automatically become a member of the Swedish State church. With some degree of generalization it can be said that the value system of the Swedish Lutheran church has had a deep societal impact, wnich has remained into the present time. Corruption is essentially non-existent, work ethics are high and honesty, justice and objectivity remain important codes of conduct. This has also benefited the Nobel Prize work.

The Huguenots were Protestants in France inspired by the Swiss reformer John Calvin. The rapid growth of people adopting this confession created major tensions in society and elicited violence. In order to protect the Huguenots, representing more than one-tenth of the French population at the time, and to ensure that they had equal rights like other citizens, the Edict of Nantes was established. This granted the Huguenots extensive religious and political independence. However, during the coming century theocracy was markedly strengthened under Louis XIV. The Edict of Fontainebleau was introduced revoking the rights granted to the Huguenots. This led to the eruption of major violence and resulted in a massive exodus of Huguenots from France. Many of them were professional people, craftsman and artisans and hence they were often welcomed in their new home country. They went to neighboring European countries, including England, where they were called Strangers but also further to the English colonies of North America and to the Dutch Cape Colony in South Africa. The passengers of the famous emigrant ship *Mayflower* were mainly Strangers and Quakers. The imprints of the Huguenots in South Africa are readily seen even today, whereas in the United States, with a democratic constitution emphasizing religious freedom of individuals as well as their integrity and integration they form only one of many religious groups with an independent profile. It should be mentioned that the motive for acquiring new territories for settlements may heavily influence the regional consequences of the emigration.

The Diasporas of religiously persecuted groups involved the movement of whole families and at least during an early phase led to inbreeding within the group. By way of contrast, when the establishment of new trading patterns and a greedy search for material wealth was the basis for annexing foreign land by force, the demographic consequences were very different. Migration to the North America involved about equal numbers of men and women, whereas the proportion of emigrants to South America was about twenty men to one woman. The long-term historical consequences are obvious. And in the case of religious wars, what have we learnt from history? Recent events sadly illustrate that history can repeat itself also in the modern era of rapidly developing globalization. One wonders when we will see a renaissance of internationally influential science in the Arab countries, such as the one that existed in the eleventh and twelfth centuries when the Greek inheritance was cultivated and further developed for a later return to Europe. In the case of the Huguenots they were not allowed to move back to France until two years before the French Revolution, when in 1787 the Edict of Versailles, also called the Edict of Tolerance was signed.

The Winding Road Towards Science

The young Monod. [Portrait by his father. From Ref. 4.]

Jacques Lucien Monod was born on February 9, 1910 in Paris. He was the fourth child of the family and had two brothers. There are many Monods in France and for generations they have had a great influence in society as clergymen, doctors, teachers, civil servants, etc. Several generations back they had been driven out of France when the Edict of Nantes was revoked. Many of them moved to Switzerland and Jacques's great-great grandfather had been a Calvinist clergyman, born in Geneva but later active in Copenhagen. After the French Revolution he was called back to Paris, where he automatically became a French citizen, but at the same time he was able to retain his Swiss citizenship. Monod's father, whose name was Lucien, did not fit into the tradition of professionals in the family. He was an artist, making his living as a portraitist. Early in life he had become severely crippled by poliomyelitis, and this disease also afflicted his son Jacques when he was only two years old. As in the case of Theorell (Chapter 4) he came out of this threatening situation with a mild, in fact a relatively much milder lameness in his lower left leg. This disability had consequences later in life when he was exempted from military service, but it certainly did not restrict his intense involvement in the Second World War during the German occupation, as we shall see. Since he developed to become an active sailor and also a skilled mountain-climber it seems that the physical limitations were not too severe.

Unusually for the Monod family the father Lucien had married an American from Milwaukee, Charlotte (Sharlie) Todd, the daughter of a Scottish clergyman, who had emigrated to the U.S. The family left for Switzerland during the First World War to live with relatives. When Jacques was eight years old and the war was over the family moved to Cannes where he spent the next ten years becoming involved in many activities. He attended the *lycée* until the age of 18, when he moved to Paris for university studies in natural sciences. There had never been a scientist in the Monod family before. According to Judson[2] there were two individuals who had a special influence on Monod's choice of a future career. One was his Greek teacher Dor de la Souchère and in particular his father who had become fascinated by Darwin and his theory of evolution. Regarding his father's enthusiasm for science Monod was quoted as saying, "He was a nineteenth-century positivist, and there's no doubt that this had great influence in attracting my attention to science. Which is amusing, because he was an artist and in fact didn't understand science at all." During his early teenage years Monod learnt to play the cello and throughout his life this remained important as a means to make a constructive interruption in his rapid flow of thoughts. He was also very involved in choral singing and honed his talents as a conductor as we shall see.

During his studies at the Sorbonne in Paris he lived with his much older brother, a lawyer. He was generally dissatisfied with the conditions of his studies, which he thought antiquated. On the other hand he himself believed with youthful enthusiasm and ignorance that problems in biology should be solved by the examination of behavior studies. However, a year into his studies he paid his first visit to the marine biology station at Roscoff, in Brittany, an institution that also played an important role in Lwoff's development, as we have seen. They met briefly but Lwoff's influence on Monod's development as a scientist came later. Other people who had a major influence on him at the time were Georges Teissier, the head of the station and a specialist on insects and marine zoology, Louis Rapkine, a biochemist, and in particular the geneticist Boris Ephrussi. Rapkine convinced Monod that in order to understand biology one must examine and describe the molecular details of processes, the living chemicals. It is worth remembering that this was before the word "biochemistry" was coined and way ahead of the introduction of the term "molecular biology." Ephrussi taught Monod the fundamentals of genetics, in particular in the context of developmental biology. Some words of explanation about this highly profiled scientist may be in order.

Unfair Omission of a Prize Candidate?

Ephrussi had a Jewish background, born in Russia, but developing his career as a professor of genetics at the University of Paris. His family had been forced to leave Russia after the Bolshevik Revolution. He first tried out the Tolstoyan way of life in Bessarabia but soon gave up agriculture and religion and moved on to France to take up science. He examined the way in which embryological processes were initiated and controlled, in part using tissue culture techniques. In 1934 he received a scholarship from the Rockefeller Foundation that allowed him to study for a year at Caltech in Pasadena in Morgan's laboratory, the scientist who the previous year had received the Nobel Prize in physiology or medicine. In this laboratory he got to know Beadle, whom we have also already met, and later in 1935 the two of them worked together in Paris. In this joint work they used eye transplants in the Drosophila insect system and received the first evidence that a particular eye color was decided by a single gene. This early observation laid the foundation for the work with the fungus Neurospora that later earned Beadle his Nobel Prize.

The 1958 prize initially made Ephrussi deeply distressed and the outspoken French press said "if the Nobel Prize is to reward finished work, the award was fair. If, however, it is to recognize those pioneers who have opened a new path towards the understanding of life, the name of Ephrussi should have been associated with that of Beadle." According to a biography of Beadle[5] in a communication between him and Ephrussi, in 1972, the former late in their life expressed the following: "I've many times thought how unjustly is credit assigned in science — and I suppose elsewhere too. You had more than your share of ideas. You and I had the basic idea of one gene — one reaction. Neurospora only confirmed it — actually only confirmed Garrod who had it all straight 40 years earlier. Ed (Tatum) and Josh (Lederberg) were incredibly lucky in being at the right place and the right time to get more than a fair share of the credit. I've many times tried to tell the story as it was but one can go only so far."

This is now history, but it can be mentioned that Ephrussi was never nominated for a Nobel Prize and hence never evaluated by the committee. He remained one of the very influential scientists in his field and his most important contributions concerned cytoplasmic inheritance in yeast. After the Second World War he became the first professor of genetics in France. He was a great lecturer and synthesizer and had the talents of a Slavic actor. According to Jacob[6] Ephrussi was a very authoritarian group leader and generally a very complex person. Jacob wrote "In a few minutes he could shift from rage to mildness, from a reasoning of wisdom to pure nonsense, from exaltation to melancholy. There was something of Ivan Karamazov in him."

Music or Science?

In the mid 1930s the Rockefeller Foundation had offered to support Ephrussi in developing genetics in France. In order to do this he asked to be allowed some research time in Pasadena and he also inquired if he could bring along a student whom he could train for future developments of the discipline in France. The Foundation accepted these proposals and Ephrussi decided to bring along Monod. This later turned out at the time not to be such a good choice. There was nothing wrong with the handsome, multi-talented young man except that he was not focused on developing his science. Originally Monod had different plans for the summer of 1936. He loved sailing and already in 1934 he had joined a research sailing ship *Pourquoi-Pas?* (Why not?) to do some Arctic work in Greenland under the leadership of Jean Baptiste Charcot. He had already signed

Boris Ephrussi (1901–1979) and Monod in California in 1936. [Photo from Barbara McClintock.]

on for another voyage with this ship in 1936, but decided to cancel this and join Ephrussi in California. Fate apparently had other plans for him. He was not to be on the ship when it became wrecked by a hurricane off Iceland, leaving only one survivor.

It could be added that normally Rockefeller scholarships were only for scientists with a Ph.D., but Monod had not yet obtained this degree. The most important lesson Monod learnt during his stay abroad was the difference in the style of science in the U.S. and in France. In contrast to the Teutonic pattern of French Science the discussions in America were much more open, with equal opportunities for senior as well as younger scientists to express their views. Dialectical critical discussions were encouraged. Monod also met the leaders in the field of genetics, but still his own science did not advance. He was simply too occupied with other matters, private love affairs and not least involvements in music. He had success in conducting orchestras and was even offered a job in Pasadena with a salary of five hundred dollars per month. This he told the deeply frustrated Ephrussi when they had returned to Paris. So what could he say? He commented that the choice of Monod's future career of course was his own, but if he wanted to choose music the recommendation was that he should not get involved as an amateur in Pasadena, but rather educate himself at the Paris Conservatory. The Monod family themselves joined in to manage the crisis and one of his brothers asked in exasperation "Is he going to become a Pasteur or a Beethoven?" Apparently there was no hesitation in appreciating his inborn capabilities.

Music remained an essential part of his life activities. His Bach choir *La Cantate* produced acclaimed performances. One of the singers in the group was Odette Bruhl, a sister of Teissier's wife. She was an orientalist at the Musée Guimet in the neighborhood of the Trocadero Gardens and had experience of archeological fieldwork as well as being a specialist in Tibetan painting. The two of them shared many interests in the arts and music and also in outdoor activities. They decided to get married and a year later, in early August 1939, their twin sons were born. The Second World War was on the verge of breaking out and this naturally upset the life of the family. Monod who originally had been exempted from military service because of his limp, volunteered for officer training. Odette and the twins stayed with her mother in Dinard on the Brittany coast. Monod's laboratory work was temporarily put on hold three years into his Ph.D. work. He was close to thirty but had still not gotten a grip on his science. Then came the German invasion and Monod was demobilized. In the autumn of 1940 he went back to work in his laboratory.

The Two-Phase Growth Curve

The original idea was that Monod should develop his science under the guidance of Ephrussi, but this did not work out after their friction-rich joint time at Caltech. Instead Monod settled in Teissier's laboratory at the Sorbonne. He was essentially all alone in his work. For his thesis he chose to study the effect of different chemicals on the growth rate of bacteria. Like Lwoff he had previously studied ciliates, but found them too complex to handle. At the time it was not apparent where the studies of the growth rates should lead, but progressively it was understood that controlled exposure of bacteria to different chemicals was a way of studying their genetics. The chemicals he used were mainly sugars and, as also previously studied by Lwoff, vitamins. In some experiments two different kinds of sugars were used and then an interesting observation was made. With certain combination of sugars a two-phase growth curve was observed. This was a puzzling result. As described in his Nobel lecture[7] he showed these data to Lwoff at Pasteur who said "That could have something to do with enzyme adaptation." Monod's reaction was "Enzyme adaptation? Never heard of it!" This was the beginning of a more serious phase in Monod's career as a scientist. However, there were a number of factors that caused a delay in the development of a full momentum.

It was in December 1940 that Lwoff gave the already mentioned eye-opening information about enzyme adaptation that put Monod on the track of 20 years of cutting edge science endeavors. Lwoff also provided him with literature to learn about earlier studies of the phenomenon. Monod called the phenomenon *diauxie*, Greek for double growth. He did in fact receive his Ph.D. late 1940, but this did not stir any interest at the Sorbonne. Thus it was at the Pasteur Institute that he developed his thinking and talent for practical laboratory work. However, it had to wait, because Monod got heavily involved in organizing resistance to the German occupants. This is a fascinating story, but outside the scope of the present book. It is well described in the already mentioned book *Brave Genius*[3] and also quite extensively in Judson's book on DNA[2], based on the interviews he made with Monod. One challenge to Monod was that the major resistance group, the Francs-Tireurs, which he planned to join, was managed by the Communist party and kept their distance from the group under Charles de Gaulle. After some difficult deliberations Monod decided to join the Communist organization, but he left it immediately after the end of the war. His code name was Marchal. Later he became a severe critic of left-wing ideology, ultimately because of the Russian invasion of Hungary in

1956, but in addition also because of other earlier incidents, as we shall see. It can be mentioned that Monod brought Lwoff back to intelligence work when the underground network to which he originally belonged had been broken by the Gestapo. These were dangerous times.

Another problem he had was that Odette was Jewish and therefore he had to send his family to safety in the southern non-occupied part of France. There was considerable turmoil during the occupation due to German uncovering of agents and the subsequent arrests. For some time Monod had to go underground, living a vagabond life. He was at high risk on many occasions but had the good fortune to manage to escape at the last moment. When the Allies landed in Normandy on June 6th, 1944 the challenges intensified. Within the French Resistance the split between Communists and Gaullists widened. Monod received new and larger responsibilities and he changed his cover name to Malivert. After the war Lwoff pointed out to him that a hero in an early Stendhal novel had that name and that he was impotent. Monod attempted to handle this embarrassing information by playing down his own role, presumably unfairly, during the late phase of the occupation. The resistance movements committed several acts of sabotage and helped foreign soldiers who because of emergency parachuting out of their aircraft or for some other reason had ended up on occupied territory. On occasions Monod sheltered American airmen in his home, putting himself at a major risk. The soldiers were later passed on to the unoccupied part of France.

Monod changed laboratory environment in 1944 when he felt that the ground was getting hot under his feet because of his deep commitment to the resistance movement. He thought that Gestapo might readily search for him at his Sorbonne laboratory, which they know about, but that it was much less likely that they would search for him at the Pasteur Institute. Monod had in fact already during the occupation performed work in Lwoff's laboratory. It was there that he was able to start the experiments completely changing the perspective on genetics in bacteria. For a long time there had been a discussion on whether the emergence of new characteristics in the laboratory workhorse bacterium *E. coli* was due to adaptations or to mutations. In collaboration with a graduate student Alice Audureau it was shown in studies of the capacity of bacteria to use the sugar lactose that *both* mechanisms were of importance. The first step was a spontaneous mutation in the bacterium that allowed it to change from not being capable to metabolize the sugar lactose to manage doing this. However, since the enzyme system needed was present in both the parental and in the mutant strain the hypothesis was that the mutation had affected a gene controlling the expression of the lactose digesting enzyme. In passing it can be mentioned that Audureau had even isolated a mutant bacterium from Lwoff himself. It was called *E. coli* ML, where the two-letter suffix stands for "*mutabile* in *Lwoffii*" and this was used for the experiments. Mutants of the bacterium grown in the presence of two sugars, glucose and lactose, retained the characteristics of the two-step growth curve phenomenon. They first grow exponentially using glucose. Then there was a delay for about an hour before they started another logarithmic growth using lactose. This was the phenomenon that occupied Monod's thinking and experimentation during the following decade. He returned, with full force, to attacking this problem when the war was over. He settled in as head of a laboratory in Lwoff's department at the Institute Pasteur in the famous attic. His research group started to grow with both French and foreign scientists of many nationalities joining in the work.

Monod had in fact towards the end of the war accidentally obtained access to some issues of the journal *Genetics* in which he found an article by Luria and Delbrück. They had demonstrated that there were bacterial phage mutants occurring spontaneously. This was an unexpected finding. The view of the time was expressed in the writings of Julian Huxley, mentioned in Chapter 3. He stated that bacteria had no genes in the sense of hereditary substance. It was considered of no use to study them as a means of learning about the genetics of more complex organisms. The Luria-Delbrück paper convinced Monod that it was time to change this attitude. Monod was so impressed by the work of the phage group that he even considered leaving his favorite field of enzyme adaptation and starting to study this kind of virus. At the time he also read the classic 1944 paper by Avery *et al.* that provided the first evidence that DNA seemed to be the carrier of the genetic information. It would take Monod, like most other leading scientists in the field time to fully appreciate the impact of Avery's findings.

The temptation felt by Monod to enter the field of phage genetics was accentuated in 1946 when together with Lwoff he participated in a meeting on "Hereditary Variation in Microorganisms" at Cold Spring Harbor, mentioned in the previous chapter. They took advantage of the recently introduced trans-Atlantic flight from Paris to New York via Ireland and Newfoundland. At this meeting he learnt that Delbrück and Hershey had demonstrated genetic interaction between bacteriophages and that Lederberg and Tatum had discovered bacterial sexuality as also referred to in the previous chapter. However, Monod overcame the temptation to enter the phage field and stayed with his problem of enzyme adaptation. In fact in writing a review in 1947 he noticed a widening interest in the scientific community in the set of problems it offered. In order to advance he had to learn more biochemistry. By contacts, not least with visiting scientists working in his group at Pasteur and by trial and error he was able to expand his knowledge, but later in life he referred to himself and his collaborators as "chemists without a license." Monod also needed resources to update his laboratory environment at Pasteur, so that it could match what he experienced in the laboratories in the U.S. He stayed in that country for two months and established important contacts with the Rockefeller Foundation and other American grant-providing organizations.

The Dawn of Molecular Biology

It should be remembered that at the time it was unclear whether a particular protein was represented by a homogenous population of molecules (cf. Chapter 5) and it was also argued by Schoenheimer (Chapter 6) that proteins could be in a dynamic state with parts being continually destroyed and rebuilt. In addition of course it remained to identify the chemical nature of the genetic material. In this state of ignorance Monod initiated studies of the control of metabolic events. Challenging and complex hypotheses were combined with ingenious experiments. The complex two-step growth curve activation seen with cultures provided with two sugars, like glucose and xylose, was speculated to be due to the fact that the bacteria preferred the glucose and hence first this kind of sugar was broken down by use of a specific enzyme to take advantage of the energy provided by this molecule. When this carbon source was exhausted the system adjusted itself, by turning off the enzyme needed to digest glucose and instead produce an enzyme that was capable of breaking down xylose. What kind of molecular switch could bring about this change?

Not knowing the chemical nature of the information-carrying molecules the approaches had to be empirical. There had to be an inducer, in the simple case the substance to be metabolized itself. But where could it act, directly on the gene or at a nearby site on the genome? It was deduced — correctly as it turned out — that the controlling part must be separate from the gene directing the synthesis of the specific enzyme. This separate control region could theoretically serve as a switch in two different ways. One way would be that it became activated by contact with a special metabolite, serving as an *activator*, a *positive* control. But potentially there could also be a *negative* control, with a critical chemical compound serving as a *suppressor*. It turned out that both these mechanisms were used by bacteria. The controlling elements later came to be called *promoter* and *operator*, respectively. These specific gene expression controlling elements will be further discussed in the last chapter. As we shall see Jacob in the mid-1950s had the ingenious idea that the system he was studying using phages and genetic transfer between bacteria and the one Monod was using might illustrate similar examples of fundamental gene controlling events. This led to some very pioneering contributions to the emerging field of molecular biology which soon became amplified and substantiated by the rapidly growing insight into the processing of information by use of different kinds of nucleic acids. This will be discussed in depth in the last chapter, where we will also meet the term *operon* coined by Jacob and Monod in 1960 to describe the system for control of gene expression allowing the synthesis of one or of a series of functionally related specific proteins.

In retrospect it can be noted that it was fortunate that these early studies of gene control were performed by use of bacterial systems. Later studies of molecular mechanisms have demonstrated that gene regulation in eukaryotic cells is much more complicated than in bacteria. The whole system is driven by transcription factors of which there is one in bacteria and five in eukaryotes. The operons found in bacteria are absent in eukaryotes and each gene must be regulated individually. In bacteria each gene is controlled by a single or a few regulatory proteins, but in eukaryotes there are many such proteins, sometimes even hundreds. Some of the latter may interact with DNA at a great distance from the gene. Finally the DNA in eukaryotes is packed with basic proteins called histons into a complexes referred to as chromosomes. There are no such structures in bacteria. This fact adds another level of control in eukaryotes, namely the regulation of the availability of DNA as it becomes dissociated from histone proteins by some, as yet poorly defined, overriding control mechanism. A famous quote by Monod is "What is true for *E. coli* is true for an elephant." This needs to be qualified. It is certainly true when it comes to the fundamental principles of flow of information as expressed by the central dogma; the replication of DNA, the transfer of information by *transcription* into RNA and the final synthesis of proteins in ribosomes guided by the messenger-RNA, the *translation*. However, further evolutionary developments have added additional levels of complexity in the eukaryotes, such as for example the processes of trimming messenger-RNA by the mechanism of splicing. The latter phenomenon was recognized by the Nobel Prize in physiology or medicine in 1993 to Roberts, mentioned in Chapter 6, and Philip A. Sharp. We used one of the shortest motivations ever, "for their discovery of split genes."

In 1949 there were important developments in Lwoff's attic at Pasteur. We have already discussed his brute force attack on lysogeny and the dramatic breakthrough. We will learn that it was at this time that Jacob made his first fruitless approaches to Lwoff. Finally Monod's group was expanded and among others the American immunologist Melvin Cohn joined him as a critical collaborator. A drift in the conceptual thinking from enzyme adaptation towards induced enzyme synthesis started, which we will soon return to. However, a year before Monod had become intensely involved in the debate on the inheritance of acquired characters, which had a major influence on the focus of his ideological orientations.

Interlude 1 — Lysenko and a Convenient Untruth

Trofim Lysenko (1898–1976).

The influence of unfounded dogmas has often held back the advance of science. When such dogmas have an ideological underpinning sad events can transpire and the question is whether or not the so-called "Lysenko affair" represents the most tragic of all such cases. For three decades Mendelian genetics was rejected in the Soviet Union and substituted by pseudoscientific ideas proposed by Trofim D. Lysenko. He was born to a peasant family in the present-day Poltava oblast, Ukraine, in 1898. After studies at the Kiev Agricultural Institute he became employed at an agricultural experimental station in Azerbaijan. He developed an interest in a phenomenon called *vernalization*, from the Latin word *vernus*, meaning spring. The idea was to prime seeds by exposing them to conditions, like moisture and cold, mimicking the ideal spring for sprouting. Treatments of this kind had been employed for a long time but in Lysenko's hands they led to amazingly increased yields. And not only that, he also proposed that the improved capacity to grow could be inherited, based on fake experiments. These findings appealed to the Communist leadership, not least to Joseph Stalin. It was very agreeable to the Communist ideology of dialectic materialism that exposure to different environmental conditions could lead to changes in the hereditary properties of plants and why not also in humans?

Lysenko became the chosen hero of Soviet agriculture and there was a major need for such a hero, because of the consequences of the shift from an economy based on traditional farming to forced collectivization into industrial farming. This change had had a disastrous effect and repeated famines occurred. He was put in charge of the agricultural affairs of the country and he took this opportunity to criticize traditional geneticists. They were not true Marxists. The 7th International Congress of Genetics was originally supposed to be arranged in Moscow in 1937, but the planning faltered badly so that eventually it was decided to move the congress forward and hold it in Edinburgh in August 1939. The Soviet delegation led by Academician Nikolai Vavilov, originally the chairman of the canceled Moscow congress, were denied travel visas just a month before the Edinburgh congress was going to be held. After that the situation rapidly grew worse for traditional Soviet geneticists. At the end of the 1940s this kind of genetics — recalling names like Mendel and Morgan — was declared "a bourgeois pseudo-science." Lysenkoism became the only correct theory. Vavilov had already been arrested in 1940 and died in prison three years later. It has been estimated that along with him some 3,000 mainstream biologists were fired, put in prison or executed due to Lysenko's use of the totalitarian political machinery to suppress his opponents. Similar developments were seen in the other Eastern Bloc countries. Considering these developments it is not surprising that Tiselius in his banquet speech in 1948 raised the issue of freedom of research as mentioned in Chapter 5.

The American Nobel Prize recipient Hermann J. Muller had some personal experiences of the development of Lysenkoism in the Soviet Union. He had a career taking many turns. It started in the laboratory of the forthcoming recipient of the 1933 Nobel Prize in physiology or medicine Thomas H. Morgan. It was in the early 1920s that Muller had been introduced to work on Drosophila, the fruit fly, for studies of fundamental genetic problems. Then he was invited by Julian Huxley to do post-doctoral work at William Marsh Rice institute (presently Rice University) and later he moved to Texas University. While there he made his breakthrough discovery of the mutagenic effects of X-rays. After the Great Depression in the early 1930s Muller became very pessimistic regarding the future of capitalism. In 1932 he moved to Berlin to work with the famous Russian expatriate Nikolay Timofeeff-Ressovsky. It was this scientist who later encouraged the physicists personified in

Delbrück, a student of Niels Bohr, to get involved in biological research. It did not take long before Muller was confronted by the rising Nazi movement and being Jewish on his maternal side he again had to move on. For political reasons he could not at the time return to the U.S., since he had been heavily involved in a leftist student newspaper, *The Spark*. Therefore he moved to the Institute of Genetics in Leningrad and developed his program for fruit fly work there. After two years his laboratory moved to Moscow. Following an early phase of expansion of the work and productive scientific contributions there were problems caused by the emerging Lysenkoism. He had to leave one more time and managed to get out, together with his 250 strains of fruit flies, via Spain to Edinburgh. He did not get back to the U.S. until 1940. Thus he was still in Edinburgh at the abovementioned 1939 congress of genetics. In this context he gave a lecture on "How could the world's population be improved more effectively genetically?" Since his youth he had been a believer in eugenics. Let us now return to Lysenko.

Hermann J. Muller (1890–1967), recipient of the 1946 Nobel Prize in physiology or medicine. [From *Les Prix Nobel en 1946*.]

When Stalin died in 1953, Lysenko, in spite of growing open criticism, managed to retain his position, due to support given him by Khrushchev. There remained a ban on studies of traditional genetics, but many biologists were released and rehabilitated. The final abolition of the ban occurred in the mid-1960s, but already in 1962, when Khrushchev was still in power, three of the most famous Soviet physicists of the time — Yakov Zel'dovich, Vitaly Ginzburg and Pyotr Kapitsa — attacked Lysenko. Two years later Andrei Sakharov said the following against Lysenko at the General Assembly of the Soviet Academy of Sciences:

He is responsible for the shameful backwardness of Soviet biology and of genetics in particular, for the dissemination of pseudo-scientific views, for adventurism, for the degradation of learning and for the defamation, firing, arrest, even death of many genuine scientists.

The whole story of the rise and fall of Lysenko has been well presented by Zhores A. Medvedev[8]. He was a man of great courage and with a strong conscience. His father was a professor at the Military-Political Academy, who was arrested in a wave of political terror in 1938 and three years later he died in one of the far eastern Kolyma Gulag camps. The surviving mother and her twin sons could eventually settle down and she managed to secure their academic education. Zhores became a biologist and his brother a historian. As a target for his biological research Zhores selected aging and in this field he pioneered the proposal that the process of senescence was dependent on accumulated errors in the production of the critical components, the proteins and nucleic acids. In 1962 he circulated as a "samizdat" — "self-publishing" — a manuscript entitled *The Rise and Fall of T. D. Lysenko*. This also found its way to the West and was published in 1969 in the U.S. As a consequence of this Medvedev was dismissed from his position as head of the Molecular Radiobiology Laboratory in Obninsk. This did not deter Medvedev who in succession published two more samizdat books, one on the need for international cooperation between scientists and another on the postal censorship in the Soviet Union. The political establishment moved into action and arrested Medvedev and subjected him to forced detention in a psychiatric hospital in Kaluga. Immediately there was a strong reaction by leading scientists, not least physicists, and well-known writers. Medvedev was released and the experiences he had gained he immediately used to write, together with his brother a book entitled *A Question of Madness*, published in London. In 1972 he accepted an offer to work at the National Institute for Medical Research in London. As soon as he and his family got there, their Soviet citizenship was taken away and their passports confiscated. However, their citizenships were finally restored by Gorbachev in 1990.

Already at the end of the 1940s there had been a powerful international reaction against Lysenko. Western Communist intellectuals were forced to take a stand and most of them renounced their original support for Marxism. One of them was Monod. Albert Camus, who was to receive a Nobel Prize in literature at a relatively young age in 1957, had become editor-in-chief of the non-communist left-wing Paris newspaper *Combat*. This had already started appearing during the war as a clandestine publication and continued for more than 30 years thereafter. Over the years many French intellectuals contributed to the newspaper. In 1948 the newspaper ran a series of articles under the heading "Mendel or Lysenko." This series was concluded by a contribution by Monod entitled "The victory of Lysenko has no scientific character whatever." It was a powerful criticism of doctrinal fanaticism and fraudulent science. It was Monod's first public attack on the communist party and it caused quite a stir in the academic establishment. The article also brought Monod in a closer contact with Camus, an intellectual encounter he treasured[3]. By way of contrast it should be added that Jean-Paul Sartre did not waver in his left-wing sympathies at the time and remained impressed by the Potemkin-like facades presented by the Soviet regime. He remained a leading pro-Communist intellectual and did not condemn the politicization of science in the Soviet Union exemplified by the Lysenko affair. Sartre's turnaround did not happen until after the Hungarian uprising in 1956, which we will return to below.

Research in most fields of science to a major extent is supported by money from governmental sources. It is natural that politicians ask about the return on the tax money that they have provided and that they also want to have some influence on the way the money is used. It is not always easy for the scientists to explain that the overall majority of major breakthroughs in their endeavors occur in fundamental, curiosity-driven research, as history tells. The interactions between politicians and scientists work best when a relationship of trust can be developed and when there is an ongoing dialogue allowing informed exchanges about the progress of the work. And still there are temptations for politicians to influence the direction of the work, in particular when it comes to more applied research. One example is the field of global warming on which it may be tempting to apply more subjective opinions. Thus right-wing American politicians may strategically and emotionally exclaim "I don't believe in global warming." However, science is not a matter of belief. The only comment the politician in this case can make is "I do not accept the data amassed by hundreds of scientists strongly indicating that our world is getting warmer and that this is related to the increasing concentration of carbon dioxide in the air." A corollary of this is also that the politician needs to define which data he does not accept and for which reason. The remarkable success story of scientific advances since the 18th century is due to the valuable neutral and reproducible data generated by scientific experiments. There is an impenetrable divide between this solidly accumulated collection of facts and the beliefs formulated by individuals or groups of them. The products of science cannot be censored and modified in their content by invoking a political ideology or for that matter a belief system assuming the existence of some kind of anonymous higher power with a priority right to formulate the "truth." As Galileo Galilei said in a muted voice *E pur si muove* — "and yet she moves (the Earth around the sun)."

The Breaking of New Ground in Science

In the early 1950s the rumor of the frontline science pursued by Monod and Lwoff at the Pasteur Institute began to disseminate in the scientific community. A number of foreign scientists started to join the group for shorter or longer visits. English, the emerging language of science, was increasingly heard in the laboratory. This was something completely new to French biological science. Many temporary collaborators could be mentioned, but one that stands out is the immunologist Melvin Cohn. He came as a postdoctoral fellow to Monod's laboratory in 1949 and stayed for five years. He then returned for

Melvin Cohn (1917–2014).

shorter periods and on the occasion of a later visit in the early 1960s he met his future wife Suzanne Bourgeois, who was a graduate student in Monod's laboratory at the time. After their marriage they moved to La Jolla to the Salk Institute, of which Cohn was one of the founders and the history of which was written by his wife after her retirement. Cohn has described Monod in the following way[9]:

> … He was the most secure scientist I have ever known …. Monod's charisma and leadership derived from his extraordinary breadth of knowledge and his interest in the social and political world around him. … He was a cellist, a philosopher, extraordinarily knowledgeable in literature, painting, etc. … He had polio and he limped on his left leg which was like a matchstick, yet he was one of the finest mountain climbers in France, he rode horseback and he played tennis. His protestant background (Huguenot) made him puritanical with respect to himself, and he was driven to overcome any infirmity. Science was only a plaything for him.

From Adaptation to Induction

It was not at all obvious in the early studies, if adaptation meant a conversion of an already preexisting immature product or a synthesis of a completely new molecule carrying the enzyme activity measured. It soon became clear that in the system of beta-galactosidase production in *E. coli*, the enzyme was synthesized *de novo*. The bacteria were fed the disaccharide lactose, which cannot be used as an energy source by the bacterium. This led to the synthesis of an enzyme that could break down the lactose into its two monosaccharide components, glucose and galactose. Since the enzyme was not present when the bacteria were fed lactose, but appeared thereafter the original term adaptation was changed to induction. It might be added that this kind of *inductive* enzyme represents an exception and not the rule in nature. Most enzymes are instead what are called *constitutive*. This means that the cell synthesizes the enzyme at a stable rate without a stimulus from any substance added to the cell. However, Monod wanted to explain the induction of enzymes originally referred to as adaptive. A lot of experiments by him, Cohn and others, not to be presented here, led to a completely new perspective. In 1953 Monod and Cohn together with three reputable biologists at the time — Martin Pollock, Sol Spiegelman and Roger Stanier, also referred to as members of the "Adaptive Enzyme's College of Cardinals" — proposed in the journal *Nature* that instead of "enzyme adaptation" one should use the term " enzyme induction." It was Pollock who from his studies of an inducible enzyme — penicillinase — that could break down penicillin in a certain kind of bacterium, first introduced the concept of an *organizer*, which in some indirect way could mediate the specific induction. This organizer served as an intermediate in an as yet undefined chain of interactive molecules.

In his studies of galactosidase induction Monod and his collaborators also made another important discovery. They were able to demonstrate that there were critical enzymes in the bacterial membrane that determined the access of the substrate to the interior of the cell, some kind of gate or pump. It was referred to as "galactosidase permease." This added one more dimension to the system studied. It took many years before the scientists could identify the nature of the elusive permease. Monod was also interested in constitutive enzyme systems. His group examined the synthesis of the amino acid tryptophan and found that in the presence of large quantities of this compound the synthesis of the enzyme needed to produce it was blocked. This spurious finding was not followed up, but it was picked up by another scientist to be examined as an example of a feedback system. It was the highly original Leo Szilard. He had a marked influence on the Pasteur group and Monod said in his Nobel lecture[7]:

> Of course I had learned, like any school boy, that two negatives are equivalent to a positive statement, and Melvin Cohn and I, without taking it too seriously, debated this logical possibility that we called the "theory of double bluff," recalling the subtle analysis of poker by Edgar Allan Poe.
>
> I see today, however, more clearly than ever, how blind I was in not taking this hypothesis seriously sooner, since several years earlier we had discovered that tryptophan inhibits the synthesis of tryptophan synthetase; also, the subsequent work of Vogel, Gorini, Maas and others (ref.) showed that repression is not due, as we thought, to an anti-induction effect. I had always hoped that the regulation of "constitutive" and inducible systems would be explained one day by a similar mechanism. Why not suppose, then, since the existence of repressible systems and their extreme generality were now proven, that induction could be effected by an anti-repressor rather than by repression by an anti-inducer? This is precisely the thesis that Leo Szilard, while passing through Paris, happened to propose to us during a seminar. We had only recently obtained the first results of the induction experiment, and we were still not sure about its interpretation. I saw that our preliminary observations confirmed Szilard's penetrating intuition, and when he had finished his presentation, my doubts about the "theory of double bluff" had been removed and my faith established — once again a long time before I would be able to achieve certainty.

In 1959 in a famous article by Arthur Pardee, François Jacob and Jacques Monod[10], historically nicknamed the "PaJaMo" paper, an acronym derived from the first two letters in their respective surname, the repeated contributions by Szilard were formally acknowledged. It was the data presented in this article that led to the formulation of two fundamentally new proposals. One was the new model for gene regulation, the two negatives cancelling out each other, already introduced above. The other was the speculation about the existence of an as yet unidentified form of RNA, which could transfer information from the genome to the protein-synthesizing machinery. In the text of this publication it was stated: "We are much indebted to Professor Leo Szilard for illuminating discussions during this work." We will return to this publication in the final chapter. So let us digress for a moment and briefly review who this Szilard was.

Interlude 2 — Leo Szilard, a Remarkable Physicist, Biologist and Peacemaker

In an earlier introduction to the 1944 recipient of the Nobel Prize in chemistry George de Hevesy[11] a book by George Marx entitled *The Voice of the Martians. Hungarian Scientists Who Shaped the 20th Century in the West* was mentioned. Among the twenty remarkable scientists who all had their upbringing in Budapest in the early part of the previous century both Hevesy and Szilard were included. It was in fact Szilard who inspired Marx to use the title of his book. It was he who suggested that "A Martian spaceship did indeed land in Budapest around 1900, then departed, and due to excess weight had to leave the less talented Martians behind." Szilard's life has been extensively described in a book entitled *Genius in the Shadows: A Biography of Leo Szilard, the Man Behind the Bomb*[12].

He has been described as a physicist, biologist and peacemaker with a Jewish background, who grew up in Budapest, but then spent most of his adult vagabond life in hotels first in Europe and then in the U.S. Because of his Jewish descent he had to leave Germany in the mid-1930s. Before this happened he had already in the 1920s presented a thesis departing from the hard-to-grasp metaphor of Maxwell's Demon, that allowed violation of the Second Law of Thermodynamics specifying a relentless increase of disorder, entropy. Using an analogy Szilard introduced the concept of this law into a new kind of information theory. He formulated the first equation on negative entropy and information. This thinking was later picked up by Claude Shannon in his influential 1948 book on information theory. In the 1940s there were extensive discussions about conditions for regulatory systems and the means of establishing a state of balance. This is of course highly relevant to biological systems and we have frequently referred to the concept of homeostasis. Norbert Wiener introduced the term "cybernetics" from a Greek work meaning governance. A fundamental part of such a system was feedback control, a concept highly applicable to gene control. The intellectual giant John von Neumann, another Hungarian polymath of Jewish extraction and also a Martian, who became the father of computer science and has been well presented in George Dyson's *Turing's Cathedral*[13], also contributed in thought experiments by the introduction of cellular automata.

Before leaving Germany Szilard had already patented the concept of a linear accelerator and cyclotron, but most importantly he had discovered the possibility of a nuclear fission process, for which he also had applied for a patent. Apparently his remarkable brain was spinning at full speed. When Szilard left Germany he went to England, where he helped to establish an organization attempting to assist refugee scholars to find new employment. During this time he became interested in biological/medical problems, which turned his interest towards isotopes. Together with a British colleague T. A. Chalmers he invented a way to concentrate isotopes. In 1949 he was in fact nominated for a Nobel Prize in physics for this finding, but the committee in a short remark noted that this contribution did not suffice to justify a Nobel Prize. This was his only nomination for a prize, but he is mentioned for his important inspiring contributions in the 1950s by Monod as already referred to.

It was in 1934 that Szilard had filed a patent on a neutron-induced nuclear chain reaction. Four years later he moved to the U.S. In that country he made fundamental contributions to the creation of a practically useful nuclear chain reaction in collaboration with Enrico Fermi. In a famous letter written by Szilard but signed by Albert Einstein, President Roosevelt was informed about the important recent achievements. This led to the initiation of the Manhattan Project, presumably the most comprehensive collective intellectual effort undertaken by mankind and of decisive influence on the surrender of Japan in the Second World War. After the war, however, Szilard spoke out against the use of nuclear weapons and participated in the first Pugwash conferences on Science and World Affairs. He also wrote a number of short stories collected in a book *The Voice of the Dolphins*. In this book he raised several moral questions associated with the situation caused by the Cold War and he also discussed the responsibilities of scientists, exemplified by the challenges of his own involvement in the development of atomic weapons. In the book, inspired by the establishment of CERN (Conseil Européen pour la Recherche Nucléaire) he sketched a similar establishment of an institution coordinating research in molecular biology in Europe. A few years later this materialized by the establishment of the European Molecular Biology Laboratory in Heidelberg, Germany, which includes a Szilard library featuring dolphins as its stamp. Szilard was also instrumental in encouraging Jonas Salk in his late 1950s efforts to establish the Salk Institute in La Jolla. According to Georg Klein, himself by birth a Jew and originally a Budapest citizen, "He (Szilard) not only accepted the duty of representing the conscience of the world, he was the conscience of the world." Szilard's intellectual-moral testimony left two questions open. How effective are scientists working inside or outside their professional domains and what about the relationship of scientists and policymakers? According to Winston Churchill the latter question can be readily answered — "scientists should be on tap, not on top"!

From his biography it becomes clear that during the latter two decades of his life — he died in 1964 — he was mostly involved in giving advice in the field of biological sciences. His thinking had a major influence on Monod's and Jacob's conceptualizing as mentioned above. Monod met him the first time in 1947 and recalling this first encounter he said:

> Many of the questions seemed very unusual, startling, almost incongruous. I was not sure I understood them all, especially since he insisted on redefining the basic problems in his own terms, rather than mine.

Their meeting occurred at a Cold Spring Harbor phage course taught by Delbrück and Luria. Inspired by his new insights Szilard teamed up with a chemist Aaron Novick. In a laboratory located in the basement of an abandoned synagogue they together invented the Chemostat. This was a device, a balanced flow bioreactor, which allowed a continuous regulated growth and metabolism of microorganisms. Szilard also developed many contacts with phage and bacterial geneticists like Hershey, Lederberg and Luria and regular

Leo Szilard (1898–1964) and Monod at the blackboard in 1961.
[From Ref. 2.]

meetings were held. A recurrent theme was the negative gene control already referred to. In a remarkable way he cross-fertilized microbial geneticists and according to Monod he "was as generous with his ideas as a Maori chief with his wives." Similarly Jacob described him as an "intellectual bumble-bee" bringing about a cross-pollination of ideas. In the discussions with Monod of the basis for the two-step growth curve when bacteria were offered two different sugars Szilard made a number of suggestions. He emphasized the role of the concentration of each of the sugars and suggested that varying these could influence the kinetics of the sugar consumption. It was these conversations that had inspired Szilard to invent the Chemostat and in parallel Monod developed a related system, which he called Biogen. All in all Szilard left many marks on the critical work by Monod and Jacob that eventually led to their recognition by a Nobel Prize.

The famous physicist Edward Teller characterized his early co-worker and later adversary concerning a possible deployment of nuclear weapons in international conflicts in the following way:

> Of all Hungarians Leo Szilard was the most Hungarian. A Hungarian is one who enters a revolving door behind you and comes out ahead. Leo Szilard was a dedicated nonconformist. He did not mind offending anybody, but never committed the sin of being boring. He had one principle which he did not violate on any occasion: never to say what was expected of him.

So, one may ask where are the Szilards of today?

Although it is somewhat distant from the main theme of this book I cannot refrain from citing what is known as Szilard's Ten Commandments. They are:

1. Recognize the connections of things and the laws of conduct of men so that you may know what you are doing.
2. Let your acts be directed toward a worthy goal but do not ask if they will reach it; they are to be models and examples, not means to an end.
3. Speak to all men as you do to yourself, with no concern for the effect you make, so that you do not shut them out from your world, lest in isolation the meaning of life slips out of sight and you lose the belief in the perfection of the creation.
4. Do not destroy what you cannot create.

5. Touch no dish except that you are hungry.

6. Do not covet what you cannot have.

7. Do not lie without need.

8. Honor children. Listen reverently to their words and speak to them with infinite love.

9. Do your work for six years; but in the seventh, go into solitude or among strangers so that the memory of your friends does not hinder you from being what you have become.

10. Lead your life with a gentle hand and be ready to leave whenever you are called.

After these pontificating statements by Szilard it is now high time to return to the intellectual giant who is the subject of this chapter, Monod.

The Unique Collaboration

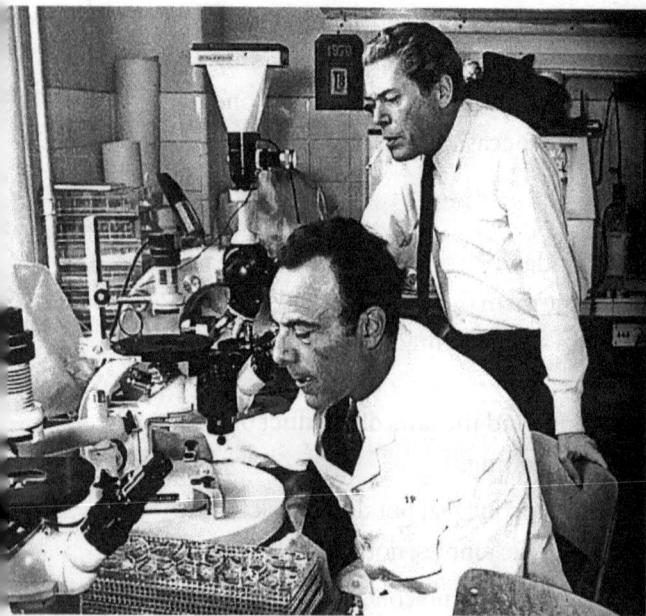

Jacob (in front) together with Monod in the laboratory. [From Ref. 20.]

In 1953 Monod was made head of a new department and moved from the attic to the bottom floor of the building. The department was called Cellular Biochemistry. Jacob remained in the attic and it was in remarkable brainstorming sessions some five years later that these two scientists, usually in Monod's office on the bottom floor, could amplify the conceptually new thinking each one of them was involved in. In this case one plus one added up to much more than two. By their remarkable achievements they came to serve as midwives of the emerging field of molecular biology. It may be appropriate to note how Jacob later in life described his collaborator[6].

Jacques Monod had a very special, very personal style. At the same time headstrong and blazing. A mixture of logic and passion. A persistent furrowing of the same line and quick probing in all directions. This puritan was a daring man. This atheist concealed a believer. Driven by a moral obligation, of a need to search for the truth about nature and make this truth known. To convince his colleagues and laymen about it. He did not only trust nature he also believed in it, its coherence and harmonization. Hence the famous aphorism: what is true for the coli bacteria is also true for the elephant. To analyze the functions of bacteria was a way to study man. What a pleasure to work together with a man to whom life was the same as research. In all fields. In all directions.

And further on he described their interaction as follows:

At this time was founded an intellectual relation between us that was extraordinary intensive, unique in its intimacy. We spent several hours together every day, mornings and evenings correcting the experimental results, analyzing and criticizing them, drawing conclusion from them; we corrected our hypotheses and formulated plans for new experiments. We sang a cantata or whistled perhaps a quartet. And we made jokes. Because all this took place in an intensive and glad atmosphere, in which we at any time could share a joke between us: about our origin, medical man or zoologist, about our "war generals," de Gaulle or Lattre de Tassigny, about our culinary, political or literary taste. All this accompanied by great laughter that filled the office to which Monod had moved ....

Their classical publication in *J. Mol. Biol.*[14], the culmination of their work, which will be discussed at length in the last chapter, included speculation as to whether the repressor might be composed of RNA or be a protein. Their studies eventually showed that it was a protein. However, later studies have shown that the nature of the repressor is dependent on the kind of bacterial system that is the object of the study. Jacob and Monod used the bacterium *E. coli*, but if they instead had chosen a so-called Gram-positive bacterium, like *B. subtilis*, the first repressor discovered would have been found to be of RNA nature — a non-coding portion of the messenger-RNA. However, this did not come about because of their selection of experimental system.

Jean-Pierre Changeux.

Monod's studies of mutants of the inductor protein yielded important insights. The essence of these studies, performed in collaboration with Jacob and Jean-Pierre Changeux, was that the interaction between the inducing substance and the inducer led to a structural change in the latter. This phenomenon was referred to as an *allosteric* effect, a naming derived from the Greek words *allos* meaning different and *stereo* meaning stiff, solid. Thus the inducer serves as a relay. Later on similar effects were identified in many different systems. Monod and his collaborators Jeffries Wyman and Changeux summarized their findings in an article entitled "On the nature of allosteric transitions: a plausible model"[15]. This is considered as one of the most influential publications in the respected journal *J. Mol. Biol.* Further studies of allosteric enzymes have demonstrated that their saturation patterns were not, as in "classic" enzymes systems, linear but often multi-molecular. An excellent illustration of this is the saturation of hemoglobin with oxygen as demonstrated by Max Perutz[9]. In the latter case the explanation is that the allosteric protein has multiple identical subunits, in the case of hemoglobin, four. Further applications of increasingly more efficient and high resolution crystallography have given innumerable examples of structural changes in proteins as they exert their operative functions, alone or as complexes. In fact there are even a fair number of proteins which have no fixed three-dimensional structure in their native polypeptide state, but take on a structure at first when they interact with their selected target molecule.

Agnes Ullmann has told[4] a charming story about the discovery of the allosteric phenomenon. Late one evening at the end of 1961 Monod walked into her laboratory. He looked concerned and somewhat exhausted. After a long delay he said "I think I have discovered the second secret of life." After having accepted the consecutive few glasses of whisky offered to him he presented the hypothesis of allostery. This included the proposal that proteins could take alternative forms and that this could be directed by the compound they were interacting with. And further that the site of interaction and the specific effect on this particular site by a defined molecule could have dramatic consequences on the structure of other distally located sites, in the simplest case as an on/off signal. In essence what Monod and his collaborators had discovered was that there was a world of potential information transfer by protein molecules on their own. We generally think of DNA and RNA as representing information transferring molecules, but we overlook the particular role of proteins in these realms. The theme of *protein information transfer* has come back with full force in discussion of prion proteins. Without elaborating this theme too much[16, 17] it has now been learnt that an incorrectly folded "sick" protein can influence its homologous "healthy" protein to change folding to become like the inducer. If this new folding includes flakes of beta-pleated sheets the outcome, by a "snow-balling" process, may be an aggregation into an amyloid-like structure. Such aggregates may cause severe cell damage. Presently major efforts in biomedical research are devoted to determining the role of amyloid formation in various kinds of diseases in the central nervous system, not only the rare Creutzfeldt-Jacob's disease, but many others, like Alzheimer's disease, Parkinson's disease, etc., and also outside the brain, like certain forms of diabetes. Protein folding and protein aggregation is a central theme in modern molecular medicine.

Boris Magasanik was a scientist of Ukrainian origin, who studied at the University of Vienna. Expelled by the Nazis he moved to the U.S. in 1938 and made an impressive career. He has been described as a brilliant polymath, who showed an unparalleled erudition including also history and the arts. After the war he obtained his Ph.D. at Columbia University. He then spent about a decade at Harvard University; before in 1960 he was recruited to MIT by Luria. At that institution he developed an impressive career that made him one of the pioneers in studies of gene

Boris Magasanik (1919–2013).

regulation. In 1959 he spent a year with Monod in Paris where on the side, he and his wife taught Agnes Ullmann to speak English. The reason for mentioning him here is partly that his chair at MIT was named the Jacques Monod Professorship and further that he pointed out to Monod that the theory of allostery was one of the most decadent (promiscuous?) in biology. It could explain essentially any phenomenon. It is said that Monod was prone to agree.

Let us now see how the Nobel committee at the Karolinska Institute evaluated Monod's contributions.

Nominations and Evaluations of Monod for a Nobel Prize in Physiology or Medicine

Monod was well aware of the procedure for nominating Nobel Prize candidates. As described in my preceding book[11] Monod in 1962 gave the critical nomination for a prize to Crick, Watson and Wilkins for their discovery of the structure of DNA. The nomination was very extensive and a considerable part of the material included was provided by a correspondence with Crick. For natural reasons there is no information available on the extent to which a nomination is the product of the nominator alone, perhaps by some assistance by colleagues with a good insight into the field concerned, or is in some way elicited by or initiated by the nominee himself. The term *nobelomania* (also *nobeliasis*) has been previously introduced and it is an impression that a not negligible number of scientists are obsessed by the idea that they are worthy of a Nobel Prize and should receive it. They may therefore mobilize their friends in smaller or larger numbers to submit nominations for their candidacy. During my time on the committee, an orchestrated large number of nominations for a certain candidate generally were viewed with a certain suspicion. A Nobel laureate, a good friend or mine, and I had a joint senior colleague, now deceased, who had a severe case of nobelomania. He pushed my friend to nominate him repeatedly for the prize, although my friend knew that the case was closed. One year he got so tired of the recurrent requests to make a proposal that he simply said "OK, I am going to write to Stockholm." He then called me and I agreed to receive a letter from him, which I later could throw into my waste basket. This saved the committee one surplus nomination. I have not told this anecdote to throw any suspicion on the nominations for a prize to Monod and Jacob. On the contrary, as we shall see these nominations were submitted only over a brief time period and reflected the rapidly increasing strength of their candidacy with time.

The first nomination of Monod for a Nobel Prize in physiology or medicine was submitted in 1960. It also included Lwoff's name as mentioned in the previous chapter. Not surprisingly it was submitted by Cohn, who knew from the inside all the whereabouts of their pioneering work at the Pasteur Institute. In fact it is an impressive nomination including a letter covering nine pages, an attached pile of all relevant references, and a complete CV taken from the recent appointment of both nominees as professors in biochemistry and microbiology, respectively, at the Sorbonne in 1959. Cohn highlighted in particular two discoveries by Monod. One was the identification of permeases and the other that the rate of enzyme synthesis was controlled by feedback mechanisms through the action of repressors. The separate identification of structural and regulatory genes in the latter context was elaborated. Cohen also described the important development of continuous cultures and the introduction of physiological time (growth) instead of absolute time.

The committee decided to let the professor of bacteriology Berndt Malmgren make a review. It is very comprehensive, covering 18 pages, and provided both a historical background and a detailed presentation of Monod's and his collaborators achievements. He noted that for some decades the transport of material, especially proteins, through the lipid-rich membrane surrounding cells had been a matter of discussion. It had been debated whether possibly specific transport mechanisms might exist. In this context Monod's discovery of a specific gene product controlling permeation of a certain sugar was very important. He also noted as a departure that bacteria were excellent model systems for investigation of the control of synthesis of inductive, later adaptive, as well as constitutive enzymes, also taking into consideration the one gene-one enzyme concept introduced by Beadle. He discussed whether the difference between the two kinds of enzyme regulation was quantitative rather than qualitative, but in the light of insights gained later this might be seen as a speculative reflection of its time.

Malmgren analyzed in more detail the specific experiments performed by Monod and his collaborators. The overall majority of the experiments were performed in *E. coli* using isotope-labeled methyl-beta-D-thiogalactoside (TMG). Competing compounds with a chemical structure similar to TMG were employed in many of the experiments. A number of new techniques were progressively introduced into the experiments performed by Monod and the increasing number of French and visiting collaborators. In some cases specific immunological reagents were used. The important collaborator Cohn was after all by his training an immunologist. It was noted that the adaptive induction was dependent on protein synthesis, since it was blocked by the inhibitor chloromycin. Similar findings were made in other systems like the use of adaptive consumption of the sugar inosite by *Aerobacter aerogenes* in studies by the abovementioned Magasanik and collaborators. Monod's group concluded that besides and *separate from* the gene-regulating protein there was a critical membrane-associated protein which had stereo-specific receptors for the critical sugar to be transported. In 1956 the name "galactoside permease" had been proposed for this protein that functioned by a highly specific recognition of TMG. The fact that the critical proteins could be demonstrated by use of genetic techniques was paradigmatic. It was this approach that allowed a distinction between the surface-active protein and the *de novo* synthesized intracellular protein involved in regulation of genome expression of the specific part of the genome directing the in enzyme production. It can be compared with Burnet's introduction of the conjectured receptor concept as a means of explaining the specificity of interaction in the first step of contact between a virus and a cell although at his time of this discovery the potential use of genetic approaches had not been revealed.

A number of different mutants of *E. coli* with different character of their membrane permeases and of the protein involved in inducing synthesis of galactosidase were identified and they could be examined by use of recombinants and transduction experiments. Jacob's merging with Monod's efforts in the mid-1950s in the studies of gene regulation, using their two respective systems, accelerated and deepened the emergence of new knowledge. Since the introduction of completely new concepts of gene recognition — the principle of receptor and operator — deriving from this collaboration will be developed in the next chapter these experiments will not be further discussed here. Malmgren's conclusions were cautious and somewhat humble. He asked for supplementary reviews allowing a deeper understanding of biochemical and genetic mechanisms, which he did not feel fully competent to grasp. He noted that the discovery of the permease system, alone would not qualify the nominee for a prize. The contributions in a wider sense were considered not amenable to a definitive conclusion regarding entitlemen to a prize. There was an unexpressed need to analyze Jacob's contributions, but he was not nominated. The committee concluded that Lwoff was worthy of a prize, but that at the time this did not apply to Monod.

In 1961 there was an impressive nomination by Jean Boyer from the Faculty of Medicine of Paris. Now all three candidates, not only Lwoff and Monod but also Jacob, were presented over 23 pages. The nomination reviewed the recent critical contributions for which Jacob in his close collaboration with Monod from 1955 and onwards had been a major driving force. This time the committee asked Gard to use his broad insights in a review including not only Lwoff, whom he had analyzed previously, but also Jacob and Monod. In fact, since the committee was so unfamiliar with Jacob's work they first asked for a preliminary investigation. Gard promptly concluded that a full investigation was strongly motivated. As already mentioned in the previous chapter he started his review with Lwoff. Since his major discoveries dated from before 1954, the analysis in essence was a repeat of the earlier evaluation. Gard noted that the follow-up work by both Lederberg using the techniques of recombina-tion and transduction and not least by Jacob and Élie Wollman, to be further discussed in the last chapter, had provided evidence for the correctness of Lwoff's proposal that the prophage DNA existing under conditions of latency was associated with the bacterial genome. This was the reason for Gard's earlier proposal to combine Lederberg and Lwoff into one prize, discussed previously. He also provided some general remarks that the presence of such a prophage could endow the bacterium with new properties, like the toxin production of *C. diphteria*, as also already mentioned.

In his evaluation of Monod Gard referred to a gene map of the Lac region in *E. coli* also included in Malmgren's presentation the previous year. This map defined the region to include (at least) three genes, referred to as y, i, and z. y and z were *structural* genes containing information that was necessary for the synthesis of permease and beta-galactosidase, respectively. By way of contrast i was a *regulatory* gene. The z gene was mapped in some detail and found to include eight loci, meaning that mutagenic modifications of eight different kinds led to absence of enzyme activity of the protein produced. Mutations in the i gene always led to inactivity. Since this gene directed a protein that served as inhibitor or repressor in Monod's terminology, its absence would lead to a full activity of synthesis of the enzyme. The fact that the synthesis of the repressor was not inhibited by chemicals that blocked protein synthesis was seen as puzzling. Could the repressor be of another chemical nature? Could it be RNA? The origin of the data that prompted this discussion is not clear at the present time. It was shown later that the repressor was a protein, but Gard's reflection that it might alternatively be RNA is clever. As briefly already discussed above it has turned out that in certain bacterial systems other than *E. coli* the repressor is in fact RNA and not protein. We will return to this kind of question when we dissect the operon concept, the formulation of which was in large part dependent on Jacob's work. Both a discussion of this and of the emerging insight into the nature of messenger-RNA will be reviewed in the next chapter.

Gard's concluding paragraphs (translated from Swedish) read as follows:

In his evaluation of 1960 Prof. Malmgren was of the view that he needed to delay a decision on the prize-worthiness of Monod's contributions; he considered that the discovery of the permease system by itself was not worthy of a prize and as for the contributions of genetic relevance, it was his view that the lines (of possible future developments) were not sufficiently clear and that the possible additional contributions by geneticists needed to be the target of a separate investigation. In my opinion the situation this year is different. Several very critical publications have been added and Jacob has entered the picture. I do not therefore hesitate to describe Monod's and Jacob's joint contributions as worthy of a prize.

It has already been emphasized that the investigations of lysogeny, transmission of genetic information and genetic control of cellular and viral activities are closely connected to each other and that they act to be mutually stimulating. It is in fact difficult to allocate individual contributions to one or the other of the groups of problems examined. It therefore seems to me justified to gather them (the three prize candidates) under one heading: the discovery of mechanisms for the transmission of genetic information and genetic control of the activities of bacterial cells and viruses and propose that a Nobel Prize in medicine or physiology is divided equally between Jacob, Lwoff and Monod. These three apparently have been the leading (ones) in the Pasteur group and there is no need to discuss any other name in this context.

Another report requested by the committee to elucidate the chemical side of Monod's contributions was submitted the same year by the biochemist Gunnar Ågren at Uppsala University, who was previously active at the Karolinska Institute. He was not impressed by Monod's early work in the 1940s on bacterial metabolism and noted that it was only in the early 1950s when Monod inspired by Lederberg took a new tack and changed focus to the beta-galactosidase in *E. coli* that major advances were made. Various aspects of the induction phenomenon were analyzed. He reviewed the techniques used; isotope labeling, preparation of antisera, preparation of chemical analogues, the use of metabolic inhibitors, etc. It was noted that the enzyme activity could be destroyed without significantly altering the capacity of the specific protein to serve as an antigen in immunological tests. Ågren then praised the pioneering studies of the permease, the brilliant identification of separate and particular systems for arranging the transport of the critical substrate beta-galactoside through the cellular membrane. Then he noted that the work performed by the biochemist — without citation marks — Monod together with the geneticist and microbiologist Jacob were of interest from a general biochemical perspective. He referred to the observation that they had made that the genetic information residing in a fraction of the genome, called cictron (consistently misspelled!), controlled by a repressor. We will return in the next chapter to the concept *cistron*. Ågren emphasized that the first scientist to introduce the concept of repressor was Henry Vogel at Yale University.

Vogel was studying the biosynthesis of the amino acid arginine in *E. coli*. This is not the place to discuss priorities in any detail, but it can be noted that Vogel is only mentioned once in Judson's book[2], and yet his critical contributions in the introduction of the concept and formulating a term for the inducting substance, originally represser, have been presented over two pages in a book by Brock[9], who leans toward giving him the priority — "Vogel clearly anticipated Monod by at least two years." As already mentioned it was Szilard who turned Monod's thinking from an instructive role of the inducer to a negative repression model. Åström discussed the various aspects of control of the expression of inductive as well as constitutive enzymes and outlined the two alternatives: (a) that the primary phenomenon was induction, in which case the repressor always should work as an antagonist of an exogenous or endogenous factor or (b) that the primary regulative factor was negative in its action, that is repressive with the inductive substance functioning as an antagonist of an endogenous factor. He then discussed the gene set i, z and y already referred to in Gard's evaluation and to be discussed in more detail in the next chapter.

Åström summarized his report by noting Monod's contributions in two different fields, the permease system and the induction/repression phenomena. He noted that the important advances in the latter field to a large extent has been dependent on his collaboration with Jacob. His final paragraph read (translated from Swedish):

> It appears to me as if Monod's contribution to the clarification of the permease system not by itself is a contribution worthy of a prize. By way of contrast it can be discussed whether or not the joint contribution by Monod in his studies of the permease and those performed (together) with Jacob have given results of such fundamental value that they could be considered for a prize. From a strictly chemical perspective it would seem that the results described at the present time are not worthy of a prize but together with the contributions in the field of biochemical genetics, (a field) in which I do not consider myself a fully competent commentator, this might be the case.

The use of biochemical techniques in the field of microbial genetics was new and a traditional researcher in this, albeit still relatively young field may not have felt comfortable with the mixed approaches used. This was to change rapidly and markedly over time. The borderline between microbial genetics and molecular biology rapidly vanished when it was discovered how the technology of the two fields could generate results cross-fertilizing each other and lead to important new discoveries. After Monod there were no more "biochemists without a license." Everyone involved had to have training in both fields. In 1965 there was in fact a nomination for a prize in chemistry for Monod together with Jacob by Michel Magat, a professor of physical chemistry in Paris. The committee of chemistry did not make a review but simply referred to that a contact with the committee on physiology or medicine would be desirable. As discussed in Chapter 5 early prizes in this field were reviewed under the heading physiology or medicine, but progressively more and more prizes illustrating the dramatic advances in the field of life sciences were recognized by prizes in chemistry.

The conclusion of the committee was straightforward. All three candidates, Lwoff and for the first time both Monod and Jacob were considered worthy of a prize. In 1962 the strength of nominations increased markedly. As mentioned in the previous chapter there were four nominations of all three candidates from three different Nobel laureates, Cournand, Lipmann and Tatum and one nomination supporting Cournand's also from New York, by Richards. In particular the nomination by Cournand was very comprehensive. The committee decided at its meeting on May 3rd that there should be another in-depth investigation, but this was not carried out in that year. All three candidates were registered as worthy of a prize by the committee in its final report. This was the year recognizing the discoverers of the structure of DNA Crick, Watson and Wilkins, which as remarked above seems to have been decided already before the start of the critical work by the committee this year. The following year there was only one outside nomination. It was by the respected French embryologist Étienne Wolff and it only included Jacob and Monod. There was a relative increased emphasis on the recently gained insights into mechanisms of gene regulation as well as on advances into insights into a critical factor in the central dogma, the messenger-RNA. When Gard noticed on the last day of nomination that there was no nomination of Lwoff he moved into action. On January 31 he nominated all three candidates and referred back to the previous evaluations performed. He wanted to secure that Lwoff was not left out of the discussions as already mentioned.

This year the three candidates were reviewed by Reichard. His analysis was very focused and clear. He first commented on Lwoff's discoveries, already referred to in the previous chapter. After this he discussed the rapid progression of Jacob's research during his first five years in Lwoff's laboratory. We will return to this in the last chapter. This was followed by summary of the already presented findings by Monod's and collaborators until 1955, the time when he initiated his uniquely productive collaboration with Jacob. Their impressive joint advances were discussed under three headings; the repressor concept; the operon and the operator concept; messenger-RNA and allosteric proteins. Reichard reflected on the concept allostery that it might be a new name for previously recognized phenomena. He referred to the U.S. enzymologist Daniel Koshland, Jr., who in 1959 had proposed that upon interaction with a substrate an enzyme might change its configuration. At the time this remained a speculation. He also mentioned the "Bohr-effect." This dates way back and was a speculation by the famous physicist Niels Bohr's father, Christian, a highly respected physiologist, that the hemoglobin in blood changed structure when it reacted with oxygen. It would take a long time until this was proven by Perutz's meticulous crystallographic work[11]. In spite of his reservation Reichard considered Monod's, Changeux's and Jacob's work important in providing more concrete insights into a change in protein structure. They used the enzyme aspartate carbamyle transferase as a model. When it reacted with a selected inhibitor, called CTP for short, it changed its configuration. However, the site of interaction with CTP was separate from the enzymatically critical site, which was suppressed.

Reichard started his summarizing comments by showing a diagram of the central dogma — DNA (=gene) gives RNA gives enzyme (the protein product, my remark) — "in the so-called molecular biology." This may be the first time that this concept was cited in the annals of the Nobel archives. It was introduced already in the 1940s by Harry Weaver at the Rockefeller Foundation as a combination of two words originating one and a half centuries earlier. The word biology — from Greek *bios*, life and Latin *logy(os)*, word, meaning, e.g. field of learning — was used for the first time in an article published in 1802 by the natural scientist Gottfried R. Treviranus. The title of his work translated from German was *Biology, or the Philosophy of the Living Nature*. The term molecule was introduced about the same time. It is the diminutive form of the Latin word *moles*, meaning mass, barrier. Molecule means an "extremely minute particle."

Reichard noted that the *qualitative* aspects of these relationships iden-tified in the central dogma had been clarified by several scientists, a number of whom had been recognized by Nobel Prizes in physiology or medicine —

Kornberg and Ochoa; Beadle, Tatum and Lederberg; Watson, Wilkins and Crick. For comparison he commented on the quantitative aspects of gene expression and referred to Lwoff's work on induction of lysogenic phages and Monod's work on beta-galactosidase in *E. coli*. Concerning the latter he noted the discovery of separate genes for the permease and for the enzyme and the existence of a separate regulator gene, as discussed above. He then stated "However, the most important and definitive results were obtained when Monod initiated the fruitful collaboration with Jacob." These advances were illustrated by a diagram and we will return in full to these impressive achievements in the final chapter. In that context we will also discuss the identification of *messenger-RNA*.

The final paragraph of Reichard's evaluation read (translated from Swedish):

Lwoff's, Jacob's and Monod's joint research has elucidated essential phenomena, which all in one way or the other concern aspects of the synthesis of proteins. It appears to me without doubt that discoveries like induction of lysogenic viruses, the repressor phenomenon and messenger-RNA are *each one on its own* (my italics) worthy of a prize. It seems to me difficult to find any research group in the field of molecular biology, which can be considered more worthy of a prize than the Pasteur group.

This was a powerful endorsement. The committee agreed that all three candidates were worthy of a prize, but the prize that year went to the discoveries of the ionic mechanisms of nerve signaling, as presented in Chapters 1–3.

In 1964 there was one nomination of Jacob and Monod by J. N. R. Grainger from Dublin and another repeat nomination of Lwoff and Monod by Lipmann. At its spring meeting the committee proposed that another full investigation should be undertaken, but this was never done during the year and the three candidates remained worthy of a prize in the summary report of the committee this year. The situation that developed the following year has already been summarized in the previous chapter. The large number of twelve separate nominations for a prize to Jacob and Monod, besides the single nomination for a prize to all three candidates, highlighted the impressive impact that their work had in the burgeoning field of molecular biology. It took a new review by Gard and extensive discussions in the committee until it finally and fortunately decided to include all three scientists.

Behavioral Sciences and Nobel Prizes in Physiology or Medicine

To digress for a moment it can be noted that many pictures of Monod and Jacob depict them with a cigarette in their mouth (p. 458). Monod continued to smoke until his death from leukemia in 1976; Jacob interrupted his smoking for periods of time and for good a few years before his death when he developed a serious disease. Lwoff never smoked. Due to active advertizing — the Marlboro Man etc. — the frequency of smoking increased in the early 1960s. U.S. citizens on average smoked 11 cigarettes per day and the figure for Frenchmen was certainly not lower. Probably most professors at the Karolinska Institute were smokers during this time. Among the professors we have met in the previous chapters, for example Gard, Hammarsten and Tiselius, were avid pipe-smokers. This increase in a severely health-threatening habit is somewhat surprising since already at this time pioneering epidemiological studies had begun to demonstrate a strong association between smoking and cancer, in particular in the lungs. These studies were initiated by Austin Bradford Hill. His earlier clinical studies had proven that streptomycin was an effective means of treating tuberculosis. Hill was skilled in statistics and pioneered the use of randomized clinical studies to demonstrate causal connections between environmental factors and disease. As a collaborator in studies of a possible role of smoking in cancer he employed Richard Doll, who during the Second World War had performed epidemiological studies.

Austin B. Hill (1897–1991).   Richard Doll (1912–2005).

Together Hill and Doll, somewhat to their own surprise, were able to demonstrate a very strong association between smoking and lung cancer. Their first study was met with suspicion, but Doll stopped smoking. The study was repeated with an enlarged group composed of essentially all physicians in the U.K. at the time. An inquiry was sent to the circa 60,000 physicians and 41,000 out of them responded. As time went on it became increasingly clear that there was a very strong relationship. During 29 months 789 physicians had died, 36 of them from lung cancer. All of them were smokers, whereas no similar correlation was found in other forms of cancer in this early study. The follow-up studies clearly demonstrated that smoking was the single most important factor primarily in lung cancer but also for many other diseases. Still today about a billion people in our world smoke and each year about a million die of lung cancer caused by this habit. Doll also made a number of other important contributions. He described the dangerous combination of smoking and exposure to asbestos and he detected the linear relationship between exposure to ionizing radiation and development of leukemia. He became a member of the Royal Society in 1966 and was knighted in 1971. He received many honors, including the United Nations prize for outstanding research into the causes and control of cancer and the Edward Jenner medal of the Royal Society of Medicine in 1981. A year before his death he was recognized by the Royal Swedish Academy of Sciences, which awarded him the Gold Medal for Radiation Protection. The motivation was "He is awarded for his pioneering work on causes of cancer. He has received special recognition for his work identifying tobacco smoking as a promoter and partly as a cause of lung cancer, and for his hypothesis on a linear dose-response relationship for ionizing radiation."

It is obvious that both Hill, who died in 1991 and Doll, who died in 2005, must have been repeatedly nominated for a Nobel Prize in physiology and medicine. In fact the first nomination had already been made in 1960 by A.W. Woodruff at the London School of Hygiene and Tropical Medicine. A preliminary evaluation was made by Nils Ringertz, professor of pathology at the Institute. He made a repeat preliminary analysis five years later when again both Hill and Doll had been nominated. His concluding paragraphs read (translated from Swedish):

Hill and Doll must be considered to belong among the foremost medical statisticians. However, one cannot point to a central biological problem that (for the solution of which) their joint contribution has represented an obvious pioneering contribution or where they have had the primary responsibility for the solving of the problem — in this context they are of course handicapped by the fact that they are statisticians, because statistical methods alone cannot provide a final solution of biological problems.

Thus it is my view that Sir Austin Bradford Hill's and Dr. W. R. S. Doll's scientific contributions as medical statisticians, although of high class, are not of such a nature that they are deserving a prize.

To this may be added that when Ringertz lectured to us during the semester long course of pathology in the third year of our medical studies he always had a packet of cigarettes in the breast pocket of his white coat.

No prize has ever been awarded for important clinical epidemiological discoveries. This is regrettable, but it may be that behavioral studies are not high up on the lists made by the committees at the Karolinska Institute. And still the importance of selected environmental factors, such as, in particular, smoking, for health and the influence of both hereditary and environmental factors discovered by studies in geographical medicine are important. A very critical part of modern preventive medicine is to encourage the choice of life styles; no smoking, exercise, appropriate eating habits, etc. The only prize for behavioral studies that comes to mind is the award in 1973. This was the first year that I was an associate member of the Nobel committee and I remember that I was somewhat surprised by the candidates finally recommended to the Medical Faculty. The prize concerned animal behavior, *ethology*. It was Karl von Frisch, Konrad Lorentz and Nikolaas Tinbergen, who were finally chosen to receive the prize "for their discoveries concerning organization and elicitation of individual and social behaviour patterns." The findings made by these scientists are fascinating and highly impressive, but what has their relevance been for modern medicine?

Two years later, in 1975, the prize in physiology or medicine recognized Baltimore, Dulbecco and Howard Temin "for their discoveries concerning the interaction between tumor viruses and the genetic material of the cell." I became deeply involved in this prize and accompanied Temin in the procession at the prize ceremony, but it was Reichard who gave the laudation address on this occasion. Temin was a truly devoted scientist, original thinker and a committed human being. He had fully understood the seriousness of the problem of smoking and health. In all the press conferences he participated in during his visit to Stockholm he took the opportunity to warn against the habit of smoking. Let us now return to the post-prize developments relating to Monod.

Chance and Necessity

In 1967 Monod was appointed professor at the Collège de France. His inauguration lecture was entitled "From molecular biology to the ethics of knowledge." This theme was developed later in a series of Robbins lectures which he gave two years later at Pomona College, Claremont in California, not far from Pasadena where Caltech is located. Pomona College is one of the highest ranked undergraduate colleges in the U.S. with an impressive endowment. In passing it may be mentioned that I have visited its charming campus many times often together with my family to see Franklin Scott, a historian writing about Sweden and Scandinavia. He and his family became good friends with my wife's family during the early 1950s when he worked in Sweden, mostly at Uppsala University. His daughter was helped in school by my future wife, who had had her first five years of schooling in the U.S.

The overriding title of Monod's Robbins lectures was "Modern Biology and Natural Philosophy" and he prepared the four lectures during one of his recurrent winter sojourns at the Salk Institute of Biological Sciences. He had contributed to the establishment of this institute in La Jolla, CA, and was a member of its board for a number of years. The four lectures had the following titles: Living beings as unnatural objects; DNA and emergence; Proteins and teleonomy; The kingdom of ideas. This is not the place to review the remarkable biological and philosophical content of the final book *Le hasard et la necessite* by Editions du Seuil, Paris, published in English the following year[1]. The book needs to be read and reread, because it contains many statements that have stood the test of time and served as catalysts for expanded discussions of Darwinism and its derivations, but it also includes statements that cause debates into the present time. In a book on EvoDevo (evolution of development)[18], written by Sean B. Carroll the same author who later wrote the already praised book on Monod and Camus[3], the admiration for the Pasteur trio, and not the least Monod shines through. It is said "The historical setting at the outset of their work, the fundamental importance of the implications of their discoveries, and the extraordinary character of these three individuals make the story of enzyme induction and gene logic in bacteria one of the most dramatic in the history of modern biology." After this the book provides a diagram illustrating the genetic switch that controls beta-galactosidase production and lactose metabolism in *E. coli*. And then towards the end there is praise for Monod's book. The text reads:

Jacques Monod captured this interplay of randomness and selection in evolution most eloquently in the title of his landmark book, Chance and Necessity (a reference to the Greek philosopher Democritus who said, "everything existing in the Universe is the fruit of chance and necessity"). Evolution is indeed a matter of chance, but in the random lottery of mutations, some numbers and combinations better meet the imperatives of ecological necessity, and they arise and are selected for repeatedly.

Besides the reference to Democritus, there is a citation of the closing passage of his friend Albert Camus' The Myth of Sisyphus in the opening epigraph of the book. The book discussed among many matters the existence of information-carrying structures and their evolution. Monod described the existence of internal autonomous determinism serving to allow the development of the very complex structures of living systems. Complexity is a matter that continues to challenge us even at the present time. A matter frequently returned to in the book is the teleonomic level of a species, a reflection of the sum of its particular information content. But teleonomy, the goal-directed emergence of structures and functions of a given organism was according to Monod secondary to reproductive invariance or expressed differently a consequence of it. Monod was a supporter of Karl Popper's ideas (see Chapter 2) and discussed the philosophical problems against the background of dualism, contrasting idealism and materialism. He wrote the foreword to the French translation of Popper's central work "The logic of scientific discovery" and his brother translated the book into French. Seeking an ethic of knowledge Monod emphasized that science itself is value-free, which however, did not prevent it from being of great value in the objective world. It was also proposed to have a central role in fostering high human values like altruism, courage, generosity, the need to create, etc. In summary Monod's style of writing was pronounced by Carroll to be "clear, elegant and incisive." However, the book was also subjected to criticism even from his closest collaborators, Lwoff and Jacob. They did not agree with Monod that one could start from molecular biology and derive a system of ethics. A parallel problem is if ethics can be derived from esthetics, as briefly discussed in the concluding section of this book. Jacob gave a very unflattering comparison with Teilhard de Chardin's book The Phenomenon of Man, which had been mentioned in Monod's book.

Chardin was a Jesuit priest trained in paleontology and geology. He was involved in the excavation of Peking man and also Piltdown man, the latter of which was later shown to be a fraud. He wrote his famous book in which he tried to make a synthesis of theology and science in the 1940s, but he was forbidden by the Church to publish it and it appeared posthumously in 1955. It included concepts like the omega point in which the maximum level of complexity and consciousness was reached and he developed the concept noosphere, a collective reflection of human thought. Monod, like many other scientists, for example Peter Medawar, was very critical to Chardin's writings at the time. Thus when Jacob compared the development of his philosophical thoughts to those of Chardin it appears as a harsh evaluation. It is possible that Jacob exaggerated his negative review of the book and that his assessment was tainted by the fact that he was losing a collaborator. There was still much more work to be done on the repressor system. Once techniques became available to produce larger quantities of repressors it was possible to determine their chemical nature. This was finally done by Walter Gilbert, Benno Müller-Hill and Mark Ptashne. Later Jacob developed his own philosophical perspectives on biology as we shall see.

Let us leave the more serious matters of philosophy and return to the quality of interaction between Monod and Jacob during the golden days when they had a lot of fun together. One needs to emphasize the importance of a lighter side to scientific work. As has been repeatedly mentioned there is a daily toil involved in scientific research with many failed or negative experiments. It is a matter of remaining an obsessionalist, to use Medawar's term, and of retaining belief in the hypothesis that is subjected to a test at the time. In order to manage this, most often an uphill battle, it helps to have a good sense of humor. Arthur Koestler discussed in his book *The Act of Creation*[19] the related conscious and unconscious underpinnings of scientific discoveries, artistic creations (see Chapter 3 and concluding section of the book) and jokes. He referred to what he called *bisociative* thinking. When two independent patterns of perception or reasoning interact with each other, several things can happen. The matrices may fuse leading to an advance of knowledge by an intellectual synthesis. Another alternative is a release of the tension that has developed by interpreting it in the form of an aesthetic creation. Finally the collision may be irresolvable and the only way out may be by a joke. Many scientists have a good sense of humor and as mentioned the intense intellectual intercourses between Monod and Jacob at the blackboard now and again was resolved by a joke or by reference to shared cultural impressions. However, there may be a certain difference between the French and the American culture in, that in the latter, the border between involving collegiality and personal friendship may be more readily transgressed. Monod's and Jacob's intense collaborative efforts led to many shared creative moments and superb personal interactions. However, this did not spill over in a warm private friendship.

In the 1960s I got to know a fellow virologist Alexander Kohn from Ness Ziona, Israel. In 1955, together with the physicist Harry J. Lipkin, he had started *The Journal of Irreproducible Results*, which survived until 1994 published by Blackwell Scientific Publications. Just a few years earlier in the 1990s a similar journal entitled *Annals of Improbable Research (AIR)* was started. This magazine is connected to the Ig Nobel Prizes, a parody of the true Nobel Prizes. The name is a pun playing on ig-noble — characterized by baseness, lowness or meanness. These prizes aim at "honor achievements that first make people laugh and then make them think." Most often there is a veiled criticism, but not infrequently there may be some useful knowledge gained by what seems like a way-out scientific achievement. The prizes are awarded at Harvard University about three months before the traditional Nobel Prize ceremony on December 10. The Ig nobel prizes are handed over by genuine Nobel laureates. We have frequently returned to the exceptional creativity of scientists with a Jewish background and the question is if not the well developed capacity to generate and appreciate jokes is of added value in this context. There is a particular form of Jewish humor. One needs only think of Woody Allen — "eternity is long in particular towards the end."

It is now overdue to bring the presentation of Monod to a conclusion.

The Phasing Out of an Intense Life Involving Science

Monod's life was guided by moral convictions, an intense curiosity to read and understand the books of nature and broad cultural involvements in general. We have already discussed his intense attack on Lysenkoism and as a consequence the progressive pronounced reduction of his left-wing sympathies. Less than ten years later the spontaneous uprising by the Hungarians and the quenching of this revolt by the Soviet army led to Monod's dedicated involvement in helping scientists who wanted to get out but were trapped in the country. In the frequently cited recent book[3] a whole chapter was devoted to this episode and to Monod's courageous efforts to bring Agnes Ullmann and her husband out of Hungary. Eventually he was successful in doing this although it took a long time and required repeated alterations of plans until they could leave hidden in a trailer. Agnes later became one of his most devoted collaborators and has written about him in a well balanced memoir[4], already cited.

Monod's visibility increased rapidly after the award of the Nobel Prize. For example, when the Reverend Dr. Martin Luther King Jr. visited Paris in 1966, two years after he had received the Nobel Peace Prize, to raise money for his forthcoming efforts it was Monod who introduced him in front of an audience close to five thousand people at the Palais des Sports. In the spring of 1968 there was a general uprising among students around the world. This also happened in May in Paris. Monod tried to establish a dialogue with the

Monod shaking hands with the 1964 Nobel Peace Prize recipient Martin Luther King's wife Loretta. To the left are actress Simone Signoret, the singer Harry Belafonte and the actor Yves Montand. [From Ref. 3.]

students, but to his regret he failed. He then attempted to negotiate a truce to get wounded students out of the zone of revolt, but again he failed. However, Monod retained his commitment to defend human values and to fight various forms of injustices, like capital punishment and lack of availability of legal abortions.

In 1971 Monod was asked whether he would consider becoming the Director of the Pasteur Institute. He was very ambivalent about how to react to this offer, not least because the institute at that time was at the brink of bankruptcy. However, feeling solidarity with the institution that had been his scientific home he accepted. He did in fact manage to restore a financial balance and at the traditional Institute competitive laboratories pursuing modern molecular sciences were able to evolve. He stayed with his heavy tasks in spite of the fact that in 1975 he was diagnosed with leukemia. A year later he died on his post, sailing into unknown waters.

In spite of his flamboyancy and extrovert character Monod was not a person to actively collect honors. He did receive some prizes from the Académie des Sciences, but he never became a member of this learned society. However, he did become foreign member of both the Royal Society and the National Academy of Sciences in the U.S. both in 1968. He also became a foreign member of the Royal Swedish Academy of Sciences. He was nominated by Gard in 1972 who wrote (translated from Swedish) in a summary statement justifying the proposed election "With his perspicacity, his analytic and outstanding associative capacity, his pregnant formal elegance, Monod takes an absolutely leading position within the science of molecular genetics, which he himself has founded." On the same occasion the Academy also admitted Hodgkin (see Chapter 3)

Monod in his sailing boat outside Cannes three days before his death. [From Ref. 4.]

following a proposal by Bernhard. Monod had earlier become an honorary doctor of the University of Chicago (1965) as well as of Rockefeller University (1970). Finally he had many military distinctions after the Second World War and in 1963 he became an Officer of the Legion of Honor.

It is now time to address the third remarkable scientist, Jacob, to give a complete picture of the exceptional trio who made decisive contributions to the early development of molecular biology at the Pasteur Institute and in the world.

Chapter 9

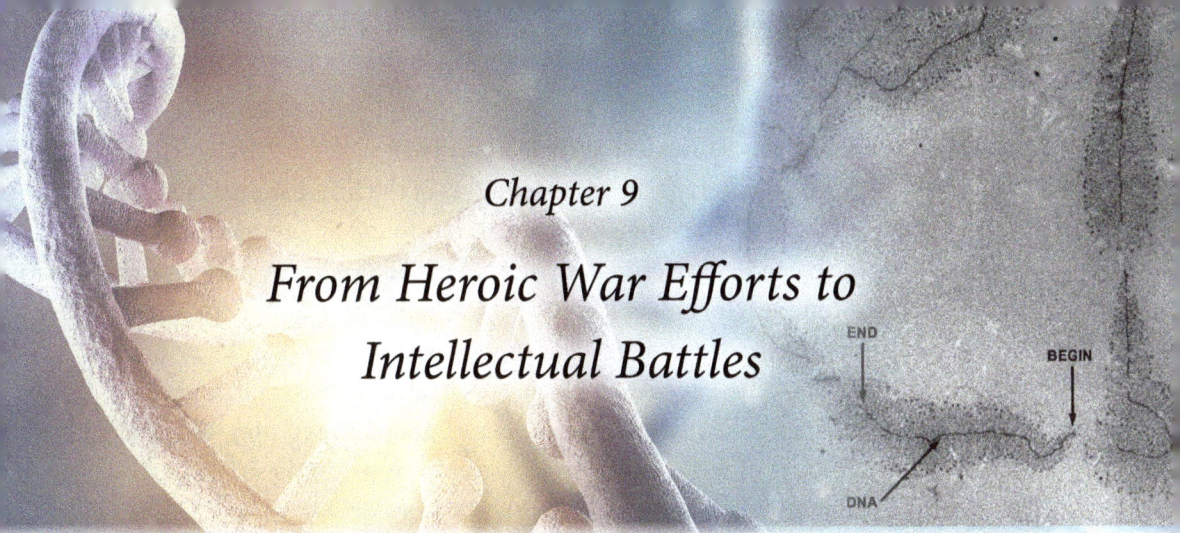

From Heroic War Efforts to Intellectual Battles

A MULTIFACETED HERO
PIONEERING NEW SCIENCE
AND WRITER OF CLASS

One dark evening in November 1988 a small group gathered at a restaurant in the old city of Stockholm. The host was Dorothea Bromberg, an eclectic and successful small-scale publisher of books in Sweden, and the guest of honor was François Jacob. His book *La statue intérieure* originally edited and published in French by his daughter Odile the year before[1] had been translated into Swedish and was to be launched by Brombergs Bokförlag AB[2]. I do not remember the other guests except for Georg Klein. He had contributed a foreword and extracted from this there were some words on the front cover, which translated into English, read "I have never read an autobiography which has given me as much, both as a literary composition, a time document and insight into the inner world of many outstanding men of science." I had previously received the French version of the book from Reichard, but regrettably my knowledge of the French language was not sufficient to appreciate the richness of the text. The Swedish version of the book, in which the dedication "Pour Erling Norrby, très amicalement, François Jacob" had been written, allowed me to enjoy his exquisite text.

On two other occasions I had the pleasure of meeting Jacob. One was in connection with his visit to Stockholm to receive his prize in 1965. As mentioned in Chapter 7, I was invited to a private dinner in Gard's home to honor Lwoff and at the coffee after dinner we moved to the neighboring house where Malmgren was the host for both Monod and Jacob. My last meeting with Jacob was in 1999 when in my function as Permanent Secretary I hosted him at the Royal Swedish Academy of Sciences as one of the representatives of the International Human Rights Network of Academies, of which he was a co-founder. This meeting was honored by the presence of Her Majesty the Queen of Sweden. A photograph of some of the founders including Jacob was shown in my previous book about Nobel Prizes[3]. In contrast to my few encounters with Jacob, Klein knew him much better. This is apparent from his, in fact poetical in its own right, foreword to Jacob's abovementioned book. He had known and admired him for some 30 years at the time of writing. His impression until then was summarized (translated from Swedish) in the following way: "I saw only the incomparable scientist in front of me — one of the best biologists of his generation." The book revealed that there was so much more to learn about him and this was brought to light by the "... unexpected vigor and intimate cordiality." Jacob had been endowed with a unique talent to formulate his thinking and express his usually hidden emotions. His use of word was admirable — "How would it be possible to reconstruct worlds which have lost the uncertainty about their future and become past time." Sometimes he invented aphorisms "I love fixed ideas on conditions that they change." Let me in all humbleness give a summary presentation of Jacob's early life and his development into an admired scientist.

A Single Child Exposed to a Rich Environment

Jacob was born in 1920 in Nancy and when he was three years old the family moved to Paris. He was the only child. On both sides of his family he had a Jewish heritage. His mother, a nurse with experience of the horrors of the First World War, came from a family of highly assimilated Jews. The relationship between mother and child grew to become very close. His grandfather on this side meant a lot to Jacob and paid considerable attention to his grandson, in particular ensuring that he received a classical education. If he could not give an immediate answer to the questions from his inquisitive grandchild he simply told him that he would search for the relevant information. The grandfather had advanced insights into mathematics and had settled for a military career. This was highly successful and he ended up being the very first four-star general of a Jewish background in the French army. It can be assumed that Jacob's deeply rooted feeling of national pride and his capacity to mobilize exceptional courage in situations of military combats was markedly encouraged by his contacts with his grandfather. On his deathbed Jacob was told that after a human being has left life there is "Nothingness. My only hope is you. You and the children you will beget." Jacob must have taken these words to his heart and carried them with him into the future. On his father's side there were no academic traditions. He was a real estate business man, who enjoyed presenting pictures of his previous experiences in life in a way that fascinated the precocious child. The father was driven by a sense of duty and followed the Jewish traditions. Hence Jacob had his Bar Mitzvah, but he seems to have kept some analytical distance from the events in the gender-separated environment of the synagogue. Religion did not come to play any role in his life. His school, Lycée Carnot in Paris, provided an intellectual environment suiting the competitive and talented young man, but in general the atmosphere of the school did not make him happy. Although not athletic he could defend himself with vigor when bullied by anti-Semitic remarks.

Jacob himself divided his life into separate phases; his carefully nursed childhood; his self-assured teenage years, full of confidence in himself and a certain preference for left-wing ideologies, but somewhat insecure in relation to the opposite sex; the years of war as a member of de Gaulle's independent forces, completely out of touch with his family; the intense and completely absorbing association with science — an impressively successful career in a short time; and the years of mature and full life with time for development of independent philosophical reflections on the logic of life and its possible and actual evolution. The baccalauréat (high school) examination took place at Sorbonne and was passed with honors. Originally he planned to enter the École Polytechnique, because he liked mathematics and physics and had performed well in both subjects. He had a likable uncle on his mother's side who was a physician. The contacts with him could have been one of the reasons for his final choice of medicine for his academic studies. His fantasy circled around the heroism of surgery, but his uncle did not encourage him to pursue this speciality. According to his uncle medicine at least required some sharpness of thinking and certain finesse, but surgery had none of that. Jacob did not accept this simplified description and asked his uncle to arrange for him to be present at an operation. This was arranged and Jacob came out of that with the firm conviction that surgery was his future. After he had entered medical school he learnt that there were a lot of facts to take in, but this came easily to him. He became acquainted with the empirical nature of medical knowledge and he also experienced the strict hierarchical structure of the medical establishment; those of a higher rank were referred to as "Monsieur" and those of a lower rank "Vieux (old friend)" or "Mon vieux (old chap)."

The Young Medic and the Gehenna of War

During the time of his medical studies frightening things were happening in Europe. "Peace was in throes." The warm and tender world of his childhood rapidly went off into the distance. On top of this his mother developed a serious illness. The approaching war was relentless as was the outcome of his mother's cancer. Coinciding with the culmination of the German offensive into France she died in June 1940. Jacob had just finished the second year of his medical studies and turned 20. Together with some friends of the same age from the faculty he set out southwest through a France that was collapsing. Everything was in chaos. Over the radio they heard Marshal Pétain instruct the French to lay down their arms. There was emotional turmoil amongst the group of young men. They were greatly upset and felt an urge to leave France to fight for their country. In the back of Jacob's head he could hear his maternal grandfather's voice. In Arcachon he met his uncle, the physician. He recommended Jacob to try to get to England and not to North Africa. But how would this be possible? Together with his friend Roger, Jacob managed to sneak into a group of Polish soldiers and unidentified, via a small fishing boat, to reach a Polish ship that finally ended up in England. During the voyage Jacob got to know about the plans that Charles de Gaulle had to form an army in parts of France not occupied by

President Charles de Gaulle.

the Germans, in particular North Africa. At the time he was based in England, where he began to assemble his army.

Jacob wanted to become a soldier and be a part of this project. After many difficulties he and his friend were enrolled in the Free French Forces and again because of the influence of his grandfather they chose to become artillerists. Dressed in his new uniform Jacob strolled the streets of London, but soon a much more brutal reality emerged. However, Jacob was not to become an artillerist. There was a shortage of doctors and because he had finished his two years of medical school he was mobilized for medical duties. With time the army unit grew to some four thousand soldiers. This was the time when the young soldier's world-view was formed. He ended up fusing the feeling for nationalism inherited from his grandfather and the naïve socialism represented by his father, to a certain extent incompatible attitudes. De Gaulle appeared to be a remarkable leader who made a deep impression on the young soldiers. His experiences of conflict in the coming years in various locations in North Africa — Fezzan, Libya, Tripolitania and Tunisia — left major imprints on the maturing young man. He contributed heroically on a number of occasions. As an example, he carried a wounded fellow soldier on his shoulder behind enemy lines, thereby saving his life. On one occasion he was himself slightly wounded. These wartime experiences included everything from the cruelest exposure to the hell of battle to occasional sweet romances. Here we shall telescope time, referring to *The Statue Within* and fast forward to the return of the allies to French soil in 1944. During this time Jacob lost his close friend Roger, who had been killed in Chad when an assegai — a short javelin — had pierced his heart.

As one of the last units of the Second French Tank Division, at the time an integral part of the American Third Army, the medical unit to which Jacob belonged was finally able to board the flat-bottomed boat that would take him and his fellow-countrymen back to France. It was an overwhelming experience to join the huge number of other similar boats approaching the sandy beaches, not in the stillness of a morning but accompanied by the sounds of airplanes and of artillery in the distance. The emotional exposure was intoxicating, but soon brutal reality set in. After one week the unit to which Jacob belonged received an order on August 8th to move towards Le Mans. The camouflage nets were removed from the cars of the train and they were just getting started when the sound of German Junker planes was heard. The sound intensified and soon bombs started to fall on the neighboring fields. Jacob and the sergeant accompanying him rapidly left their car and sought protection in a neighboring ditch. The Gehenna-like situation got worse — whistles turned into thunder, light flashes lit up the darkness, there were cries of wounded soldiers. Jacob was hiding motionless in the ditch. Then he saw close by Lieutenant B. with a twisted face lying by his car bleeding profusely. Jacob left his protected position and moved close to him — "Don't leave me." Both of them pressed themselves into the ground and then the thunder increased, the ground started to shake and clods started to fly through the air. Then Jacob felt an impact and soon realized that he was bleeding and could not move his right hand. He tried to raise himself but then the pain hit him and he lost consciousness. After this there followed many months of care at different hospitals, and repeated operations, and it took time before he could gradually orientate himself and start to assimilate the world around him. Jacob detested the complete dependence on others and being enclosed in an extensive plaster cast, but he was of course also grateful for all the efforts made to help him heal the extensive damage he had suffered.

And still he felt the final return to Paris after three days and nights very cumbersome, with his train journey interrupted by alerts caused by the ongoing war with the allies pressing towards Paris the most humiliating. His dream, after all the years in Africa, coming back to an independent France was shattered. He had imagined the many ways he would be part of the return to a free Paris, but being carried on a stretcher was not one of them. He could follow the developments by radio and he could finally after the many years of no contact see his father and other relatives and friends. Piece by piece his originally extensive cast was removed and bit-by-bit his different physical faculties returned. A lot of shrapnel had been removed in repeated operations, but many metal splinters remained in his body throughout the rest of his life. This was a cause of occasional pain and caused him problems in later life when sitting too long in a fixed position. However, he was finally able to progressively return to a life outside the hospital. But he would have to come back repeatedly for follow-up operations, one of which took place on May 8, 1945, the day when Germany finally capitulated. Not even that great day could he be a part of! Walking the streets of his beloved Paris he could reflect on his future. He was grateful for having been given a chance to return to life. Out of his 40 officer comrades in the 1940 Company in Dakar, only three managed to return home. He was one of them!

Back to the School Bench

It was time to restart his medical studies five years after he had left them in 1940. The return to the school bench and all the exams were managed with variable success. After this followed a number of periods of clinical practice, but the question remained what specialty he should focus on when he could not, because of his physical limitations, become a surgeon. He was 26 years old living on his accumulated salary as a lieutenant during his long period of hospitalization. Should he try something completely different, become a journalist, a politician?? He lived in a world of arrogance and shyness. Eventually he became involved in a small scale factory producing penicillin. He learnt to grow bacteria and was even able to generate some new information in his experiments. He accumulated a sufficient amount of new knowledge on the moderately important antibiotic tyrothricin to produce a dissertation and receive his diploma proving that he had become a medical doctor. He even participated in a congress of microbiology in Copenhagen in 1947, only to realize that scientific research was not for him. He knew too little! So of course he started to read the rich world literature providing insights into the human condition — Camus, Proust, Kafka, Dostoevsky, Faulkner, etc. but also books on science, J. Huxley on evolution, Brachet on embryological differentiation, Schrödinger on a physicist's perspective on life. His mood alternated between apathy and nervousness. He walked his Paris — "It was the hour when the colors started to hesitate, slowly releasing themselves from the sun and blending with the vastness of the night." And then everything changed. He found the woman of his life, Lise, a pianist, who with her background was able to widen his insights into music and cultural enrichment in general. After a short acquaintanceship they married.

Through a new acquaintance Jacob learnt that it was possible to move from medicine to biology. His new friend told him about his work with Ephrussi, whom we have already met, studying mutations in yeast cells. However, again it was realized that additional knowledge would have to be accumulated and that required a return to the school bench. The decision was taken, biology it would be, but what kind of biology? The soon-to-be 30 years old Jacob established the appropriate contacts with leading science administrators in the field. Upon his third contact, which was with Professor Jacques Tréfouël at the Pasteur Institute, he got some encouragement. He would receive a scholarship and start learning about bacteriology, virology and immunology the coming autumn, 1949. Then he could start at one of the laboratories at the Institute. His choice was biased, since he had met Lwoff already when he worked with antibiotics and also Monod, but in a different context, namely at a debate about Lysenkoism. Monod had a lower rank in the academic hierarchy and was younger so Jacob tried to contact him first. However, Monod immediately sent him to Lwoff, the head of the department. Thus it came that Jacob made his first visit to the famous attic, but the contact was not encouraging. After Jacob had presented his ignorance and his ambitions Lwoff simply told him that his laboratory was full and that there was no vacancy. The stubborn Jacob did not give up. He made repeated visits and a few months after the first contact he found Lwoff's "… glance bluer than normal, his head-shaking more intense, but his reception more heartwarming." And then he told Jacob "I have just discovered the induction of the prophage." And the answer was "isn't that interesting" with Jacob thinking "what is a prophage, what does it mean to induce" and a turmoil of other confusing thoughts born of ignorance. When the question was raised "Would you like to work on the phage?" there was of course only one answer, "That is exactly what I want." The embarrassed and baffled Jacob was instructed to come back after the vacation and start in the laboratory on the first of September. A new life was going to start and it would lead to a remarkable development of an influential scientist over the coming ten years. Jacob would become one of the founding father of the emerging discipline of molecular biology, pioneering the field of gene regulation.

What It Meant to Become a Scientist

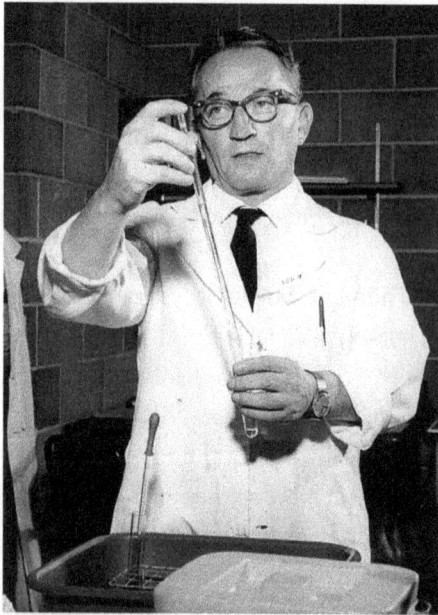

Sol Spiegelman (1914–1983).

It was a completely new world that Jacob was confronted with. He was eager to learn its culture — its rules and patterns of interactions, its lore and language. One of his first experiences was to listen to a seminar given by Sol Spiegelman, a particularly narcissistic geneticist. He later became very influential, not least because of his introduction of nucleic acid hybridization techniques to compare the genetic information in different kinds of nucleic acids. Klein has described him as a person running to the top of a mountain and then announcing "I am here!" and this agrees with my own impressions of him. Jacob was amazed by the intensity of the attacks by the eight to ten scientists participating in the conference. Teasing, provocations and picking at each other were accepted behavior. The most Olympian was Monod, who in flawless English attempted to convince Spiegelman that he had not asked the right questions and that he should have done different experiments. Other scientists present made their separate attacks. Lwoff guided the session in an unperturbed manner. Spiegelman enjoyed making his counterattacks. Later in the evening the discussions continued at a good French restaurant with a complete pulverization of the new theory that Spiegelman had proposed at the end of his seminar. Jacob was amazed at the search for intellectual dominance of the gathering of scientists. He was perplexed and fascinated at the same time, but in fact he had missed most of the argumentations because of his limited acquaintance with the English language. This was soon to be remedied, but it took somewhat longer time to become fully acquainted with the hands-on work in the laboratory.

The contacts with Lwoff were colored by a certain paternal distance. Jacob addressed him as "monsieur." However, Monod did not like to be called "monsieur." Jacob had to choose between Jacques or Monod. He preferred the former. On one occasion the shy Jacob asked Lwoff about the possible importance of joining some of the courses at the Sorbonne. The answer was, "If you are aiming at a university career, you should definitely not come here." He then added "Don't worry your head about this. Do the experiments and the rest will follow suit." A few weeks later when Jacob showed him his first results Lwoff said with a smile, "There you are. Research is first and foremost a matter of flair. But also about strategy and tactics. Character flaws are something I have come to understand that you lack, judging from your achievements during the war." Lwoff had become Jacob's father figure. Progressively

The young Jacob and Lwoff at Cold Spring Harbor in 1953. [Photo from Karl Maramorosch.]

Jacob learnt that a crucial part of experimentation was to first formulate a question — a hypothesis — which could then be experimentally tested. The secrets of biology were unraveled by a mixture of fantasy and reality. The bacterial system was ideal as an object of experimentation because of the speed of replication of bacteria. An experiment initiated in the morning one day often provided results the following day. Altogether Jacob felt that he had come to the right place

All three scientists in the Pasteur laboratory at the time of their Nobel Prize. *From left*: Jacob, Monod and Lwoff. [© Institut Pasteur]

at the right moment. A unique friendship developed furthered by the luncheon meetings in the attic, between the three strongly profiled scientists.

Lwoff's laboratory offered a unique academic environment in France in that the major respect for it came not from fellow French scientists but from coworkers of foreign nationalities. Important members of the American phage group, Seymor Benzer and Günther Stent came for a longer or shorter stay. Élie Wollman had just returned from a two-years stint in Delbrück's laboratory at Caltech in Pasadena. Lwoff had suggested that Jacob should study the occurrence of lysogeny in another kind of bacterium than the one he had used. The pus bacterium *Pseudomonas pyocyanea* was chosen. Out of thirty different strains studied, ten were found to carry prophages as identified by UV irradiation. Jacob was also able to demonstrate that some bacterial strains carried two prophages, only one of which was inducible. Only three months after entering the laboratory Jacob submitted his first scientific article. Later on he and Wollman together switched to studies of *E. coli*. There were two reasons for choosing this alternative system. One was that the Lederbergs together had demonstrated transfer of genetic material between different strains of this bacterium and the other that Esther Lederberg had discovered that bacteriophage *lambda* could establish a state of lysogeny in the K12 strain of the bacterium. The idea was to use genetic tools to examine in more detail the state of lysogeny.

During 1951–52 Benzer and Jacob shared an office. There were no close contacts since they had completely different working hours. Jacob, who had a growing family with a son born in 1949 and twins to be born in 1952, a boy and a girl, generally worked regular daytime hours. Benzer had completely different habits taking advantage of the freedom of scientific pursuits. We have met him briefly before[3] since in 1962 there were two nominations for Crick together with Benzer for a prize in physiology or medicine. In the coming years until 1965 there were no more nominations of Benzer and thus he was not subjected to any review. The reason for the joint nomination with Crick was that Benzer around 1960 had generated over 2,300 mutants in a single gene, rII in the *E. coli* phage T4. Using this large array of mutants he had proven the linearity of this gene by mapping their exact positions. This was important work.

Benzer was a physicist by training. Like Picasso his scientific life could be divided into three distinct periods. He obtained his Ph.D. from Purdue University. In this work he studied metallic germanium, a metal of critical importance for the development of semiconductors for transistors. He then became interested in biophysics and was encouraged by Luria to take the phage course at Cold Spring Harbor, which he did in 1948. This led to somewhat more than 10 years' work on phages which included his two years at the Pasteur Institute and his time with Crick and Brenner in Cambridge in 1957 and 1958. In the early 1960s Benzer shifted research field one more time. He became excited about embryological differentiation and after another course at Cold Spring Harbor he entered the field of fruit fly research. Very soon he advanced his science to become a leader also in this field. Important discoveries were made, but it is beyond the scope of this book to present these in more detail. His important achievements in these different fields were recognized by many prizes, but not by a Nobel Prize. It is a truism to note that there will always be a large number of impressive scientists who come close but who in the end will never be selected to "make a trip to Stockholm." However, this is not completely true in the case of Benzer. In 1993 he did in fact come to Stockholm but this was to receive the Crafoord Prize in biology.

The Crafoord Prize was established in 1980 by the promise of a yearly donation by the Swedish industrialist Holger Crafoord, the inventor of the artificial kidney that led to the establishment of the

Holger Crafoord and then permanent secretary of the Royal Swedish Academy of Sciences, Carl Gustaf Bernhard, a neurophysiologist presented already in Chapter 1, discussing the arrangements for the forthcoming Crafoord Prizes.

Gambro Company, and his wife Anna-Greta. It was decided that each year the Crafoord Trust in Lund should provide a sum of money corresponding roughly to half a Nobel Prize, plus overhead for the prize work, to recognize discoveries in disciplines *not* covered by the traditional Nobel Prizes. The recipient of the yearly donation should be the Royal Swedish Academy of Sciences, which is responsible for the Nobel Prizes in physics and chemistry. The selected fields were mathematics, astronomy, geological sciences, biological sciences, with an emphasis on ecology, and also a particular branch of medicine, polyarthritis research. The latter discipline should only be recognized if the class of medical sciences of the Academy could identify worthy recipients of the prize, and it was specifically included since Holger Crafoord himself had severe arthritis. In 1993 the selected discipline was biosciences and in this year the famous British evolutionary biologist William C. Hamilton and Benzer were awarded a prize. Hamilton was cited for his contributions in theoretical biology leading to the formulation of the concept of reciprocal altruism, whereas Benzer was awarded for his groundbreaking genetic and neurophysiological studies of the fruit fly. The prize was handed over to Benzer by His Majesty the King. The Crafoord Prizes are less well known than the Nobel Prizes, because they have been awarded only since 1982 and furthermore each individual discipline is recognized only about every fourth year. However, the list of scientists who have received this prize is impressive. Woese was a towering recipient as mentioned in Chapter 5. After this digression it is time to return to Jacob.

In 1952 Jacob first traveled abroad as an active scientist. There was a meeting on viruses arranged by the Society for General Microbiology in Oxford. The organizers were two colorful virologists, Bawden and Pirie, whom we have met previously[4] and in Chapter 5. They had been the first to demonstrate, as early as the late 1930s that tobacco mosaic virus particles contain about 5% RNA. It had been planned that Luria should

Seymor Benzer (1921–2007) on the occasion of receiving the Crafoord Prize from His Majesty the King. [Photo by Lars Falck.]

be the leading keynote speaker, but he did not receive a passport for traveling abroad because of the prevailing era of McCarthyism in the U.S. His student Watson had been delegated to present his data, but he instead preferred to give data from a recent letter by Hershey, which described the paradigmatic experiment he had performed together with Martha Chase as mentioned in my previous books[3,4]. They had labeled the phage nucleic acid with radioactive phosphorous and its proteins with sulfur. To their great surprise it appeared that only the nucleic acid entered the bacteria. The phage protein shell was separated from the target bacteria by gentle treatment of the infected culture by use of a Waring blender. At later meetings the same year Jacob got an opportunity to meet the many important scientists in the field, not only Luria but in addition the pontiff of phage genetics, Delbrück. Jacob also realized that he had a competitor in Bertani, working in Luria's laboratory at the time as already mentioned in Chapter 7.

During the autumn of the same year Lwoff had arranged a meeting for the international phage group at Royaumont in the northeastern part of France, a charming place as I can testify from my own experience. All the main actors in the field participated in the meeting. Jacob was able to enlarge his professional network and also describe his first results to a wider audience. There was one problem, however, and that was that as yet he did not feel comfortable using English. A colleague a physicist from Geneva and a member of Delbrück's phage group at Caltech, Jean Weigle, promised to help him. Jacob was allowed to rehearse his performance in his hotel room in the presence of Weigle. Everything seemed to be under control, but when Jacob stood at the rostrum to give his speech he could not find his manuscript! He had to memorize the text and extemporize. When it all was over Weigle came to him and said "I am sorry that I played this prank on you, but I thought this was the best way of teaching you that one should never read a paper. You did excellently." A few months later Jacob was on his first trip to the U.S. He had been invited by Delbrück himself to participate in a meeting at Cold Spring Harbor. His career as a scientist was indeed developing very rapidly. At the meeting he heard the great news about the structure of DNA from Watson himself. A revolution in molecular biology was on its way, but it took some time for the momentous shift towards prioritizing DNA over proteins as the central actor in genetics. On the side Jacob had in fact taken some formal courses at the Sorbonne and in May 1954 he presented his thesis at the same institution. He became formally licensed to become a scientist, but the truth is that he was already well on his way.

The Sexology of Bacteria

Jacob and Élie Wollman decided to make an attack on the phenomenon of transfer of genetic material between bacteria, an issue that had become very hot at the time. It had been found that the transfer of such material in certain systems could only go in one direction, from one particular strain of bacteria to another — from a "male" to a "female." This system was used to examine the relationship between the prophage genetic material and the bacterial host cell genome. It could then be demonstrated that a transfer of the lysogenic trait could only occur when there was an effective transfer of genetic material from a male to a female, but it did not work in the other direction. Furthermore it was discovered that the prophage was often expressed in the recipient bacterium leading to a lytic destruction. These observations demonstrated the association between the phage and host cell genetic material. Following the French cultural tradition the phenomenon was referred to as the *erotic induction* of the prophage, but when presented in a scientific publication it was rebaptised to *zygotic induction*.

Jacob's wife Lise was not too excited about the American Waring blender he had given her after his first trip to the U.S. Therefore he took it to his laboratory. It came to unexpected good use. The idea was to let male and female bacteria interact for a predetermined time and then separate them by the mixer — by the French terminology of course a *coitus interruptus*. This turned out to be a very simple and informative experiment. The mixer did not only separate the two strains of bacteria, it also cut the male chromosome in a time-wise fashion, as it was introduced into the female. This made it possible to follow how one property after the other was transferred with time. A simple proof that the genes in the genome were linearly ordered was obtained. These different observations put a new light on the phenomenon of conjugation. Monod disrespectfully referred to these studies as "spaghetti"-experiments, a term not to Élie Wollman's liking. In fact this simple approach, further developed by labeling the DNA with radioactive phosphorus, allowed the first bacterial gene maps to be constructed. The newly-developed techniques of electron microscopy also became important in the studies and finally it could be demonstrated that unexpectedly the *E. coli* chromosome was not linear but circular. This structure was postulated by Jacob but experimentally confirmed by John F. Cairns, an influential molecular biologist of British origin, for extensive periods active in Australia and in the U.S.

It was during this time that Monod received his chair and moved to the bottom floor of the building. Jacob and Monod loved to challenge each other at the blackboard, but this now took place in the latter's new office. As described by Jacob himself[1, 2]:

> I had come down to describe the fresh results from the morning. Results that were still uncertain. Only seen once. But I needed to discuss them, to share my excitement. In order to think and move on I needed to debate them. Test ideas and let them bounce back on me. And no one was as good at this game as Jacques. He listens. Looks at me. Let his cheek rest in his hand. Walks to the blackboard. Draws a scheme. Comes back to me. Ask brusquely if I have considered making some confirmatory experiments, without which my findings do not have the slightest value.

It is easy to enliven this scene and it would be repeated many times over the coming 8–10 years.

Jacob rapidly enlarged his contacts in the international scientific community by presenting his and Élie Wollman's new data in the U.S. and in the U.K. They were generally well received. At a meeting in London in 1957 Jacob for the first time learnt to understand the greatness of Crick. He rapidly came to appreciate his unique capacity to synthesize data accumulated mostly by others and to give a direction to the forthcoming developments. Step by step the central dogma was established, but it took time to appreciate the existence of different forms of RNA. Some critical role of RNA in the synthesis of proteins had been suggested already by the early experiments by Torbjörn Caspersson and Jean Brachet in the late 1940s, but it was first when effective isotope labeling techniques were developed that different kinds of RNA could be identified. During the 1950s microsomes later named *ribosomes*, because of their high content of RNA, were morphologically identified in the cytoplasm of nucleated cells and were shown to be central in protein synthesis. Originally Watson and Crick offered a very vague speculation that protein synthesis might take place in the major groove of the double-helical structure of DNA they had presented in 1953. However, since in eukaryote cells DNA is located in the cell nucleus, where it also is replicated, and protein synthesis occurs in the cytoplasm it rapidly became clear that there must be some other explanation. Crick postulated that besides the protein synthesizing ribosomes in the cytoplasm some kind of adaptor molecules should also be present. Progressively it was identified that the critical actors were low molecular weight RNA, originally referred to as soluble RNA and later named *transfer RNA*. But it remained to identify how the genetic information could be transported from the nucleus to the cytoplasm.

An Interrupted Movie Viewing

In the summer of 1958 Jacob had received a number of flattering invitations to present the data emerging from his group, including the prestigious Harvey lecture in New York. He tried to focus his intellect on preparing an exciting and attractive presentation of his new data, but inspiration did not come to him. He walked restlessly in his study and his wife Lise therefore proposed that they should go to the movies. This he accepted but the movie did not catch his attention. His thoughts continued to meander when suddenly in the middle of the movie it struck him that there might be hitherto unappreciated parallels in the phage system he and Élie Wollman were studying and the lactose enzyme induction system in *E. coli* that was the center of all Monod's involvements. Why had this not struck him before? In both cases it was a matter of controlling the gene expression of DNA, the single central actor in the process of information transfer. Jacob himself described this moment of conceptual discovery in the following way[1,2]:

> These so far rough, sloppily sketched, badly formulated hypotheses pushed and jostled in me. Hardly had they emerged when I was filled with an intense pleasure, a wild satisfaction. And in addition a feeling of power and strength. As if I had climbed a mountain, reached a peak from where I could see far away an inviting landscape. No longer did I feel mediocre or even mortal. I needed air.

Lise noticed his state of mind with curiosity and asked the logical question "Have you had enough? Shall we go?" A soon as they had reached Rue Montparnasse there came the urge to tell. Jacob's defenseless wife had to try to assimilate a number of bewildering, conjectural thoughts, some logical and comprehensible, others more speculative and vague. He needed to communicate with his closest scientific partners, Lwoff and Monod, but this was France and August. Thus everyone was away on vacation — Lwoff painting in the countryside and Monod sailing in the Mediterranean. In early September Jacob, who had just had a sleepless return flight from New York, entered Monod's office, tense and excited. He presented his new ideas. The reaction was what Jacob had anticipated knowing Monod's sharp and inquisitive intellect. A huge number of counterarguments were fired at him and the hypothesis seemed to be in shambles. But the next day it was another close collaborator who greeted him. He showed the other side of his Janus face. Monod's schizophrenic personality has already been introduced in the previous chapter. Jacob described him on the one hand as "Jacques," the empathic and generous person asking the right questions using his incisive intellect, and on the other hand "Monod," a narcissistic, dogmatic, domineering person, who did not hesitate on the spot to turn from friend to foe, by a sharp comment. Taking the Jacques attitude the dialogue between the two could restart and now the dialectics opened new insights and potential syntheses in a good Hegelian tradition. Theses and antitheses challenged each other and a new thinking emerged, new hypotheses, which could be tested in the laboratory, were formulated. Possibilities for examining new perspectives on gene regulation emerged. This renewed and intensified collaboration turned out to be a turning-point in Jacob's rapidly evolving career. He had just finished the long time and fruitful collaboration with Élie Wollman, who had traveled to Caltech in Berkeley to work with Stent. The two of them had summarized their joint findings in a book entitled *Sexuality and the Genetics of Bacteria*[5].

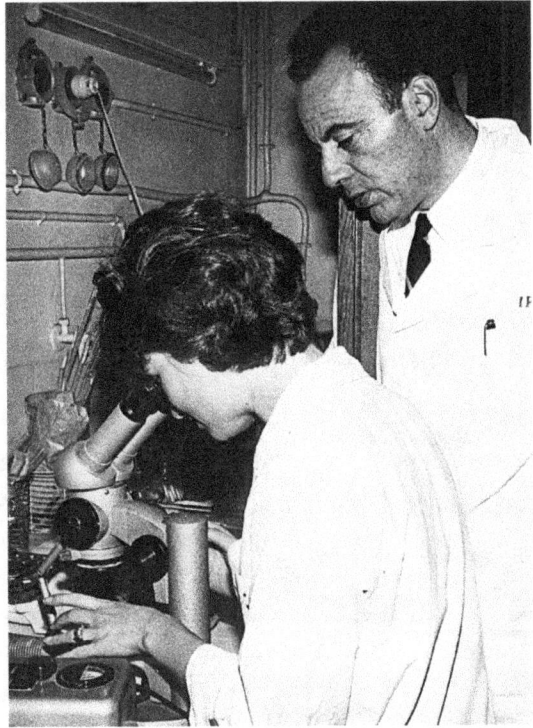

Jacob in the laboratory. [From Ref. 24.]

The fresh new start with Monod led to a dynamic and unique collaboration that lasted for some seven years and resulted in 22 joint publications, most of which had a seminal value. A critical component in this work was the PaJaMo experiment, briefly alluded to in the previous chapter. So let us discuss this publication in more detail. First a few words on Arthur Pardee, the first-named

Arthur Pardee, the critical collaborator in the PaJaMo experiment. [From Ref. 25.]

author of the publication. He was one year younger than Jacob and had received his training in science with Pauling. Early on he made his mark on science by being a co-discoverer of the already mentioned microsomes (ribosomes) in the early 1950s. His stay in Paris, which was a sabbatical from the Virus laboratory in Berkeley, California, meant an important step forward in his scientific career, which later came to involve many different subfields of the growing discipline of molecular biology. During the latter part of his scientific life he was active at the Dana-Farber Cancer institute and at Harvard Medical School, both in Boston.

The genes y, i and z identified in the Lac region of E. coli were referred to in Malmgren's and Gard's reviews cited in the previous chapter. y and z were structural genes producing proteins with identified catalytic functions in the bacteria, whereas i was a regulatory gene. In the classic experiment one additional structural gene had been identified, the gene a for the enzyme transacetylase. The key outset for the PaJaMo experiment was that male bacteria carrying z^+ and i^+ were conjugated with female bacteria lacking functional forms of these two genes. After mating the time-wise appearance of beta-galactosidase was measured. This is a very clever experiment since in their original state none of the bacterial strains could produce the enzyme; the male strain because of the absence of an inducer and the female because of the lack of a functional gene for the enzyme. After crossing enzyme synthesis began within a few minutes but after somewhat more than an hour the synthesis stopped. At this stage the inducer was added and the synthesis was resumed. There was also one additional variant of the experimental conditions[16]. When Pardee had returned to California he performed some important studies with his student Monica Riley. They labeled the DNA of the male bacterium with ^{32}P prior to letting it conjugate with the female bacteria. DNA labeled in this way had a shortened life span. As a result of the decay of the isotope there were breaks in the DNA which interrupted its function as a mold for synthesis of RNA.

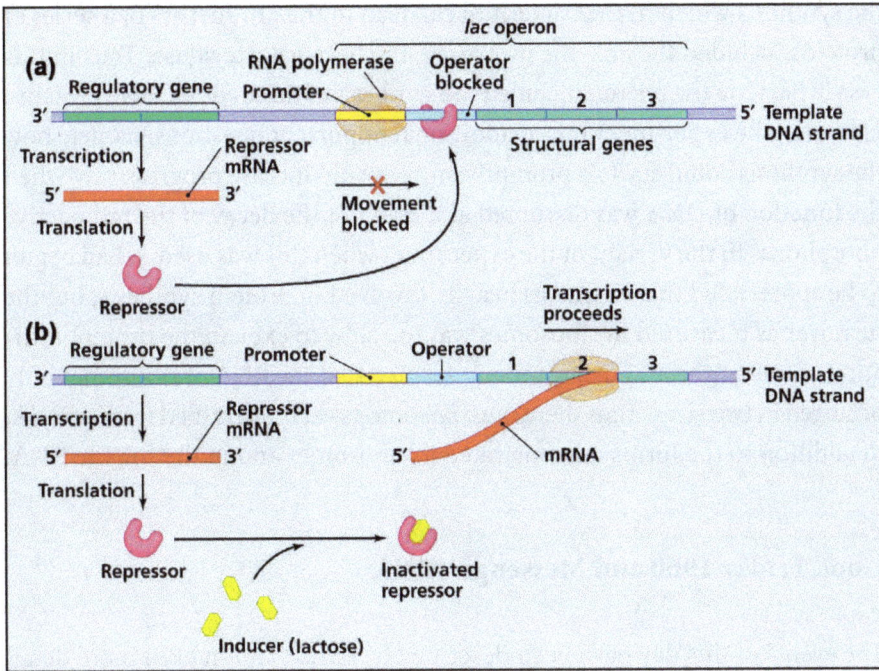

(a) In the lac operon, the operator (structural) gene is blocked by a repressor protein when there is no lactose present. (b) In the presence of lactose it acts as an inducer by binding to the repressor protein. This leads to that this protein no longer can attach to the operator, allowing instead RNA polymerase to proceed, leading to transcription of all three structural genes, 1, 2 and 3. [From Ref. 26.]

Hence the outcome of the experiment was that production of the enzyme beta-galactosidase in the female bacterium to which the labeled DNA had been transmitted ceased after some time. It was these findings that later led to the identification of a previously unidentified form of RNA, a messenger to be further discussed.

The sequence of experiments performed by Monod and Jacob and their collaborators in the late 1950s and early 1960s led to the final insight into the existence of negative regulatory mechanisms already discussed in the previous chapter and they finally led to the formulation of the *operon* concept. Although I generally do not use diagrams in this kind of book an exception has been made here, simply because it would take an extensive text to present a description without the aid of a picture. Briefly what the diagram illustrates is a repressed state of the operon on the top and an induced state at the bottom. The repression which is a natural state is broken when the inducer reacts with the Lac i product because of the added lactose. Blocking the repressor allows an activation of the operator and a formation of RNA that can direct the synthesis which in the bacterium can lead to the production of a series of proteins included the enzyme measured, the beta-galactosidase. Thus in this case in bacteria the operon included the synthesis of three consecutive proteins. However, the experiment raised additional important questions, namely how the synthesis could start so promptly and why did it cease progressively when the function of DNA was disrupted as a result of the decay of the radioactive phosphorus in the variant of the experiment when this was used. It had begun to be appreciated that RNA was heavily involved in protein synthesis, but the turnover as measured in ribosomes was too slow to explain the prompt initiation of the synthesis. Furthermore it seemed that the RNA of ribosomes only occurred in two sizes. Thus there *must be* some as yet unidentified form of RNA, in addition to the forms demonstrated in ribosomes and in the soluble RNA.

Good Friday 1960 and Messenger-RNA

The events of this day have been described by Crick[6] and in even more detail in the already referred to book by Judson and of course they are also referred to in Jacob's biography[1,2]. He was visiting Cambridge and a range of local as well as other visiting molecular biologists took this opportunity to discuss the details of the PaJaMo experiment. At the time Sydney Brenner and Crick were planning experiments to isolate different ribosome populations and see if they produced different kinds of proteins. However, technological obstacles prevented them from performing the experiments and in addition there were conceptual objections. The ribosomes seemed only to contain two different size classes of RNA, as mentioned, and how could this tally with the fact that proteins had varying lengths? The "grilling" of Jacob started in the morning. Since the laboratory was closed on Good Friday the meeting took place in the Gibbs building at King's College of which Brenner was a Fellow. Six or seven people including influential scientists like Leslie Orgel (a qualified organic chemist, good friend of Crick) and Ole Maaloe (a Danish microbiologist of high international standing) participated. In parenthesis it can be added that already six months earlier Jacob had given a presentation at a meeting arranged by Maaloe in Copenhagen. It was a small but highly select group; Crick, Watson, Monod, Benzer, Brenner and even the famous physicist Niels Bohr. During all the presentations Watson read his newspaper and therefore when it was his turn to give a presentation, as a sweet revenge, everyone in the audience took up a newspaper and read it. In a presentation on this occasion Jacob did in fact hint at a possible need for a labile informa-tion-carrying molecule, substance X, during protein synthesis. But no one paid attention and Watson continued to read his newspaper.

Before returning to the April meeting in Cambridge some addi-tional comments on Orgel may be in order. He received his early training as a biochemist at Oxford. He was one member of a particular group of scientists including Brenner, Jack Dunitz, Dorothy Hodgkin and Beryl M. Oughton who traveled to Cambridge in April 1953 to see Watson's and Crick's model of DNA. Everyone in the group was very

Leslie Orgel (1927–2007).

impressed. Later during the 1950s Orgel worked in Cambridge and in 1964 he was appointed head of the Chemical Evolution Laboratory at the Salk Institute in La Jolla and developed a wide network of collaborators not least Crick, after he had also moved to the Institute. He had a particular interest in the origin of life and summarized the state of the art in this field in 1973 in a book entitled *The Origins of Life on the Earth*[7]. Later he was one of the first to argue for the potential importance of an RNA world in the origin of life. He died at the age of 80 in 2007. I had the pleasure of meeting him and his wife a couple of times in Gustaf and Jenny Arrhenius' home in La Jolla. He was a charming scientist and is remembered partly because of Orgel's rules of which number two states "Evolution is cleverer than you are."

Now back to the April meeting in Cambridge with Jacob as the central figure. Crick was leading the discussion and started the cross-examination of him. According to Crick it was Brenner who first had a premonition of the right track to take. He remembered an experiment performed four years earlier by a biochemist from the Oak Ridge National Laboratory, Elliot Volkin, who together with a colleague Lazarus Astrachan had made some particular observations in their studies of the production of nucleic acids during the latent period of phage infection in *E. coli*. They had found a rapid rise in a population of large-size RNA. Similar observations had been made by others, but Volkin and Astrachan took the experiment one step further. They examined the presence of different nucleotides in the RNA and found that the representation did not correlate with the even presence of the four nucleotides as found in *E. coli*, but instead the proportion was similar to the very different representation of nucleotides found in the virus DNA. The discussion concerned whether this RNA could be a precursor of DNA, but this did not fit with the findings being made at the time by Arthur Kornberg as discussed previously[4]. The data remained unexplained until Brenner came up with the idea that maybe there could be an as yet unidentified form of RNA, with a rapid and transitory appearance which could be of importance for transferring information from DNA to the ribosomes, a *messenger*.

In my upbringing Good Friday was one of the most boring days of the year. No movies, no dancing, only a gloomy day characterized by the color purple in the church. However, such aspects were not considered by the heretic Crick family. The parties they arranged at their home the Golden Helix were famous for their great atmosphere, reflecting the enthusiastic and charming host couple. As formulated by Crick himself "the molecular biologists' parties were considered to be the liveliest in Cambridge." The particular party on this Good Friday was a little different. The major part of the group were just thoroughly enjoying themselves. As Jacob has described it "It was a very British evening with gratin à la Cambridge, an excess of sweet girls, a lot of different liquor and pop music." And further: "It was hard to isolate oneself during such a splendid party, surrounded by people crowding around us. People who talked, screamed, laughed, danced and sang. Like on an isolated island we (Brenner was the counterpart, my remark) sat squeezed in at a little table and continued our discussion of our new model." The question was how it might be possible to demonstrate experimentally the presence of a short-lived heterogeneous population of RNA, the conjectural messenger. Many ideas were brought to the table. It was noticed that coincidentally it had already been planned for both Brenner, invited by Matt Meselson and Jacob invited by Delbrück, to spend the coming month of June at Caltech in Pasadena, California. The former of the two hosts was highly respected in the scientific community because of his responsibility for what had become the classic Meselson-Stahl experiment. This experiment, using isotope-labeled N, provided strong evidence for the semi-conservative replication of DNA. It was possible for the two visiting scientists to jointly test their ideas during their brief Californian sojourn. Jacob has described how their collaboration in California developed.

When Brenner, Meselson and Jacob got started with their experiments they expected they would soon strike gold. But everything seemed to be against them. All experiments failed. Disappointed Meselson took off — to get married. A female biologist in the laboratory registered their distress and made some plans to encourage them to think about other things. She took them to a charming part of the California coast, where they could relax in the sand listening to the never quiet Pacific Ocean. Brenner was in a miserable mood. Very exceptionally for him he was quiet. All of us who have had the privilege of getting to know him have taken note of his never-ending verbosity. He can speak on any subject and he is a superb storyteller. When he stays at a hotel it is of course "in a room with a Jew." The mood was truly very low. Why should they continue, why be stubborn, what use was it? The entertaining stories told by their considerate host fell on deaf ears.

Then Brenner cried out and started to jump around like a devil. He ululates "It must be the magnesium! It is our magnesium!" The company quickly jumped into the car and drove back at impermissible speed to the laboratory. And now the experiments worked a charm. An excess of magnesium was needed to keep the ribosomes together and allow them to produce new proteins. The messenger-RNA had finally been identified, but a number of follow-up experiments had to be performed by Brenner when he was back in Cambridge. This experience highlights the sometimes enormous importance of employing the proper experimental conditions. A very small difference in the concentration of one of the chemical reagents used can be decisive. In this case it was the divalent cation magnesium that was critical. In another case it may be something else. In order to control that the proper conditions are used scientists have to pursue often boring experiments in which the concentration of the different reagents and the physical conditions used are adjusted to be optimal to observe the phenomenon searched for. The discovery of messenger-RNA illustrated one more thing namely that unknowingly a seminal discovery may be made simultaneously in two different laboratories.

One of the most exciting discoveries I have made was the identification in 1964 of never before seen projections at the vertices of the icosahedral adenovirus particles. Later on this structure was found to be critical in the first step of virus replication, the attachment of the virus particles to specific receptors on cells. Completely unknown to me the same discovery was made in Cambridge by Robin Valentine and Helio Pereira as already told[3]. In the case of messenger-RNA an independent identification of this previously unidentified RNA was made in Watson's laboratory at Harvard University. Another scientist from the Pasteur Institute François Gros, a specialist in RNA, had gone, at the same time as Jacob, to the U.S. to work in Watson's laboratory. They also found a previously unidentified form of RNA that was a candidate for transferring information from DNA to ribosomes. The results from the two groups demonstrating the existence of messenger-RNA were published back to back in the journal *Nature*[8,9] as referred to in Jacob's Nobel lecture[10]. The story of how Gros and the Harvard group independently discovered messenger-RNA, a term by the way introduced first in 1961 by Jacob and Monod in the famous paper in *J. Mol. Biol.* that year, which we will return to, has been excellently described by Watson in his historically impressive 1962 Nobel lecture[11]. It provided a high quality overview of the early events under the title "The involvement of RNA in the synthesis of proteins."

Gros had discovered that when bacteria were grown in the presence of the atypical nucleotide base 5-fluorouracil, this was rapidly incorporated into the newly formed RNA and caused it to make mistakes in the positioning of amino acids in proteins. Not only were dysfunctional proteins produced but there was also a general shut-down of protein synthesis. This was an argument against the existence of a stable messenger with the inference that it must have a short lifespan. In the late 1950s the abovementioned Spiegelman and his colleagues demonstrated the importance of magnesium for the formation of stable ribosomes and chains of them. Jacob and Brenner using proper magnesium concentrations had discovered the messenger-RNA in virus-infected bacteria, but Gros and his collaborators in Watson's laboratory successfully searched for it in uninfected bacteria by short pulse labeling with isotopes and isolation of the RNA by dissociation of polyribosomes in the presence of low concentrations of magnesium. Not long thereafter Spiegelman and collaborators managed to form hybrids between DNA and the recently identified messenger-RNA proving that they had matching sequences of nucleotides. This made the story complete and also introduced the new technology that allowed a comparison of nucleotide sequences in two populations of nucleic acids, RNA with DNA or DNA with DNA. This technique of nucleic acid hybridization came to good use during the continued development of the burgeoning field of molecular biology. Finally about ten years after the scheme DNA to RNA to protein, which Watson had already written on the wall of his temporary office in Cambridge already in the early 1950s, had been consolidated, with Crick as the number one standard-bearer, into the central dogma; DNA by transcription gives RNA, which by translation gives protein.

Gros has left many imprints on biological science in France during his long and active life. At the time in the late 1950s and early 1960s when both he and Jacob worked at the Pasteur Institute there was a problem. Both had the given name François. This was resolved, according to Agnes Ullmann[12] by referring to Jacob as "le Grand François" because of his more impressive physiognomy. Gros became Director of the Pasteur Institute succeeding Monod in 1976 and in 1991 he became the *secretaire perpetuel* of the Académie des Sciences, a position he retained for ten years. I have had the pleasure of interacting with him on a number of occasions since my time as Permanent Secretary of the Royal Swedish Academy of Sciences partly overlapped with his. Over the years it has been possible for me to make a number of visits to

François Gros dressed in the ornamentary tail dress of the Académie des Sciences in which he served as *secretaire perpetuel* for ten years.

the Institute de France, the charming complex of historical buildings on the left bank of the Seine in Paris. It hosts five academies of which the most famous is the Académie Française, founded in 1635. The Académie des Sciences initiated in 1666 is another of the five academies located at the learned quarters. Some of the buildings are open to the public and a visit to the Bibliothèque Mazarine is breathtaking. Cardinal Mazarin, who was of Italian origin, was Prime Minister of France on two occasions. The precious cumulation of antique books was first collected between 1642 and 1653 by Mazarin's qualified librarian and it remains one of the four most famous libraries in Europe. A visit to the library allows one to travel back in time.

Christmas Eve and a Remarkable Article

The *Journal of Molecular Biology* was started in 1959 and rapidly became the leading journal in its field. The articles it contained generally did not exceed about a dozen pages, but in early 1961 a remarkable exception to the length of an article was made. Jacob and Monod published an article covering 39 pages[13]! The title was "Genetic Regulatory Mechanisms in the Synthesis of Proteins." It was an impressive *tour de force* and provided a great mixture of their accumulated discoveries, including the demonstration of messenger-RNA, which as yet had not been published, and a number of prophetic hypothetico-deductive reasonings. The two scientists had worked on this article for the whole of the autumn and the final version of the extensive paper was finished on Christmas Eve. It represents a milestone in the field of molecular biology. The text elaborates the development of the thinking and experimentation that lead to the model of gene regulation already presented above in the introduction of the operon concept (p. 499). The text is uniquely logical and lucid. It departs from the conviction that the basic mechanisms for gene regulation are the same for simple organisms like bacteria and for the more complex vertebrate organism, like ourselves, motivating the already mentioned famous quip, "What is true for the bacterium is also true for the elephant."

After the Introduction the authors discussed Inducible and Repressible Enzyme Systems, Regulator Genes, The Operator and the Operon, The Kinetics of Expression of Structural Genes, The Nature of the Structural Message and finally Conclusions. In the latter they emphasized "The property attributed to the structural messenger of being an unstable intermediate is one of the most specific and novel implications of this scheme." The final paragraph of the paper read:

> According to the strictly structural concept, the genome is considered as a mosaic of independent molecular blue-prints for the building of individual cellular constituents. In the execution of these plans, however, *co-ordination is evidently of absolute survival value* (my italics). The discovery of regulator and operator genes, and of repressive regulation of the activity of structural genes, reveals that the genome contains not only a series of blue-prints, but a coordinated program of protein synthesis and the means of controlling its execution.

This was written in 1960 and of course there have been major developments since then, not least in the world of RNA. It would take too long to summarize these developments and a reference is therefore given to the 2011 book by James Darnell[14]. A considerable part of this book discusses all the RNA synthesized from DNA that is not translated into proteins, the so-called "non-coding RNA" — in many senses a misnomer. Many classes of such RNA have been identified, but the field continues to grow. Two discoveries, as already mentioned, have been recognized by Nobel Prizes; the prizes in chemistry in 1989 to Altman and Cech "for their discovery of catalytic properties of RNA" and in physiology or medicine in 2006 to Fire and Mello "for their discovery of RNA interference — gene silencing by double-stranded RNA." In spite of all the progress it is obvious that a lot remains to be learnt before we can start to make a coherent interpretation of the comprehensive picture of the balancing mechanisms that keeps a complex organism in order under variable environmental conditions — the homeostasis. Which are the secret mechanism (the hidden algorithms?) that allow this many-instrument orchestra to play without a conductor? To talk about the end of science in this context is hypocritical.

The Nobel Committee Reviews Jacob

Jacob's analysis by the Nobel committee at the Karolinska Institute was of short duration, encompassing only four years, 1961–65. This can be compared with the evaluation by the chemistry committee of Watson and Crick which, because of the surprisingly late first nomination of their 1953 discovery, only extended over three years, 1960–62. Jacob was always nominated in tandem with Monod, and hence was already alluded to in the previous chapter. The comprehensive nomination by Jean Boyer, Paris in 1961 included all three candidates Lwoff, Monod and, for the first time, Jacob. The thorough review made by Gard included Jacob after Gard in a preliminary investigation had promptly noted that he should be included together with the other two in a complete review. The three of them were evaluated one by one and the remarks on Lwoff and Monod were already quoted in the previous two chapters. The last two pages on Monod discussed in some detail the discovery of messenger-RNA. This is somewhat misleading, since although this work is included, prior to the original publication, in the 1961 overview article in *J. Mol. Biol,* it is Jacob's discovery together with Brenner and Meselson. Hence it should have been placed under the heading François Jacob.

Under this rubric Gard first discussed Jacob's early work on lysogeny initiated by Lwoff. He then noted that Jacob soon started to find his own way in particular in the studies together with Élie Wollman examining the progressive transfer of genetic material from "male" *E.coli* bacteria to "female" bacteria. It was these studies that created the first map of a part of the linear arrangement of genes in the bacterial genome. The concepts of episomes and sexduction were then introduced in Gard's review. He also described in quite some detail the data demonstrating that the hereditary material of lysogenic phages was physically associated with the bacterial genome as demonstrated by the experiments showing zygotic induction. The final paragraph under Jacob concerned the genetics of protein synthesis. Very briefly Gard concluded that all the important publications in this field had been signed by both Monod and Jacob, but he noted that "Jacob's part in both the experimental work and the theoretical analysis most likely have been very significant." He then referred back to the discussion of these data in the part of the review that concerned Monod's contributions. Gard's conclusions (translated from Swedish) started:

> Jacob's contributions have made it possible to integrate Lwoff's virus work, Monod's studies of the protein synthesis of the normal cell and his own work on the fertility factor into a wider context and the joint contribution by the group as a consequence has been developed to reach a much wider general biological significance than each one of the individual presentations.

Hereafter in an attempt to summarize and evaluate the results Gard in fact borrowed a citation from Jacob, which he in his turn had adapted from Tocqueville. It read "Generalizations can only give incomplete descriptions; what one gains in understanding, one loses in accuracy. When it comes to a subject like the genetics of viruses there are rich opportunities to be alternatively inexact and incomprehensible." Gard liked this citation so much that he used it in the beginning of his laudatory speech at the prize ceremony. He then continued his summary "The contributions spread over large areas within biochemistry, genetics and microbiology; if one does not have a wide personal experience in all these fields it is difficult to distinguish the main messages among the throng of detailed observations. However, in my opinion the deductions made by the authors themselves in most cases are brilliant and throughout they are characterized by a logical acuity and on the same time scientific imagination."

The continuation of the extended summary mostly discussed Jacob's discoveries, in particular his characterization of episomes. This term was used to define genetic material independent of the chromosomal hereditary material. Following Jacob's schematic relationships Gard showed in a diagram how episomes could be either non-viral or viral and that there were conjectural possibilities for a change of one to the other and he also discussed the relationship to plasmids, independent genetic material that allows the extra-chromosomal exchange of for example antibiotic resistance between bacteria, and (obligatory) viruses. There was a question mark against the potential possibilities of the latter two forms of extrachromosomal genetic material to be converted one into the other. Gard's final conclusion was that Jacob's characterization of episomes was worthy of a Nobel Prize. He noted that Jacob's discoveries in this area would not have been possible without the identification of the fertility factors of bacteria by the Lederbergs and Tatum, the demonstration of the transduction phenomenon by Lederberg and finally the induction phenomenon demonstrated by Lwoff. He then noted that both Lederberg and Tatum had already received Nobel Prizes and also stressed that Jacob could not be recognized by a prize without the inclusion of Lwoff.

In a brief paragraph Gard then discussed the significance of the discovery of messenger RNA. He noted that this finding was interesting and important, but that it did not seem to be of the magnitude that it could be considered worthy of a prize. This is a view that might be contested. Discovering the messenger RNA was in fact absolutely crucial to the definite establishment of the central dogma. After this Gard discussed the phenomenon of repression and it was emphasized that the differentiation of genes into structural, operative and regulatory and also of the need to widen the simplified one gene — one enzyme concept were highly important contributions. He expressed a curiosity about the potential application of Jacob's and Monod's newly introduced concept also on eukaryotic cells. As already mentioned in the previous chapter Gard did not hesitate to declare Monod's and Jacob's joint contributions as (very) worthy of a prize. He finally emphasized that Lwoff absolutely should be included in an award to them. The committee agreed and listed all the three nominees as deserving of a prize.

The situation in 1962 and in 1963 was already described in the preceding chapter. All three candidates were proposed in three independent nominations from Nobel laureates, Cournand, Lipmann and Tatum and the nomination by Cournand was supported by another nomination by Richards, also from New York. The extensive text of the nomination submitted by Cournand, to a large extent followed the disposition of content in Jacob's and Monod's impressive 1961 publication in *J. Mol. Biol.* referred to above. No review was made and the three candidates remained listed as worthy of a prize. In 1963 there was an extensive nomination of Jacob and Monod by Wolff and a brief supplementary last-minute nomination by Gard to ensure that Lwoff would be available as a candidate. Reichard was commissioned to carry out a review, as also already discussed in the previous chapter. In the first half of the review he discussed the three individual candidates in the order Lwoff, Jacob and Monod, but in the second half he discussed the two latter scientists together. He once again reviewed the repressor concept, the operon and the operator concept, messenger-RNA and finally the added finding of allosteric proteins. Concerning the chemical nature of the repressor, Reichard noted that Jacob and Monod first speculated that it might be composed or RNA, but that additional experimental evidence indicated that it would be a protein in the experimental system using *E. coli*.

In the earlier review by Gard in 1961 the possible existence of messenger-RNA had just been referred to in Jacob's and Monod's comprehensive *J. Mol. Biol.* article[13] but the original articles on the messenger-RNA discovery appearing later in 1961[8,9] were not accessible to him. The new information on allostery was published first in 1963[15] and hence could only be commented on by Reichard. He gave a broad analysis of the history of the discovery of messenger-RNA. It was remarked that the first important observations were made in the 1940s by Brachet and Caspersson indicating that RNA had some role in protein synthesis. He also noted that this synthesis had now been clearly identified as occurring on ribosomes located in the cytoplasmic environment. Hence a central problem identified was the transfer of information from DNA in the nucleus of eukaryote cells to the protein-synthesizing machinery, the ribosomes in the cytoplasm. Citing the PaJaMo experiment and the follow-up publication by Riley et al. a year later[16] Reichard noted the first evidence for a short-lived population of RNA of a previously unidentified nature. He then referred to the identification of messenger-RNA by two different groups, Brenner et al.[8] and Gros et al.[9] and he also alluded to the first hybridization experiments mixing RNA and DNA by Spiegelman and his collaborators. He even alluded to the first experimental evidence for the existence of a triplet genetic code by Marshall Nirenberg and Severo Ochoa. The final discussion in Reichard's review concerned allostery and this was already commented on in the previous chapter. He was somewhat less impressed by the introduction of this concept and questioned if it was a truly new observation.

The final paragraph of the review truly gave praise to the three scientists. After an introductory mention of the central dogma he described all three nominees as true groundbreakers in the young field of molecular biology. He emphasized in particular that the early complex experiments that led to the repressor hypothesis — and implicitly also the introduction of the operon concept — in addition led to the identification of messenger-RNA. It is worth repeating the statements presented in the last two sentences of the review.

The first one concluded that the discoveries of the induction of lysogenic viruses, of the repressor phenomenon and of messenger-RNA, could each (!) be considered worthy of recognition by a prize. The second one emphasized that there seemed to be no other group of scientists in the field of molecular biology that could be considered more worthy of a prize than the trio of scientists at the Pasteur Institute.

In 1964 the focus of the committee was on Bloch and Lynen, as described in Chapter 6. The three French scientists were nominated pair-wise, including Monod in both nominations. No new review was carried out, but all three remained on the list of worthy candidates for a prize. The way the situation evolved in 1965 was explained in Chapter 7. For Jacob and Monod it was smooth sailing but the central question was whether Lwoff should also be included. The main advocates for his candidature were Gard and Klein and they finally managed to convince the committee that all three Pasteur scientists should be recognized. This may not have been an easy decision to take since twelve out of the thirteen nominations submitted only proposed Jacob and Monod.

It was clear that the strength of the candidature of the three French candidates was increasing rapidly and probably they themselves could not be unaware of this. In early October 1965 there was a symposium in Naples with a broad international representation including Jacob. Others present were Tiselius and Agnes Ullmann. Her husband had had a previous experience as a young scientist to work with Tiselius and Svedberg in Uppsala for two years. Tiselius therefore invited Agnes for a dinner at which he informed himself about the three molecular biologists at the Pasteur Institute. Tiselius, the doyen of the Nobel work, was apparently well informed about the candidacy for a prize in physiology or medicine of the three French scientists. Furthermore, in addition to the repeated nominations of Monod and Jacob for a prize in this discipline there was in fact, as already mentioned, a nomination of Jacob and Monod for a prize in chemistry in 1965. The committee did not initiate a review and citing Tiselius it wrote (translated from Swedish) in its summarizing section:

What concerns Jacob and Monod, they too (this year Tiselius had made a first review of Marshall W. Nirenburg — one of the discoverers of the genetic code to be recognized by the prize in physiology or medicine [in 1968, my remark]) have been nominated for the first time for a prize in chemistry — which — as the committee has learnt — also applies to a prize in medicine. In order to make a decision on this matter it would seem recommendable to establish a contact between the two committees.

It was clear to Jacob that something was cooking. When Agnes returned to her hotel she was promptly invited by him for a drink. He wanted to be informed about what had been discussed at the dinner. When Agnes diplomatically related that the discussions had concerned her husband and the Hungarian revolution Jacob seemed most disappointed. Scientists too have emotions of anticipation and possible fulfillment. Still, Jacob's unconfirmed premonitions did not last long. The announcement of his shared prize was made within two weeks.

Time for Nobel Festivities

The three French scientists who arrived in Stockholm in December 1965 were exceptional luminaries. Not only did they represent prototypes of successful scientists but in addition they identified with the best in human culture; curiosity, creativity, commitment, critical intelligence, cooperative amplifications of efforts, compatriots in crises, etc. And they enjoyed being together in the limelight. The prize recipients were addressed by Gard at the prize ceremony. He gave a very lucid presentation of the important advances in the field of molecular biology and the large influence "the French group,"

Lwoff, Monod and Jacob receive their Nobel Prizes from the hands of his Majesty the King. [From Ref. 24.]

a term he used repeatedly, had had. He noted at the start that it had been discovered that the answer to question of life mysteries could only be answered by achieving an understanding about the structure and function of genes. He described as a background the function of DNA and the nature of the genetic code. He also noted that "A chain of genes can contain from a few hundred to many thousand links and within this frame can be included the individual pattern for the total number of well over a million genes, which one estimates that a cell may contain." This is an interesting reflection of its time. We now know that the total number of protein-coding genes in, for example,

Jacob, Lwoff and Monod with their Nobel diploma. [From Ref. 24.]

the human genome is of the order of 20,000 and that only a very small fraction of the total genome is involved in directing the synthesis of proteins. In addition a lot of information is transferred from what is called "non-coding DNA," implicitly "non-(protein)coding DNA" to a large number of different categories of RNA of variable sizes. It still remains to demonstrate the function of a number of these different forms of RNA[14]. Gard then highlighted the importance of the identification of the short-lived messenger-RNA, the discovery of which in part was based on Jacob's contribution. A good part of the talk was devoted to an attempt to describe in an easy-to-follow form the complicated circuits for gene control elucidated in the work by the French group. The second to last paragraph prior to the concluding invitation in musical French for the scientists to receive their prizes from the hands of His Majesty the King read (translated from Swedish) as follows:

Lwoff represents microbiology, Monod biochemistry, Jacob cell genetics. Their decisive discoveries would not have been possible (to make) without knowledge and technical skills in all these fields and without a close collaboration between the three researchers. But the mystery of life is not resolved simply by access to knowledge and to technical skills. In addition to this there is a requirement for a keen capacity to observe, a logical intellect, capacity to combine scientific imagination and intuition, qualities that all the three scientists personify in ample measure.

Crown Prince Carl Gustaf at his first Nobel Prize ceremony.

The three prize recipients, one at a time received their gold medal and diploma from the hands of the King, Gustaf VI Adolf. There were some changes in the representations from the Royal Family. The Queen Louise had died the same year and the 19-year-old Crown Prince Carl Gustaf participated in the Nobel Prize Ceremony for the first time. In 1973 he would take over the responsibilities from his grandfather to hand over the prizes, as mentioned earlier.

At the banquet in the City Hall all the three laureates expressed philosophical thoughts in elegant French. It is tempting to cite from all of them, but I choose to quote only from Lwoff:

It is the search that led me here and research, as we know, is a form of play, noted already by Montaigne, who said: "the studious passion that amuses us to the pursuit of things leading to acquisitions that make us desperate." In general, it must be said, it is not the discovery that engenders despair, but rather its absence.

And later:

The Grand Inquisitor has dug into the past of the researcher ....
The victim takes obvious pleasure in the ceremony and there are many candidates for martyrdom. For every scientist, deep inside, wants

to become recognized. However, fame conferred on a recipient in the form of a rare prize, so envied and so spectacular, somewhat arbitrarily separates him from his peers. It forces him to think and to reflect about himself and it also obliges him to meditate on prizes in general on the generosity of fate, on the charms and constraints of notoriety.

And then there is something more serious. Chrysippus, a Greek stoic philosopher, recommended us to attach importance only to things that depend on us. This is a very wise counsel which I have always tried to follow. And here there is an event that does not depend on me, suddenly occupying a place in my life that it would be pointless to deny the importance of and which, moreover, is a happy event. Happy for my country, for the institution to which I belong, and for the discipline I cultivate. Happy also for my family and no doubt also for myself. According to Spinoza, joy is the transition from a lower to a higher perfection, that is something important, and despite my wise principles I feel great joy. I say this with regret, but without much astonishment. Moreover, if it is true that regret is a form of sadness, it should be dismissed as being reduced.

Lwoff finished his speech by quoting his countryman Camus, who received the Nobel Prize in literature in 1957, only 44 years old. This quote could equally well have been used by Monod, who had a special intellectual relationship to Camus, as highlighted by the recent book[17], referred to frequently in the previous chapter. Lwoff's concluding comments were:

A few years ago a great writer came to sit at your table to break the bread of friendship (compare *cum pane* — etymological origin of companion, my remark). It was a sensitive artist and a generous man, whose life had been a struggle for freedom and truth. These were his words "It remains for me to thank you publicly, as a personal token of gratitude, the same ancient pledge of fidelity, that every true artist makes to himself every day in silence.

Following the rule specified by Nobel's will that would still be adhered to for a long time to come, Nobel lectures in 1965 were given the day after the Prize Ceremony. As already mentioned Jacob made a compromise to at least partially accommodate the requirement that Frenchmen in official functions should use their native language. He gave half his lecture in French and the other half in English. As a preamble he said:

If I find myself here today, sharing with André Lwoff and Jacques Monod this very great honor which is being bestowed upon us, it is undoubtedly because, when I entered research in 1950, I was fortunate enough to arrive at the right place at the right time. At the right place, because there, in the attic of the Pasteur Institute, a new discipline was emerging in an atmosphere of enthusiasm, lucid criticism, nonconformism, and friendship. At the right time, because then biology was bubbling with activity, changing its ways of thinking, discovering in microorganisms a new and simple material, and drawing closer to physics and chemistry. A rare moment, in which ignorance could become a virtue.

He then went on to describe the milestones, or rather aggregates of star dust that he had passed during his meteoric career. He started with lysogeny and bacterial conjugation leading to the identification of genome-independent pieces of genetic material, the episomes, and also his success of showing the linear map of the bacterial chromosome. Then he described the hunt for messenger-RNA, the third species of RNA identified. The bulk of the lecture focused on the complicated matters of gene control, reviewing the recently introduced concept, the operon and the replicon. The latter was discussed in the perspective of how the division of the circular DNA of the *E.coli* bacterium could be synchronized with the division of the cell and there was speculation in relation to some recent data about the possible role of structural changes in membranes as a means of coordinating genome and cellular duplication. It was inferred that evolution had conserved a system of molecular communication between the cell surface and DNA both in bacteria and in eukaryotic cells. This remark was just a foretaste of the rapidly expanding field of cell signaling via a multitude of receptors at the surface of cells during the following decades.

In his summary he widened the perspective and said "The message inscribed in the genetic material thus contains not only the plans for the architecture of the cell, but also a program to coordinate the synthetic process." And further on "What the study of regulatory circuits has shown is that the compounds in question serve only as simple stimuli: they act as signals to initiate a synthesis whose mechanism and final rejection remain entirely determined by the nucleotide sequence of the DNA" and as a summary "In the expression of the nucleic message, as well as in its reproduction, adaptation results from an elective rather than an instructive effect of the environment."

And finally he acknowledged that much more work remained to be done. The chemical nature of the repressor had not as yet been identified. Eventually this was to be done by others.

A Life in Full

Jacob had a long life dying in 2013 at the age of 93. The narrative of his famous autobiography finished in 1961 although the book was published 26 years later. The selected endpoint was Christmas time when he and Monod had finally submitted their magnum opus summarizing their advances in understanding the regulation of gene expression and announcing prior to the final publication the identification of messenger-RNA. The last paragraph of the book read:

> The snow had started to fall again on the Luxembourg Garden. The light was fading, at first it turned into shabby white, then dark grey. As if one had folded away the day to put it into a box. To leave room for the night, for ghosts, dreams, horrors. When I left the garden I suddenly had an idea about an experiment about cell division to perform. A rather simple experiment. It would be enough to ....

Science is a never ending process of searching for new knowledge. All those involved in this enterprise "stand on the shoulders of others" and at some stage the committed endeavors will be taken over by the next generation of scientists. After the mid-1960s Jacob did in fact part with Monod and he changed his focus in science. In 1964 he had become professor of cellular genetics at Collège de France. He shifted to studies of animal systems, in particular mice in the early 1970 with a focus on differentiation. He developed a particular interest in tumors of germ cells, teratocarcinomas. The target was well chosen, but techniques to examine embryonic stem cells and evaluate their potential use in regenerative medicine were developed first at a much later stage. In parallel with his science he developed his special capacity to use the written language, as reflected in the already repeatedly referred to autobiography. This was not published until 1987, but prior to that Jacob had put his mark on the history of biology leading up to its molecular phase. Thus he published a book entitled (in English) *The Logic of Living Systems: A History of Heredity*[18], mapping the growing understanding of the fundamental levels of organizations in organisms representing different branches of the tree of life. The broad approach taken combines the description of historical events with discussions of the philosophy and sociology of science. He attempted to outline the relationship between science and ethics. His second book in this category may have had an even larger impact. It is entitled *The Possible and the Actual*[19] and appeared in 1981. The central theme is that evolution acts as a tinkerer (bricoleur) providing strong arguments against any form of intelligent design. As mentioned in the previous chapter, Jacob, as well as Lwoff, expressed a certain criticism of Monod's book *Chance and Necessity*[20]. A part of this criticism may have been the reservation about Monod's ranking of *teleonomy* as a secondary property deriving from invariance.

The term teleonomy was introduced in 1958 to create a contrast to the already established term *teleology*. These two terms have a joint root in Greek *telos*, meaning end purpose, but differ in their suffixes; *nomos* meaning law or rather purpose-law and *logos* meaning reason, explanation. In everyday terms language teleology is often called "intelligent design." Since Lwoff and Jacob did not accept that knowledge about molecular mechanisms could in any way be used to make predictions about evolutionary developments they clearly favored the use of the concept tinkering over teleonomy. Jacob's second book also dealt with the scientists as tinkerers, the way they play with models and ideas. In his final book providing philosophical and historical perspectives, *Of Flies, Mice and Men*[21] the description of scientists as one group critically furthering societal advances was compared to other influential groups, involved in the exploitation of the arts and myths. Jacob always brought to light the particular value of scientific knowledge. In his books he elaborated on the unique scientific culture in which experimental testing of the hypothesis was critical and the enormous value of the large jury of other scientists scrutinizing presented data. Altogether it was interpreted to provide a very healthy self-controlling culture. Anyone attempting to falsify data will sooner or later be revealed. To Jacob, doing and defending science were two sides of the same thing. This also led to Jacob becoming an active defender of the rights of scientists in any political system. He became heavily involved in attempting to help scientists who had been imprisoned in the Soviet Union. In 1993 Jacob and a number of fellow leading scientists established the International Human Rights Network of Academies and Scholarly Societies, administered by the National Academy of Sciences in the U.S., already described in my previous book on Nobel Prizes[3].

The apex of Jacob's writing was the creation of his autobiography. It was this book which led the man who had become a member of The French

Academy of Sciences in 1976 in addition to becoming a member of the foremost academy, the academy of language, Académie Française. The latter academy is based on Plato's original Academy with 40 life members. In 1995 the French novelist Jean-Louis Curtiz, using as a pseudonym Louis Lafitte, had died. Jacob succeeded him in chair number 38.

A man of Jacob's stature of course received many honors. He was an exceptional war hero, something he himself never wanted to discuss. Still in his heart he appreciated the recognition he received for his unselfish acts. The one award he appreciated the most was the Croix (Compagnon) de la Libération, the highest-ranking French military decoration of the Second World War, created by Charles de Gaulle. It was awarded only to 1,056 French citizens, who had distinguished themselves exceptionally during the war. He wore this award with pride when he dressed in tails for the Nobel Prize ceremony. Later in life, 2007–2011, he was chancellor of this unique order. His wartime contributions were also recognized by many other medals; The Grand Cross of the Legion of Honor, The Great Official of the National Order of Merit, The Cross of the Qar with 5 citations, The Colonial Medal with Memories "Fezzan-Tripolitaine" and "Tunisia" and more. His membership of Académie Française has already been mentioned. In the academic world of the natural sciences he became a member of not only the French Academy of Sciences but also the Royal Society and the U.S. National Academy of Sciences as well as the American Philosophical Society. He was also a Foreign Member of the Danish Royal Academy of Arts and Sciences, the Académie Royale de Médicine de Belgique, the Hungarian Academy of Sciences and the Royal Academy of Sciences (Madrid). His list of honorary doctorates is too long to give here as are his invitations to give a number of the most prestigious lectureships, like the Harvey Lecture and the Dunham Lectures. In 2012, a year before his death, he was proud to stand alongside the

Jacob and Professor Alice Dautry, President of the Pasteur Institute at the inauguration of a new research building for emerging diseases named the François Jacob building in 2012. [© Institut Pasteur]

President of the French Republic and inaugurate a new building at the Pasteur Institute that carries his name. He was pleased that in addition to buildings at the Institute carrying Lwoff's and Monod's name there was also one carrying his name. He retained his sharp intellect until the end of his life. Jacob described General de Gaulle, after he had met him for the first time in London as a "gothic cathedral." Interestingly some of Jacob's younger students saw their mentor himself as a "gothic cathedral." They did not feel fully at ease with this tall shy man with his many accomplishments. It was sometimes with regret that they noted the impossibility to get beyond the "statue without."

Exit Lwoff, Monod and Jacob. What Next?

In the early 1960s Gard published an article entitled *Exit poliomyelitis. What next?* This was the time when one could register the impressive impact of polio vaccination. Would it be possible to develop a world free from infections with poliovirus and would there be something taking its place in the panorama of viral diseases? During his long life, like Jacob he lived to become 93 years old, dying in 1993, Gard could register impressive developments in the science he had contributed to and I helped to keep him informed about the most recent advances in the field. The consequences of the broad applications of polio vaccines in our world have been described in my two previous books on Nobel Prizes[3, 4]. Since 2013 there have been further encouraging developments. Following the eradication of polio from the huge and highly populated Republic of India it has now been possible to eliminate the single remaining source of polio in Africa, by also extinguishing the disease from Nigeria. This is a major achievement and in the whole world there are at present only two countries in which polio is now endemic, meaning that the infection as yet has not been possible to eliminate completely. These countries are Pakistan and Afghanistan. It is not hard to imagine the difficulties in administering general vaccinations in such countries with major social unrest and without complete control by the central government. In 2015 there were only 57 cases of polio in the whole world, all of them occurred in Pakistan. Tragically there have been as many as 70 killings of vaccinators and other health personnel fulfilling their duties within the framework of immunization programs directed by the World Health Organization (WHO) in collaboration with local health authorities. The so-called end-game of the global polio eradication program continues to be pursued with vigor. Once the infection has been fully eradicated from the world it will be possible to phase out vaccinations, after a transitory phase of use of inactivated instead of live vaccines. In fact one of the three types of poliovirus, type 2, has already been eliminated from the world and the consequence therefore is that at present new vaccines, only containing two of the originals three types of the viruses causing polio, types 1 and 3, have been introduced.

Gard could not have foreseen these developments and his question instead concerned whether there would be an evolutionary void once an infection had been regionally eliminated. His answer of course was that this should not be the case. The filling out of an ecological niche in a population is dependent on a number of factors that influence the urge for and possibility of the selfish genes carried by a virus or any kind of cellular forms of life to manage its survival. It should be reminded that acute virus infections have only accompanied the human population since major cohorts of humans living together in urbanized areas became established some 3,500 years back. When the human species lived in hunter-gatherer communities there were no acute infections, no common cold or measles, only infections with potentially persistent viruses, like herpes viruses or retroviruses. Gard's appropriate answer to the question he posed was that there are no natural mechanisms to fill up niches that have been cleared of a certain kind of virus infection. There is no *horror vacui*. Similarly once smallpox, a disease that had killed hundreds of millions of individuals throughout the time of settled human civilizations, had been eradicated by the laudable efforts of the WHO in 1978, no niche to be filled by another infection was established.

Evolution is a matter of interplay between the mutable agents or organisms and the environment. The history of evolvement of different life forms on earth is highly complex due to the continuously changing environments and sometimes dramatically altered conditions of life due to major incidents as discussed in Chapter 7. It is a hallmark of evolution that the way new forms of life develop, generally as a response to new challenges posed by the environment, is unpredictable. It is a matter of tinkering as repeatedly emphasized by Jacob, and a characteristic feature is the reuse of structures originally serving one function for a completely different purpose. No one could have predicted that what were originally gill pouches in fish could eventually end up as the small hearing bones in our ears. And who could have predicted that the original evolution of feathers to keep dinosaur eggs warm eventually could be of use to endow the only surviving extensively developed derivate of the dinosaurs, the birds, with a capacity to fly. There are in fact a number of alternative structures allowing animals to fly, as already repeatedly exploited by evolution.

As mentioned the environmental changes on Earth have sometimes been very dramatic. A good example is the enormous impact of the large meteor that hit the Yucatan peninsula in present day Mexico some 65 million years ago. The wiping out of all the dinosaurs opened up possibilities for some negligible small mammals to develop. This led to the wide representation of this kind of animals that we see today, including one species — us — with an advanced consciousness and a capacity to reflect on, analyze and in recent times take control over both the environment and the genetic properties of life forms as repeatedly referred to above.

We have over and over again taken note that the major factors guiding the advance of human civilizations are science and technology. Due to the accelerated accumulation of new knowledge this influence has increased markedly during the last hundred years. The development from a residential rural agrarian society into the modern global, ever more urbanized society has occurred within a remarkably short time. The impact of factors favoring globalization becomes increasingly more evident. It was noted above that the control of infectious diseases is a global responsibility. Infectious agents do not respect the artificial cultural constructs of national borders. Attempts to globally eradicate certain virus infections, and also to control selected other infectious diseases will be continued by the WHO. The epidemic spread of measles and rubella in the world is continuously being reduced and it should be possible to eradicate these two infectious diseases within perhaps a decade. Viruses with a capacity to continuously change their antigenic properties cannot be completely eliminated but unceasing global monitoring of the appearance of new strains of virus, like influenza virus, or a previously unencountered form of virus infection, allows early epidemiological interventions and development of effective vaccines or suitable drugs. This is another example of the global needs and benefits of collaboration and sharing of knowledge.

The Millennium Development Goals defined by the United Nations at the turn of the present century to promote dignity, equality, equity and to reduce extreme poverty has become a considerable success, not least in reducing extreme poverty in the world. In 2015 these goals were replaced by a list of 17 sustainable developmental goals. The fulfillment of these ambitious goals is set for 2030. At the time of writing a conference in Paris has led to an agreement about how to take action to manage global warming with a goal of reaching an increase of less than 2 degrees Celsius during the rest of the present century and the World Trade Organization (WTO) has against all odds reached an agreement at a meeting in Nairobi, Kenya. It was the first time this organization had met in Africa and the results were referred to as a "historic" package for Africa and the world. There is no doubt that we live in a global world and there seem to be great possibilities for the future of our five grandchildren, to whom this book is dedicated. An important part of this optimism is based on the global advances in making schools available to all children including also, of course, all girls. The spread of existing knowledge, facilitated by the new electronic media, and the accelerated generation of new knowledge and new discoveries by the increasing number of scientists representing all corners of the world will improve the human condition. *La condition humaine* is the name of a famous book by André Malraux, one of Jacob's favorite authors.

It would be naive not to acknowledge that there are threats to future developments. What about violence? The shadow of the two world wars during the previous century still hangs over the oldest generation of today. In addition there has been an enormous development of arms technology. The dual use of scientific knowledge remains a fact throughout history. And still the amount of violence in the world has been continuously *decreasing* as described in the book *The Better Angels of Our Nature* by Steven Pinker[22]. This may not be the impression one gets from the information provided by the daily news reports. Regrettably negative news sells better than positive information. How should humans as individuals and as groups represented by democratically elected politicians respond to the threat of violence including the risk, albeit very small, of a terrorist attack? In this context we may reflect on national pride and patriotism versus pacifism. This is a large topic only to be touched on in the present context. It might be mentioned that the German philosopher, Johann G. Herder, a friend of Goethe and a man of the enlightenment once said "… and wars are mere crimes. This is so because all large wars are essentially civil wars, since men are brothers, and wars are a form of abominable fratricide." In view of what we know today about our close genetic relationships and of our common ancestors moving out of Africa some 60,000 years ago this statement shows a remarkable foresight.

Jacob was a true patriot and a outstanding war hero as already emphasized. But he was also a staunch advocate for human rights and together with his two fellow prize recipients, Lwoff and Monod, he argued strongly against capital punishment. As a medical officer Jacob was of course armed but one wonders whether during the African experiences where he was a part of the action in Fezzan, Libya, Tripolitania and Tunisia and in particular during the invasion of Normandy, he ever used his weapon. As a medical officer he might have relied on the armband with the red cross he carried. This does not in any way detract from the fact that Jacob risked his life in defending the fundamentals of democracy and freedom when he earlier in Africa and finally as a member of the French Second Armored Division participated in the recapturing of his home country. There must have been occasions when Jacob during his visits to Cambridge met Sanger. This could have allowed for a discussion of patriotism versus pacifism, but I feel quite sure that this issue was never raised by these two outstanding and relatively shy scientists.

As has been repeatedly stressed individual researchers can sometimes catapult the advance of science by their discoveries, but its broad advance is a collective effort. In this context it is tempting to raise the counter-factual question, what would have happened if Jacob had not been one of the three survivors among the group of 200 officers that de Gaulle gathered in London. How much would this have slowed down the introduction of important aspects of the early developments of molecular biology? And could Lwoff and Monod have carried a Nobel Prize on their own? It is in fact a principle that among a group of prize recipients, each one represented should be capable of carrying a prize on his own. This is certainly applicable to Lwoff and Monod, but the competitive strength of their contributions was markedly amplified by the presence of Jacob. They did in fact form a unique triad in science and each one of them contributed different profiled creative qualities. Their joint appearance at the prize ceremony was a tribute to their collective competitiveness.

Science and the Arts Revisited

Molecular model building is critical for the solving of structural problems by crystallographers as mentioned in Chapter 3. The structures deduced not infrequently are referred to as esthetically pleasing. In my previous book on Nobel Prizes[3] the chapter describing the discovery of the double-helix structure of DNA by Watson and Crick was entitled "It is so beautiful, you see, so beautiful." This was the only way Watson could conclude a particular lecture about the recent discovery. An event occurring at the time of writing reminds about the fact that medallions can also be esthetically pleasing and have a value. Watson just sold his Nobel gold medallion at a Christie's auction. He pocketed 4.1 million U.S. dollars. This was the first time a living Nobel laureate has sold his medal. The price markedly exceeded the 2.2 million dollars paid a year earlier for the gold medal awarded to his late colleague, Crick.

Seeing the Invisible World and a Biased Proposal

As discussed in Chapter 3 and above we need to visualize phenomena to create a form of rationalized understanding. In their description of the operon Jacob and Monod inferred the existence of molecules of certain reactive capacity, presented as arrows in a diagram (p. 499). However, they did not chemically identify for example the critical repressor. It was too elusive for the Pasteur group. Thus Monod and Jacob could not discuss the structure and function of the macromolecules they had conceptually identified. This was for others to do later. In modern chemistry based on the identification and characterization of molecules with specific activities, we use model building employing different kinds of modules, space filling or sticks, the latter emphasizing the significance of the void. This was briefly discussed in Chapter 3. Today we can technically increase the resolution to visualize individual molecules. Thus we can "see" things that are several orders of magnitude smaller than what we discern by the naked eye. The possibilities for understanding the correlation between structure and function in the world of macromolecules, prime actors in biology, have increased dramatically. To this should be added the fantastic tools that modern digital techniques have put in our hands. We can now readily rotate molecules to view them from all sides and we can dock them to other molecules and notice how they change configuration as a consequence of the interaction. In fact there are almost no limits to the new techniques.

We can make an autopsy of the corpse of a human being or an animal by non-invasive techniques. We can wake up dinosaurs from their long sleep and make them run across our movie screen. There is of course also a flip side. The power of the violence entertainment industry has become dramatically expanded. Space travel is a simple business and massive killings in virtual situations — and even in real life in the control of drones — has become a part of the cozy corner home entertainment. There has been a dramatic development since the time of the movies of cowboys and Red Indians that I grew up with. One question is how it may influence our attitude to violence, but that is way beyond the scope of this book to discuss. What should be discussed however, is how the animations may disregard the laws of natural science, for example the denial of gravity, and introduce many other supernatural phenomena. Regrettably this unconsciously may foster anti-scientific attitudes. And still, as already emphasized, the illustrations we use generally must for didactic reasons allow certain violation of physical facts; depict realities that do not exist or at best are rough approximations. It is only by using various metaphorical forms of visualization, that we can grasp the principles of structure and function. When doing this we need to keep in mind that in the world of molecules and even more so when it comes to the building stones of atoms the absolute majority of what there is, is void. And there are no colors, but we use them in an artificial manner to distinguish different entities, like different kinds of atoms.

Often when we feel we are close to the optimal depiction of the conditions we have deduced we call what we see beautiful. In our attempts to illustrate complex phenomena we are forced to simplify and generalize. This needs to be done with consideration taken to the limits of perception of our senses. We can identify a certain spectrum of visible lights, like the colors of the rainbow, but other animals can use part of the wavelength spectrum out of reach of our eyes, like ultraviolet light. According to Goethe "colors are the deeds and suffering of light." The colors of the rainbow provide a good illustration of the colors that we can perceive, as mentioned. It was defined by Isaac Newton as having seven colors, but this was not a categorical scientific fact based on distinctions of the range of wavelengths. Instead it was subjectively based on what the human eye could discern. Newton originally included five colors, but then he added orange and indigo. It can be argued, as it has been done by Isaac Asimov, that indigo is a form of deep blue and should not be included, giving a total subjective number of six colors. However, Newton searched for an analogy for the musical scale that has seven notes and furthermore he might have been influenced by cabalistic thinking, the prime number seven was more appealing than the trivial two times three that is six. Interestingly this biased proposal has stayed with the human cultural inheritance since Newton's day. One of the most important challenges in science is to avoid subconscious biases of this kind. And on this theme there is an aphorism which says — "science is spectrum analysis; art is photosynthesis." To which can be added a statement by T. H. Huxley (see Chapter 3) who said "every great advance in natural knowledge has involved the absolute rejection of authority."

Searching for Metaphors

In the making of scientific knowledge available to a broader audience we need to use metaphors, as has repeatedly been emphasized. However, in our choice of metaphors we have to ensure that there is a reasonable agreement among the public about the interpretation of the concept used. One dark August night a few years ago, our oldest son, Jacob, and I were comfortably cooling off outdoors at the bridge of our summer-house after a pleasant sauna. The weather was powerful, combining a close to full moon and rapidly passing showers of rain. All of a sudden we saw a beautiful silvery gray arch being formed in the sky. After some reflection we came to the conclusion that it was a rainbow without colors formed by the moon rays diffracted by the densely falling drops. I checked this in the 1908 encyclopedia that we have in our summer-house. It has the charming title (in English) *The Nordic Family Book; Conversation Lexicon and Reality Encyclopedia (The Owl Version)*. The book gave a full description of conventional rainbows, but then in the final paragraph it said "Moon rainbow, a very rare phenomenon." In English it is referred to as "moon bow." It struck me that a rainbow without colors might serve as a powerful metaphor and that one could compose a piece of music entitled "Over the colorless rainbow." But then upon further reflection I had to note that although attractive the moon rainbow would not serve well as a metaphor, because almost no one has seen it. The metaphors we use need to strike a chord of ready identification in order to allow their effective use. This is particularly challenging when we reach the limits of what can be made visible. How are we to reliably illustrate the innermost nature of matter, such as the Higgs' particle or the different kinds of neutrinos, highlighted by the Nobel Prizes in physics in 2014 and 2015?

As mentioned our son's name is Jacob, a name used way back in our family and which I also carry. This name has its origin in the Old Testament and the fantastic stories about Jacob can be read in Genesis. Jacob the grandson of Abraham, living close to two millennia BCE, had a special covenant with God, who later in his life bestowed on him the name Israel. Thus Jacob was the original ancestor of the Jewish people, whom we encounter repeatedly in presentation of Nobel laureates, because of their remarkable over-representation among them. The biblical Jacob was a complex character and he deceived his blind father Isaac who gave him the blessing as the first-born. Jacob was in fact born after his twin brother Esau and during delivery he held his foot. This is the source of the English saying "to pull someone's leg," a connection rarely recognized. This name suits our son, who likes pranks and mischief, well, but it was probably less applicable to the main character of this chapter, who had Jacob as a surname. However, he was a remarkable example of a creative scientist, again, not surprisingly, of Jewish extraction.

As we move from the atomic to the subatomic world the lack of specific structures begins to dominate and a massive emptiness prevails. And still there are certain rules for the interaction among the subatomic components within this extended void. This is truly the invisible world that needs to be made visible. We do this by use of abstractions and metaphors as mentioned. They represent higher generic units, giving our consciousness an opportunity to increase its speed of processing and hence its effectiveness. We accept that a particle and a wave phenomenon represent two expressions of the same entity and we add a durability factor talking about the space-time warp. Physicists attempting to provide a coherent theoretic framework for their discipline unpretentiously refer to the theory of everything (ToE). It is critical that many of the relationships in the world of physics are mathematically defined. If that had not been the case it would not have been possible to unravel the relationships and predict the occurrence of certain units in the subatomic world. There is a challenge in the application of mathematics when we want to give an updated explanation of the nature of matter which in modern string theory is presented as including nine spatial and two time dimensions. By our senses we are limited to naturally conceive three dimensions of space and a single time reference. We also need, as repeatedly emphasized here, to find appropriate metaphors of a nature allowing a congruent interpretation of the majority of humans and we need to reflect on the complex relative entity of time. It is beyond the framework of this chapter to discuss this complex issue. Perhaps it would suffice to use a citation attributed to Mark Twain; "Time was invented so that not everything would happen on the same occasion."

Symbols and Esthetics

Of course scientists frequently use visual impressions as a way of conceptualizing dramatic new pieces of knowledge. Thought (*Gedanke*) experiments need to be anchored in symbolic or integrated pictorial impressions. Since way back in human civilization development of symbols have been critical in advancing, by quantum leaps, culture at large and science in particular. In the historical development of numbers and letters we have truly transgressed the border between esthetics and functionality. Mathematics could take the first step to advanced exploitation by the introduction of symbols of numbers. Historically several approaches were taken. We still honor Latin numbers in noting the year a classical building was erected, but this numbering system was not useful for example in multiplication and division. The introduction of the Arabic number symbols 1 to 9 and the later critical introduction of 0 from India allowed the access to radical new combinations and interactions. Because we have ten fingers we settled for a base of ten. However, since embryological developments in their original design are binary, certain modifications were introduced to finally get the five fingers on one hand, which apparently provided some survival advantages. If instead our fourth proximal finger, the ring finger, had not been a fusion of two rudiments our number system base would have been twelve. This would have given several mathematical advantages. It would have allowed us to multiply more quickly in our head. However, human culture has not overlooked the potential of using the number 12. Astronomers of far back have taught us that there are celestial bodies that by their representation accentuate the numbers five and twelve. Therefore we have chosen to have two times twelve hours per day plus night and these are broken up into 60 minutes and further into 60 seconds. In addition the number of months per year is twelve. In the same vein the compass is divided into 360 degrees. The digital revolution has reduced the numbers used to only two, 0 and 1. The year 2016, when this book was published becomes MMXVI in Latin numbers and expressed in a binary form 00000111111, neither of which is useful for further mental calculations. The reduction of the symbols to only two of course limits the possibilities for wider esthetic modulations of their use.

Another very central group of symbols potentially combining esthetics and functionality are used to represent words. Originally they were recorded in the form of ideograms — a simplified picture of reality representing an idea or concept, and this still remains in use in some languages in the Far East, as in China and Japan. Then the Phoenicians converted some of the ideograms to letters — the inverted head of an oxen A; the outline of a house — B; a camel — C, etc. — and an extremely useful combinatorial message formulation was created. The first languages only had symbols for consonants and the vowels were indicated by different diacritical marks. Later vowels in many languages became represented by their own symbols and the number of letters in different languages varies both for this reason and also because of other idiosyncratic additions like the å, ä and ö in the Scandinavian languages. The writings were often developed into forms of art, calligraphy. Regrettably this beautiful handwriting may become obsolete in the future. The personal signature certainly will disappear. Electronic media do away with handwriting and instead we will use non-forgeable signatures like our fingerprints or eye bottom patterns or why not our unique DNA? Thus technological developments will lead to a disappearence of potentially esthetically pleasing texts, but on the other hand electronic media will open up new modes for producing various forms of art, ArtSci. These developments are extensively discussed in the book *Colliding Worlds*, as already referred to in Chapter 3.

The "beauty" of numbers and letters is their potential for combinatorial use. Of course the number of letters needed for effective use could be markedly reduced from the number employed in most languages. However, languages are dynamic "live" structures that continuously change by evolutionary adaptive modifications. It is the need of the day that decides the way a language is structured. A Swede of today, for example, would have difficulties in understanding a fellow countryman of his from the 18th century. Since we now live in a global world, languages also need to transcend borders. This has lead to a progressive extinction of languages spoken by too small groups of people. Out of the originally about 6,000 languages it has been estimated that about 200 will survive in the future. In the natural sciences, English has developed to become the language of use. As a consequence this language has come to have a major influence on the advancement of human global culture at large.

Beside the use of numbers and letters, symbolism or different forms or metaphorical presentations have been critical in the advance of science. The list of scientific epiphanies is long and continues into the future. Berzelius introduced a letter system to identify individual chemical elements and combinations of them. In the mid-1900s the German August Kekulé had a dream of a snake biting his own tail. This led him to speculate that six atoms of carbon can form a ring with alternating single and double bonds. This is a very important structure in organic chemistry. In physics there are many examples of epiphanies related to symbolic thinking. Albert Einstein sitting on a tram noticed the clock on the wall of a house. He thought "If I transport myself with the speed of light away from that clock, the time will remain unchanged." This led to $e=mc^2$, probably the most famous equation in the world. A grey morning in 1933 Leo Szilard (Chapter 8) was standing on a street corner in London. The stop light changed to green and he stepped off the curb. When crossing the street it dawned on him that a nuclear chain reaction may be conceived. The rest is history. The unpredictable moment of synthesis may occur in any situation in life. The role of visualization in understanding quantum physics has been amply expressed by Richard Feynman, the 1965 Nobel laureate in physics. He said "What I cannot see (construct) I cannot understand."

The Magic of Interplay of Numbers

Mathematics is a particular branch of science. It was originally conceived by particularly resourceful and profiled brains. The outcome of their reflections was a sorting out of axioms and proposals, often in the form of a theorem. One of the most famous theorems is the one conjectured by Pierre de Fermat in 1637 in the margin of his copy of Diophantus's *Arithmetica*. He claimed to have the necessary proof but noted that the margin of the book was too small to accommodate it. It took 358 years until Andrew Wiles managed to develop a successful proof. This is just one out of many theorems and its associate, in this case particularly challenging, solution in mathematics to which the term "beautiful" can appropriately be attached. Another challenging theorem was the four color map theorem. The first conjecture of this theorem was made as early as 1852 when a British mapmaker, Francis Guthrie, empirically found in mapping counties of England that four colors sufficed. In his maps no two regions of the same color met. It would take until 1976 before Kenneth Appel and Wolfgang Haken at the University of Illinois could prove that the theorem was correct. Interestingly they needed to use a computer to validate the proof. A number of mathematicians were hesitant at first to accept that the assistance of a computer was necessary, but with time this opposition has waned. Other theorems have also now been proven by the use of computers. Here then is joint territory for the human brain and for the additional intelligence provided by a computer. A recurrent topic of discussion is whether mathematics is a human construct or if the way numbers interact represents a reflection of an inherent world order. Many astrophysicists and elementary particle physicists would argue that their disciplines *are* mathematics. Although seemingly capricious evolution would also seem to make mathematical sense, since it is anchored in the success of reproduction, viz. in flow of information. The periodic table definitely has an inherent mathematical structure and it is not unlikely that new mathematical algorithms will be discovered that may explain the complex regulation of gene expression and also the still, in parts, enigmatic principles of protein and RNA folding.

As Marston Morse from Princeton has said, "There is a center and final substance in mathematics whose perfect beauty is rational, but rational 'in retrospect.'" In a conference entitled *Mathematics and the Arts* honoring the poet Robert Frost Morse furthermore used the following emotion-rich formulation:

It is science without its penumbra or its radiance, science after birth, without intimations of immortality. The creative scientist lives in the "wildness of logic" where reason is the handmaiden and not the master. I shun all monuments that are coldly legible. It is the hour before the break of the day when science turns in the womb, and, waiting, I am sorry that there is between us no sign, no language, except by mirrors of necessity. I am grateful for the poets who suspect the twilight zone.

It is a truism to note that interpretation of a piece of art is in the eye of the beholder. It is the combined perception of form and color, anchored in prior conscious or subconscious experiences, that decides if the message by the artist strikes a responsive chord in the spectator. Today it is possible to measure the processing of impressions by the brain using MRI, a technique mentioned in Chapter 3. There is an increased tendency for certain parts of the prefrontal cortex to become activated when a viewer interprets a piece of art to be emotionally engaging. One of the proponents of the new field of neuroesthetics, Semir Zeki, a brain researcher at University College London, has proposed that an artist "in a sense is a neuroscientist."

The term beauty is very difficult to define. It is used by artists and by scientists, but in many different contexts and covering many kinds of impressions. A mathematician finding a satisfactory and harmonious solution to a problem refers to it as beautiful. A biologist may take notice of the preference of Nature to favor symmetrical structures, like in the human face. Therefore such structures are often interpreted to be beautiful. Symmetry in different mathematical forms may contribute to our appreciation of what we see in nature. The Fibonacci sequence of numbers 1, 2, 3, 5, 8, 13, 21, 34 …, where each new number in the series is the sum of the two previous numbers, is decisive in many structure we enjoy in nature, be they pine cones, sunflowers or something else. In addition the relation between a higher and a lower number as the series advances asymptotically approaches the Golden Ratio, empirically used in many designs of open spaces that humans enjoy. Our brains are also preprogrammed to appreciate certain kinds of landscapes. The highly social primate who climbed down from the tree to use the open savanna together with his group of similar animals appreciated the openness and hence even modern man enjoys the open landscape. In other cases the forms — often characterized by simplicity and still with high information content — associated with functionality may be described as beautiful, as cited above in the case of DNA. The great French mathematician Henri Poincaré once wrote "The scientist does not study nature because it is beautiful; he studies it because he delights in it, and he delights in it because it is beautiful." And additionally to tie the sciences and the arts loosely together we may cite the device used by the Danish astronomer Tycho Brahe who said "Neither power nor richness, only the dominions of the arts and science last."

The Act of Creation

And to end let us return once more to science as a form of art. The process of searching, often in an obsessive way, for a particular truth, sometimes has a reward, a discovery is made. Knowledge that no one else has earlier retrieved, has come to light and this elicits euphoria and a major happiness in the scientist. The observation is "beautiful" and the flooding of the brain with reward transmitter substances is the same in the scientist, who is in this particular and rare situation, and an artist, who has just managed to create a unique piece of art that he has doggedly worked towards for a long time. The two professions or rather vocations of scientists and of artists share many common traits. Their obsessive involvements are acted out in situations of a high degree of freedom used to gain new, sometimes revolutionary heterodox insights or to produce a piece of art reflecting an experience of increased understanding amplified by esthetic qualities. The word esthetic comes from *aisthetikos*, a Greek word meaning a "fully integrated perception by the senses." Both these groups of creative humans often like to refer to intuition, a general means of describing the final synthesis. However, the true meaning of this word is difficult to define. It is sometimes said by the artist that the final piece of art can only be intuited and not explained. In science this would not be an acceptable position to take. Scientists work with factual findings which can be confirmed by other scientists in their *research*. They read the books of Nature, as has been said. Of course there are (a few?) examples of artists who using the mindset of a scientist truly make a factual contribution to the ever-growing enterprise of knowledge accumulation. Goethe was proud of his discovery of the intermaxillary bone in humans too, a synthetic rather than analytical contribution of importance to the field of anatomy.

It may be appropriate in this context to cite Tiselius from his banquet speech in 1948. A part of his address read:

To what extent is progress in literature and science lined with the personality of poets and scientists? That this is indeed the case is beyond doubt, but there is a difference between these two forms of endeavor — one cannot judge them by the same standards. Poets, like all creative artists, impress the stamp of their own personality on their work far more than scientists can. The work of the poet bears his individual mark. Thus, as an individual, he is indispensable for the development of culture and civilization. The scientist, however, seeks objective truth, which is and must be completely free from all traces of his individual personality. It can be said with certainty of all scientific discoveries that if they are not made by one scientist, they will sooner or later be made by another. Naturally there is a strong emphasis on the "sooner or later." The enormous progress made in science and in medicine seems today to demonstrate a process of organic growth which proceeds according to its own laws, and in which it is frequently difficult to distinguish between individual achievements. It not infrequently occurs that the same discovery is made in different parts of the world at more or less the same time, with perhaps a few days or weeks between them. Rapid communication of all new discoveries and intensive correspondence between scientists in all parts of the world has contributed to the advancement of science in a spirit of team work in numerous fields. This can only be of benefit to science and thus to civilization as a whole. These thoughts should serve as a reminder for the individual scientist when considering his own role in assisting development, and should deter him from any false conception of his own importance.

It is of course correct that the scientist in his creed contributes to the unique building enterprise of knowledge and is often a part of an important working collective as well as a member of an invisible guild that continuously scrutinizes his contributions. And still in the end he carries his independent responsibility for his work. In her book *Mémories d'Hadrien* Marguerite Yourcenar, the first woman to become a member of the Académie Française, has reflected on the contributions by the individual departing from the Latin expression *Trahit sua quemque voluptus* — each one is led by his own liking. She reflected: "Each and everyone has his own mind, each and everyone has his own goal, or if you like, his own ambition, his secret preference and his pure ideals. Mine (Hadrian's, my remark) is included in the word beauty, which is difficult to define, in spite of the convincing evidence of the senses and the eyes."

Towards the end of my preceding book on Nobel Prizes[3] I gave a quote from Ronald Dworkin's book *Justice for Hedgehogs*, but the origin of the title was not explained. Among the fragments of the Greek lyric poet Archilochus, who lived during the seventh century BCE, has been found the line "The fox knows many things, but the hedgehog knows one big thing." The meaning of these words has been discussed among scholars, but the general interpretation is the following; in spite of all his cunning the fox is outsmarted by the single defense used by the hedgehog. Taken figuratively this simple statement has been extensively used to illustrate different groups of thinkers and writers and why not also scientists. The difference lies between those who follow many leads, sometimes even contradictory, without referring to any common principle and those who depart from a single moral, esthetic or scientific system. The former tries to accommodate a wide variety of experiences, without attempting to harmonize them into a coherent concept, whereas the latter are searching for a unity of vision, in the case of science, for a structured and coherent system of knowledge. In an essay on Lev Tolstoy the philosopher and historian Isaiah Berlin has used Archilochus's division. He could readily identify among Russian writers representatives of both foxes and hedgehogs. Pushkin was classified as a fox but Dostoevsky as a hedgehog. Berlin did, however, have difficulties in classifying Tolstoy.

Joseph Brodsky, the Russian Nobel Prize recipient in literature in 1987, presumably also a hedgehog, has emphasized as a central theme that esthetics gives a guidance to ethics. This is a very complex and demanding statement, since the esthetic dimension is defined by the perceptions of the reader/viewer whereas ethics is a collective cultural product defined to improve interactions between the individuals forming the society. Following Brodsky one could argue that the reading of the books of life (evolution), just to select one scientific discipline, could also give guidance to the optimized ethics of science, concerning both the forms of its pursuit and the application of the new knowledge it generates. Again it seems difficult to derive one from the other. The discussion of teleology in Chapter 8 addressed some of these problems. Consequently there remains a need for continuous discussions of the ethics to apply in performing science and to allow appropriate use of the fruits of its labor. We have seen that the introduction of new and advanced technology may raise new ethical problems. But to return to the saying by Archilochus, which incidentally is also a name given to a genus of hummingbirds, it would seem that the overall majority of successful scientists are hedgehogs. This trait may unite them with at least some artists, and what is science if not a form of art? The greatest moment for a scientist is when he becomes part of his discovery — the unique art — he is creating.

References

Chapter 1

1. Norrby, E. (2010) *Nobel Prizes and Life Sciences*. World Scientific, Singapore.
2. Mörner, K.A.H. (1907) Nobelpriset i fysiologi och medicin (in Swedish and associated German translation). In *Les Prix Nobel en 1904*. Imprimerie Royale, P.A. Norstedt & Söner, Stockholm, pp. 24–30.
3. Pavlov, I.P. (1907) Physiology of digestion (in German). In *Les Prix Nobel en 1904*. Imprimerie Royale, P.A. Norstedt & Söner, Stockholm.
4. Babkin, B.P. (1949) *Pavlov, a Biography*. University of Chicago Press, Chicago.
5. Granit, R.G. (1983) *Hur det kom sig. Forskarminnen och motiveringar* (in Swedish). P.A. Norstedt & Söner, Stockholm.
6. Mörner, K.A.H. (1908) Nobelpriset i fysiologi och medicin (in Swedish and associated French translation). In *Les Prix Nobel en 1906*. Imprimerie Royale, P.A. Norstedt & Söner, Stockholm, pp. 27–31.
7. Gregory, P.R. (2007) *Lenin's Brain and Other Tales from the Secret Soviet Archives*. Hoover Institution Press, Stanford, CA.
8. Golgi, C. (1908) *La Doctrine du neurone. Théorie et Faits* (in French). In *Les Prix Nobel en 1906*. Imperimerie Royale, P.A. Norstedt & Söner, Stockholm.
9. Nansen, F. (1887) *The Structure and Combination of the Histological Elements of the Central Nervous System*. John Grieg Forlag, Bergen.
10. Jansen, J.K.S. (2001) Fridtjof Nansen som nevrobiolog (in Norwegian). *Tidsskr Nor. Laegeforen.* 121:210–211.
11. Cajal, S.R. (1908) Structure et connexions des neurones (in French). In *Les Prix Nobel en 1906*. Imperimerie Royale, P.A. Norstedt & Söner, Stockholm.
12. Partsalis, A.M., Blazquez, P.M. and Triarhou, L.C. (2013) The renaissance of the neuron doctrine: Cajal rebuts the rector of Granada. *Translational Neuroscience* 4:104–114.
13. Shepherd, G.M. (1991) *Foundations of the Neuron Doctrine*. Oxford University Press, New York.
14. Liljestrand, G. and Bernhard, C. G. (1972) The Prize in Physiology or Medicine. In *Nobel, the Man and His Prizes,*. 3rd ed., Odelberg, W. (ed.). American Elsevier Pub. Co., New York, pp. 139–278.

537

15. Sherrington, C. S. (1906) *The Integrative Action of the Nervous System*. Charles Scribner's Sons, New York.
16. Granit, R. (1967) *Charles Scott Sherrington. A Biography of the Neurophysiologist*. Doubleday, Garden City, New York.
17. Sherrington, C.S. (1934) Inhibition as a coordinative factor. In *Les Prix Nobel en 1932*. Imprimerie Royale, P.A. Norstedt & Söner, Stockholm.
18. Adrian, E.D. (1932) *Mechanism of Nerve Action*. Oxford University Press, Oxford.
19. Liljestrand, G. (1934) 1932 års Nobelpris i fysiologi och medicin (in Swedish and associated English translation). In *Les Prix Nobel en 1932*. Imprimerie Royale, P.A. Norstedt & Söner, Stockholm, pp. 25–29.
20. Adrian, E.D. (1934) The activity of the nerve fibres. In *Les Prix Nobel en 1932*. Imprimerie Royale, P.A. Norstedt & Söner, Stockholm.
21. Norrby, E. (2013) *Nobel Prizes and Nature's Surprises*. World Scientific, Singapore.
22. Liljestrand, G. (1937) 1936 års Nobelpris I fysiologi eller medicin (in Swedish and associated German translation). In *Les Prix Nobel en 1936*. Imprimerie Royale, P.A. Norstedt & Söner, Stockholm, pp. 34–38
23. Dale, H. (1937) Some recent extensions of the chemical transmission of the effects of nerve impulses. In *Les Prix Nobel en 1936*. Imprimerie Royale, P.A. Norstedt & Söner, Stockholm.
24. Medawar, J. and Pyke, D. (2000) *Hitler's Gift. The True Story of the Scientists Expelled by the Nazi Regime*. Arcade Publishing, New York.
25. Loewi, O. (1937) Die chemische Übertragung der Nervenwirkung (in German). In *Les Prix Nobel en 1936*. Imprimerie Royale, P.A. Norstedt & Söner, Stockholm.
26. Feldberg, W. (1976) The early history of synaptic and neuromuscular transmission by acetylcholine: Reminiscences of an eye witness. In Hodgkin, A. L. *et al.*, *The Pursuit of Nature. Informal Essays on the History of Physiology*. Cambridge University Press, Cambridge.
27. Granit, R. (1946) 1944 års Nobelpris I fysiologi och medicin (in Swedish and associated English translation). In *Les Prix Nobel en 1944*. Imprimerie Royale, P.A. Norstedt & Söner, Stockholm, pp. 29–31.
28. Gasser, H.S. (1946) Mammalian nerve fibers. In *Les Prix Nobel en 1944*. Imprimerie Royale, P.A. Norstedt & Söner, Stockholm, pp. 128–141.
29. Erlanger, J. (1949) Some observations on the responses of single nerve fibers. In *Les Prix Nobel en 1947*. Imprimerie Royale, P.A. Norstedt & Söner, Stockholm, pp. 173–195.
30. Stolt, C-M. (2002) Moniz, lobotomy and the 1949 Nobel Prize. In *Historical Studies in the Nobel Archives. The Prizes in Science and Medicine*, Crawford, E. (ed.). Universal Academy Press, Tokyo.
31. Olivecrona, H. (1950) Nobelpriset i fysiologi och medicin för år 1949 (in Swedish with an associated English translation). In *Les Prix Nobel en 1949*. Imprimerie Royale, P.A. Norstedt & Söner, Stockholm, pp. 35–38.
32. Uvnäs, B. (1958) Nobelpriset i fysiologi eller medicin för år 1957 (in Swedish with an associated French translation). In *Les Prix Nobel en 1957*. Imprimerie Royale, P.A. Norstedt & Söner, Stockholm, pp. 31–33.
33. Bovet, D. (1958) Relations d'isostérie et phénomenes compétitifs dans le domaine des médicaments du système nerveux végétatif et dans celui de la transmission neuromusculaire. In *Les Prix Nobel en 1957*. Imprimerie Royale, P.A. Norstedt & Söner, Stockholm, pp. 134–161.

34. Granit, R. (1945) The electrophysiological analysis of the fundamental problem of colour reception. *Proc. Phys. Soc. of London* 57:337–463.
35. Ohlmarks, Å. (1969) *Nobelpristagarna* (in Swedish), Forssell, G. B. (ed.). F. Beck & Son, Stockholm.

Chapter 2

1. Norrby, E. (2013) *Nobel Prizes and Nature's Surprises*. World Scientific, Singapore.
2. Olivecrona, H. (1950) Nobelpriset i fysiologi och medicin för år 1949 (in Swedish with an associated English translation). In *Les Prix Nobel en 1949*. Imprimerie Royale, P.A. Norstedt & Söner, Stockholm, pp. 35–38.
3. Liljestrand, G. and Bernard, C.G. (1972) The Prize in Physiology or Medicine. In *Nobel, the Man and His Prizes*, 3rd ed., Odelberg, W. (ed.). American Elsevier Pub. Co., New York, pp. 139–278.
4. Startzl, T.E. (1992) *The Puzzle People*. University of Pittsburgh Press, Pittsburgh.
5. Sargent, W. (1957) *Battle for the Mind. A Physiology of Conversion and Brain-washing*. Heinemann, London.
6. Sacks, O. (1973) *Awakenings*. Duckworth & Co., London.
7. Bentivoglio, M. and Kristensson, K. (2014) *Trends in Neurosciences* 37:325–333.
8. Kalsbeek, A., *et al.* (Eds.) (2012) *The Neurobiology of Circadian Timing. Progress in Brain Research*, Vol. 199.
9. Stolt, C-M. (2002) Why did Freud never receive the Nobel Prize? In *Historical Studies in the Nobel Archives. The Prizes in Science and Medicine*, Crawford, E. (ed.). Universal Academy Press, Tokyo
10. Norrby, E. (2010) *Nobel Prizes and Life Sciences*. World Scientific, Singapore.
11. Prusiner, S.B. (2014) *Madness and Memory. The Discovery of Prions — A New Biological Principle of Disease*. Yale University Press, New Haven.
12. Barker, C.F. and Markmann, J.F. (2013) *Historical overview of transplantation*. Cold Spring Harbor Perspect. Med. DOI:10.1101/cshperspect.a014977.
13. Magoun, H.W. (1958) *The Waking Brain*. Charles C. Thomas, Springfield.
14. Marshall, L.H. (2004) *Horace Winchell Magoun: A Biographical Memoir*. National Academy of Sciences, Washington, D.C.
15. Curtis, D.R. and Andersen, P. (2001) John Carew Eccles. *Biographical Memoirs of Fellows of the Royal Society* 47:159–187.
16. Kandel, E.R. (2006) *In Search of Memory. The Emergence of a New Science of Mind*. W.W. Norton, New York.
17. Stuart, D.G. and Brownstone, R.M. (2011) The beginning of intracellular recording in spinal neurons: Facts, reflections, and speculations. *Brain Res.* 1409:62–92.
18. Eccles, H.C. and Popper, K.R. (1977) *The Self and Its Brain. An Argument for Interactionism*. Springer Verlag, New York.
19. Perutz, V. (Ed.) (2009) *What a Time I Am Having. Selected Letters of Max Perutz*. Cold Spring Harbor Laboratory Press, New York.
20. Stent, G.S. (2002) *Paradoxes of Free Will*. American Philosophical Society, Philadelphia.
21. Hodges, A. (2014) *Alan Turing: The Enigma*. Princeton University Press, Princeton.

22. Kandel, E. R. (2012) *The Age of Insight: The Quest to Understand the Unconscious in Art, Mind and Brain, from Vienna 1900 to Present*. Random House, New York.
23. Denton, D. (1993) *The Pinnacle of Life. Consciousness and Self-Awareness in Humans and Animals*. HarperCollins, New York.
24. Changeux, J-P. (1985) *Neuronal Man. The Biology of Mind*. Pantheon Books, New York.
25. Liljefors, M., Lundin, S. and Wiszmeg, A. (2012) *The Atomized Body. The Cultural Life of Stem Cells, Genes and Neurons*. Nordic Academic Press, Lund.
26. Ottoson, D.G.R. (1982) The introductory speech to the 1981 Nobel Prize in physiology or medicine. (English translation from Swedish). In *Les Prix Nobel en 1981*. Almqvist & Wiksell International, Stockholm, pp. 26–28.
27. Sperry, R. (1982) Some effects of disconnecting the cerebral hemispheres. In *Les Prix Nobel en 1981*. Almqvist & Wiksell International, Stockholm, pp. 209–219.
28. Ohlmarks, Å. (1969) *Nobelpristagarna* (in Swedish; ed. Forsell, G. B.). F. Beck & Son, Stockholm.

Chapter 3

1. Hodgkin, A. (1992) *Chance & Design. Reminiscences of Science in Peace and War*. Cambridge University Press, Cambridge.
2. Norrby, E. (2013) *Nobel Prizes and Nature's Surprises*. World Scientific, Singapore.
3. Norrby, E. (2010) *Nobel Prizes and Life Sciences*. World Scientific, Singapore.
4. Granit, R. (1964) Nobelpriset i fysiologi eller medicin (in Swedish with an associated English translation). In *Les Prix Nobel en 1963*. Imprimerie Royale, P.A. Norstedt & Söner, Stockholm, pp. 34–37.
5. Hodgkin, A.l. (1964) Banquet address. In *Les Prix Nobel en 1963*. Imprimerie Royale, P.A. Norstedt & Söner, Stockholm, pp. 52–54.
6. Eccles, J.C. (1964) The ionic mechanism of postsynaptic inhibition. In *Les Prix Nobel en 1963*. Imprimerie Royale, P.A. Norstedt & Söner, Stockholm, pp. 261–283.
7. Hodgkin, A.L. (1964) The ionic basis of nervous conduction. In *Les Prix Nobel en 1963*. Imprimerie Royale, P.A. Norstedt & Söner, Stockholm, pp. 224–241.
8. Huxley, A.F. (1964) The quantitative analysis of excitation and conduction in nerve. In *Les Prix Nobel en 1963*. Imprimerie Royale, P.A. Norstedt & Söner, Stockholm, pp. 242–260.
9. Finch, J. (2008) *A Nobel Fellow on Every Floor. A History of the Medical Research Council Laboratory of Molecular Biology*. Icon Books, Cambridge.
10. Gann, A. and Witkowski, J. (Eds.) (2012) *The Annotated and Illustrated Double Helix by James D. Watson*. Simon & Schuster, New York.
11. Skou, J.C. (1998) The identification of the sodium-potassium pump. In *Les Prix Nobel en 1997*. Almqvist & Wiksell International, Stockholm, Sweden, pp. 245–260.
12. Skou, J.C. (2013) *Om Heldige Valg. Eller vad fröer, krabber og hajer også kan brukes till* (in Danish). Aarhus Universitetsforlag, Aarhus.

13. Lindkvist, K. and Sundling, S. (1993) *Xylocain – en uppfinning – ett drama – en industri*. (in Swedish) Stellan Ståls Tryckerier AB, Södertälje.
14. Grillner, S. (1992) The Nobel Prize in physiology or medicine. Introductory speech translated from the Swedish text. In *Les Prix Nobel en 1991*. Almqvist & Wiksell International, Stockholm, pp. 24–25.
15. Neher, E. (1992) Ion channels for communication between and within cells. In *Les Prix Nobel en 1991*. Almqvist & Wiksell International, Stockholm, pp. 120–135.
16. Sakmann, B. (1992) Elementary steps in synaptic transmission revealed by currents through single ion channels. In *Les Prix Nobel en 1991*. Almqvist & Wiksell International, Stockholm, pp. 141–169.
17. Malmström, B.G. (1989) The Nobel Prize in chemistry. Introductory speech translated from the Swedish text. In *Les Prix Nobel en 1988*. Almqvist & Wiksell International, Stockholm, pp. 20–21.
18. MacKinnon, R. (2004) Potassium channels and the atomic basis of selective ion conduction. In *Les Prix Nobel en 2003*. Almqvist & Wiksell International, Stockholm, pp. 214–235.
19. Yu, F.H., Yarov-Yarovoy, V., Gutman. G.A. and Catterall, W.A. (2005) Overview of molecular relationships in the voltage-gated ion channel superfamily. *Pharmacological Reviews* 57:387–395.
20. Miller, A.I. (2014) *Colliding Worlds. How Cutting-Edge Science is Redefining Contemporary Art*. W.W. Norton, New York.
21. Bernhard, C.G. (1968) Nobelpriset i fysiologi eller medicin. Introductory speech in Swedish with an accompanying English translation. In *Les Prix Nobel en 1967*. Imprimerie Royale, P.A. Norstedt & Söner, Stockholm, pp. 59–62.
22. Ottoson, D.G.R. (1982) The introductory speech to the 1981 Nobel Prize in physiology or medicine (English translation from Swedish). In *Les Prix Nobel en 1981*. Almqvist & Wiksell International, Stockholm, pp. 26–28.
23. Hubel, H.H. (1982) Evolution of ideas on the primary visual cortex, 1955–1978: A biased historical account. In *Les Prix Nobel en 1981*. Almqvist & Wiksell International, Stockholm, pp. 224–256.
24. Hubel, D.H. and Wiesel, T.N. (2005) *Brain and Visual Perception. The Story of a 25-Year Collaboration*. Oxford University Press, Oxford.
25. Wiesel, T.N. (1982) The postnatal development of the visual cortex and the influence of environment. In *Les Prix Nobel en 1981*. Almqvist & Wiksell International, Stockholm, pp. 261–283.
26. Kandel, E. (2003) *In Search of Memory. The Emergence of a New Science of Mind*. W.W. Norton, New York.
27. Uvnäs, B. (1971) Nobelpriset i fysiologi eller medicin. Introductory speech in Swedish with an accompanying English translation. In *Les Prix Nobel en 1970*. Almqvist &Wiksell International, Stockholm, pp. 56–57.
28. Ungerstedt, U. (2001) The introductory speech to the 2000 Nobel prize in physiology or medicine (English translation from Swedish). In *Les Prix Nobel en 2000*. Almqvist & Wiksell International, Stockholm, pp. 24–26.
29. Prusiner, S.B. (2014) *Madness and Memory. The Discovery of Prions — a New Biological Principle of Disease*. Yale University Press, New Haven.

30. Lefkowitz, R.J. (2013) A brief history of G protein coupled receptors. In *The Nobel Prizes (Les Prix Nobel) 2012*. Watson Publishing International LLC, Sagamore Beach, pp. 159–194.
31. Ohlmarks, Å. (1969) *Nobelpristagarna* (in Swedish; ed. Forsell, G.B.). F. Beck & Son, Stockholm.

Chapter 4

1. Norrby, E. (2010) *Nobel Prizes and Life Sciences*. World Scientific, Singapore.
2. Ferry, G. (1998) *Dorothy Hodgkin: A Life*. Granta Books, London.
3. Norrby, E. (2013) *Nobel Prizes and Nature's Surprises*. World Scientific, Singapore.
4. Liljestrand, G., revised by Bernhard, C.G. (1972) The Prize in Physiology or Medicine. In *Nobel, the Man and His Prizes*, 3rd ed., Odelberg, W. (ed.). American Elsevier Pub. Co., New York, pp. 139–278.
5. Wallin-Levinovitz, A. and Ringertz, N. (2001) *The Nobel Prize. The First 100 Years*. Imperial College Press, London and World Scientific, Singapore.
6. Olovsson, I., Liljas, A. and Lidin, S. (2015) *From a Grain of Salt to the Ribosome. The History of Crystallography as Seen through the Lens of the Nobel Prize*. World Scientific, Singapore.
7. Lagerkvist, U. (1999) *Karolinska Institutet och Kampen mot Universiteten* (in Swedish). Gidlunds Förlag, Hedemora.
8. Friedberg, E.C. (2014) *A Biography of Paul Berg. The Recombinant DNA Controversy Revisited*. World Scientific, Singapore.
9. Theorell, H. (1977) *Växlande Vindar* (in Swedish). Natur och Kultur, Stockholm.
10. Teachers and Public Officials at the Institute (1960) *Karolinska Mediko-Kirurgiska Institutet. Forskning och Undervisning inom Olika Ämnen* (in Swedish). Almqvist & Wiksell, Uppsala.
11. Hofmann, K. (1987) *Vincent du Vigneaud 1901–1978. A Biographical Memoir*. National Academy of Sciences, Washington, D.C., pp. 543–595.
12. De Duve, C. (1975) Exploring cells with a centrifuge. In *Les Prix Nobel en 1974*. Imprimerie Royale, P.A. Norstedt & Söner, Stockholm, pp. 142–160.
13. Theorell, H. (1956) The nature and mode of action of oxidation enzymes (English translation of a presentation given in Swedish). In *Les Prix Nobel en 1955*. Imprimerie Royale, P.A. Norstedt & Söner, Stockholm, pp. 132–146.
14. Boyer, P.D. (1998) Energy, life and ATP. In *Les Prix Nobel en 1997*. Almqvist & Wiksell International, Stockholm, pp. 120–141.
15. Björk, R. (2007) Nobelsystemet. Karolinska Institutet och Nobelpriset i medicin till Hugo Theorell 1955 (in Swedish). *Lychnos*, pp. 43–62.
16. Tiselius, A. (1968) Reflections from both sides of the counter. *Ann. Rev. Biochem.* 37:1–23.
17. Vincent du Vigneaud (1956) A trail of sulfur research: from insulin to oxytocin. In *Les Prix Nobel en 1955*. Imprimerie Royale, P.A. Norstedt & Söner, Stockholm, pp. 112–131.
18. Prusiner, S. B. (2014) *Madness and Memory. The Discovery of Prions — a New Principle of Disease*. Yale University Press, New Haven.

19. Fredga, A. (1956) Nobelpriset i kemi för år 1955. Introductory speech in Swedish with an associated English translation. In *Les Prix Nobel en 1955*. Imprimerie Royale, P.A. Norstedt & Söner, Stockholm, pp. 25–27.
20. Hammarsten, E. (1956) Nobelpriset i fysiologi eller medicin för år 1955. Introductory speech in Swedish with an English translation. In *Les Prix Nobel en 1955*. Imprimerie Royale, P.A. Norstedt & Söner, Stockholm, pp. 31–33.
21. Ohlmarks, Å. (1969) *Nobelpristagarna* (in Swedish), Forsell, G.B. (ed.). F. Beck & Son, Stockholm.

Chapter 5

1. Norrby, E. (2010) *Nobel Prizes and Life Sciences*. World Scientific, Singapore.
2. Norrby, E. (2013) *Nobel Prizes and Nature's Surprises*. World Scientific, Singapore.
3. Kekwick, R.A. and Pedersen, K.O. (1974) Arne Tiselius. 1902–1971. *Biogr. Mems. Fell. R. Soc.* 20:401–428.
4. Tiselius, A. (1968) Reflections from both sides of the counter. *Ann. Rev. Biochem.* 37:1–23.
5. Kay, L.E. (1993) The Intellectual Politics of Laboratory Technology. The Protein Network and the Tiselius Apparatus. In *Center on the Periphery. Historical Aspects of 20th-Century Swedish Physics* (Ed. Svante Lindqvist). Science History Publications, Watson Publishing International, Canton, MA.
6. Tiselius, A. (1947) Introduction to the Nobel Prize in Chemistry in 1946 (in Swedish with an English translation). In *Les Prix Nobel en 1946*, Imprimerie Royale, P.A. Norstedt & Söner, Stockholm, pp. 24–28.
7. Tunlid A. (2008) Den nya biologin. Forskning och politik i tidigt 1960-tal (in Swedish). In *Vetenskapens Sociala Strukturer* (Ed. Sven Widmalm). Nordic Academic Press, Lund.
8. Westgren, A. (1949) Introduction to the Nobel Prize in Chemistry in 1948 (in Swedish with a French translation). In *Les Prix Nobel en 1948*, Imprimerie Royale, P.A. Norstedt & Söner, Stockholm, pp. 27–30.
9. Tiselius, A. (1949) Electrophoresis and adsorption analysis as aids in investigations of large molecular weight substances and their breakdown products (English translation of a presentation in Swedish). In *Les Prix Nobel en 1948*, Imprimerie Royale, P.A. Norstedt & Söner, Stockholm, pp. 102–121.
10. Brownlee, G.G. (2014) *Fred Sanger. Double Nobel laureate. A Biography*. Cambridge University Press, Cambridge.
11. Tiselius, A. (1959) Introduction to the Nobel Prize in Chemistry in 1958 (in Swedish with an English translation). In *Les Prix Nobel en 1958*. Imprimerie Royale, P.A. Norstedt & Söner, Stockholm, pp. 21–24.
12. Sanger, F. (1959) The chemistry of insulin. In *Les Prix Nobel en 1958*. Imprimerie Royale, P.A. Norstedt & Söner, Stockholm, pp. 134–146.
13. Sanger, F. (1988) Sequences, sequences, and sequences. *Ann. Rev. Biochem.* 57:1–28.
14. Finch, J. (2008) *A Nobel Fellow on Every Floor. A History of the Medical Research Council laboratory of Molecular Biology*. Icon Books, Cambridge.

15. Malmström, B.G. (1981) Introduction to the Nobel Prize in Chemistry in 1980 (translated into English from Swedish). In *Les Prix Nobel en 1980*. Almqvist & Wiksell International, Stockholm, pp. 21–22.

16. Sanger, F. (1981) Determination of nucleotide sequencing in DNA. In *Les Prix Nobel en 1980*. Almqvist & Wiksell International, Stockholm, pp. 143–159.

17. Hargittai, I. (2002) Frederick Sanger. In *Candid Science II. Conversations with Famous Biomedical Scientists*, Imperial College Press, London, pp. 72–83.

18. Judson, H.F. (1996) *The Eighth Day of Creation. Makers of the Revolution in Biology.* Cold Spring Harbor Laboratory Press, New York.

19. Bremer, B. (2007) Linnaeus' sexual system and flowering plant phylogeny. *Nord. J.Bot.* 25:5–6.

20. Pääbo, S. (2014) *Neanderthal man: In Search of Lost Genomes.* Basic Books, New York.

21. Venter, J.C. (2013) *Life at the Speed of Light. From the Double Helix to the Dawn of Digital Life.* Brockman, New York.

22. Hutchison, Clyde A. III *et al.* (2016) Design and synthesis of a minimal bacterial genome. *Science* 351:6253–63.

23. Hodgkin, A. (1992) *Chance & Design. Reminiscences of Science in Peace and War.* Cambridge University Press, Cambridge.

24. Ohlmarks, Å. (1969) *Nobelpristagarna* (in Swedish), Forsell, G.B. (ed.). F. Beck & Son, Stockholm.

Chapter 6

1. Norrby, E. (2010) *Nobel Prizes and Life Sciences*, World Scientific, Singapore.

2. Söderbaum, H.G. (1929) Nobelprisen i kemi för åren 1927 och 1928 (in Swedish with an accompanying German translation). In *Les Prix Nobel en 1928*. Imprimerie Royale, P. A. Norstedt & Söner, Stockholm, pp. 14–17.

3. Wieland, H. (1929) Die Chemie der Gallensäuren (in German). In *Les Prix Nobel en 1928*. Imprimerie Royale, P. A. Norstedt & Söner, Stockholm, pp. 1–12 under *Les Conférences Nobel*.

4. Windaus, A. (1929) Nobel-Vortrag (in German). In *Les Prix Nobel en 1928*. Imprimerie Royale, P. A. Norstedt & Söner, Stockholm, separate pp. 1–19 under *Les Conférences Nobel*.

5. Brown, M.S. and Goldstein, J.L. (2009) Cholesterol feedback: From Schoenheimer's bottle to Scap's MELADL. *J. Lipid Res.* April Suppl. pp. 15–27.

6. Norrby, E. (2013) *Nobel Prizes and Nature's Surprises.* World Scientific, Singapore.

7. Medawar, J. and Pyke, D. (2000) *Hitler's Gift. The True Story of the Scientists Expelled by the Nazi Regime.* Arcade Publishing, New York.

8. Pääbo, S. (2014) *Neanderthal Man: In Search of Lost Genomes.* Basic Books, New York.

9. Bergström, S. (1983) The prostaglandins: From the laboratory to the clinic. In *Les Prix Nobel en 1982*. Almqvist & Wiksell International, Stockholm, pp. 129–148.

10. Friedberg, E.C. (2014) *A Biography of Paul Berg. The Recombinant DNA Controversy Revisited.* World Scientific, Singapore.

11. Meyers, M. A. (2012) *Prize Fight. The Race and the Rivalry to be the First in Science.* Palgrave Macmillan, New York.

12. Hargittai, I. (2002) *Candid Science II. Conversations with Famous Biomedical Scientists*. Imperial College Press, London.
13. Konstantinov, I.E., Mejevoi, N. and Anichkov, N.M. (2006) Nicolai N. Anichkov and his theory of atherosclerosis. *Tex. Heart Inst. J.* 33:417–423.
14. Friedman M. and Friedland, G.W. (1998) *Medicine's 10 Greatest Discoveries*. Yale University Press, New Haven.
15. Bergström, S. (1965) Nobelpriset i fysiologi eller medicin för 1964 (in Swedish with an accompanying English translation). In *Les Prix Nobel en 1964*. Imprimerie Royale, P.A. Norstedt & Söner, Stockholm, pp. 34–37.
16. Bloch, K. (1965) The Biological Synthesis of Cholesterol. In *Les Prix Nobel en 1964*. Imprimerie Royale, P. A. Norstedt & Söner, Stockholm, pp. 179–203.
17. Lynen, F. (1965) Der Weg von der "Activieren Essigsäure" zu den Terpenen und den Fettsäuren (in German). In *Les Prix Nobel en 1964*. Imprimerie Royale, P. A. Norstedt & Söner, Stockholm, pp. 205–243.
18. Brown, M. S. and Goldstein, M.S. (1986) A receptor-mediated pathway for cholesterol homeostasis. In *Les Prix Nobel en 1985*. Almqvist & Wiksell International, Stockholm, pp. 166–206.
19. Mutt, V. (1986) Introductory speech to the Nobel Prize for Physiology or Medicine (English translation of a speech given in Swedish). In *Les Prix Nobel en 1985*, Almqvist & Wiksell International, Stockholm, pp. 23–24.
20. Goldstein, J.L. and Brown, M.S. (2015) A century of cholesterol and coronaries: From plaques to statins. *Cell* 161:161–172.
21. Brown, M.S. and Goldstein, J.L. (2004) A tribute to Akira Endo, discoverer of a "Penicillin" for cholesterol. *Atherosclerosis Supplements* 5:13–16.
22. Ohlmarks, Å. (1969) *Nobelpristagarna* (in Swedish), Forsell, G.B. (ed.). F. Beck & Son, Stockholm.

Chapter 7

1. Norrby, E. (2010) *Nobel Prizes and Life Sciences*. World Scientific, Singapore.
2. Norrby, E. (2013) *Nobel Prizes and Nature's Surprises*. World Scientific, Singapore.
3. Judson, H.F. (1996) *The Eighth Day of Creation. Makers of the Revolution in Biology*. Cold Spring Harbor Laboratory Press, New York.
4. Jacob, F. (1995) *The Statue Within. An autobiography*. Cold Spring Harbor Laboratory Press, New York.
5. Brock, T.D. (1990) *The Emergence of Bacterial Genetics*. Cold Spring Harbor Laboratory Press, New York.
6. Lwoff, A. (1965) The specific effectors of viral development. (The First Keilin Memorial lecture, September 17, 1964). *Biochem. J.* 96:289–301.
7. Lwoff, A. (1953) Lysogeny. *Bact. Rev.* 17:239–253.
8. Bertani, G. (2004) Lysogeny at mid-twentieth century: P1, P2, and other experimental systems. *J. Bacteriol.* 186:595–600.
9. Lwoff, A. (1957) The concept of a virus. *J. Gen. Microbiol.* 17:239–253.
10. Lwoff, A., Horne, R.W. and Tournier, P. (1962) A system of viruses. *Cold Spring Harbor Symp. Quant. Biol.* 27:51–55.

11. Jacob, F. and Girard, M. (1998) André Michel Lwoff. 8 May 1902–30 September 1994. *Biogr. Mems. Fell. R. Soc.* 44:255–263.
12. Lwoff, A. (1966) Interaction among Virus, Cell, and Organism (in French). In *Les Prix Nobel en 1965*. Imprimerie Royale, P. A. Norstedt & Söner, Stockholm, pp. 233–243.
13. Hargittai, I. (2002) *Candid Science II. Conversations with Famous Biomedical Scientists.* Imperial College Press, London.
14. Ullmann, A. (1995) André Lwoff (1902–1994): remembrances. *EMBO Journal* 14:3289–3291.
15. Lwoff, A. (1981) *Jeux et combats.* Fayard, Paris.
16. Friedberg, E.C. (2014) *A Biography of Paul Berg. The Recombinant DNA controversy revisited.* World Scientific, Singapore.
17. Yi, D. (2015) *The Recombinant University. Genetic Engineering and the Emergence of Stanford Biotechnology.* The University of Chicago Press, Chicago.
18. Cohen, S.N. (2013) DNA cloning: A personal view after 40 years. *Proc. Natl. Acad. Sci.* 110:15521–15529.
19. Baltimore, D. (1971) Expression of animal virus genomes. *Bacteriol. Rev.* 35:235–241.
20. Koonin, E.V. and Dolja, V.V. (2013) A virocentric perspective on the evolution of life. *Current Opinion in Virology* 3:546–557.
21. Koonin, E.V., Dolja, V.V. and Krupovic, M. (2015) Origins and evolution of viruses of eukaryotes: The ultimate modularity. *Virology* 479–480:2–25.
22. Greene, S.E. and Reid, A. (2014) *Viruses Throughout Life & Time: Friends, Foes, Change Agents.* A report on an American Academy of Microbiology Colloquium, San Francisco, 2013.
23. Ohlmarks, Å. (1969) *Nobelpristagarna* (in Swedish), Forsell, G.B. (ed.). F. Beck & Son, Stockholm.

Chapter 8

1. Monod, J. (1971) *Chance and Necessity. An Essay on the Natural Philosophy of Modern Biology.* Vintage Books, a Division of Random House, New York.
2. Judson, H.F. (1996) *The Eighth Day of Creation. Makers of the Revolution in Biology.* Cold Spring Harbor Laboratory Press, New York.
3. Carroll, S. B. (2013) *Brave Genius. A Scientist, a Philosopher and Their Daring Adventures from the French Resistance to the Nobel Prize.* Broadway Books, New York.
4. Ullmann, A. (2011) In Memorian: Jacques Monod (1910–1976). *Genome Biol. Evol.* 3:1025–1033.
5. Berg, P. and Singer, M. (2003) *George Beadle. An uncommon farmer. The Emergence of Genetics in the 20th Century.* Cold Spring Harbor Laboratory Press, New York.
6. Jacob, F. (1995) *The Statue within. An autobiography.* Cold Spring Harbor Laboratory Press, Cold Spring Harbor, New York.
7. Monod, J. (1966) De l'adaptation enzymatique aux transitions allosteriques. In *Les Prix Nobel en 1965*. Imprimerie Royale, P.A. Norstedt & Söner, Stockholm, pp. 244–263.

8. Medvedev, Z. (1969) *Rise and Fall of T. D. Lysenko*. Columbia University Press, New York.
9. Brock, T.D. (1990) *The emergence of bacterial genetics*. Cold Spring Harbor Laboratory Press, New York.
10. Pardee, A.B., Jacob, F. and Monod, J. (1959) *J.Mol.Biol.* 1:165–178.
11. Norrby, E. (2013) *Nobel Prizes and Nature's Surprises*. World Scientific, Singapore.
12. Lanouette, W. with Szilard, B. (2013) *Genius in the Shadows. A Biography of Leo Szilard, the Man Behind the Bomb*. Skyhorse Publishing, New York.
13. Dyson, G. (2012) *Turing's Cathedral: The Origins of the Digital Universe*. Pantheon Books, New York.
14. Jacob, F. and Monod, J. (1961) Genetic regulatory mechanisms in the synthesis of proteins. *J. Mol. Biol.* 3:318–356.
15. Monod, J., Wyman, J. and Changeux, J.P. (1965) *J.Mol.Biol.* 12:306–320.
16. Norrby, E. (2010) *Nobel Prizes and Life Sciences*. World Scientific, Singapore.
17. Prusiner, S.B. (2014) *Madness and Memory. The discovery of prions — A New Biological Principle of Disease*. Yale University Press, New Haven.
18. Carroll, S.B. (2005) *Endless Forms Most Beautiful. The New Science of Evo Devo*. W.W. Norton, New York
19. Koestler, A. (1964) *The Act of Creation*. Hutchinson & Co., London.
20. Murphy, F.A. (2012) *The Foundations of Virology*. Infinity Publishing, West Conshohocken, PA.

Chapter 9

1. Jacob, F. (1987) *La statue intérieure*. Éditions Odile Jacob, Paris.
2. Jacob, F. (1988) Den inre gestalten (in Swedish).Brombergs Bokförlag AB, Stockholm.
3. Norrby, E. (2013) *Nobel Prizes and Nature's Surprises*. World Scientific, Singapore.
4. Norrby, E. (2010) *Nobel Prizes and Life Sciences*. World Scientific, Singapore.
5. Jacob, F. and Wollman, E.L. (1961) *Sexuality and the Genetics of Bacteria*. Academic Press, New York.
6. Crick, F. (1988) *What Mad Pursuit*. Basic Books, New York.
7. Orgel, L.E. (1973) *The Origins of Life : Molecules and Natural Selection*. John Wiley & Sons, New Jersey.
8. Brenner, S., Jacob, F. and Meselson, M. (1961) *Nature* 190:576–580.
9. Gros, F. *et al*. (1961) *Nature* 190:581–585.
10. Jacob, F. (1966) Genetique de la Cellule Bacterienne. In *Les Prix Nobel en 1965*. Imprimerie Royale, P.A. Norstedt & Söner, Stockholm, pp. 212–243.
11. Watson, J.D. (1963) The involvement of RNA in the synthesis of proteins. In *Les Prix Nobel en 1962*. Imprimerie Royale, P.A. Norstedt & Söner, Stockholm, pp. 155–178.
12. Ullmann, A. (2014) Le "Grand" François. *Research in Microbiology* 165:327–330.
13. Jacob, F. and Monod, J. (1961) *J. Mol. Biol.* 3:318–356.
14. Darnell, J. (2011) *RNA. Life's Indispensible Molecule*. Cold Spring Harbor Laboratory Press, New York.
15. Monod, J., Changeux, J. P. and Jacob, F. (1963) *J. Mol. Biol.* 6:306–329.
16. Riley, M., Pardee, F., Jacob, F. and Monod, J. (1961) *J.Mol.Biol.* 2:216–230.

17. Carroll, S. B. (2013) *Brave Genius. A Scientist, a Philosopher and Their Daring Adventures from the French Resistance to the Nobel Prize.* Broadway Books, New York.
18. Jacob, F. (1974) *The Logic of Living Systems: A History of Heredity.* Allen Lane, a Division of Penguin Books, London.
19. Jacob, F. (1981) *The Possible and the Actual.* Reprinted in 1994 by University of Washington Press, Washington.
20. Monod, J. (1971) *Chance and Necessity. An Essay on the Natural Philosophy of Modern Biology.* Vintage Books. A Division of Random House, New York.
21. Jacob, F. (1998) *Of Flies, Mice and Men.* Harvard University Press, Cambridge, MA.
22. Pinker, S. (2011) *The Better Angels of Our Nature. The Decline of Violence in History and Its Causes.* Allen Lane, a Division of Penguin Books, London.
23. Ullmann, A. (2011) In Memorian: Jacques Monod (1910–1976). *Genome Biol. Evol.* 3:1025–1033.
24. Ohlmarks, Å. (1969) *Nobelpristagarna* (in Swedish), Forsell, G.B. (ed.). F. Beck & Son, Stockholm.
25. Judson, H.F. (1996) *The Eighth Day of Creation. Makers of the Revolution in Biology.* Cold Spring Harbor Laboratory Press, New York.
26. Murphy, F.A. (2012) *The Foundations of Virology.* Infinity Publishing, West Conshohocken, PA.

Index

A

acetylcholine, 40, 42, 43, 45–47, 58, 59, 108, 123

acquired characters debate, 446–451

action potential, 128, 355
    conduction velocity, 121
    ions and, 52, 121, 122, 136, 164
    measurement and, 31, 51–52, 95, 121, 138
    neurons and, 31, 52, 91, 161. *See also* neurons
    resting potential and, 122, 128, 132, 149
    saltatory conduction and, 52, 73, 138
    sodium and. *See* sodium

adenosine triphosphate (ATP), 149–150, 225

adrenal glands, 328

adrenaline, 39–42, 43, 58, 328

Adrian, Edgar D., 14t, 22–35, 31f, 32f, 50, 123, 134

AIDS, 386, 424–426, 428

allostery, 386, 460, 469, 510

Alzheimer's disease, 88, 177, 461

amines, 58, 62, 333. *See also specific types, topics*

amino acids, 282, 297, 307
    disulphide bonds and, 214, 283, 287–291
    DNA and, 283. *See also* deoxyribonucleic acid
    DNP method, 285
    enzymes and, 159, 287. *See also* enzymes
    essential, 280
    insulin and, 187, 277, 281–290, 298
    MELADL, 377
    nucleic acids and, 302. *See also* nucleic acids
    polypeptides and, 298. *See* polypeptides
    proteins and, 276, 281, 283–291, 294, 298,

    303, 370, 430. *See also specific topics*
    RNA and, 206, 302, 419, 504. *See also* ribonucleic acid
    sequencing, 283, 425, 430
    See also genetics; *specific types, topics*

anesthesia, 47, 108

Anfinsen, Christian B., 224

Äng, Gröder, 274–275. *See* Tiselius, Arne

animal experiments, 11–12, 26, 72, 85, 105, 107, 124, 473

Anitschkow, Nikolay, 359–361

antibodies, 253

Apáthy, Stephan von, 21

APG system, 311

aphasia, 16

apoptosis, 246

archaea, 308–309, 310, 311, 424

Archilochus, 535

Aristotle, 4

art, science and, 167–168, 533

artificial intelligence, 85, 109

Asimov, Isaac, 526

Astrachan, Lazarus, 501–502

atherosclerosis, vii, 336, 360, 373, 380. *See also* cholesterol

ATP. *See* adenosine triphosphate

atrophine, 59

Auer, J., 90

autoimmune disorders, 81, 82, 386, 424, 425, 428

Avery, Oswald, T., 242, 243

Awakenings (Oliver Sacks), 80

Axelrod, Julius, 61

axons
    action potential and. *See* action potential

axons
  axoplasm, 126
  cell body of, 98, 147
  electricity and, 32, 52, 92. *See also*
    electricity
  giant, 124, 124f, 125, 128, 130f, 131, 138,
    150, 159
  measurement of, 51, 142
  myelin and, 19, 52–53, 124, 138
  neuronism and, 20
  neurons and, 2, 33. *See also* neurons
  number of, 117
  viruses and, 30, 72

B

bacteria, 498
  archaea and, 308
  bacteriophages, 381, 387, 392–399, 405,
    407, 443, 490, 493
  classification of, 423–424
  DNA and, 414, 445
  enzymes and, 498. *See also* enzymes
  eukaryotes and, 409. *See also* eukaryotes
  fertility factors of, 509
  genetics and, 408, 494. *See also* genetics
  groups of, 308
  induction and, 494. *See also* induction
  lysogeny and, 392–398, 407, 410, 414, 490.
    See also lysogeny
  transduction, 408
  viruses and. *See* viruses
  See also specific topics
Baker, P. F., 138
Baltimore, David, 404, 428, 473
basal brain, 67
Bayliss, William M., 10
behavioral studies, 473
Békésy, Georg von, 33, 48, 53, 60, 69, 99, 140
Benzer, Seymor, 490, 492f
Berg, Paul, 355
Berger, Hans, 35–37, 35f
Berger rhythm, 35, 123
Bergman, Ingmar, 147
Bergstrand, Hilding, 39f, 52f
Bergström, Sune, 62, 341–345, 343f, 356, 357
Berlin, Isaiah, 107, 535
Bernhard, Carl Gustaf, 63, 64, 64f, 66, 96,
  170, 491f
Bernhard, Claude, 53
Bernstein, Julius, 121, 130, 131, 149
Bertani, Giuseppe, 397, 397f

Berzelius, J. Jacob, 5, 190–194, 191f, 251, 333
Bethe, Albrecht von, 19, 21
biochemistry, 186, 306, 318. *See also specific
  topics*
biotin, 208, 212, 349, 358, 362
Blackburn, Elisabeth, 305
Blalock, Alfred, 87
Bloch, Konrad, 101, 347–350, 347f, 354–362,
  365–368, 365f
blood, 45
  adrenaline and, 41
  brain and, 3, 55, 67, 79–82, 177. *See also* brain
  cholesterol and, 336, 362, 368, 372. *See also*
    cholesterol
  circulation, 9
  clotting, 6, 178, 201
  electrophoresis, 268, 270
  energy and, 246, 330, 469
  hormones. *See* hormones
  human groups, 281
  MRI and, 185
  salinity of, 213, 328
  sedimentation, 219
  sleep and, 79
  staining and, 80
  See also specific topics
body, as microbiome, 426
Bohr, Niels, 105
Bovet, Daniel, 14t, 58, 58f
Boyer, Paul D., 224, 225
Brahe, Tycho, 533
brain
  artificial intelligence and, 85, 109
  asymmetry, 115, 116
  basal parts, 71f
  blood and, 3, 55, 67, 79–82. *See also* blood
  Broca's area, 6
  cerebral cortex, 15, 70, 75, 102, 103, 172
  choice and, 105
  consciousness and. *See* consciousness
  corpus callosum, 117, 118
  cranial nerves, 3
  electricity and, 35, 72. *See also* electricity
  function of, 3–4
  hemispheres, 116, 117–119
  hormones and, 211. *See also* hormones
  illustration of, 5f
  information and, 2, 22, 67, 76, 85, 110, 118,
    426
  integrative centers, 72
  lateralization, 4, 6
  memory and, 85

consciousness
  choices and, 105
  concept of, 106
  dualism and, 102
  emotions and, 76
  evolution and, 76, 111, 112, 113
  interpretation of, 15
  mind and, 119
  Mörner on, 14–15
  nervous system and, 2, 15
  self and, 107
  subconscious and, 79
  wakefulness and, 75
contraceptives, 417
Cornforth, John W., 358, 359
countercurrent distribution technique,
  209–211
Crafoord, Holger, 491, 491f
Crafoord Prize, 491, 492, 492f
Crick, Francis H. C., 70, 109, 111, 129, 137,
  301f, 412, 495, 500, 501, 502, 524
CRISPR spaces, 422
crystallography, 189, 190
curare, 59
Curtis, Howard, 124
cybernetics, 454
cysteine, 194, 208
cytochromes,, 223, 245, 246, 248, 249

D

Dale, Henry H., 14t, 37, 38f, 40, 41, 45, 46, 47,
  51, 61, 92, 123
Damasio, Antonio, 110
Darwin, Charles, 108, 133, 153
Dautry, Alice, 519f
de Gaulle, Charles, 483, 484, 484f
de Hevesy, George, 47–48, 103, 158, 337, 339
Dean, Robert B., 156
degradation methods, 214, 289
Delbrück, Max, 414
Demerec, Milislav, 397f
Democritus, 475
dendrons, 103–104
Dennett, Daniel, 110, 114
Denny-Brown, Derek, 25
Denton, Derek, 112, 119, 120
deoxyribonucleic acid (DNA)
  amino acids and, 283. See also amino acids
  bacteria and, 445
  disulphide bonds and, 283
  enzymes and, 204, 510. See also enzymes

evolution and, 516
genetics and, 242, 313, 316, 392, 422. See also
  genetics
human, 312, 313
induction and, 496. See also induction
information and, 307
mitochondria and, 306, 310
nucleic acids and, 198, 402, 427. See also
  nucleic acids
purified, 243
recombinant technology, 162, 199, 356, 419,
  420, 464
RNA and, 155, 300, 302, 303, 305, 419, 427,
  428, 445, 504, 513. See also ribonucleic acid
sequencing, 304, 305, 307–319, 400
structure of, 100, 223, 412, 493, 524
dermatomes, 29, 74, 429
Descartes, 71, 93, 102, 105, 107, 108
dextran, 263
dideoxy method, 306
digestion, 8, 10
discovery, Nobel prizes and, 23, 27, 215–216, 503
disulphide bonds, 283, 287, 288, 289, 291
DNA. See deoxyribonucleic acid
DNP method, 285
Doll, Richard, 471, 471f
Domagk, Gerhard, 36
dopamine, 62, 175, 176, 180
double bluff, theory of, 453
du Bois-Reymond, Emil, 31, 121
du Vigneaud, Vincent, 186, 205–217, 206f, 250,
  250f
dualism, 71, 102, 103, 119
Dulbecco, Renato, 17, 473
Dworkin, Ronald, 534

E

Eccles, John C., 8, 25, 27, 40, 69, 71, 91–101, 92f,
  102, 123, 145, 145f, 171
Edelman, Gerald, 85, 110
Edman, Pehr V., 289
Edman degradation technique, 289
EEG. See electroencephalography
Ehrlich, Paul, 20, 80, 83
Einstein, Albert, 530
electricity, 30–35, 147
  action potential and, 31, 91, 121, 136. See also
    action potential
  axons and, 32. See also axons
  brain and, 4, 5. See also brain

F

Fåhraeus, Robin, 200
Falck, Bengt, 63, 63f
families, prizes and, viii, 17, 128, 152, 153, 155, 156f, 196, 276, 278, 364
Feldberg, Wilhelm, 45, 46, 47, 123
Fernel, Jean, 30
Feynman, Richard, 530
Fichtelius, Karl Erik, 107–108
finger-printing method, 302
Fischer, Emil, 295
Florey, Howard W., 25, 343f
Flynn effect, 183
folding, of proteins, 168, 223, 243, 283, 299, 461, 531
Folkers, Karl, 350–354, 354f, 357
forebrain, 70, 71–72
forensic processes, 313
Foster, Michael, 24
Foucault, Michel, 167
fractionation methods, 282
Frankenhaeuser, Bernhard, 66, 96, 135, 136, 136f, 138, 150
Franklin, Benjamin, 5
French, Jack, 88
Freud, Sigmund, 110
Fulton, John F., 54, 88–89

G

Gage, Phineas, 54
galactosidase, 452, 463, 464, 466, 470, 498, 500
Galen, 4
Galilei, Galileo, 451
Galvani, Luigi, 4, 5, 121, 191
Gard, Sven, 69, 520
Garrod, Archibald, 370
Gasser, Herbert S., 14t, 33, 47, 48f, 50–51
Gaule, Justus, 71
gene regulation, 488
genealogy, 153–156
genetically modified organisms (GMOs), 420–421, 422
genetics, 312, 314, 418
    acquired characters, 446–451
    amino acids. See amino acids
    bacteria and, 428, 494. See also bacteria
    DNA. See deoxyribonucleic acid
    evolution and, 313, 419
    forensic processes and, 313
    gene regulation, 488

genetic code, 300, 303, 306, 510
genetic disease, 422
genetic engineering, 420–421, 422
genetic properties, 408
    genome and, 305, 307–308, 311–313, 316, 430
    information and, 443
    Lysenko affair, 326, 446–451
    mechanisms in, 442
    metagenomics, 308
    nucleic acids. See nucleic acids
    recombination and, 392
    synthetic cells, 316
    of viruses, 430
Giauque, William F., 270, 271f
global warming, 450, 522
glucosteroids, 328
GMOs. See genetically modified organisms
God, 114
Goethe, J. W., 24, 526, 533
Goldmann, Edwin E., 80
Goldstein, Joseph L., 368–371, 369f
Golgi, Camillo, 6, 13, 14t, 15, 15f, 16, 18–22, 27, 31
Goltz, Friedrich, 24
GPS systems, 169–170, 182
Granit, Ragnar, 13, 25, 27, 34, 48, 49f, 66, 67, 89, 96, 97, 98, 100, 133
grid cells, 170
Grillner, Sten, 160
Gros, François, 503–505, 505f
Grünewald, Isaac, 57
Guillain-Barré disease, 81
Gustaf V, 28f, 108, 272
Gustaf VI Adolf, 188f, 250f, 251f, 252, 272, 273, 273f, 365f, 512f, 514
Gustavson, Karl H., 293
Guthrie, Francis, 531

H

Haber, Fritz, 109
Haldane, B. S., 154
Haldane family, 155, 156f
Hammarsten, Einar, 194–205, 194f, 233–234, 251, 251f, 269, 288
Hammarsten, Olof, 330, 331f, 333
Hanson, Jeanne, 152
Hansson, Göran, x
Hartline, Haldan K., 34, 49, 67
heart, nerves and, 40
Helmholz, Hermann von, 32, 121

heme component, 223, 223f
hemoglobin, 245
Henschen, Folke, 359
Henschen, Salomon E., 16
hepatitis virus, 387, 407, 431
Herder, Johann G., 523
Herophilus, 4
herpes family, 407, 415, 429, 521
Hess, Walter R., 14t, 53, 70f, 71, 72, 89
Heymans, Corneille J. F., 61
Hill, Archibald V., 61, 124
Hill, Austin Bradford, 471, 471f
Hillarp, Nils-Åke, 63, 63f
hindbrain, 70
histamine, 41, 58, 59
Hitler, Adolf, 17, 36, 45, 125, 220, 279, 338
HIV infections, 386, 424–426, 428
Hjertén, Sigrid, 57
HLA antigens, 82
HMG CoA reductase, 374, 378, 379
Hodgkin, Alan L., 8, 52, 67, 69, 90, 98, 99,
    122–156, 122f, 132t, 146f, 159
Hodgkin, Dorothy Crowfoot, 122, 187–189,
    326, 353
Hodgkin-Huxley theory, 164
Holmgren, Frithiof, 31, 67
homeostasis, 67, 71, 72, 316, 454
hormones
    brain and, 102, 211, 328
    cortisone, 328
    discovery of, 10
    human growth hormone, 215
    insulin. See insulin
    nerves and, 10
    oxytocin, 210, 211, 215
    polypeptides and, 214, 216
    as proteins, 208
    vasopressin, 211, 215
    See also specific types, topics
Horsfall, Frank Jr., 262
Houssay, Bernardo A., 52f, 61
Hubel, David H., 85, 171
Huguenots, 435
humoral transfer, 92, 93, 99, 101–102, 171
Huxley, Aldous, 109, 154, 155
Huxley, Andrew F., 8, 69, 90, 97–101, 121,
    122–156, 125f, 135t, 146f, 160, 164
Huxley, Hugh, 151–152, 151f, 301f
Huxley, Julian, 443
Huxley, Thomas H., 93, 153
Huxley family tree, 153, 153f
Hwasser, Israel, 192

hybridization, 488, 504, 510
hypothalamus, 71, 80, 81, 211

I

Ig Nobel Prizes, 477
immune system, 82, 83, 84
induction
    adaptation and, 452, 463
    discovery of, 406, 407
    DNA and, 496
    enzymes and, 434, 452–453, 466, 467
    galactosidase and, 452
    lysogeny and, 406, 407, 409, 414, 470, 511
    phages and, 406–407
    prophages and, 488, 494
    repression and, 453, 467
    viruses and, 396–397. See also viruses
    zygotic, 494, 508
    See also specific persons, topics
influenza virus, 81
information, 264–323
    allostery and, 510
    biology and, 307
    brain and, 2, 22, 67, 76, 85, 110, 118, 426
    cistron and, 466–467
    DNA and, 307, 377, 443, 445, 461, 510, 513
    entropy and, 86, 454
    epigenetic phenomena, 84
    evolution and, 475, 531
    genetics and, 242, 264–323, 295, 413, 418,
        466
    human genome, 312, 316, 425
    immune system and, 83
    information theory, 105–106, 454
    nucleic acids and, 266, 287, 295, 487
    operons and, 445
    prions and, 461
    RNA and, 419, 430, 461, 465, 496, 501, 502,
        504, 510
    storage of, 113, 182–183, 264–323
insecticides, 108
insulin, 187, 208, 277, 281–290, 283f, 298
Internet, 113, 182, 183
invention, discovery and, 215–216
ion exchanges, 131–138, 160, 162–165, 167. See
    also sodium
ionists. See soupers/sparkers
isoprenes, 351
isotope methods, 134, 137, 158, 165, 166, 208,
    318, 338–341, 357, 397, 497

Matthews, Brian H. C., 35, 123
McLuhan, Marshall, 183
measurement, methods of, 33, 91, 98, 128–129, 150, 165. *See also specific types, topics*
Mechnikov, Ilya I., 83
Medawar, Peter B., 69, 83, 90, 150, 476
Medvedev, Zhores A., 449
memory, 77, 84, 85, 107, 110, 118, 165, 176, 182
Mendel, B., 47
Mendel, Gregor, 418, 446, 449
mental disorders, 16, 47, 54–57, 78, 109
mentalism, 119
Merck & Co., 379
Meselson, Matt, 502
metabolism, 246, 355, 370, 409, 444
methionine, 208, 302
method, scientific, 104, 317, 319, 423, 503, 526
mevalonate, 353, 358, 378
Meyer, Hans H., 42, 133
Meyer-Overton theory, 133
midbrain, 70–77, 79
Millennium Development Goals, 522
Miller, Steve, 166
mind. *See* consciousness
mind-body problem, 104–105. *See also* dualism
mirror neurons, 111
mirror test, 107
mitochondria, 225, 306, 309, 310, 426
molecular biology, 186, 266, 287, 434, 444–446, 458, 469, 488, 493, 504, 506, 525. *See also specific persons, topics*
Moniz, António, 53, 55, 55f, 56, 57
Monod, Jacques, 108, 381, 381f, 391f, 392, 433–479, 439f, 456f, 478f, 479f, 488, 490f, 497, 512f, 513f, 516
Moore, Stanford, 209, 209f
Mörner, Karl A. H., 9, 13, 14, 15, 193–194, 193f
Morse, Marston, 531
Moruzzi, Guiseppe, 74, 75f, 86, 90, 99
Mountcastle, Vernon, 172, 173
MRI. *See* magnetic resonance imaging
MRS. *See* magnetic resonance spectroscopy
Muller, Hermann J., 418, 447, 448f
multiple sclerosis, 82, 83
Murray, Joseph E., 87f, 88
music, 217–219, 225–230
Mutt, Viktor, 371, 371f
myelin, 19, 52, 53, 124, 138
myoglobin, 243

Myrbäck, Karl, 228, 228f

N

Nansen, Fridtjof, 19
narcolepsy, 82
naturalism, 114
Nazis, 43, 44, 72, 338, 442, 448, 461
Neher, Erwin, 160, 161f
nerves, 2, 3, 108
  acetylcholine, 40–47, 58, 59, 108, 123
  action potential, 31, 51–52, 91, 161. *See also* action potential
  all-or-nothing in, 142
  axons. *See* axons
  brain and. *See* brain
  C fibers, 60
  chemical transmission. *See* chemical signals
  conditioned reflexes, 8–13
  conduction rates, 50, 52, 60, 98
  cross sections, 50
  cyclic AMP, 176
  cytokones and, 85
  depolarization, 34, 136
  differentiated functions, 48
  diseases, 16. *See also specific topics*
  DNA in, 84. *See also* deoxyribonucleic acid
  electrical signals, 30–35, 37, 39, 51, 98. *See also* electricity
  G proteins, 176
  generation of, 20
  heart and, 40
  hormones and, 10. *See also* hormones
  humoral transmitters, 101–102, 175
  inhibition in, 98
  instruction and, 84
  ionic exchange, 129, 131–133, 138, 158
  IPSP/EPSP, 148
  major groups of, 50
  measurement in, 51
  myelination and, 19, 52, 53, 124, 138
  nervous system. *See* nervous system
  neurons. *See* neurons
  permeability, 98, 139
  reflex and, 22–30
  saltatory conduction, 53
  second messengers, 176
  silver impregnation technique and, 15
  soupers/sparkers, 92, 93, 99, 171
  synapse and, 29

nerves
  vagus nerve, 11
  viruses and, 73, 74
nervous system, 16
  action potential and, 91. *See also* action
    potential
  autonomic, 39
  brain. *See* brain
  diseases of, 16. *See also specific topics*
  electricity and, 31, 121. *See also* electricity
  environment and, 84
  functions of, 14
  immune system and, 83
  integrative functions, 23, 26
  major parts, 2
  motor system, 39
  nerves and. *See* nerves
  neurons. *See* neurons
  opistothonus and, 26
  parasympathetic system, 39
  peripheral part of, 3
  sympathetic system, 39
  See also specific persons, topics
neurons, 22, 27
  acetylcholine and, 40–47, 58, 108, 123
  action potential and, 31, 52, 91, 161
  afferent, 3, 28
  axons and. *See* axons
  cell body, 147
  cholinergic, 47
  contiguity/continuity, 20
  efferent, 3, 28
  electricity and, 32, 33, 101–102. *See also*
    electricity
  firing of, 33
  functions of, 32
  humoral transfer and, 101–102, 175
  inhibitory mechanisms, 98
  link theory and, 20
  mirror neurons, 111
  neuron theory, 7, 13, 16–22
  neuropharmacology and, 56, 57–59, 80, 82
  neurotransmitters, 62. *See also specific*
    topics
  noncholinergic, 47
  reciprocal relationships, 15
  reflex and, 28. *See* reflexes
  synapse. *See* synapses
Newton, Isaac, 526
niacin, 212
Nobel, Alfred, 12, 185, 252f, 296, 320
Nobel Foundation, 64–65

Nobel Institutes, 64, 65–66. *See also specific*
  persons, topics
noradrenaline, 40, 58, 62
Norrby, Erling, v–xi, 116, 159, 163, 183, 186,
  194, 195–196, 221, 227–228, 229, 264–265,
  315, 319, 399, 403–404, 462, 473, 474, 476,
  481–482, 505, 515, 527
Norrby, Johannes, 227
nuclear weapons, 325, 457
nucleic acids, 300, 304, 445, 488
  bacteria and. *See* bacteria
  DNA and. *See* deoxyribonucleic acid
  evolution and, 424. *See also* evolution
  genetics and. *See* genetics
  hybridization, 504
  RNA and. *See* ribonucleic acid
  viruses and, 424. *See also* viruses
  See also specific topics

O

O'Keefe, John, 170
Oldstone, Michael, 317
olfactory system, 7, 7f, 34, 174
Olivecrona, Herbert, 54
One Flew Over the Cuckoo's Nest (Kesey), 56
one gene/one enzyme, 305, 374, 463, 509
operons, 445, 465, 469, 499–500, 510, 516, 525
optogenetics, 179
Orgel, Leslie, 500, 501, 501f
Ossietzky, Carl von, 36
Ottoson, David G. R., 116, 116f, 117
Overton, C. Ernest, 133
oxidation, 79, 221, 232, 242, 245, 310, 343
oxytocin, 209–211, 210f, 213, 214

P

Pääbo, Svante, 342
pain, 34, 50, 60, 72
PaJaMo experiment, 453, 497, 498, 500, 510
parasites, 389, 390, 431
Pardee, Arthur, 497–498, 498f
Parkinson's disease, 176
particles, elementary, 103
Pasteur, Louis, 383, 384, 385
Pasteur Institute, 383–388, 384f
patch-clamp technique, 160–162
Pauling, Linus, 254
Pavlov, Ivan P., 8–13, 8f, 11f, 14t, 78
pellegra, 212
Penfeld, Wilder D., 25–26

permeases, 452, 462–467, 470
Perutz, Max F., 70, 103, 162, 295, 301f
pharmacology, 59, 62. *See also specific topics*
phrenology, 5
physics, 103, 104, 167, 528, 533
pineal gland, 5, 71, 102
Pinker, Steven, 523
pituitary gland, 71, 187, 209, 210, 211, 220,
  328
Poincaré, Henri, 532
polio, 73, 400, 520, 521
politics, science and, 450, 518
polymaths, 153–156
polypeptides, 214, 286, 287, 298
Popper, Karl R., 93, 93f, 102, 103, 475
porphyrin, 223
Porter, Rodney R., 281–282, 281f
Possible and the Actual, The (Jacob), 518
posture, 26, 27, 31
potassium, 133–134, 136, 137, 160, 164. *See
  also* sodium
Priestley, Joseph, 5
Prigogine, Ilya, 86
prion proteins, 84, 176
prokaryotes, 309, 329, 355, 395, 419, 422, 426,
  427
prostaglandins, 62, 344
protein synthesis, 84, 504, 508
psychiatric disorders, 16, 47, 54–57, 78, 109
Purkinje cells, 6
Purkyně, Jan E., 6
purpose, 103, 518
Putnam, Tracy J., 36

Q

Quakers, 278–279, 280, 320, 321, 322, 434, 435
quantum theory, 103, 104

R

rabies virus, 73
radiation, 158, 165, 166, 338
Ramakrishnan, Venkatraman, 295
Ramón y Cajal, Santiago, 6, 13, 14t, 15, 15f,
  16–19, 27, 31
Ranson, Stephen W., 73, 89
Ranvier, Louis-Antoine, 52, 53, 150
recombinant technology, 162, 199, 356, 419,
  420, 464
reflexes, 22–30, 34. *See also specific topics*

Reichard, Peter, 203, 411, 411f, 412
religion, 114. *See also specific topics*
repressor system, 476, 499f
  chemical nature of, 517, 525
  discovery of, 411
  enzymes and, 463, 465
  genetics and, 467
  induction and, 453, 467
  operons and, 510
  protein synthesis and, 465
  RNA and, 411, 459, 465, 470, 499, 510
respiration, cellular, 109, 223, 232, 248
resting potential, 122, 128, 132, 149. *See also*
  action potential
reticular formation, 16, 20–21, 75–76, 76f
retroviruses, 428, 429, 432
ribbon model, 168
ribonuclease, 210, 283, 299
ribonucleic acid (RNA), 198, 300, 430, 495
  catalytic activity, 253
  DNA and, 302–305, 445, 504, 513. *See also*
    deoxyribonucleic acid
  genetics and. *See* genetics
  messenger, 500–505, 506, 509, 510, 513,
    516
  nucleic acids and, 402. *See* nucleic acids
  reverse transcriptase and, 428
  ribosomes and, 500
  sequencing, 400
  transcription and, 295, 309, 377, 419, 428,
    445, 499f, 504
  See also specific persons, topics
ribosomes, 253, 495, 498, 500, 502, 503
Ringertz, Nils, 472
Rizzolatti, Giacomo, 111
RNA. *See* ribonucleic acid
Roberts, Richard J., 421
robots, 107, 109
Rose, Caleb, 24
Rous, Peyton, 91, 125, 127, 145, 371
Runnström, John, 249
Rutherford, Ernest, 146

S

Sakmann, Bert, 160, 161f
saltatory conduction, 53, 138
Samuelsson, Bengt, 62, 244, 317, 344, 363, 364
Sanger, Frederick, 187, 276–307, 281f, 296f,
  297f, 301f, 304f, 306, 314, 317, 319, 322f
Santesson, Carl Gustaf, 26
Sargent, William, 78

SARS. *See* severe acute respiratory syndrome
Sartre, Jean-Paul, 450
schizophrenic, 55
Schmiedeberg, Oswald, 42
Schoenheimer, Rudolph, 335, 338–341, 338f
Scolnick, Edward, 379
science, art and, 167–168, 450, 518, 533
scientific method, 104, 317, 319, 423, 503, 526
secretin, 10, 42
self-awareness, 107. *See also* consciousness
separation techniques, 209, 214, 232, 263, 270, 285
sequencing techniques, 283, 425, 430
severe acute respiratory syndrome (SARS), 424, 425
sexuality, evolution and, 112
Shaw, T. I., 138
Sherrington, Charles S., 14t, 22–30, 23f, 29f, 31, 32f, 49, 91
shingles, 74
shock treatments, 78
Shumway, Norman, 87f
silver impregnation technique, 15, 16
Skoglund, Carl Rudolf, 66
Skou, Jens C., 156, 157f, 158
sleep, 75, 77–82, 108, 111
"sleeping" sickness, 81
smell, 7, 7f, 34, 174–175
smoking, 471, 472, 473
Söderbaum, Henrik G., 331, 331f
sodium
    action potential and, 91, 98, 122, 130, 130F, 133
    ATP and, 157
    ionic hypothesis, 160, 166f
    nerve impulse and, 41, 131, 136, 137, 151, 159
    potassium and, 133, 134, 136, 137, 149
    pump, 156–159
    sodium hypothesis, 149
    voltage-gating, 176
Sola, Eduardo Garcia, 21
soul, 105
soupers/sparkers, 92, 93, 99, 171
space-filling model, 167–168
Sperry, Roger W., 115, 115f, 117
Spiegelman, Sol, 488, 488f
spinal cord, 2–3, 27–28, 64, 70, 96, 98, 142
squalene, 351, 353, 358, 378
staining techniques, 6, 18, 63, 152
Stalin, Josef, 11
Stanley, Wendell, 265
Steitz, Thomas A., 295

starch column technique, 209
Starling, Ernest H., 10
Starzl, Thomas E., 86–88, 87f
statins, vii, 375, 377–379. *See also* cholesterol
Stefánsson, Kári, 312
Stein, Gertrude, 155
Stein, William H., 209, 209f
Stent, Günter, 104
stereotaxic scalpel, 72
sterols, 327, 332, 334, 337, 367, 377. *See also specific types, topics*
stick-ball model, 168
Stockholm Concert Hall, 226, 226f, 227
streptomycin, 351, 352
Sunge, Richard L. M., 282f
Sutherland, Earl W., 176
Svedberg, The, 62, 259f
Swedish language, 364
symbolism, 528–530
synapses, 22–30, 37–44
Szilard, Leo, 453, 454–458, 456f, 467, 530

T

Tasaki, Ichiji, 53
teleonomy, 518
Temin, Howard, 473
tetrodotoxin, 160
Theorell, Hugo, 62, 66, 185, 186, 216–255, 219f, 226f, 231f, 251f, 343f
thermodynamics, 86
Thomas, E. Donnall, 88
Tigerstedt, Robert, 10
Tinbergen, Nikolaas, 473
Tiselius, Arne, 62, 69–70, 232, 257–275, 260f, 262f, 273f, 275f, 276f, 295, 511, 533–534
tobacco mosaic virus, 265, 405
Tocqueville, Alexis de, 508
transacetylase, 498
transduction, 408, 509
transmethylation, 208, 212
trypanosomiasis, 80–81
Turing, Alan, 105–106
Turing test, 109
Twain, Mark, 528
Tydén, Thomas, 276

U

Ullmann, Agnes, 477
uncertainty principle, 104